Lipids in Aquatic Ecosystems

Michael T. Arts • Michael T. Brett
Martin J. Kainz
Editors

Lipids in Aquatic Ecosystems

Springer

Editors
Michael T. Arts
Aquatic Ecosystems Management
Research Division
National Water Research Institute –
Environment Canada
P.O. Box 5050, 867 Lakeshore Road
Burlington, ON, Canada L7R 4A6
Michael.Arts@ec.gc.ca

Michael T. Brett
Department of Civil & Environmental
Engineering
University of Washington
Box 352700, 301 More Hall, Seattle
WA 98195-2700, USA
mtbrett@u.washington.edu

Martin J. Kainz
WasserKluster Lunz
Biologische Station
Dr. Carl Kupelwieser Promenade 5
3293 Lunz am See, Austria
martin.kainz@donau-uni.ac.at

QH
541.5
.W3
L56
2009

ISBN: 978-0-387-88607-7   e-ISBN: 978-0-387-89366-2
DOI: 10.1007/978-0-387-89366-2
Springer Dordrecht Heidelberg London New York

Library of Congress Control Number: 2008942065

© Springer Science+Business Media, LLC 2009
All rights reserved. This work may not be translated or copied in whole or in part without the written permission of the publisher (Springer Science+Business Media, LLC, 233 Spring Street, New York, NY 10013, USA), except for brief excerpts in connection with reviews or scholarly analysis. Use in connection with any form of information storage and retrieval, electronic adaptation, computer software, or by similar or dissimilar methodology now known or hereafter developed is forbidden.
The use in this publication of trade names, trademarks, service marks, and similar terms, even if they are not identified as such, is not to be taken as an expression of opinion as to whether or not they are subject to proprietary rights.

The artwork depicted in the small inset on the front cover is a collaboration between the three editors and the artist, Andrew Turnbull (www.turnbullsculpture.com), with subsequent modifications by graphic artist Lucas Neilson.

Printed on acid-free paper

Springer is part of Springer Science+Business Media (www.springer.com)

b2514546

Foreword

The direction of science is often driven by methodological progress, and the topic of this book is no exception. I remember sitting with a visitor on the terrace of a hotel overlooking Lake Constance in the early 1970s. We were discussing the gravimetric method of measuring total lipids in zooplankton. A few years later, as a visitor in Clyde E. Goulden's lab, I was greatly impressed by the ability of an instrument called an Iatroscan to discriminate and quantify specific lipid classes (e.g., triacylglycerols, polar lipids, wax esters). At that time, food web analysis was mainly concerned with bulk quantitative aspects. For example, lipids, because of their high energy content, were considered mainly as an important food source and storage product.

Nearly a decade ago, when Michael Arts and Bruce Wainman edited the first volume entitled "Lipids in Freshwater Ecosystems" (Springer), the focus had already changed. Fatty acid analysis had become more mainstream, because new, less expensive, instruments had become available for ecological laboratories and because ecology, in general, was diversifying and integrating with other disciplines. Hence, there was increased emphasis on studies which dealt with the qualitative aspects of lipid composition. The concept of lipids in ecosystems was no longer restricted to just providing fuel; lipid composition had, by then, already been recognized as a factor controlling the flow of matter and the structure of food webs. In his foreword to the first book, Robert G. Wetzel defined a rapidly evolving field that he called "biochemical limnology" and identified lipid research as one of its facets. Judging from the ever increasing numbers of published papers and congress contributions the field is presently evolving even more rapidly.

However, progress was not restricted to limnology. In fact, methods of lipid and fatty acid analysis were probably more advanced in marine ecology, and essential fatty acids were an important factor in marine aquaculture. Lipid research in aquatic organisms profited also from the growing connections to human nutrition science interested in the importance of highly unsaturated fatty acids (HUFA; fatty acids with ≥ 20 carbons and ≥ 3 double bonds) originating from fish and shellfish. This became very evident at the 2002 summer meeting of the American Society of Limnology and Oceanography in Victoria, British Columbia, when Michael A. Crawford delivered an unusual, but fascinating plenary lecture entitled

"The evolution of the human brain." Consequently, this new volume has broadened its scope from freshwater to "aquatic ecosystems." It is, thus, a contribution to finding common principles in marine and freshwater systems.

My personal interest in fatty acids has been stimulated again in recent years by the controversy over food quality factors controlling the growth of zooplankton, which used to be more a topic in limnology than in marine ecology. Two schools developed at about the same time, one proposing that zooplankton growth was limited by the availability of essential fatty acids, the other one developing the concept of zooplankton growth controlled by inorganic nutrient stoichiometry. In principle, both groups of resources can be limiting as they must be taken up with the same food package and cannot be completely synthesized by the consumer itself. Unfortunately, the empirical data were contradictory, and there was support for both concepts. As usual, this resulted in a heated debate; however, we are now on the way to a concept incorporating both groups of resources as limiting factors. The controversy had a striking effect on aquatic lipid research; it stimulated discussion, created new ideas, and fostered methodological progress. Lipids and fatty acids are now regular topics of special sessions at aquatic science conferences.

Robert Wetzel's statements in the earlier foreword are still valid and up-to-date, but the field has broadened considerably in the past decade. The "classical" studies on lipids as storage products and carriers of lipophilic contaminants are continuing. Research on lipids as nutritional factors now concentrates on the role of essential components, e.g., polyunsaturated fatty acids (PUFA) and sterols, in modifying the growth and reproduction of animals. This includes studies on biosynthesis and metabolic pathways in food organisms and the characterization of fatty acid profiles in organisms at the base of food webs and in allochthonous material. Spatial and temporal variations in lipid composition need to be investigated to reach the goal of a mechanistic prediction of food web structures under changing environmental conditions. Finally, specific fatty acids and ratios of fatty acids are being developed as biomarkers to aid in the identification of key food web connections.

Evolutionary ecology is beginning to explore adaptations of organisms to the changing availability of essential fatty acids in their food, e.g., the evolution of life histories, provision of offspring with PUFA, and the timing of diapause. However, lipid production may also be considered as an adaptation by algae and bacteria against their consumers. Evidence is accumulating indicating that not all fatty acids are beneficial to consumers. Some are toxic or are precursors of toxic products, and the question therefore now arises as to why organisms produce such costly products.

Finally, lipid and fatty acid research has gained considerable applied importance as humans are often "top predators" and also depend on essential dietary nutrients. Public awareness of healthy nutrition is increasing, and this relates to both acquiring necessary food compounds and avoiding toxic contaminants. Lipids play a key role in these processes.

The past 10 years have seen a rapid increase in our knowledge about the ecological importance of lipids. As with all progressive scientific initiatives this new knowledge has also generated new questions. It is thus time for a new synthesis.

This book addresses most of the topics mentioned above; hence it is a timely book. I am sure it will not only summarize the status quo; it will also stimulate new research within the important and exciting field of biochemical aquatic ecology as well as foster new and fruitful connections with the field of human nutrition.

Plön, Germany   Winfried Lampert

Contents

Introduction xv
Michael T. Arts, Michael T. Brett, and Martin J. Kainz

**1 Algal Lipids and Effect of the Environment
on their Biochemistry** 1
Irina A. Guschina and John L. Harwood

**2 Formation and Transfer of Fatty Acids in Aquatic
Microbial Food Webs: Role of Heterotrophic Protists** 25
Christian Desvilettes and Alexandre Bec

3 Ecological Significance of Sterols in Aquatic Food Webs 43
Dominik Martin-Creuzburg and Eric von Elert

4 Fatty Acids and Oxylipins as Semiochemicals 65
Susan B. Watson, Gary Caldwell, and Georg Pohnert

**5 Integrating Lipids and Contaminants in
Aquatic Ecology and Ecotoxicology** 93
Martin J. Kainz and Aaron T. Fisk

6 Crustacean Zooplankton Fatty Acid Composition 115
Michael T. Brett, Dörthe C. Müller-Navarra, and Jonas Persson

**7 Fatty Acid Ratios in Freshwater Fish, Zooplankton
and Zoobenthos – Are There Specific Optima?** 147
Gunnel Ahlgren, Tobias Vrede, and Willem Goedkoop

**8 Preliminary Estimates of the Export of Omega-3
Highly Unsaturated Fatty Acids (EPA+DHA) from
Aquatic to Terrestrial Ecosystems** 179
Michail I. Gladyshev, Michael T. Arts, and Nadezhda, N. Sushchik

9	**Biosynthesis of Polyunsaturated Fatty Acids in Aquatic Ecosystems: General Pathways and New Directions**	211
	Michael V. Bell and Douglas R. Tocher	
10	**Health and Condition in Fish: The Influence of Lipids on Membrane Competency and Immune Response**	237
	Michael T. Arts and Christopher C. Kohler	
11	**Lipids in Marine Copepods: Latitudinal Characteristics and Perspective to Global Warming**	257
	Gerhard Kattner and Wilhelm Hagen	
12	**Tracing Aquatic Food Webs Using Fatty Acids: From Qualitative Indicators to Quantitative Determination**	281
	Sara J. Iverson	
13	**Essential Fatty Acids in Aquatic Food Webs**	309
	Christopher C. Parrish	
14	**Human Life: Caught in the Food Web**	327
	William E. M. Lands	
Name Index		355
Subject Index		367

Contributors

Gunnel Ahlgren
Department of Ecology and Evolution (Limnology), Uppsala University,
P.O. Box 573, 751 23 Uppsala, Sweden
gunnel.ahlgren@ebc.uu.se

Michael T. Arts
Aquatic Ecosystems Management Research Division, National Water Research
Institute – Environment Canada, P.O. Box 5050, 867 Lakeshore Road,
Burlington, ON, Canada L7R 4A6
michael.arts@ec.gc.ca

Alexandre Bec
Laboratoire de Biologie des Protistes, Université Blaise Pascal,
Clermont-Ferrand II, Campus des Cézeaux, 63177 Aubiere Cedex, France
alexandre.bec@univ-bpclermont.fr

Michael V. Bell
Institute of Aquaculture, University of Stirling, Stirling, Stirlingshire FK9 4LA, UK
mvb1@stir.ac.uk

Michael T. Brett
Department of Civil & Environmental Engineering, University of Washington,
Box 352700, 301 More Hall, Seattle, WA 98195-2700, USA
mtbrett@u.washington.edu

Gary Caldwell
School of Marine Science and Technology, Newcastle University, Ridley
Building, Rm 354, Claremont Road, Newcastle upon Tyne NE1 7RU, UK
gary.caldwell@newcastle.ac.uk

Christian Desvilettes
Laboratoire de Biologie des Protistes, Université Blaise Pascal,
Clermont-Ferrand II, Campus des Cézeaux, 63177 Aubiere Cedex, France
christian.desvilettes@univ-bpclermont.fr

Aaron T. Fisk
Department of Biology (Great Lakes Institute for Environmental Research),
University of Windsor, 2990 Riverside Drive West, Windsor, ON,
Canada N9B 2P3
afisk@uwindsor.ca

Michail Gladyshev
Institute of Biophysics, Siberian Branch of the Russian Academy of Sciences,
660036 Krasnoyarsk, Akademgorodok, Russia
glad@ibp.ru

Willem Goedkoop
Department of Environmental Assessment, Swedish University of Agricultural
Sciences, Box 7050, 750 07 Uppsala, Sweden
willem.goedkoop@ma.slu.se

Martin Graeve
Pelagic Ecosystems/Marine Chemistry and Marine Natural Products,
Alfred Wegener Institut für Polar- und Meeresforschung, Am Handelshafen 12,
27570 Bremerhaven, Germany
mgraeve@awi-bremerhaven.de

Irina A. Guschina
School of Biosciences, Cardiff University, P.O. Box 911, Cardiff CF10 3US,
Wales, UK
guschinaia@cf.ac.uk

Wilhelm Hagen
Marine Zoology (FB2), Universität Bremen, P.O. Box 330440,
28334 Bremen, Germany
whagen@uni-bremen.de

John L. Harwood
School of Biosciences, Cardiff University, P.O. Box 911, Cardiff CF10 3US,
Wales, UK
harwood@cardiff.ac.uk

Sara Iverson
Department of Biology – Life Sciences Centre, Dalhousie University,
1355 Oxford Street, Halifax, NS, Canada B3H 4J1
sara.iverson@dal.ca

Martin J. Kainz
WasserKluster Lunz – Biologische Station, Dr. Carl Kupelwieser Promenade 5,
A-3293 Lunz am See, Austria
martin.kainz@donau-uni.ac.at

Gerhard Kattner
Pelagic Ecosystems/Marine Chemistry and Marine Natural Products,
Alfred Wegener Institut für Polar- und Meeresforschung, Am Handelshafen 12,
27570 Bremerhaven, Germany
gkattner@awi-bremerhaven.de

Christopher C. Kohler
Director, Fisheries and Illinois Aquaculture Center, Southern Illinois University,
Carbondale, IL 62901-6511 USA,
ckohler@siu.edu

Winfried Lampert
Max Planck Institute for Limnology, Plön, Germany
lampert@mpil-ploen.mpg.de

William E. M. Lands
6100 Westchester Park Drive, Apt. #1219, College Park, MD 20740, USA
wemlands@att.net

Dominik Martin-Creuzburg
Limnological Institute, Universität Konstanz, Mainaustrasse 252,
78464 Konstanz, Germany
dominik.martin-creuzburg@uni-konstanz.de

Dörthe Müller-Navarra
Aquatic Ecology, Universität Hamburg, Zeiseweg 9, 22609 Hamburg, Germany
doerthe.mueller-navarra@uni-hamburg.de

Christopher C. Parrish
Ocean Sciences Centre, Memorial University of Newfoundland,
St. John's, NF, Canada A1C 5S7
cparrish@mun.ca

Jonas Persson
Department of Ecology and Evolution, Uppsala University, Husargatan 3,
75 123 Uppsala, Sweden
jonas.persson@ebc.uu.se

Georg Pohnert
Laboratory of Chemical Ecology – LECH, Ecole Polytechnique Fédérale
de Lausanne, EPFL SB ISIC LECH – BCH 4306, 1015 Lausanne, Switzerland
georg.pohnert@epfl.ch

Nadezhda N. Sushchik
Institute of Biophysics, Siberian Branch of the Russian Academy of Sciences,
660036 Krasnoyarsk, Akademgorodok Russia,
labehe@ibp.ru

Douglas R. Tocher
Institute of Aquaculture, University of Stirling, Stirling, Stirlingshire FK9 4LA, UK
d.r.tocher@stir.ac.uk

Eric von Elert
Institute of Zoology, Universität zu Koeln, Weyertal 119, 50923 Koeln, Germany
evelert@uni-koeln.de

Tobias Vrede
Department of Ecology and Environmental Sciences, Umeå University,
90187 Umeå, Sweden
tobias.vrede@emg.umu.se

Susan B. Watson
Aquatic Ecosystems Management Research Division, National Water Research
Institute – Environment Canada, P.O. Box 5050, 867 Lakeshore Road,
Burlington, ON, Canada L7R 4A6
sue.watson@ec.gc.ca

Lipids in Aquatic Ecosystems

Michael T. Arts, Michael T. Brett, and Martin J. Kainz

Introduction

Life began as a process of self-organization within a lifeless environment. For single and, subsequently, multicellular organisms to differentiate themselves from the outside world, they needed an effective, adaptable barrier (i.e., the cell/cytoplasmic membrane). The modern cell membrane is mainly composed of phospholipids, proteins, and sterols, which in unison regulate what goes into and out of the cell. Some have hypothesized that spontaneously formed phospholipid bilayers played a key role in the origin of life. The precise structure and composition of these biochemical groups have an enormous influence on the integrity and physiological competency of the cell. It should not be surprising that this organizational and functional specificity at the cellular level readily translates into profound systemic effects at the macroscopic level. Thus, cellular lipid composition and organization orchestrate both subtle and obvious effects on the health and function of organisms → populations → communities → ecosystems.

Ecology is, by its very nature, an integrative field of inquiry that actively promotes the examination of processes that span both cellular and macroscopic levels of organization. Modern ecologists are challenged and motivated to put their research into a broader perspective; ecology thrives at the intersections of disciplines! Lipids provide an effective platform for this mandate because they are a global energy currency and because of their far-reaching physiological roles in aquatic and terrestrial biota. Two previous, comprehensive efforts to examine the role of lipids in aquatic environments exist. The first (Gulati and DeMott 1997) arose as the proceedings of an international workshop held at Nieuwersluis, the Netherlands in 1996. The objective of this workshop was "to take stock of the state of the art in food quality research, to address factors that determine food quality" and "to integrate the available information into a coherent and consistent view of

---

M.T. Arts (✉), M.T. Brett, and M.J. Kainz
Aquatic Ecosystems Management Research Division, National Water Research Institute –
Environment Canada, P.O. Box 5050, 867 Lakeshore Road, Burlington, ON, Canada L7R 4A6
e-mail: Michael.Arts@ec.gc.ca

food quality for the zooplankton." A second, more extensive publication followed 2 years later (Arts and Wainman 1999). That publication set about to "establish a general reference and review book for those interested in aquatic lipids" and to "demystify lipid research." Its focus was mainly on freshwater ecosystems. Since these two publications in the late 1990s, the field has advanced considerably, most notably in such areas as:

- Refining the understanding of the essentiality of specific lipids
- Biochemical pathways and controls on PUFA synthesis and degradation
- Fatty acid as trophic markers
- Importance/essentiality of sterols
- Integrating contaminant and lipid pathways
- Trophic upgrading by protists, heterotrophic flagellates, and zooplankton
- Role of fatty acids and other lipids in the maintenance of membrane fluidity
- Role of fatty acids in cell signaling
- Effect of essential fatty acids (EFAs) on human health and behavior (e.g., n-3 deficiency)
- EFAs as seen from a conservation perspective

Advances such as these convinced us that, nearly a decade after the first edition, a second book project should be undertaken. We envisioned that this book should (a) have a much broader mandate than the original; for example, it should encompass both freshwater and marine ecosystems, (b) touch on several of the recent advances highlighted above, and (c) break new ground by interconnecting the fields of lipid research with other highly topical areas such as climate change, conservation, and human health.

A survey of the literature clearly shows that interest in lipids within environmental sciences is increasing almost exponentially. As more detailed and informative experiments and observations are made, it is becoming clear that some lipids (e.g., the long chain, polyunsaturated, omega-3 fatty acid "docosahexaenoic acid" or "DHA" for short, 22:6n-3) have a critical role to play in maintaining the health and functional integrity of both aquatic and terrestrial organisms. Thus, the more general interest in lipids as structural components and as purveyors of energy is increasingly being coupled with this deeper understanding resulting in a parallel increase in publications dealing specifically with individual lipid molecules such as DHA.

The chapters in this book are broadly organized so as to elaborate and synthesize concepts related to the role of lipids from lower to higher trophic levels up to and including humans – an objective that has seldom been attempted from an ecological perspective. A précis of the book's 14 chapters follows:

In Chap. 1, "Algal Lipids and Effect of the Environment on Their Biochemistry," Irina Guschina and John Harwood explore the origins and synthesis of a wide variety of algal lipids (glycolipids, phospholipids, betaine lipids, and nonpolar glycerolipids) and provide important clues as to how environmental signals (temperature, light, salinity, and pH) may influence the production of specific lipids and lipid classes. Their chapter concludes with a concise summary of how nutrients and nutrient regimes affect the production of lipids in algae.

Introduction

The second chapter, "Formation and Transfer of Fatty Acids in Aquatic Microbial Food Webs: Role of Heterotrophic Protists," by Christian Desvilettes and Alexandre Bec provides details on the biosynthesis pathways for polyunsaturated fatty acids in heterotrophic protists and, in so doing, demonstrates that protists may perform an ecologically important service by "trophically upgrading" some fatty acid molecules to more physiologically active forms for zooplankton and eventually fish consumers. They also showcase the variability in lipid profiles among protists.

In "Ecological Significance of Sterols in Aquatic Food Webs" (Chap. 3), Dominik Martin-Creuzberg and Eric von Elert demonstrate that sterols play key roles in the physiological processes of all eukaryotic organisms. Their chapter provides details on the occurrence and biosynthesis of sterols followed by an informative summary of the physiological properties and nutritional requirements for sterols. These authors use an ecological perspective to demonstrate how sterols affect herbivorous zooplankton, trophic interactions, and food web processes.

In Chap. 4, "Fatty Acids and Oxylipins as Semiochemicals," Susan Watson, Gary Caldwell and Georg Pohnert showcase the subtlety of chemical communication in aquatic ecosystems. In so doing, they expose a "darker" side of lipids and demonstrate that, under some conditions, certain lipids (e.g., aldehydes derived from polyunsaturated fatty acids) can induce a range of negative effects in aquatic organisms. They also reveal that aquatic organisms are capable of avoidance behaviors, detoxification, and other adaptive strategies to either avoid or deal with exposure to toxic lipids.

"Integrating Lipids and Contaminants in Aquatic Ecology and Ecotoxicology" (Chap. 5) is a relatively new area being pioneered by Martin Kainz and Aaron Fisk. They show that the uptake of contaminants, both lipophilic and hydrophilic, and EFAs can be coupled in aquatic organisms but that, sometimes with the appropriate ecological foreknowledge, actions and procedures can be instituted to minimize risk and maximize benefit. They stress the ecotoxicological need to understand how potential contaminants are linked with lipids and their specific structural and/or storage compounds at the cell, tissue, and, eventually, at the food web level.

The subject of biomarkers has received a great deal of attention in the last decade. Zooplankters, such as members of the herbivorous genus *Daphnia*, provide excellent opportunities to test the veracity of the biomarker concept. Thus, in Chap. 6, "Crustacean Zooplankton Fatty Acid Composition," Michael Brett, Dörthe Müller-Navarra, and Jonas Persson provide a state-of-the-art summary of what is known about how taxonomic affiliation and diet influence the fatty acid composition of freshwater and marine zooplankton. This chapter also explores the literature on reproductive investments in essential lipids, as well as temperature and starvation impacts on zooplankton fatty acid profiles.

Clearly essential or growth regulating fatty acids must be supplied in appropriate proportions. This is especially true of the highly physiologically active fatty acids such as arachidonic, eicosapentaenoic, and docosahexaenoic acids. Gunnel Ahlgren, Tobias Vrede, and Willem Goedkoop have, in their chapter (Chap. 7) "Fatty Acid Ratios in Freshwater Fish, Zooplankton and Zoobenthos – Are There

Specific Optima?," integrated a large body of information which suggests that specific optima between specific omega-3 and omega-6 fatty acids do indeed exist for aquatic biota.

Establishing a more formal link between aquatic and terrestrial ecosystems, with respect to the fate and distribution of EFAs, requires that "Preliminary Estimates of the Export of Omega-3 Highly Unsaturated Fatty Acids (EPA + DHA) from Aquatic to Terrestrial Ecosystems" be conducted. Michail Gladyshev, Michael Arts, and Nadezhda Sushchik (Chap. 8) demonstrate the strengths and inherent weaknesses of this approach, and call for more studies to fill in the current gaps in our knowledge. They also highlight the new concept that aquatic ecosystems, in addition to their previously established roles, should now also be seen as key purveyors of essential PUFA to terrestrial ecosystems.

A clear understanding of the pathways of synthesis is a prerequisite to understanding the potential limitations faced by aquatic organisms in nature. In Chap. 9, "Biosynthesis of Polyunsaturated Fatty Acids in Aquatic Ecosystems: General Pathways and New Directions," Michael Bell and Douglas Tocher provide a succinct summary of what we know about the biosynthesis of fatty acids in fish. They also provide a stimulating section on potential future directions of research on the biosynthesis of fatty acids by aquatic organisms.

In Chap. 10, "Health and Condition in Fish: The Influence of Lipids on Membrane Competency and Immune Response," Michael Arts and Christopher Kohler comment on the role that specific fatty acids play in maintaining the health and condition of teleost cell membranes especially in terms of temperature adaptation and on the close association between EFAs and healthy immune system function.

Global warming is currently a center stage issue in science. In Chap. 11, "Lipids in Marine Copepods: Latitudinal Characteristics and Perspective to Global Warming," Gerhard Kattner and Wilhelm Hagen showcase the enormous diversity in marine copepod lipid profiles and demonstrate that these profiles have evolved in response to the specific habitats and temperature regimes occupied by the various copepod species. They sugggest that the effects of climate change on species shifts and consequently lipid profiles may not be straightforward and predictable.

Researchers interested in using fatty acid trophic markers to explore food web dynamics have begun to realize that the "honeymoon phase" is over. There is a real need for more quantitative methods to determine the impact of particular diet organisms on the lipid profiles of consumers. Sara Iverson (Chap. 12), "Tracing Aquatic Food Webs Using Fatty Acids: From Qualitative Indicators to Quantitative Determination," introduces us to the underlying assumptions, concepts, and development of the quantitative fatty acid signature analysis (QFASA) approach and elaborates both the strengths and weaknesses of this tool.

The concept of essentiality of fatty acids is discussed in detail by Christopher Parrish in Chap. 13 – "Essential Fatty Acids in Aquatic Food Webs." The chapter starts with a definition of what constitutes an EFA and then highlights some of the key effects of EFAs on aquatic organisms. He concludes by making the case that particular n-6 fatty acids (e.g., 22:5n-6) should also be included in the list of EFAs.

Humans occupy a singularly unique position in the global food chain. We are at once free from the "rules" that govern the population dynamics of other species and yet we are also constrained by many of the same biochemical requirements. So then, why are algae and human brains linked by the fact that docosahexaenoic acid is the most prevalent fatty acid in brain tissue (which is ~60% lipid by dry weight), but DHA is produced de novo primarily by algae and some fungi? And what is the connection between this knowledge and the fact that fish have had, and continue to have, a deeply embedded cultural significance in our psyche (Reis and Hibbeln 2006)? In an effort to address these questions William (Bill) Lands' thought-provoking Chap. 14, "Human Life: Caught in the Food Web," examines the position of humans in the global food web and highlights our requirements for essential omega-3 fatty acids, thereby underscoring the urgency of protecting and enhancing the aquatic food web → human nutrition connection.

This book should appeal to a broad audience from divergent fields. Our readers are expected to include academics/graduate students, government researchers, and resource managers interested in understanding how these essential compounds affect the function and dynamics of aquatic ecosystems in their sphere of influence. Specific audiences likely to have an interest in this book include:

- *Plankton ecologists and physiologists* – interested in (a) the relationship between lipid production in algae and various environmental variables including nutrient concentrations, nutrient ratios, underwater light climate, and temperature and (b) the dynamics of transfer and retention and synthesis of EFAs in zooplankton because such an understanding is a prerequisite to a better understanding of fish production, cold tolerance, and fitness in both marine and freshwater ecosystems.
- *Nutritionists* – It is now well recognized that EFAs play a critical role in the health and well-being of all vertebrates including humans. What is less clear, given global declines in fish stocks, is how we can maintain sustainable EFA production at the base of the food chain for ultimate incorporation into the human diet stream and also what alternatives exist to ensure our continued access to these essential compounds.
- *Aquaculturists* – It is now well established, from both laboratory and field studies, that EFAs contribute to the somatic growth and productivity of invertebrates and fish. Thus, the burgeoning field of aquaculture has a strong interest in understanding the role of lipids and, in particular, the role of EFAs in optimizing/maximizing the EFA content of commercially raised and harvested species, while, simultaneously, minimizing the bioaccumulation of potential contaminants.
- *Toxicologists* – It is now clear that a more thorough knowledge of the distribution, type, concentrations, and pathways of lipids within and amongst organisms in aquatic systems is crucial for understanding how heavy metals (e.g., the neurotoxin methyl mercury) and lipophilic contaminants (e.g., PCBs) are accumulated in aquatic organisms and eventually in humans. Thus, environmental managers, working in consultation with health professionals, have a strong

interest in providing environmental management solutions that minimize contaminant loads while simultaneously maximizing EFA availability in fish.
- *Environmental chemists* – Environmental chemists will gain a deeper understanding of the more holistic, ecological effects that lipids have on living organisms and, by extension, on the relationships between lipids and higher scale processes (biochemical and ecological) at the population and ecosystem level.
- *Environmental managers* – It is anticipated that policy makers, charged with overseeing either degraded and/or pristine ecosystems, will profit from a deeper understanding of the role that EFAs play in maintaining the health and ecological integrity of aquatic ecosystems. Superimposed over this, and of imminent concern to policy makers, is the specter of climate change with its, as yet largely unappreciated, potential to alter EFA production at the base of the food web.

The global ecosystem faces many threats (e.g., climate change, cultural eutrophication, contaminants, invasive species, declining fish stocks, UV radiation, and overpopulation). The study of lipid dynamics is germane to understanding the consequences of many of these threats because lipids are sensitive, and both specific and broad, indicators of stress and change. The study of lipids in aquatic ecosystems also provides an effective vehicle for bringing different disciplines together. This is important because, in order to better define the consequences of global threats to ecosystem sustainability, we need integrative interdisciplinary science that allows us to scale up from the very specific biochemical and physiological roles that lipids have to their broader effects on energy flow in food webs, fisheries production, contaminant accumulation and, ultimately, human health at a global scale.

The editors and contributors of this book are greatly indebted to the many people who made this book possible. In particular, we extend our heartfelt appreciation to our external anonymous reviewers.

References

Arts, M.T. and B.C. Wainman (eds.), 1999. Lipids in Freshwater Ecosystems. Springer, New York, 319 pp.

Gulati, R.D. and W.R. DeMott (eds.), 1997. The role of food quality for zooplankton: remarks on the state-of-the art, perspectives and priorities. Freshwater Biology 38: 455–768.

Reis, L.C. and J.R. Hibbeln. 2006. Cultural symbolism of fish and the psychotropic properties of omega-3 fatty acids. Prostaglandins Leukotrienes Essential Fatty Acids 75: 227–236.

Chapter 1
Algal Lipids and Effect of the Environment on their Biochemistry

Irina A. Guschina and John L. Harwood

1.1 Introduction

Lipids play a number of roles in living organisms and can be divided into two main groups: the nonpolar lipids (acylglycerols, sterols, free (nonesterified) fatty acids, wax, and steryl esters) and polar lipids (phosphoglycerides, glycosylglycerides). Polar lipids and sterols are important structural components of cell membranes which act as a selective permeable barrier for cells and organelles. These lipids maintain specific membrane functions providing the matrix for a very wide variety of metabolic processes and participate directly in membrane fusion events. In addition to a structural function, some polar lipids may act as key intermediates (or precursors of intermediates) in cell signalling pathways (e.g. inositol lipids, sphingolipids, oxidative products) and play a role in responding to changes in the environment. Of the nonpolar lipids, the triacylglycerols are abundant storage products, which can be easily catabolised to provide metabolic energy (Gurr et al. 2002). Waxes are common extracellular surface-covering compounds but may act (in form of wax esters) as energy stores especially in organisms from cold water habitats (Guschina and Harwood 2007). Sterols of algae have been studied extensively and a number of comprehensive reviews are already available on these nonpolar lipids (e.g., Patterson 1991; Volkman 2003; see also Chap. 3).

Algae are important constituents of aquatic ecosystems, accounting for more than half the total primary production at the base of the food chain worldwide. Algal lipids are major dietary components for primary consumers where they are a source of energy and essential nutrients. The role of algal polyunsaturated fatty acids (including the human essential fatty acids linoleic (LIN; 18:2n-6) and α-linolenic (ALA; 18:3n-3) as well as eicosapentaenoic acid (EPA; 20:5n-3) and docosahexaenoic acid (DHA; 22:6n-3) in aquatic food webs is well documented (e.g., see Chaps. 6 and 13). They provide a substantial contribution to the food quality for invertebrates and are vital for maintaining somatic and population growth, survival, and

---

I.A. Guschina and J.L. Harwood (✉)
School of Biosciences, Cardiff University, Museum Avenue, Cardiff CF10 3US, Wales, UK
harwood@cardiff.ac.uk

reproductive success. Not only are they important membrane components, but polyunsaturated fatty acids (PUFA) are involved in the regulation of physiological processes by serving as precursors in the biosynthesis of bioactive molecules such as prostaglandins, thromboxanes, leukotrienes, and resolvins, which may affect egg-production, egg-laying, spawning and hatching, mediating immunological responses to infections, and have a wide range of other functions (Brett and Müller-Navarra 1997). The fatty acids are constituents of most algal lipids and rarely occur in the free form. They are mainly esterified to glycerolipids whose main classes in algae are the phosphoglycerides, glycosylglycerides and triacylglycerols. In the present chapter, we will give an overview of lipid composition in algae with a special emphasis on how environmental factors may affect algal glycerolipid biochemistry.[1]

1.2 Lipid Composition of Algae

1.2.1 *Polar Glycerolipids*

In general, algae have a glycerolipid composition similar to that of higher plants, although some species also contain unusual lipids. The basic structure of glycerolipids is a glycerol backbone metabolically derived from glycerol 3-phosphate to which the hydrophobic acyl groups are esterified at the *sn*-1 and *sn*-2 positions, and there are three main types. Glycosylglycerides are characterized by a 1,2-diacyl-*sn*-glycerol moiety with a mono- or oligosaccharide attached at the *sn*-3 position of the glycerol backbone. Phospholipids have phosphate esterified to the *sn*-3 position with a further link to a hydrophilic head group. Betaine lipids contain a betaine moiety as a polar group, which is linked to the *sn*-3 position of glycerol by an ether bond. There are no phosphorus or carbohydrate groups in betaine lipids.

1.2.1.2 Glycolipids

In algae (as in higher plants and cyanobacteria), glycolipids (glycosylglycerides) are located predominantly in photosynthetic membranes. The major plastid lipids, galactosylglycerides, are uncharged. They contain one or two galactose molecules linked to the *sn*-3 position of the glycerol corresponding to 1,2-diacyl-3-*O*-(β-D-galactopyranosyl)-*sn*-glycerol (or monogalactosyldiacylglycerol, MGDG) and 1,2-diacyl-3-*O*-(α-D-galactopyranosyl-(1,6)-*O*-β-D-galactopyranosyl-*sn*-glycerol (or digalactosyldiacylglycerol, DGDG) (Fig. 1.1). MGDG and DGDG represent

---

[1]For comprehensive descriptions of the biosynthesis of algal and plant lipids see Harwood and Jones (1989), Guschina and Harwood (2006a) and Murphy (2005) and references therein.

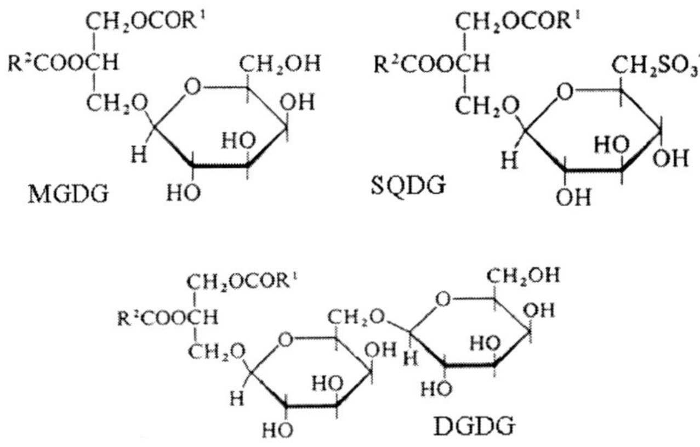

Fig. 1.1 The main glycosylglycerides of algae. R1 and R2 are the two fatty acyl chains. *MGDG* monogalactosyldiacylglycerol; *DGDG* digalactosyldiacylglycerol; *SQDG* sulfoquinovosyldiacylglycerol

40–55 and 15–35% of thylakoid lipids, respectively. Another class of glycosylglyceride, which is present in appreciable amounts (e.g., up to 29% of total lipids in the red tide alga *Chattonella antique* and 22% in the bladder wrack seaweed *Fucus vesiculosus* (Harwood and Jones 1989)) in both photosynthetic and in non-photosynthetic algal tissues, is the plant sulfolipid, sulfoquinovosyldiacylglycerol, or 1,2-diacyl-3-*O*-(6-deoxy-6-sulfo-α-D-glucopyranosyl)-*sn*-glycerol (SQDG) (Fig. 1.1). This lipid is unusual because of its sulfonic acid linkage. It consists of monoglycosyldiacylglycerol with a sulfonic acid in position 6 of monosaccharide moiety. The sulfonoglucosidic moiety (6-deoxy-6-sulfono-glucoside) is described as sulfoquinovosyl. The sulfonic residue carries a full negative charge at physiological pH giving the sulfolipid distinct properties.

A unique feature of plastid galactolipids is their very high content of PUFA. Similar to higher plants, MGDG of fresh water algae contains ALA as the major fatty acid, and ALA and palmitic acid (16:0) are dominant in DGDG and SQDG. The glycolipids from some algal species, e.g. green algae *Trebouxia* spp., *Coccomyxa* spp., *Chlamydomonas* spp., may also be esterified with unsaturated C16 acids, such as hexadecatrienoic (16:3n-3) and hexadecatetraenoic (16:4n-3) acids (Guschina et al. 2003; Arisz et al. 2000). The plastidial glycosylglycerolipids of marine algae contain, in addition to 18:3n-3 and 16:0, some very-long-chain PUFA, e.g. arachidonic (ARA; 20:4n-6), EPA, DHA as well as octadecatetraenoic acid (18:4n-3). In contrast, a complex mixture of SQDG has been identified in an extract of the marine chromonad *Heterosigma carterae* (Raphidophyceae) with the main fatty acyl residues consisting of 16:0, 16:1n-7, 16:1n-5, 16:1n-3, and EPA (Keusgen et al. 1997). MGDG from the marine diatom *Skeletonema costatum* contained another unusual fatty acid (18:3n-1) in significant amounts (~25%) (D'Ippolito et al. 2004). Table 1.1 shows some examples of the fatty acid distribution

Table 1.1 Some examples of the fatty acid composition (% of total fatty acids) of glycosylglycerides from different algal species

Lipid Class	16:0	16:1	16:2 n-6	16:3 n-3	16:4 n-3	18:0	18:1 n-9	18:1 n-7	18:2 n-6	18:3 n-6	18:3 n-3	18:4 n-3	20:4 n-6	20:5 n-3	22:6 n-3
Chlorophyta	*Chlamydomonas moewusii* (Arisz et al. 2000)														
MGDG	2	1[a]	2	4	36	tr.	2	–	9	–	43	–	–	–	–
DGDG	28	2[a]	8	11	2	–	19	2	10	–	18	–	–	–	–
SQDG	81	–	–	–	–	1	3	2	5	–	9	1	–	–	–
	Parietochloris incisa (Bigogno et al. 2002a)														
MGDG	2	1[b]	9	21	–	tr.	4	1	15	1	32	–	14	1	–
DGDG	16	2[c]	1	2	–	2	6	4	26	2	19	–	18	1	–
SQDG	36	tr.	–	–	–	2	4	13	21	1	19	–	3	–	–
Haptophyta	*Pavlova lutheri* (Eichenberger and Gribi 1997)														
MGDG	8	9	–	–	–	–	–	–	3	4	4	26	tr.	44	–
DGDG	19	10	–	–	–	–	–	–	tr.	2	2	13	tr.	49	1
SQDG[d]	45	10	–	tr.	–	tr.	6[e]	–	tr.	tr.	tr.	1	tr.	3	tr.
	Isochrysis galbana (Alonso et al. 1998)														
MGDG	8	20	–	–	–	1	2	1	2	–	4	15	–	28	5
DGDG	15	18	–	–	–	tr.	1	2	1	–	3	12	–	25	7
SQDG[d]	29	20	–	–	–	1	1	4	1	–	1	3	–	8	1
Bacillariophyta	*Phaedactylum tricornutum* (Alonso et al. 1998)														
MGDG	7	20	13	18	4	1	tr.	tr.	1	–	–	–	2	31	tr.
DGDG	12	22	13	17	5	tr.	tr.	tr.	2	–	–	–	2	22	tr.
SQDG[d]	40	31	3	–	–	2	1	–	–	–	–	–	1	11	–
Rhodophyta	*Porphyridium cruentum* (Alonso et al. 1998)														
MGDG	30	2	–	–	–	7	3	1	8	–	–	–	21	16	–
DGDG	40	3	–	–	–	2	2	3	7	–	–	–	18	20	–
SQDG	48	–	–	–	–	2	1	2	5	–	–	–	14	19	–

The positions of double bonds were assigned following capillary gas-liquid chromatography but were not confirmed by other methods. For lipid abbreviations see text. In *Phaedactylum tricornutum* 16C unsaturated fatty acids were reported as 16:2n-4, 16:3n-4 and 16:4n-1. *Dashes* mean none detected, *tr:* trace. For other examples refer to Harwood and Jones (1989) and Gunstone et al. (2007)

[a]*n*-9 isomer
[b]*n*-11 isomer
[c]The sum of n-11 and n-7
[d]Significant amount of 14:0 also detected
[e]The sum of two isomers of 18:1 given

in algal glycolipids from various taxonomical groups. More examples may be found in Harwood (1998a).

In addition to these common glycolipids, a few unusual lipids have been reported in some algal species. Trigalactosylglycerol has been identified in *Chlorella* (cited by Harwood and Jones 1989). In some red algae, glycolipids may contain sugars other than galactose (e.g. mannose and rhamnose) (Harwood and Jones 1989). From the marine red alga, *Gracilaria verrucosa*, a new glycolipid, sulfoquinovosylmonogalactosylglycerol (SQMG) has been isolated (Son 1990).

A carboxylated glycoglycerolipid, diacylglyceryl glucuronide (DGGA) has been described in *Ochromonas danica* (Chrysophyceae) and in *Pavlova lutheri* (Haptophyceae) (Eichenberger and Gribi 1994, 1997). In *O. danica*, this glycolipid accounted for ~3% of the glycerolipids of the alga with the predominant molecular species being a 20:4/22:5(*sn*-1/*sn*-2)-combination. In *P. lutheri*, this lipid was enriched in 22:5n-6 and DHA (44.4 and 18.9%, respectively) (Fig. 1.2).

A new glycoglycerolipid bearing the extremely rare 6-deoxy-6-aminoglucose moiety (avrainvilloside) has been isolated from marine green alga *Avrainvillea nigricans* and its structure was established on the basis of spectroscopic data and methanolysis/GC-MS analysis (Andersen and Taglialatela-Scafati 2005). As has been shown recently, six minor new glycolipids were present in crude methanolic extracts of the red alga, *Chondria armata* (Al-Fadhli et al. 2006). These included 1,2-di-*O*-acyl-3-*O*-(acyl-6′-galactosyl)-glycerol (GL$_{1a}$) and the sulfonoglycolipids 2-*O*-palmitoyl-3-*O*-(6′-sulfoquinovopyranosyl)-glycerol and its ethyl ether derivative. GL$_{1a}$ was the first example of the natural occurrence of an acyl glycolipid acylated at the *sn*-1, *sn*-2 of glycerol and 6′ positions of galactose (Al-Fadhli et al. 2006).

1.2.1.3 Phospholipids

The major phospholipids (phosphoglycerides) in most algae species are phosphatidylcholine (PC), phosphatidylethanolamine (PE), and phosphatidylglycerol (PG) (Fig. 1.3). In addition, phosphatidylserine (PS), phosphatidylinositol (PI), and diphosphatidylglycerol (DPG) (or cardiolipin) may be also found in substantial amounts.

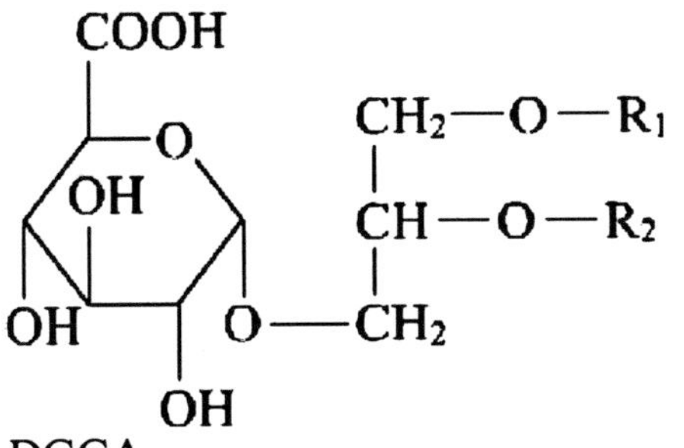

Fig. 1.2 Structure of diacylglyceryl glucuronide (DGGA). R1 and R2 are C_{18}, C_{20} and C_{22} polyunsaturated fatty acids in *P. lutheri* (Eichenberger and Gribi 1997)

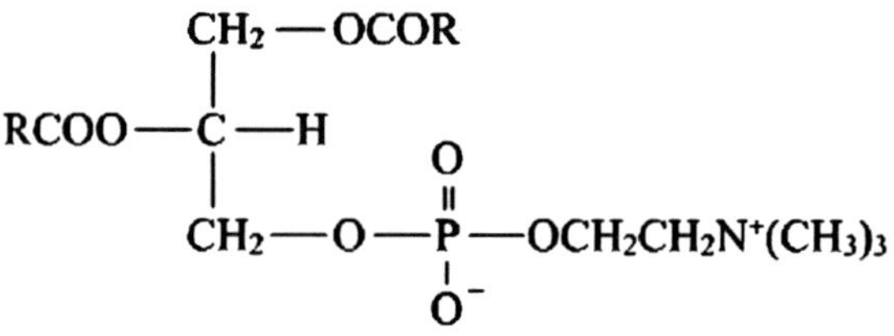

Fig. 1.3 Major phosphoglycerides of algae. *PC* phosphatidylcholine; *PE* phosphatidylethanolamine; *PG* phosphatidylglycerol; *PI* phosphatidylinositol

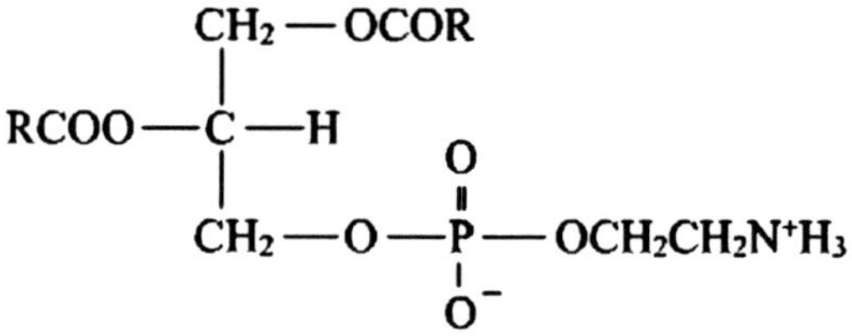

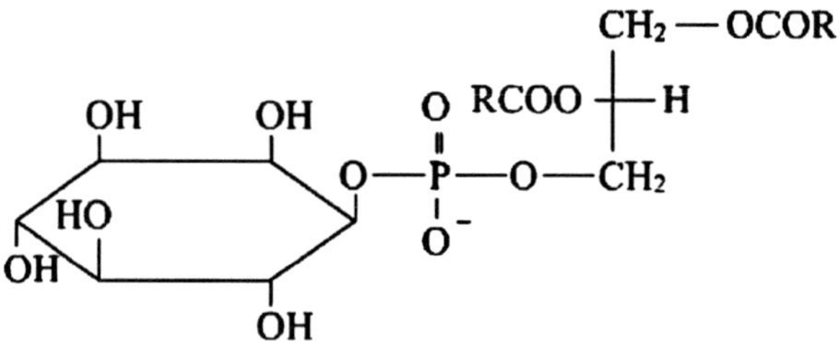

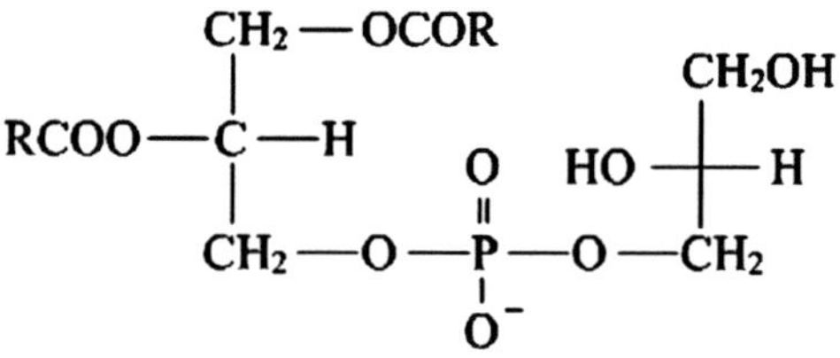

Phosphatidic acid is noted as a minor compound. Their structure is characterized by a 1,2-diacyl-3-phospho-*sn*-glycerol, or phosphatidyl moiety, and a variable headgroup linked to the phosphate.

The phospholipids are located in the extra-chloroplast membranes with the exception of PG, which is the only phospholipid present in significant quantities in the thylakoid membranes. PG represents between 10 and 20% of the total polar glycerolipids in eukaryotic green algae. An unusual fatty acid, Δ^3-*trans*-hexadecenoic acid (16:1(3*trans*)), usually esterified to the *sn*-2 position of PG, is found in all eukaryotic photosynthetic organisms (Tremolieres and Siegenthaler 1998). It is interesting to note that both the *trans*-configuration and the Δ^3 position of the double bond are very unusual for naturally occurring fatty acids. A sulfonium analog of phosphatidylcholine has been described in diatoms (Anderson et al. 1978a, b; Bisseret et al. 1984). This lipid contains a sulphur atom replacing the nitrogen atom of choline: $- S^+(CH_3)_2$ instead of $- N^+(CH_3)_3$. In a non-photosynthetic diatom, *Nitzschia alba*, this phosphatidylsulfocholine (PSC) completely replaces PC, whereas in four other diatom species both lipids were present and their total relative amount varied from 6 to 24% of the total lipids. Trace amounts of PSC (less than 2%) were also found in diatoms *Cyclotella nana* and *Navicula incerta* as well as in a *Euglena* sp. (Bisseret et al. 1984).

From brown algae, a novel lipid constituent was isolated and identified as phosphatidyl-*O*-[*N*-(2-hydroxyethyl)glycine] (PHEG) (Eichenberger et al. 1995). This lipid contains glycine as a headgroup ($-CH_2-CH_2-NH-CH_2-COOH$) and was present at between 8 and 25 mol% of total phospholipids in the 30 algal species analysed. It has been shown that this common lipid component of brown algae was accumulated in the plasma membrane of gametes of *Ectocarpus* species. With its fatty acid composition rich in ARA (80%) and EPA (10%), PHEG is considered a potential acyl donor for pheromone production and hence, possibly involved in the fertilization process of these algae (Eichenberger et al. 1995).

1.2.1.4 Betaine Lipids

Three types of betaine lipids have been identified, 1,2-diacylglyceryl-3-*O*-4'-(*N,N,N*-trimethyl)-homoserine (DGTS), 1,2-diacylglyceryl-3-*O*-2'-(hydroxymethyl)-(*N,N,N*-trimethyl)-β-alanine (DGTA), and 1,2-diacylglyceryl-3-*O*-carboxy-(hydroxymethyl)-choline (DGCC) (Dembitsky 1996) (Fig. 1.4). Table 1.2 shows some examples of the distribution of DGTS and DGTA in algae. These betaine lipids are all zwitterionic at neutral pH since they have a positively-charged trimethylammonium group and a negatively-charged carboxyl group (Fig. 1.4).

Betaine lipids are not found in higher plants, either gymnosperms or angiosperms, but are quite widely distributed in algae (as well as in ferns, bryophytes, lichens, some fungi and protozoans). The distribution of betaine lipids in various taxonomic groups of algae has been thoroughly reviewed by Dembitsky (1996) and Kato et al. (1996). On the basis of an obvious structural similarity between betaine lipids and phosphatidylcholine and on their taxonomical distribution (namely, their reciprocal relationship in many species), it has been suggested that betaine lipids, especially DGTS, are more evolutionarily primitive lipids which, in the lower plants, play the same functions in membranes that PC does in higher plants and animals (Dembitsky 1996).

In many algal species analyzed, the fatty acid composition of DGTS has been shown to vary significantly between freshwater and marine species. In freshwater algae, mainly saturated fatty acids (14:0 and 16:0) are found at the *sn*-1 position of the glycerol backbone and 18C (18 carbon atoms) unsaturated fatty acids

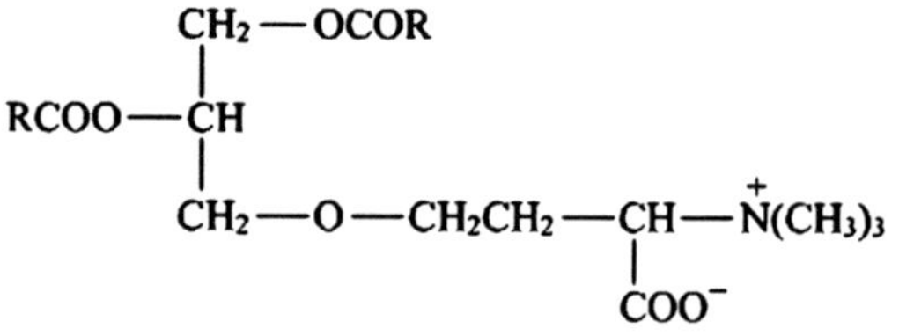

$$\begin{array}{c} CH_2-OCOR \\ | \\ RCOO-CH \\ | \\ CH_2-O-CH_2CH_2-CH-\overset{+}{N}(CH_3)_3 \\ | \\ COO^- \end{array}$$

1,2-diacylglyceryl-3-O-4'-(N,N,N-trimethyl)-homoserine (DGTS)

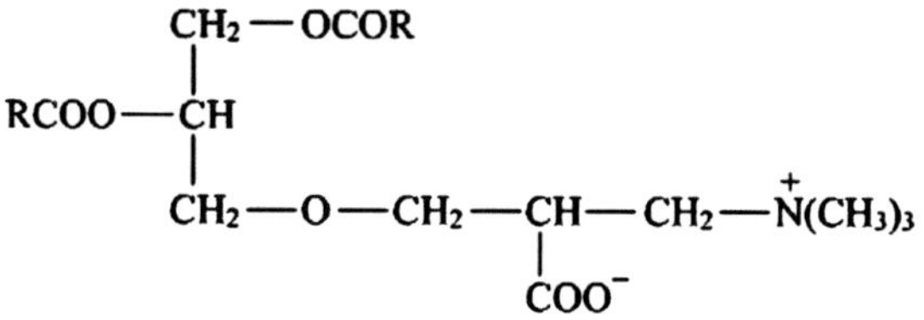

$$\begin{array}{c} CH_2-OCOR \\ | \\ RCOO-CH \\ | \\ CH_2-O-CH_2-CH-CH_2-\overset{+}{N}(CH_3)_3 \\ | \\ COO^- \end{array}$$

1,2-diacylglyceryl-3-O-2'-(hydroxymethyl)-(N,N,N-trimethyl)-β-alanine (DGTA)

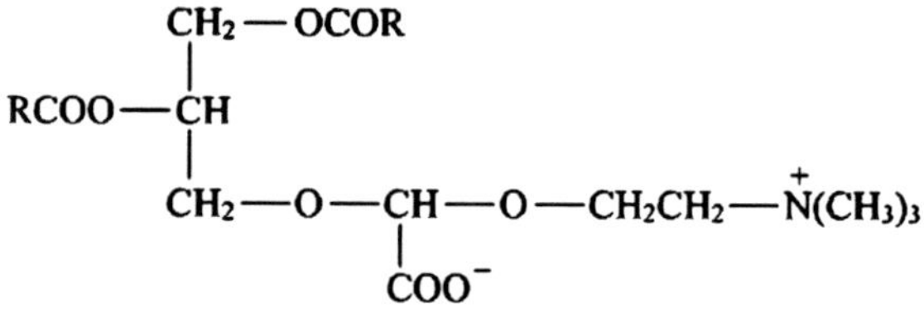

$$\begin{array}{c} CH_2-OCOR \\ | \\ RCOO-CH \\ | \\ CH_2-O-CH-O-CH_2CH_2-\overset{+}{N}(CH_3)_3 \\ | \\ COO^- \end{array}$$

1,2-diacylglyceryl-3-O-carboxy-(hydroxymethyl)-choline (DGCC)

Fig. 1.4 The main betaine lipids of algae

Table 1.2 Glycerolipid composition of selected species of algae

Composition of lipid classes (% of total)

	MGDG	DGDG	SQDG	PG	PC	PE	PI	DGTS	DGTA	MAG	DAG	TAG
Chlorophyta *Chlamydomonas moewusii* (Arisz et al. 2000)												
	35	15	10	9	–	8	6	16	–			
Chlorophyta *Parietochloris incisa* (Bigogno et al. 2002a)												
	22	14	4	2	6	3	1	4	–	–	–	43
Haptophyta *Pavlova lutheri*[a] (Eichenberger and Gribi 1997)												
	19	12	9	1	–	–	–	–	6	–	–	42
Haptophyta *Chrysochromulina polylepis* (John et al. 2002)												
	7	32	–	4[b]	43	4[b]	1	–	–	–	–	14
Haptophyta *Isochrysis galbana* (Alonso et al. 1998)												
	18	13	6	1	9	2	4	–	–	6	12	25
Bacillariophyta *Phaedactylum tricornutum* (Alonso et al. 1998)												
	21	12	2	2	6	4	1	–	–	4	7	40
Rhodophyta *Porphyridium cruentum* (Alonso et al. 1998)												
	14	24	5	–	5	–	2	–	–	3	19	25

[a]Contained DGGA and DGCC (2 and 5% of total lipids, respectively)
[b]The sum of PE and PG given. For lipid abbreviations see text and MAG is monoacylglycerol

(predominantly 18:2 and 18:3) at the *sn*-2 position. DGTS in marine algae can be esterified with long chain PUFA at both the *sn*-1 and *sn*-2 positions. As an interesting example, the marine alga *Chlorella minutissima* has been shown to produce DGTS at unusually high levels, varying from a low of ~10% to a high of ~44% of total lipids (Haigh et al. 1996). DGTS from *C. minutissima* was highly unsaturated at both positions of glycerol, and was exceptionally rich in EPA (>90% of its total fatty acids) (Haigh et al. 1996).

1.2.2 Nonpolar Glycerolipids

In many algal species, nonpolar triacylglycerols (TAG) (Fig. 1.5) are accumulated as storage products. The level of TAG accumulation is very variable (e.g., from ~2% of total lipids in *Fucus serratus* (Harwood and Jones 1989) to 77% in stationary phase *Parietochloris incisa* (Bigogno et al. 2002a)) (Table 1.2) and may be stimulated by a number of environmental factors (see below). In general, TAG is mostly synthesized in the light, stored in cytosolic lipid bodies, and then reutilized for polar lipid synthesis in the dark (Thompson 1996). Nitrogen deprivation has a major impact on TAG synthesis, and many algae show a two to threefold increase in lipid content, predominantly TAG, under nitrogen limitation (Thompson 1996). Algal TAG are generally characterized by saturated and monounsaturated fatty acids. However, some oleaginous species have demonstrated a capacity to accumulate high levels of long chain PUFA in TAG (Table 1.3). A detailed study on both accumulation of ARA in TAG of the green alga *Parietochloris incisa* and mobilization of arachidonyl moieties from storage TAG into chloroplastic lipids (following recovery from nitrogen starvation) led the authors to suggest that TAG may play an additional role beyond being an energy storage product in this alga (Bigogno et al. 2002a; Khozin-Goldberg et al. 2000, 2005). Thus, the PUFA-rich TAG were

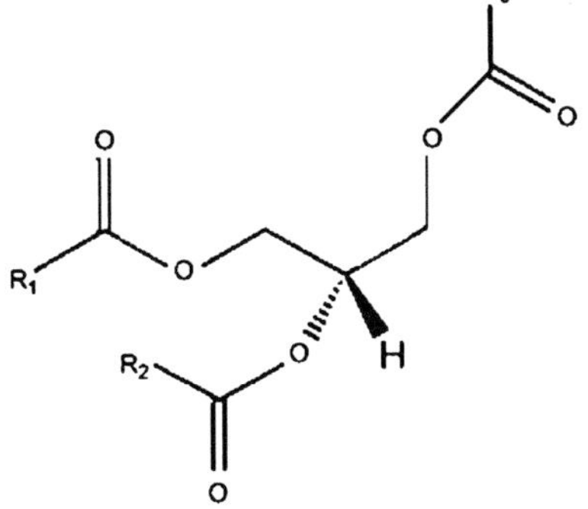

Fig. 1.5 Structure of a triacyl-*sn*-glycerol. R1, R2 and R3 are fatty acyl residues

Table 1.3 Fatty acid distribution in TAG from selected algae species

Fatty acids	Nannochloropsis sp. Eustigmatophyta	Parietochloris incisa Chlorophyta	Pavlova lutheri Haptophyta	Isochrysis galbana Haptophyta	Phaeodactylum tricornutum Bacillariophyta	Porphyridium cruentum Rhodophyta
14:0	18.8	–	8.6	4.8	4.3	1.6
16:0	41.6	8.4	39.2	12.3	13.3	21.1
16:1	33.7	0.4[a]	32.9	21.7	17.4	1.5
16:2n-4	–	–	–	–	4.8	–
16:3n-4	–	–	–	–	2.1	–
16:4n-1	–	–	–	–	0.5	–
18:0	1.0.	3.1	tr.	1.2	2.5	3.7
18:1n-9	3.8	18.0	2.1[b]	4.9	1.4	4.0
18:1n-7	–	4.0	1.2	0.1	0.9	–
18:2n-6	1.1	14.1	3.7	2.2	1.0	12.2
18:3n-3	–	0.4	–	1.5	–	–
18:3n-6	–	0.7	–	–	–	–
18:4n-3	–	–	1.1	7.2	–	1.0
20:2	–	–	4.3	–	–	1.1
20:3n-6	–	1.1-	–	–	–	24.2
20:4n-6	–	47.1-	–	–	3.7	15.9
20:5n-3	–	0.7	7.0	25.6	35.5	–
22:5n-6	–	–	–	1.2	–	–
22:6n-3	–	–	1.1	8.1	1.2	–

Positions of double bonds were assigned following capillary gas-liquid chromatography but were not confirmed by other methods. *Nannochloropsis* sp. (grown under low light conditions) (Sukenik et al. 1993); *Parietochloris incisa* (stationary phase culture analysed) (Bigogno et al. 2002a); *Pavlova lutheri* (Eichenberger and Gribi 1997); *Isochrysis galbana*, *Porphyridium cruentum*, *Phaeodactylum tricornutum* (Alonso et al. 1998)
[a] 16:1 reported as the 16:1n-11 isomer
[b] Sum of two isomers present

metabolically active and were suggested to act as a reservoir for specific fatty acids. During acclimation to sudden changes in environmental conditions, when the de novo synthesis of PUFA may be slower, PUFA-rich TAG may donate specific acyl groups to MGDG and other polar lipids to enable rapid adaptive membrane reorganization (Khozin-Goldberg et al. 2005; Makewicz et al. 1997).

Alternative forms of storage lipid, which are not glycerolipids, have been identified in some algal species. For example, when *Euglena gracilis* is grown in the dark and in aging cultures of the marine cryptomonad *Chroomonas salina*, wax esters were accumulated (Thompson 1996). A number of polyunsaturated long-chain (C_{37}–C_{39}) alkenes, alkenones, and alkenoates are synthesized by *Isochrysis galbana* and *Emiliania huxleyi*, as well as some other haptophyte algae (Eltgroth et al. 2005). A physiological role for these compounds as an energy store has been suggested based on the cellular localization of these polyunsaturated compounds and their metabolic behavior (Eltgroth et al. 2005).

1.3 Effects of the Environment on Algal Lipid Biochemistry

1.3.1 General Growth Conditions

Light intensity and temperature are probably the most important and best-studied environmental factors affecting the lipid and fatty acid composition of photosynthetic tissues or organisms (Harwood 1998b; Guschina and Harwood 2006a, b; Morgan-Kiss et al. 2006). It is generally accepted that many of the lipid changes alter the physical properties of the membrane bilayer so that normal functions (e.g., ion permeability, photosynthetic and respiratory processes) can continue unimpaired. The most commonly observed change in membrane lipids following a temperature shift is an alteration in fatty acid unsaturation (Harwood 1998b).

1.3.2 Temperature Effects

The green alga, *Dunaliella salina* has been extensively analyzed for low temperature modification of lipid composition (Thompson 1996). A temperature shift from 30°C to 12°C increased the level of lipid unsaturation in this alga significantly (Thompson 1996). This cooling regime also led to a number of ultrastructural changes. The chloroplast membrane lipid content increased by 20% while that of microsomes (mainly endoplasmic reticulum) rose by 280% (Thompson 1996). In the latter fraction, retailoring the molecular species of preexisting PE and PG has been noted as the most immediate response to temperature shift. Especially noteworthy was the increase in molecular species with two unsaturated fatty acids over the first 12 h at 12°C (Thompson 1996). The transformation in chloroplast phospholipids was shown to occur only after 36 h, and then the changes were similar to those seen in microsomes. In addition, a rise in ALA/16:1(3*trans*)-PG from 48 to 57% and a concomitant decrease in 18:2-16:1(3*trans*)/-PG from 34 to 26% of total chloroplast PG between 36 and 60 h was correlated with a significant alteration in the threshold temperature of thermal denaturation of the photosynthetic apparatus (Thompson 1996). In the microalgae, *Chlorella vulgaris* and *Botryococcus braunii*, increased temperature resulted in a decrease in the relative content of the more unsaturated intracellular fatty acids, especially the trienoic fatty acids, while the composition of fatty acids secreted into the medium was unchanged (Sushchik et al. 2003). A decrease in culture temperature from 25 to 10°C led to an elevation in the relative proportion of oleate but a decrease in linoleate and stearidonic acid (18:4n-3) in the green alga *Selenastrum capricornutum* (McLarnon-Riches et al. 1998). In cultures of *I. galbana* grown at 15 and 30°C, lipids and fatty acids were analyzed and compared (Zhu et al. 1997). At 30°C, total lipids accumulated at a higher rate with a slight decrease in the proportion of nonpolar lipids, an increase in the proportion of glycosylglycerides but no change in the proportion of phospholipids. Higher levels of ALA and DHA with a corresponding decrease in saturated, monounsaturated, and linoleic fatty acids were found in the cells grown at 15°C.

Four tropical Australian microalgal species (a diatom *Chaetoceros* sp., two cryptomonads, *Rhodomonas* sp. and *Cryptomonas* sp. and an unidentified haptophyte) cultured at five different temperatures showed similar trends in their fatty acid composition (Renaud et al. 2002). EPA was identified in all species with its highest concentration in the haptophyte. In this species, the level of EPA was lower at higher temperatures. Similarly, percentages of DHA were lower in all species cultured at higher temperatures. In contrast, moderate amounts of ARA were found in *Chaetoceros* sp. and the haptophyte and accumulated at cultivation temperatures within the range 27–30°C. Moreover, the EPA and PUFA content of the marine diatom *Phaeodactylum tricornutum* has been shown to be higher at lower temperatures within the range of 10–25°C (Jiang and Gao 2004). The highest yields of PUFA and EPA per unit dry mass were 4.9 and 2.6%, respectively, when temperature was shifted from 25 to 10°C for 12 h, with both being raised by 120% compared with the control (Jiang and Gao 2004).

In some plants, the resistance to low temperature (reduced chilling susceptibility) has been shown to be closely associated with a high proportion of the 16:0/16:0 and 16:0/16:1(3*trans*) molecular species of PG (Murata 1983). These two molecular species were considered to trigger the formation of gel phases in the membrane bilayer because of their high T_c values[2] : they undergo a liquid crystalline to gel phase transition even at room temperature. A similar relationship between the chilling sensitivity and the content of 16:0 and 16:1(3*trans*) has been demonstrated for *Chlorella ellipsoidea* (Joh et al. 1993). For PG, the content of 16:0 was 52% in the chilling-sensitive strain and 36% in the chilling-resistant strain. The content of 16:1(3*trans*) at the *sn-2* position of PG was 8% in the chilling-sensitive and 16% in

---

[2] T_c is the transition temperature, at which the acyl chains change from the gel to the liquid phase. Above the T_c the membrane is normally in a functional form with the lipids being "liquid" or, more correctly, showing low order.

the chilling-resistant strain. Moreover, the chilling-resistant strain also contained more ALA and, therefore, more unsaturation in its PG (Joh et al. 1993).

There was no effect of temperature shift on the content of the acidic thylakoid lipids, SQDG and PG, in *C. reinhardtii* (Sato et al. 2000). However, in the marine haptophyte *Pavlova lutheri*, significant changes in acidic lipid and fatty acid composition have been reported for cultures grown at 15°C compared with 25°C (Tatsuzawa and Takizawa 1995). Lower temperatures resulted in an increased relative amount of EPA and DHA. The relative percentage of betaine lipids, PG and SQDG increased when algae were cultivated at 15°C with a concomitant decrease in the levels of TAG and MGDG. The relative percentage of 16:1 in MGDG increased, but was almost unchanged in other membrane lipids. A similar response to lowered temperature is characteristic for cyanobacteria where the rapid $\Delta 9$-desaturation of palmitate to palmitoleate in MGDG has been shown to be an important thermo-adaptation mechanism (Tatsuzawa and Takizawa 1995). In the red microalga *Porphyridium cruentum*, a reduction in the growth temperature led to an increase in the proportion of the eukaryotic molecular species of MGDG, especially EPA/EPA MGDG (Adlerstein et al. 1997). These molecular species (like "eukaryotic species" of lipids in general) have been suggested to play a special role in adaptation of PUFA-rich algae to low temperatures (Adlerstein et al. 1997). In those types of oleaginous algae which accumulate high levels of ARA in TAG (e.g. *Parietochloris incisa* and *P. cruentum*), ARA can be transferred to membrane lipids as a quick response mechanism to cold-induced stress (see above) (Khozin-Goldberg et al. 2000; Bigogno et al. 2002b). Thus, more subtle alterations are often seen in many algae rather than a simple correlation of increased unsaturation with lower temperatures.

To summarise, exposure to lower environmental temperatures generally causes algae to increase their relative amount of fatty acid unsaturation. However, the details of these changes vary from organism to organism and will, naturally, be influenced by the variety and activity of those fatty acid desaturases present.

1.3.3 Light Effects

Light has been reported to produce many effects on algal lipid metabolism and therefore lipid composition (Harwood 1998b). In general, high light usually leads to oxidative damage of PUFA. Nevertheless, high light is required for the synthesis of 16:1(*3trans*) and alters the level of this fatty acid in algae. Moreover, qualitative changes in lipids as a result of various light conditions are associated with alterations in chloroplast development (Harwood 1998b). In *Nannochloropsis* sp., the degree of unsaturation of fatty acids decreased with increasing irradiance, especially the percentage of total *n*-3 fatty acids (from 29 to 8% of total fatty acids) mainly due to a decrease of EPA (Fabregas et al. 2004). In other EPA-producing algae (*Phaeodactylum tricornutum* and *Monodus subterraneus*), a similar tendency was noticed when increasing light intensity caused a reduction in EPA accumulation (cited by Adlerstein et al. 1997). High light exposure (300 µmol photons m^{-2} s^{-1}) decreased the total phospholipid content and increased the level of nonpolar lipid (namely TAG) in the filamentous green alga *Cladophora* spp. (Napolitano 1994). In low light conditions (6 µmol photons m^{-2} s^{-1}), the concentrations of two acetone-soluble fractions (most likely, MGDG and DGDG) were significantly enhanced. Low light also decreased the relative percentage of 16:0 and increased the percentage of 16:1n-7 and ALA (Napolitano 1994).

Variations in lipid composition have been studied in the marine red alga *Tichocarpus crinitus* exposed to different solar irradiance levels (Khotimchenko and Yakovleva 2005). Light intensity caused significant alterations in both storage and structural lipids. Exposure to low light intensity (shade at 8–10% of the incident photosynthetically active radiation (PAR)) resulted in increased levels of some cell membrane lipids, especially SQDG, PG, and PC, whereas higher light intensities (70–80% of PAR) increased the level of TAG. Although the total fatty acid content in *T. crinitus* did not alter, there were changes in the fatty acid composition of individual lipids. Low light exposure increased the content of EPA in MGDG and PG, while high light exposure increased the content of 16:1(3*trans*) in PG in *T. crinitus* (Khotimchenko and Yakovleva 2005).

In the green alga *Ulva fenestrate*, growing under different solar irradiances in field experiments, MGDG, SQDG and PG increased 2–3.5 times when grown at 24% of PAR compared with algae cultured at 80% of PAR (Khotimchenko and Yakovleva 2004). The contents of DGDG and betaine lipid, as well as the relative proportions of fatty acids in TAG, MGDG, and SQDG, were not affected by light intensity. Changes in the amounts of different lipid classes together with variations in the fatty acid compositions in DGDG and PG determined the differences in the total fatty acid composition under various light conditions. Palmitate and 16:4(n-3) exhibited the biggest changes (Khotimchenko and Yakovleva 2004).

Light/dark cycles also have a significant effect on algal lipid composition. A detailed study on various light regimes on lipids of the diatom *Thalassiosira pseudonana* provides a good example (Brown et al. 1996). The light regimes used were 100, 50, and 100 µmol photons m^{-2} s^{-1} on a 12:12, 24:0, and 24:0 h light/dark (L:D) cycle, respectively. A high accumulation of TAG and a reduced percentage of total polar lipids were found for cells grown under 100 µmol photons m^{-2} s^{-1} continuous light. The fatty acid composition (weight %) of algae in the logarithmic growth stage under the two continuous light regimes showed no significant differences, whereas cells grown under 100 µmol photons m^{-2} s^{-1} 12:12 h L/D conditions contained a higher proportion of PUFA and a lower proportion of saturated and monounsaturated fatty acids (Brown et al. 1996). With the onset of stationary phase, algae grown in continuous light showed increased proportions of saturated and monounsaturated fatty acids and decreased amounts of PUFA at 100 µmol photons m^{-2} s^{-1} light intensity in comparison to 50 µmol photons m^{-2} s^{-1} light intensity. As to fatty acid concentrations expressed as % of dry weight, those of myristate, palmitate, palmitoleate, EPA, and DHA were found to increase during stationary phase in all cultures (Brown et al. 1996).

The role of lipids in low light acclimation or acclimation to darkness has been studied also in some algae species. The lipid and fatty acid compositions of three species of sea ice diatoms grown in chemostats have been analysed and compared when cultivated at 2 and 15 µmol photons $m^{-2} s^{-1}$ (Mock and Kroon 2002). Growing cultures at 2 µmol photons $m^{-2} s^{-1}$ resulted in 50% more MGDG containing EPA than those grown at 15 µmol photons $m^{-2} s^{-1}$. EPA was suggested to ensure the fluidity of the thylakoid membrane (although other polyunsaturated fatty acids are just as effective) and, therefore, the velocity of electron flow, which was indicated by increasing rate constants for the electron transport between Q_A (first stable electron acceptor) and Q_B (second stable electron acceptor of photosystem II), making photosynthesis at low light levels more efficient. 2 µmol photons $m^{-2} s^{-1}$ resulted in higher amounts of nonlipid bilayer forming MGDG in relation to bilayer forming lipids, especially DGDG (the ratio of MGDG:DGDG increased from 3.4 to 5.7) than in cultures grown at 15 µmol photons $m^{-2} s^{-1}$ (Mock and Kroon 2002).

Dark treatment caused a decrease in the relative proportion of oleate and an increase in that of linoleate in the green alga *Selenastrum capricornutum* (McLarnon-Riches et al. 1998). In the dinoflagellate *Prorocentrum minimum*, dark exposure led to the reduced content of TAG and galactosylglycerides, while the total content of phospholipids changed little with decreased PC, PE, and PG, and increased PS, PA, and PI levels. The decrease of TAG and galactosylglycerides was in parallel to an increase in the activity of β-oxidation and isocitrate lyase indicating that TAG and galactosylglycerides were utilized as alternative carbon sources by the cells under nonphotosynthetic growth conditions (McLarnon-Riches et al. 1998).

In three seaweeds, *Ulva pertusa* (Chlorophyta), *Grateloupia sparsa* (Rhodophyta), and *Sargassum piluliferum* (Phaeophyta), the effect of different levels of light has been studied in combination with salinity (Floreto and Teshima 1998). In *U. pertusa*, exposure to a combination of high light and low salinity led to a significant decline in the total (mg g^{-1} dry weight) fatty acid content. Incubation under high light resulted in an increased content of most saturated fatty acids found in this alga (myristate, pentadecanoate, palmitate, and *iso*-heptadecanoate). In *G. sparsa*, low light and high salinity increased the content of all classes of fatty acids compared with normal salinity levels. The levels of myristate, oleate, vaccinate, EPA, and total *n*-3 fatty acids were elevated under high light conditions. In *S. piluliferum*, high light intensity decreased the content of almost all fatty acids while higher salinity increased the levels of 18:4n-3, ARA and EPA as well as total *n*-3 fatty acids (Floreto and Teshima 1998).

In conclusion, light will normally stimulate fatty acid synthesis, growth, and the formation of (particularly chloroplast) membranes. Therefore, the overall lipid content of algae will reflect such morphological changes.

1.3.4 Salinity Effects

Some algae are exceptional in the plant kingdom for their ability to tolerate high salt concentrations, the genus *Dunaliella* being an excellent example. The ability of *Dunaliella* species to proliferate over practically the entire range of salinities makes them useful models to study mechanisms that determine this capacity (Azachi et al.

2002). It has been shown that, in the cells of *D. salina* transferred from 0.5 to 3.5 M NaCl, the expression of β-ketoacyl-coenzyme A (CoA) synthase (KCS) (which catalyzes the first and rate-limiting step in fatty acid elongation) was induced (Azachi et al. 2002). This was commensurate with a considerably higher ratio of 18C (mostly unsaturated) to 16C (mostly saturated) fatty acids in the cells grown in 3.5 M NaCl compared with those grown at 0.5 M NaCl. The authors suggested that salt-induced KCS, together with fatty acid desaturases, may play a role acclimating the intracellular membrane compartment to function in the high internal glycerol concentrations used to balance the external osmotic pressure created by high salt (Azachi et al. 2002). (However, it should be noted that such a proposal assumes that the KCS is responsible for 18C fatty acid production rather than fatty acid synthase). An increase of the initial salt concentration from 0.5 M NaCl to 1.0 M followed by further addition of 1.0 M NaCl during cultivation of *Dunaliella tertiolecta* resulted in an increase in intracellular lipid content and a higher percentage of TAG (Takagi et al. 2006).

1.3.5 pH Effects

Lipids also react to extreme pH. Thus, alkaline pH stress led to TAG accumulation and a proportional decrease in membrane lipids (and, most likely, membranes) (Guckert and Cooksey 1990). The effects of pH on the lipid and fatty acid composition of a *Chlamydomonas* sp., isolated from a volcanic acidic lake, and *C. reinhardtii*, obtained from an algal collection (Institute of Applied Microbiology, Tokyo), have been studied and compared (Tatsuzawa et al. 1996). In the unidentified *Chlamydomonas* sp., fatty acids in the polar lipids were more saturated than those in *C. reinhardtii*. The relative proportion of TAG (as % of total lipids) was higher in *Chlamydomonas* sp. grown at pH 1 than that in the cells cultivated at higher pH. The increase in saturation of fatty acids in membrane lipids of *Chlamydomonas* has been suggested to represent an adaptive reaction at low pH to decrease membrane lipid fluidity (Tatsuzawa et al. 1996).

1.4 Nutrients and Nutrient Regimes

1.4.1 General Nutrient Effects

Nutrient availability has a significant impact and broad effects on the lipid and fatty acid composition of algae. Nutrient limitation almost invariably causes a steadily declining cell division rate. Surprisingly, active biosynthesis of fatty acids is maintained in some species of algae under such conditions (Thompson 1996). When algal growth slows down and there is no requirement for the synthesis of new membrane compounds, the cells instead divert and deposit fatty acids into triacylglycerols before conditions improve. For example, certain nutrient limited green algae more than double their lipid content (Thompson 1996).

Pronounced effects of nutrient-limitation on lipid composition have been shown for the freshwater diatom *Stephanodiscus minutulus* (Lynn et al. 2000). This alga was grown under silicon, nitrogen, or phosphorus limitation. Similarly, an increase in TAG accumulation and a decrease of polar lipids (as % of total lipids) was noticed in all of the nutrient-limited cultures (Lynn et al. 2000). An increase in TAG levels (from 69 to 75% from total lipids) together with phospholipids (from 6 to 8%) was reported for the microalga *Phaeodactylum tricornutum* as a result of reduced nitrogen concentration (Alonso et al. 2000). Conversely, the level of galactolipids decreased from 21 to 12% in these nitrogen-starved cells.

In *Chlamydomonas* spp., the concentration of PUFA decreased when the cultivation conditions changed from photoautotrophic via mixotrophic to heterotrophic (Poerschmann et al. 2004). In *Chlamydomonas moewusii*, nutrient-limitation resulted in alterations in the fatty acid composition of the chloroplast lipids, PG, and MGDG (Arisz et al. 2000). The PUFA, 16:3, 16:4, and 18:3, which were present in the plastidic galactolipids, and 16:1(3*trans*), specific for plastidic PG, decreased under nutrient-limited conditions. The synthesis of storage lipids has been suggested to be stimulated by depletion of nutrients, and this was consistent with a rise in the overall levels of 16:1 and 18:1 which were prominent in storage lipids (Arisz et al. 2000).

The photosynthetic flagellate *Euglena gracilis* has been cultivated under various conditions of autotrophy and photoheterotrophy to estimate the contribution of lactate (a carbon source) and ammonium phosphate (a nitrogen source) to its metabolism (Regnault et al. 1995). Effects of increasing ammonium phosphate concentration on lipid composition were noticed only when lactate was depleted. Such conditions increased the content of galactolipids rich in polyunsaturated 16C and 18C fatty acids as well as the ratio of MGDG/DGDG. Excess of nitrogen did not change the content of medium chain (12–14C) acids but induced a reduction of 22C acids. When ammonium phosphate was absent in the cultural medium, increasing the lactate concentration led to a decrease in all plastid lipids, whereas the accumulation of storage lipids (enriched with 14:0 and 16:0) increased, while biosynthesis of 18C PUFA was reduced as indicated by the accumulation of 18:1n-9 (Regnault et al. 1995).

The crude lipid content (as a percentage of dry weight) of the seaweed *Ulva pertusa* was increased when grown under nitrogen starvation and, surprisingly, also under very high levels of nitrogen (15 mM) (Floreto et al. 1996). Increased nitrogen concentrations led to a decrease in proportion of the major PUFA; 16:4n-3 and 18:4n-3, and a rise in the proportion of palmitate, 18:1n-7 and 18:2n-6. By contrast, phosphorus starvation decreased the proportion of 16:0 and increased that of 16:4n-3 with no effect on the total lipid content of the seaweed (Floreto et al. 1996).

Overall, nutrient limitations which cause reduced cell division rates, and therefore population growth, typically result in increased cellular production of storage lipid, primarily TAG. Because the fatty acid composition of TAG often differs between algae from different taxonomic groups, the resultant fatty acid compositions of the algae may also differ, leading to considerable variation between taxa. However, these studies of many very different species have shown that because more fatty acyl groups of TAG tend to be saturated and monounsaturated relative to those of the polar glycerolipids, the increase in TAG with nutrient limitation typically resulted in decreased proportions of polyunsaturated fatty acids in most algae.

1.4.2 Specific Phosphorus or Sulphur Effects

In the green alga *C. reinhardtii* those thylakoid lipids with a negative charge (PG and SQDG) have been studied under sulphur and phosphorus-starved cultivation (Sato et al. 2000). Sulphur-limited cells lost most of their SQDG when compared with normal conditions. Concomitantly, PG content increased by twofold, representing a compensatory mechanism for the reduced level of the other anionic lipid, SQDG. When *C. reinhardtii* was grown in a media with limited phosphorus it showed a 40% decrease in PG and a concomitant increase in the SQDG content. Thus, mechanisms that keep the total sum of SQDG and PG concentrations constant under both phosphorus and sulphur-limiting conditions appear to occur. Moreover, it has been suggested that SQDG may substitute for PG to maintain the functional activity of chloroplast membranes (Sato et al. 2000).

In general, the replacement of membrane phospholipids by non-phosphorus containing glycolipids and betaine lipids under phosphate limitation has been demonstrated in many organisms, including higher plants, photosynthetic bacteria and algae (e.g., Benning et al. 1995; Härtel et al. 2000; Andersson et al. 2003). This replacement has been suggested to represent an effective phosphate-conserving mechanism. However, only minor alterations in lipid metabolism were noticed when four green algal–lichen photobionts were exposed to low-phosphate conditions and examined by labeling with [1-^{14}C]-acetate (Guschina et al. 2003). Although growth and total lipid labeling were impaired in low phosphate media, there were only minor changes in the relative rates of phosphoglyceride labeling and hardly any decrease in the relative labelling of PG. X-ray probe electron microscopy revealed significant stores of endogenous phosphorus in the algae, which might be used to maintain normal synthesis of phosphoglycerides in these photobionts (Guschina et al. 2003).

In a study of seven species of marine algae cultured in phosphorus-limiting conditions, lipid contents increased in *P. tricornutum*, *Chaetoceros* sp., and *P. lutheri*, but decreased in the chlorophyte flagellates, *Nannochloris atomus* and *Tetraselmis* sp. (Reitan et al. 1994). Severely nutrient-limited cultures had a higher relative content of 16:0 and 18:1n-9 and lower levels of 18:4n-3, EPA, and DHA (Reitan et al. 1994). In contrast, for phosphorus-starved cells of the green alga *Chlorella kessleri*, an elevated level of unsaturated fatty acids in all identified individual lipids, namely PC, PG, DGDG, MGDG, and SQDG were found (El-Sheek and Rady 1995).

Incubation of the fresh water eustigmatophyte *Monodus subterraneus* in media with decreasing phosphate concentrations (175, 52.5, 17.5, and 0 μM) resulted in a gradual decrease in EPA concentration with a concomitant increase in 18:0, 18:1n-9, and 16:1n-7 fatty acids (as % of dry weight and as % of total fatty acids) whereas the cellular total lipid content increased, mainly due to TAG accumulation (Khozin-Goldberg and Cohen 2006). In phosphate-depleted cells, the proportion of phospholipids declined from 8.3 to 1.4% of total lipids. Among other polar lipids, cellular contents of DGDG (fg cell^{-1}) and DGTS increased while that of MGDG was not significantly changed. But the relative content (as % of total lipid) of these lipids was reduced. The proportion of EPA in DGDG, where it was located exclusively at the *sn-1* position, increased from 11.3 to 21.5%. In contrast, the proportion of this fatty acid in MGDG, SQDG, and PC did not change and decreased in all other polar lipids (PE, PG, DGTS) and TAG. The reported accumulation of free 18:0 indicated that no polar lipid can replace PC, which is apparently the only substrate for C18 desaturation in this algal species. DGTS has been suggested to be a source of 20C acyl-containing diacylglycerols under phosphate starvation (Khozin-Goldberg and Cohen 2006).

1.4.3   Carbon Availability

A general reduction in the degree of fatty acid unsaturation as a response to elevated CO_2 concentration has been reported for several species of green algae (Thompson 1996). For example, *C. kessleri* grown under low CO_2 (0.04% compared with 2% CO_2) showed elevated contents of ALA, especially at both *sn-1* and *sn-2* positions of MGDG and DGDG, and also at the *sn-2* position of PC and PE (Sato et al. 2003). The higher unsaturation levels in low-CO_2 cells has been proposed to be (at least partly) due to repressed fatty acid synthesis, which allowed desaturation of preexisting fatty acids (Sato et al. 2003).

The effect of CO_2 concentration on fatty acid composition has been studied in wild-type *C. reinhardtii* and its cia-3 mutant strain, which is deficient in a CO_2-concentrating mechanism (Pronina et al. 1998). In both strains, there was some increase in PUFA biosynthesis as a result of the decrease in CO_2 concentration from 2 to 0.03%. However, in the mutant, when compared with the wild type, an increase in PUFA was less pronounced and some fatty acids (e.g., 16:4n-3) did not change, which may indicate a correlation between the induction of the CO_2-concentrating mechanism and an acceleration of fatty acid desaturation (Pronina et al. 1998). The CO_2 concentration has also been shown to change the content and composition of fatty acids and chloroplast lipids in the unicellular halophilic green alga *Dunaliella salina* (resistant to CO_2 stress) (Muradyan et al. 2004). The response was seen after an increase in CO_2 concentration from 2 to 10% and resulted in an increase of 30% in the total amount of fatty acids on a dry weight basis. Alterations in fatty acids indicated increased fatty acid synthesis de novo but an inhibition of their elongation and desaturation. The MGDG/DGDG ratio increased fourfold while the ratio of n-3/n-6 fatty acids, as well as the proportion of 16:1(3*trans*) in PG increased significantly. These changes have been suggested to represent an adaptation of the photosynthetic membranes to ensure effective photosynthesis in *D. salina* under the experimental conditions (Muradyan et al. 2004).

In contrast to the results with *Chlamydomonas* spp., elevated CO_2 or added organic carbon sources significantly enhanced EPA production in *Nannochloropsis* sp. (Hu and Gao 2006).

1.5 Conclusions

Overall, the main trends from the studies presented show that any change in an environmental factor that affects photosynthesis and/or the production of lipids (e.g., supply of essential nutrients) will affect the amount of the lipid classes (and the main characteristic fatty acid moieties for each lipid class) involved in that process, thereby altering the lipid content and composition of the algae being studied.

In addition to the myriad of changes reported above, anthropogenic factors have also been studied (Einicker-Lamas et al. 1996, 2002). Like the general environmental stresses of light, temperature, and nutrients, such factors produce various changes in different organisms. Even for the best-studied stress, temperature, the detailed biochemical response of individual algal species varies and, apart from one species of cyanobacterium (Gombos and Murata 1998), we have rather little idea of the molecular mechanisms involved.

Perhaps the most obvious way to advance our understanding of how the environment can alter lipid metabolism in algae is to study one species under controlled laboratory conditions. The whole battery of different experimental analytical methods (including lipidomics, genomics, proteomics, microscopy, and physiological functions) should then be applied. Only that way will we begin to unravel in detail the molecular mechanisms involved in adaptation.

From studying and thereby gaining an understanding of how one organism reacts to a single stress, we can then examine further organisms, using a combination of stresses and "natural" conditions. That way we will build up our knowledge of acclimation in a sequential manner, which may, ultimately, be capable of extrapolation to other species. Obviously, there is plenty of work to do!

References

Adlerstein, D., Bigogno, C., Khozin, I., and Cohen, Z. 1997. The effect of growth temperature and culture density on the molecular species composition of the galactolipids in the red microalga *Porphyridium cruentum* (Rhodophyta). J. Phycol. 33:975–979.

Al-Fadhli, A., Wahidulla, S., and D'Souza, L. 2006. Glycolipids from the red alga *Chondria armata* (Kütz.) Okamura. Glycobiology 16:902–915.

Alonso, D.L., Belarbi, E.H., Rodriguez-Ruiz, J., Segura, C.I., and Gimenez, A. 1998. Acyl lipids of three microalgae. Phytochemistry 47:1473–1481.

Alonso, D.L., Belarbi, E.H., Fernandez-Sevilla, J.M., Rodriguez-Ruiz, J., and Grima, E.M. 2000. Acyl lipid composition variation related to culture age and nitrogen concentration in continuous cultures of the microalga *Phaeodactylum tricornutum*. Phytochemistry 54:461–471.

Andersen, R.J. and Taglialatela-Scafati, O. 2005. Avrainvilloside, a 6-deoxy-6-aminoglucoglycerolipid from the green alga *Avrainvillea nigricans*. J. Nat. Prod. 68:1428–1430.

Anderson, R., Livermore, B.P., Kates, M., and Volcani, B.E. 1978a. The lipid composition of the non-photosynthetic diatom *Nitzschia alba*. Biochim. Biophys. Acta 528:77–88.

Anderson, R., Kates, M., and Volcani, B.E. 1978b. Identification of the sulfolipids in the non-photosynthetic diatom *Nitzschia alba*. Biochim. Biophys. Acta 528:89–106.

Andersson, M.X., Stridh, M.H., Larsson, K.E., Liljenberg, C., and Sandelius, A.S. 2003. Phosphate-deficient oat replaces a major portion of the plasma membrane phospholipids with the galactolipid digalactosyldiacylglycerol. FEBS Lett. 537:128–132.

Arisz, S.A., van Himbergen, J.A.J., Musgrave, A., van den Ende, H., and Munnik, T. 2000. Polar glycerolipids of *Chlamydomonas moewusii*. Phytochemistry 53:265–270.

Azachi, M., Sadka, A., Fisher, M., Goldshlag, P., Gokhman, I., and Zamir, A. 2002. Salt induction of fatty acid elongase and membrane lipid modifications in the extreme halotolerant alga *Dunaliella salina*. Plant Physiol. 129:1320–1329.

Benning, C., Huang, Z.H., and Gage, D.A. 1995. Accumulation of a novel glycolipid and a betaine lipid in the cells of *Rhodobacter sphaeroides*. Arch. Biochem. Biophys. 317:103–111.

Bigogno, C., Khozin-Goldberg, I., Boussiba, S., Vonshak, A., and Cohen, Z. 2002a. Lipid and fatty acid composition of the green oleaginous alga *Parietochloris incisa*, the richest plant source of arachidonic acid. Phytochemistry 60:497–503.

Bigogno, C., Khozin-Goldberg, I., and Cohen, Z. 2002b. Accumulation of arachidonic acid-rich triacylglycerols in the microalga *Parietochloris incisa* (Trebouxiophyceae, Chlorophyta). Phytochemistry 60:135–143.

Bisseret, P., Ito, S., Tremblay, P.A., Volcani, B.E., Dessort, D., and Kates, M. 1984. Occurrence of phosphatidylsulfocholine, the sulfonium analog of phosphatidylcholine in some diatoms and algae. Biochim. Biophys. Acta 796:320–327.

Brett, M.T. and Müller-Navarra, D.C. 1997. The role of highly unsaturated fatty acids in aquatic foodweb processes. Freshwater Biol. 38:483–499.

Brown, M.R., Dunstan, G.A., Norwood, S.J., and Miller, K.A. 1996. Effects of harvest stage and light on the biochemical composition of the diatom *Thalassiosira pseudonana*. J. Phycol. 32:64–73.

Dembitsky, V.M. 1996. Betaine ether-linked glycerolipids: chemistry and biology. Prog. Lipid Res. 35:1–51.

D'Ippolito, G., Tucci, S., Cutignano, A., Romano, G., Cimino, G., Miralto, A., et al. 2004. The role of complex lipids in the synthesis of bioactive aldehydes of the marine diatom *Skeletonema costatum*. Biochim. Biophys. Acta 1686:100–107.

Eichenberger, W. and Gribi, C. 1997. Lipids of *Pavlova lutheri*: cellular site and metabolic role of DGCC. Phytochemistry 45:1561–1567.

Einicker-Lamas, M., Soares, M.J., Soares, M.S., and Oliveira, M.M. 1996. Effects of cadmium on *Euglena gracilis* membrane lipids. Braz. J. Med. Biol. Res. 29:941–948.

Einicker-Lamas, M., Mezian, G.A., Fernandes, T.B., Silva, F.L.C., Guerra, F., Miranda, K., et al. 2002. *Euglena gracilis* as a model for the study of Cu^{2+} and Zn^{2+} toxicity and accumulation in eukaryotic cells. Environ. Pollut. 120:779–786.

Eltgroth, M.L., Watwood, R.L., and Wolfe, G.V. 2005. Production and cellular localization of neutral long-chain lipids in the haptophyte algae *Isochrysis galbana* and *Emiliania huxleyi*. J. Phycol. 41:1000–1009.

El-Sheek, M.M. and Rady, A.A. 1995. Effect of phosphorus starvation on growth, photosynthesis and some metabolic processes in the unicellular green alga *Chlorella kessleri*. Phyton 35:139–151.

Fabregas, J., Maseda, A., Dominquez, A., and Otero, A. 2004. The cell composition of *Nannochloropsis* sp. changes under different irradiances in semicontinuous culture. World J. Microbiol. Biotechnol. 20:31–35.

Floreto, E.A.T. and Teshima, S. 1998. The fatty acid composition of seaweeds exposed to different levels of light intensity and salinity. Bot. Mar. 41:467–481.

Floreto, E.A.T., Teshima, S., and Ishikawa, M. 1996. Effects of nitrogen and phosphorus on the growth and fatty acid composition of *Ulva pertusa* Kjellman (Chlorophyta). Bot. Mar. 39:69–74.

Gombos, Z. and Murata, N. 1998. Genetic engineering of the unsaturation of membrane glycerolipid: effects on the ability of the photosynthetic machinery to tolerate temperature stress, pp. 249–262. In P-A. Siegenthaler and N. Murata (eds.), Lipids in Photosynthesis: Structure, Function and Genetics, Kluwer, Dordrecht.

Guckert, J.B. and Cooksey, K.E. 1990. Triglyceride accumulation and fatty acid profile changes in *Chlorella* (Chlorophyta) during high pH-induced cell cycle inhibition. J. Phycol. 26:72–79.

Gunstone, F.D., Harwood, J.L., and Dijkstra, A.J. 2007. The Lipid Handbook, 3rd ed. Taylor and Francis, Boca Raton, FL, 1447 pp.

Gurr, M.I., Harwood, J.L., and Frayn, K.N. 2002. Lipid Biochemistry. An Introduction, 5th ed. Blackwell, Oxford, 320 pp.

Guschina, I.A. and Harwood, J.L.2006a. Lipids and lipid metabolism in eukaryotic algae. Prog. Lipid Res. 45:160–186.

Guschina, I.A. and Harwood, J.L. 2006b. Mechanisms of temperature adaptation in poikilotherms. FEBS Lett. 580:5477–5483.

Guschina, I.A. and Harwood, J.L. 2007. Complex lipid biosynthesis and its manipulation in plants, pp. 253–279. In P. Ranalli (ed.), Improvement of Crop Plants for Industrial End Use. Springer, Dordrecht.

Guschina, I.A., Dobson, G., and Harwood, J.L. 2003. Lipid metabolism in cultured lichen photobionts with different phosphorus status. Phytochemistry 64:209–217.

Haigh, W.G., Yoder, T.F., Ericson, L., Pratum, T., and Winget, R.R. 1996. The characterisation and cyclic production of highly unsaturated homoserine lipid in *Chlorella minutissima*. Biochim. Biophys. Acta1299:183–190.

Härtel, H., Dörmann, P., and Benning, C. 2000. DGD1-independent biosynthesis of extraplastidic galactolipids following phosphate deprivation in *Arabidopsis*. Proc. Natl Acad. Sci. U. S. A. 97:10649–10654.

Harwood, J.L. 1998a. Membrane lipids in algae, pp. 53–64. In P-A. Siegenthaler and N. Murata (eds.), Lipids in Photosynthesis: Structure, Function and Genetics. Kluwer, Dordrecht.

Harwood, J.L. 1998b. Involvement of chloroplast lipids in the reaction of plants submitted to stress, pp. 287–302. In P-A. Siegenthaler and N. Murata (eds.), Lipids in Photosynthesis: Structure, Function and Genetics. Kluwer, Dordrecht.

Harwood, J.L. and Jones, A.L. 1989. Lipid metabolism in algae. Adv. Bot. Res. 16:1–53.

Hu, H. and Gao, K. 2006. Response of growth and fatty acid compositions of *Nannochloropsis* sp. to environmental factors under elevated CO_2 concentration. Biotechnol. Lett. 28:987–992.

Jiang, H. and Gao, K. 2004. Effects of lowering temperature during culture on the production of polyunsaturated fatty acids in the marine diatom *Phaeodactylum tricornutum* (Bacillariophyceae). J. Phycol. 40:651–654.

Joh, T., Yoshida, T., Yoshimoto, M., Miyamoto, T., and Hatano, S. 1993. Composition and positional distribution of fatty acids in polar lipids from *Chlorella ellipsoidea* differing in chilling susceptibility and frost hardening. Physiol. Plantarum 89:285–290.

John, U., Tillmann, U., and Medlin, L.K. 2002. A comparative approach to study inhibition of grazing and lipid composition of a toxic and non-toxic clone of *Chrysochromulina polylepis* (Prymnesiophyceae). Harmful Algae 1:45–57.

Kato, M., Sakai, M., Adachi, K., Ikemoto, H., and Sano, H. 1996. Distribution of betaine lipids in marine algae. Phytochemistry 42:1341–1345.

Keusgen, M., Curtis, J.M., Thibault, P., Walter, J.A., Windust, A., and Ayer, S.W. 1997. Sulfoquinovosyl diacylglycerols from the alga *Heterosigma carterae*. Lipids 32:1101–1112.

Khotimchenko, S.V. and Yakovleva, I.M. 2004. Effect of solar irradiance on lipids of green alga *Ulva fenestrate* Postels et Ruprecht. Bot. Mar. 47:395–401.

Khotimchenko, S.V. and Yakovleva, I.M. 2005. Lipid composition of the red alga *Tichocarpus crinitus* exposed to different levels of photon irradiance. Phytochemistry 66:73–79.

Khozin-Goldberg, I. and Cohen, Z. 2006. The effect of phosphate starvation on the lipid and fatty acid composition of the fresh water eustigmatophyte *Monodus subterraneus*. Phytochemistry 67:696–701.
Khozin-Goldberg, I., Yu, H.Z., Adlerstein, D., Didi-Cohen, S., Heimer, Y.M., and Cohen, Z. 2000. Triacylglycerols of the red microalga *Porphyridium cruentum* can contribute to the biosynthesis of eukaryotic galactolipids. Lipids 35:881–889.
Khozin-Goldberg, I., Shrestha, P., and Cohen, Z. 2005. Mobilization of arachidonyl moieties from triacylglycerols into chloroplastic lipids following recovery from nitrogen starvation of the microalga *Parietochloris incisa*. Biochim. Biophys. Acta 1738:63–71.
Lynn, S.G., Kilham, S.S., Kreeger, D.A., and Interlandi, S.J. 2000. Effect of nutrient availability on the biochemical and elemental stoichiometry in freshwater diatom *Stephanodiscus minutulus* (Bacillariophyceae). J. Phycol. 36:510–522.
Makewicz, A., Gribi, C., Eichenberger, W. 1997. Lipids of *Ectocarpus fasciculatus* (Phaeophyceae). Incorporation of [1-^{14}C]oleate and the role of TAG and MGDG in lipid metabolism. Plant Cell Physiol. 38:952–960.
McLarnon-Riches, C.J., Rolph, C.E., Greenway, D.L.A., and Robinson, P.K. 1998. Effects of environmental factors and metals on *Selenastrum capricornutum*. Phytochemistry 49:1241–1247.
Mock, T. and Kroon, B.M.A. 2002. Photosynthetic energy conversion under extreme conditions-II: the significance of lipids under light limited growth in Antarctic sea ice diatoms. Phytochemistry 61:53–60.
Morgan-Kiss, R.M., Priscu, J.C., Pocock, T., Gudynaite-Savitch, L., and Huner, N.P.A. 2006. Adaptation and acclimation of photosynthetic microorganisms to permanently cold environments. Microbiol. Mol. Biol. Rev. 70:222–252.
Muradyan, E.A., Klyachko-Gurvich, G.L., Tsoglin, L.N., Sergeyenko, T.V., and Pronina, N.A. 2004. Changes in lipid metabolism during adaptation of the *Dunaliella salina* photosynthetic apparatus to high CO_2 concentration. Russ. J. Plant Physiol. 51:53–62.
Murata, N. 1983. Molecular species composition of phosphatidylglycerols from chilling-sensitive and chilling-resistant plants. Plant Cell Physiol. 24:81–86.
Murphy, D.J. (ed.) 2005. Plant Lipids: Biology, Utilisation and Manipulation. Blackwell, Oxford, 403 pp.
Napolitano, G.E. 1994. The relationship of lipids with light and chlorophyll measurement in freshwater algae and periphyton. J. Phycol. 30:943–950.
Patterson, G.W. 1991. Sterols of algae, pp. 118–157. In G.W. Patterson and W.D. Nes (eds.), Physiology and Biochemistry of Sterols. AOCS Press, Urbana, IL.
Poerschmann, J., Spijkerman, E., and Langer, U. 2004. Fatty acid patterns in *Chlamydomonas* sp. as a marker for nutritional regimes and temperature under extremely acidic conditions. Microbiol. Ecol. 48:78–89.
Pronina, N.A., Rogova, N.B., Furnadzhieva, S., and Klyachko-Gurvich, G.L. 1998. Effect of CO_2 concentration on the fatty acid composition of lipids in *Chlamydomonas reinhardtii* cia-3, a mutant deficient in CO_2-concentrating mechanism. Russ. J. Plant Physiol. 45:447–455.
Regnault, A., Chevrin, D., Chammai, A., Piton, F., Calvayrac, R., and Mazliak, P. 1995. Lipid composition of *Euglena gracilis* in relation to carbon-nitrogen balance. Phytochemistry 40:725–733.
Reintan, K.I., Rainuzzo, J.R., and Olsen, Y. 1994. Effect of nutrient limitation on fatty acid and lipid content of marine microalgae. J. Phycol. 30:972–977.
Renaud, S.M., Thinh, L.V., Lambrinidis, G., and Parry, D.L. 2002. Effects of temperature on growth, chemical composition and fatty acid composition of tropical Australian microalgae grown in batch cultures. Aquaculture 211:195–214.
Sato, N., Hagio, M., Wada, H., and Tsuzuki, M. 2000. Environmental effects on acidic lipids of thylakoid membranes, pp. 912–914. In J.L. Harwood and P.J. Quinn (eds.), Recent Advances in the Lipid Biochemistry of Plant Lipids. Portland Press, London.
Sato, N., Tsuzuki, M., and Kawaguchi, A. 2003. Glycerolipid synthesis in *Chlorella kessleri* 11h II. Effect of CO_2 concentration during growth. Biochim. Biophys. Acta 1633:35–42.
Son, B.W. 1990. Glycolipids from *Gracilaria verrucosa*. Phytochemistry 29:307–309.

Sukenik, A., Yamaguchi, Y., and Livne, A. 1993. Alterations in lipid molecular species of the marine eustigmatophyte *Nannochloropsis* sp. J. Phycol. 29:620–626.
Sushchik, N.N., Kalacheva, G.S., Zhila, N.O., Gladyshev, M.I., and Volova, T.G. 2003. A temperature dependence of the intra- and extracellular fatty acid composition of green algae and cyanobacterium. Russ. J. Plant Physiol. 50:374–380.
Takagi, M., Karseno, B., and Yoshida, T. 2006. Effect of salt concentration on intracellular accumulation of lipids and triacylglycerols in marine microalgae *Dunaliella* cells. J. Biosci. Bioeng. 3:223–226.
Tatsuzawa, H. and Takizawa, E. 1995. Changes in lipid and fatty acid composition of *Pavlova lutheri*. Phytochemistry 40:397–400.
Tatsuzawa, H., Takizawa, E., Wada, M., and Yamamoto, Y. 1996. Fatty acid and lipid composition of the acidophilic green alga *Chlamydomonas* sp. J. Phycol. 32:598–601.
Thompson, G.A.J. 1996. Lipids and membrane function in green algae. Biochim. Biophys. Acta 1302:17–45.
Tremolieres, A., and Siegenthaler, P.A. 1998. Role of acyl lipids in the function of photosynthetic membranes in higher plants, pp. 145–173. In P-A. Siegenthaler and N. Murata (eds.), Lipids in Photosynthesis: Structure, Function and Genetics. Kluwer, Dordrecht.
Volkman, J.K. 2003. Sterols in microorganisms. Appl. Microbiol. Biotechnol. 60:495–506.
Zhu, C.J., Lee, Y.K., and Chao, T.M. 1997. Effects of temperature and growth phase on lipid and biochemical composition of *Isochrysis galbana* TK1. J. Appl. Phycol. 9:451–457.

Chapter 2
Formation and Transfer of Fatty Acids in Aquatic Microbial Food Webs: Role of Heterotrophic Protists

Christian Desvilettes and Alexandre Bec

2.1 Introduction

2.1.1 Heterotrophic Protists

The term protist was first coined by Haeckel in 1866 for diverse microorganisms including bacteria (Haeckel 1866). However, in 1925 in a paper on an amoeboid parasite of *Daphnia*, Chatton (1925) highlighted for the first time the fundamental difference between prokaryotic and eukaryotic organisms and the term protist to be now used to describe unicellular eukaryotes, which do not differentiate into tissues (see Adl et al. 2005).

In aquatic systems, the most widely known protists are the phytoplanktonic microalgae and most students are familiar with *Euglena*, which survive and grow even in the absence of light. Protists such as *Euglena* absorb soluble organic compounds via osmotrophy whereas other protists have the ability to obtain energy and nutrients via both osmotrophic and phagotrophic pathways (mixotrophy). Mixotrophy (autotrophic and/or heterotrophic) has been recognized as giving a major competitive advantage in changing environmental conditions especially in nutrient-limited systems (Stoecker 1998). In fact the photoautotrophic protists are relatively rare whereas the non-photosynthetic protists are numerous and highly diversified. The latter can be free-living, symbiotic or parasitic, and are called protozoa or more technically heterotrophic protists (HP). In the older scientific literature HP are also named as 'colourless' or 'non-pigmented' protists. The 'colourless' protists can appear pigmented from undigested pigmented material, which is present in food vacuoles. Some protozoa (such as the heterotrich ciliate *Climacostomum virens*) can sometimes, but not always, contain a small chlorophyte alga as an endosymbiont. Other protists such as

---

C. Desvilettes (✉) and A. Bec
Laboratoire de Biologie des Protistes, Université Blaise Pascal, Clermont-Ferrand II, UMR CNRS 6023, Campus des Cézeaux, AUBIERE cedex, 63177 Aubière, France
christian.desvilettes@univ-bpclermont.fr

Mesodinium rubrum can sequester plastids from their algal prey and by karyoklepty gain full access to photosynthetic products (Johnson et al. 2007).

As HP are ubiquitous and abundant in all types of habitats, they are important components of food webs (Sherr and Sherr 2002) especially in extreme habitats such as sea ice (Scott et al. 2001), solar salterns with high salinity (Park et al. 2007), and anoxic environments (Brugerolle and Müller 2000). In aquatic environments, the populations of HP often are diverse, not only in species richness but also in feeding behaviour and ecological roles. Ciliates (*Ciliophora*) and heterotrophic flagellates are the most important grazers of bacteria and picoplanktonic microorganisms in many aquatic ecosystems (Sherr and Sherr 2002) (Fig. 2.1). HP can feed on a wide size range of particles ranging from viruses (Bettarel et al. 2005) to fish (Burkholder and Glasgow 1997). The phagotrophic species can ingest particles several times larger than themselves (Arndt et al. 2000) because of particular properties of their cytoskeleton and internal membranes. Other species of HP feed on large prey by sucking out (myzocytosis) internal contents. In freshwater environments, the oligotrich ciliates *Oligotrichia*, *Choreotrichia*, and *Stichotrichia* dominate in oligotrophic waters, whereas the scuticociliates (*Scuticociliata*), haptorids (*Litostomatea*) and ubiquitous oligotrichs dominate in eutrophic waters. Marine species

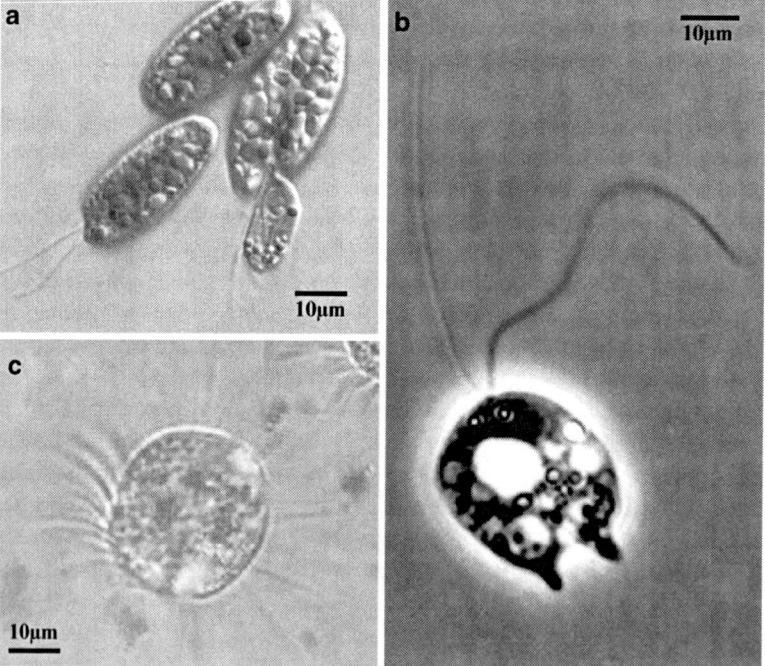

Fig 2.1 A composite image of aquatic heterotrophic protists. (**a**) *Colpodea perforans* a biflagelled alveolate predator attacking the Cryptomonadale *Chilomonas paramecium*. (**b**) *Diphylleia rotans* (syn. *Aulacomonas submarina*) a free-living algivorous flagellate (still remains as *incertae sedis*). (**c**) *Halteria* sp. an algivorous ciliate (*Oligotrichia*)

are frequently dominated by tintinnids (*Tintinnida*) and naked cells of oligotrichs (*Choreotrichia, Stichotrichia*). The available taxonomic information on non-pigmented flagellates in pelagic microbial communities shows that similar groups of nano and microflagellates are present in quite different environments (Arndt et al. 2000; Laybourn-Parry and Parry 2000). Marine, brackish and limnetic pelagic communities are dominated by chrysomonads (Stramenopiles – Chrysophyceae), bicosoecids (Stramenopiles – Bicosoecida), dinoflagellates (Alveolates – Dinoflagellata), and choanoflagellates (Opisthokonta – Choanomonada) (Arndt et al. 2000). In these environments, the non-pigmented cryptomonads (Chromalveolata – Cryptomonadales) and euglenids often are present and while they can be dominant they usually comprise only a minor fraction of the flagellate biomass. Benthic taxa are primarily dominated by euglenids (Euglenozoa – Euglenida), bodonids (Euglenozoa – Kinetoplastida), apusomonads (*protista incertae sedis*) and cercomonads (Cercozoa – Cercomonanida) (Vørs et al. 1995).

Because some heterotrophic flagellates can rapidly change to an amoeboid form (Burkholder and Glasgow 1997) and heterotrophic flagellates display few distinctive morphological features, it is difficult to determine the taxonomic composition of natural assemblages from an examination of cell morphology. This is especially true for the smallest heterotrophic flagellates, which are usually classified as unidentified picoflagellates and usually represent the most abundant component of the heterotrophic flagellate community (Carrias et al. 1998; Wieltschnig et al. 2001). The application of molecular ecology methods such as fluorescence in situ hybridization (FISH), which targets whole cells with species-specific rRNA-targeted probes, now enables simultaneous identification and quantification of specific species or genera. This has shown that the picoflagellate fraction contains a wide range of different species belonging to the *Stramenopiles, Alveolates* and *Cercozoa*, and small flagellate reproductive cells of the *Chytridiomycota* (fungi) (Diez et al. 2001; Lefèvre et al. 2007). These picoflagellates are now considered to be fundamental components of both marine and freshwater planktonic systems.

2.1.2 Heterotrophic Protists at Trophic Interfaces

Although HP were often described in aquatic systems by early investigators (Fauré-Fremiet 1924), their ecological importance was not appreciated until new methods (epifluorescence microscopy and flow cytometry) were developed. Up to that time, the notion of a classical linear food chain consisting of phytoplankton, zooplankton and fish predominated. This was replaced by the concept of trophic web in which consumption of DOM (dissolved organic matter) and POM (especially picoplanktonic particles such as detritus, heterotrophic bacteria, autotrophic picoplankton) by HP outlines a major carbon-flow pathway: i.e., the microbial food web (Pomeroy 1974; Azam et al. 1983; Sherr and Sherr 1988). It is now well accepted that picoplankton are responsible for the bulk of primary production in large parts of the open ocean and in many lakes (Li et al. 1983; Stockner and Antia 1986; Weisse

1993; Fogg 1995; Callieri and Stockner 2002). Picoplankton are largely unavailable for direct consumption because the filter apparatus of many crustacean zooplankton is too course to retain picoplanktonic-sized particles (the exception is a few species of grazing cladocerans, which can dominate in lakes). However, HP are larger and thus more efficiently grazed by crustacean zooplankton. Because they repackage their picoplanktonic prey into particles accessible to crustacean grazers, the phagotrophic protists are a crucial link between picoplankton production and higher trophic levels (Sherr et al. 1986; Sherr and Sherr 1988; Gifford 1991). In regard to carbon transformations in aquatic systems, the assimilation of picoplankton by HP leads to substantial losses via respiration. This has led to a debate about the quantitative significance of the transfer of picoplanktonic materials to higher trophic levels via protists (Sherr et al. 1987). In systems in which metazoan zooplankton, such as *Daphnia*, feed directly on picoplankton, trophic repackaging may be regarded as a sink for carbon (Stockner and Shortreed 1989). This 'link or sink' debate should, however, also be considered from a qualitative point of view because both the quantity and quality of carbon determine trophic transfer efficiency (Brett and Müller-Navarra 1997). McManus (1991) has emphasized that the nutritional value of protozoa as food for metazoan zooplankton needs to be addressed, and Sanders and Wickham (1993, and references therein) have shown that HP are an important food source for metazooplankton especially when phytoplankton abundances are low or when phytoplankton quality is reduced. Since the 1990s, considerable research has focused on phytoplankton food quality (Ahlgren et al. 1990; Hessen 1990), but there have been few studies on HP nutritional value (perhaps because of the difficulty in providing metazooplankton with monospecific HP diets in sufficient quantities to complete a growth experiment). In a review, Sanders and Wickham (1993) emphasized that the nutritional value of HP as a food for metazooplankton is highly variable. The C/N ratio of these protists is often lower than that of microalgae and consequently HP may be a source of nitrogen-rich compounds (Stoecker and McDowell Capuzzo 1990), and the bacterivorous protozoa, which obtain phosphorus from bacteria, could be an important source of phosphorus-containing compounds for zooplankton (Caron et al. 1990). However, because the mineral content of HP does not predict their nutritional value for *Daphnia* (Sanders et al. 1996), it has been suggested that variability in the nutritional value of HP is due to variability in their polyunsaturated fatty acid (PUFAs) content (Bec et al. 2003a, b). This has been supported by more recent studies showing that the levels of PUFAs, sterols and sterol-like compounds can determine the nutritional value of HP for metazooplankton (Martin-Creuzburg et al. 2005, 2006; Boëchat et al. 2007).

The nutritional value of HP for metazooplankton is variable. Bacterivorous HP have been reported to be a lower quality food for daphnids than algae such as *Cryptomonas* or *Scenedesmus* (Sanders et al. 1996). However, other protozoa, especially those grazing on nanophytoplankton, can be of higher food quality than their algal prey (Klein Breteler et al. 1999; Bec et al. 2003a). Using protozoal interactions as a model, Klein Breteler et al. (1999) developed the 'trophic-upgrading' concept, which is now understood to be a key process structuring food chains. Heterotrophic protists not only repackage their food, but they may also upgrade it in some cases. More recently, it has been demonstrated that trophic upgrading of picoplankton food quality was largely explained by the lipid composition of HP (Bec et al. 2006).

2.2 Biosynthesis Pathways of Polyunsaturated Fatty Acids in Heterotrophic Protists

Reports on the fatty acid metabolism of HP have often been controversial, especially since many studies date back several decades (see Erwin 1973) and did not focus on the most representative or numerous taxa in aquatic systems. Photosynthetic organisms, the basis of aquatic food webs, have attracted the most attention. The fatty acids of phytoplankton and microalgae have been extensively investigated, and there is an abundant literature on their lipid composition and lipid metabolism (Volkman et al. 1989; Dunstan et al. 1994; see also Chap. 1). Progress in cloning genes encoding fatty acid desaturases and elongases has lead to detailed insights on fatty acid biosynthesis in higher plants (Hashimoto et al. 2006). Now, these molecular tools are being extended to studies on fatty acid biosynthesis and lipid metabolism in protists. At present, fatty acid biosynthesis pathways are better documented in photosynthetic species than in ciliates or non-pigmented (NP) flagellates. Knowledge is increasing on fatty acid metabolism, genes and enzymes in diatoms (*Stramenopiles, Bacillariophyta*), eustigmatophyceae (*Stramenopiles, Eustigmatales*), haptophyceae (Chromalveolata – Haptophyta), euglenids (*Euglenida*) and several species of green microalgae (now classified as *Chloroplastida*) (Wallis and Browse 1999; Domergue et al. 2003; Tonon et al. 2003; Kajikawa et al. 2006). These studies as well as those on the parasitic flagellates *Kinetoplastida* and *Apicompexa* and on the *Amoebozoa* and the ciliates such as *Tetrahymena* have revealed that much of the fatty acid biosynthetic pathways found in protists can be understood based on their phylogenetic histories (Nakashima et al. 1996; Tripodi et al. 2005; Sayanova et al. 2006; Venegas-Calerón et al. 2007). All protists have a cellular structure, which was derived from several endosymbiotic events (Adl et al. 2005). Their genome was compartmentalized as the result of evolution and complex intracellular rearrangements following incorporation of the symbiotic partners into the progenitor host cell. In the case of the genes involved in fatty acid synthesis, this includes the loss, gain and intracellular transfer of genetic material (Domergue et al. 2003), which explains why the potentials for fatty acid and PUFA synthesis are closely related to phylogenetic lineages, and why the pathways can be deduced once phylogenies have been determined. Information on evolutionary history (taxonomic position) makes it possible to propose possible mechanisms of fatty acid synthesis in some aquatic NP flagellates and ciliates. In the large majority of protists, PUFAs are produced by a series of aerobic desaturations and elongations of the 16:0 and 18:0 acids produced by fatty acid synthase (FAS). Another pathway that does not require such desaturation/elongation steps has been found in the marine Thraustochytrid *Schizochytrium* (Ratledge 2004). This *Stramenopile* HP synthesizes PUFAs, including docosahexaenoic acid (DHA, 22:6n-3) using a polyketide synthase (PKS) pathway, which until recently, was only known in deep sea bacteria such as *Shewanella* and *Moritella marina* (formerly *Vibrio marinus*) (Metz et al. 2001). Therefore, *Schizochytrium*, which is well distributed in the marine environment, may be an important source of long chain PUFAs at the base of marine food webs (Parrish et al. 2007). Figure 2.2 shows different routes followed by several protist species for the synthesis of major fatty acids (Domergue et al. 2003; Tripodi et al. 2005; Kajikawa et al. 2006; Sayanova et al. 2006). The pathways illustrated are all related to the classical bioconversion of FAS products into unsaturated fatty acids via the so-called 'eukaryotic' and 'prokaryotic' reactions, which take place in the endoplasmic reticulum and the chloroplast, respectively (Domergue et al. 2003). NP flagellates belonging to lineages dominated by autotrophic species secondarily lost their photosynthetic ability (Sekiguchi et al. 2001). This includes the genera *Paraphysomonas* and *Spumella* (Stramenopiles – Chrysophyceae), *Chilomonas* (Chromalveolata – Cryptomonadales), *Pteridomonas* and *Ciliophrys* (Stramenopiles – Dictyochophyceae), *Polytoma* (Chloroplastida), and several species of colourless euglenids (*Euglenida*) (Mignot 1977; Sekiguchi et al. 2001). All these flagellates have retained the chloroplast in a vestigial form with a reduced plastid DNA of half the size of a complete plastid DNA (Gockel and Hachtel 2000; Sekiguchi et al. 2001). Therefore, it is not known whether the 'prokaryotic' pathway found in the functional chloroplast of photosynthetic protists is present in the vestigial chloroplast (leucoplast) of non-pigmented flagellates (Fig. 2.2). This pathway, inherited from the first endosymbiosis, is according to Domergue et al. (2003), Behrouzian et al. (2001), Riekhof et al. (2005) and Kajikawa et al. (2006), responsible for the entire synthesis of 16:2n-4 and 16:3n-4 in diatoms (and probably in other autotrophic *Stramenopiles*) and for the synthesis of 16:3n-3 in the *Chloroplastida* and higher plants (Fig. 2.2). The occurrence of the 'prokaryotic' pathway in the heterotrophic dinoflagellates, which have no trace of leucoplast, although they possess 18 of the plastid genes in their genome (Sanchez-Puerta et al. 2007), is not known.

According to the theory of secondary endosymbiosis, all members of the Chromista have a common ancestor, which engulfed a rhodophyte (Bodyl 2005). This common origin of *Cryptomonadales* and *Stramenopiles* (diatoms, *Chrysophyceae*, *Eustigmatale*) explains why the pathways of PUFA synthesis are quite similar in these protists (Sargent et al. 1995; Khozin-Goldberg et al. 2002; Domergue et al. 2003). Eicosapentaenoic acid (EPA, 20:5n-3) is mostly synthesized via the 'eukaryotic' pathway located in the endoplasmic reticulum (ER) (see Fig. 2.2). As noted by Domergue et al. (2003), in diatoms $\Delta 9$, $\Delta 12$, n-3 ($\Delta 15$) desaturases, $\Delta 6$ and $\Delta 5$ desaturases operate together with elongases to convert 18:0 to 20:5n-3. Curiously, *Monodus subterraneus* (*Eustigmatale*) exhibits significant variation in the synthesis of EPA by converting (n-6) 20-carbon PUFA into (n-3) 20-carbon PUFA. Subsequently, 18:2n-6 is desaturated by $\Delta 6$ desaturase to 18:3n-6 and latter elongated to 20:3n-6. This fatty acid is further desaturated by the $\Delta 5$ desaturase and the n-3 desaturase to EPA (Khozin-Goldberg et al. 2002). *Chrysophyceae* and *Cryptomonadales* complete the full range of PUFA biosynthetic reactions by forming

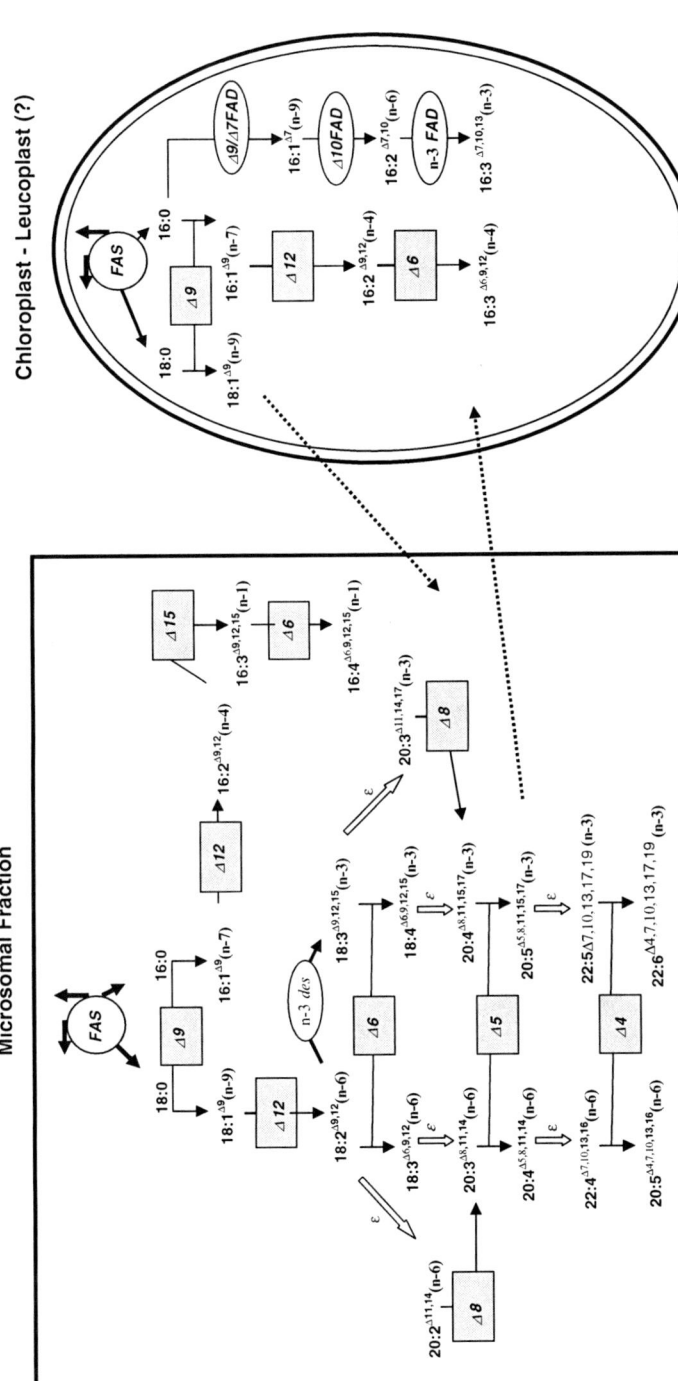

Fig. 2.2 Diagrammatic representation of the proposed pathways of PUFA synthesis in several groups of protists. Synthesis of 16:4(n-1) from 16:1(n-7) via Δ12, Δ15 and Δ6 desaturations was found in amoebozoa (Sayanova et al. 2006). The pathways for 20C and 22C PUFA synthesis, using Δ6, Δ5 and Δ4 desaturases, were depicted in *Kinetoplastida* (Tripodi et al. 2005). Δ4 desaturase is also functional in euglenids and dinoflagellates (Meyer et al. 2003). Δ8 desaturases were identified in euglenids (Meyer et al. 2003). The mode of 16:3(n-3) and 16:3(n-4) formation in chlorophyta and diatom plasts was elucidated by Domergue et al. (2003) and Kajikawa et al. (2006). However, a possible synthesis of fatty acids in vestigial plastids of non-pigmented flagellates is not proved. For clarity, FA formed by retroconversion such as 18:5(n-3) or 16:4(n-3) are not represented. Δ4, Δ5, Δ6, Δ8 desaturases are front-end desaturases (double bond is introduced between a pre-existent bond and the carboxyl end). Δ12, Δ15, n-3 *des*, n-3 *FAD* are methyl-end desaturases (double bond is introduced towards the methyl end of the aliphatic chain). *FAS* fatty acid synthase; *ε* elongation; *Dashed arrows* FA trafficking between ER and plasts (simplified)

DHA (22:6n-3) via elongation of EPA to 22:5n-3 and further desaturation by the Δ4 desaturase. The so-called "Sprecher pathway" in which DHA is formed by β-oxidation of 24-carbon PUFA seems not to be functional in protists. This may be due to the occurrence of the Δ4 desaturase identified in different species of autotrophic and heterotrophic protists (Tripodi et al. 2005). The fatty acid composition of *Paraphysomonas* sp., *Spumella* sp., *Chilomonas* sp. (Erwin 1973; Zhukova and Kharlamenko 1999; Boëchat 2005; Bec et al. 2006) clearly shows that the microsomal elongases and desaturases involved in PUFA production are conserved among those non-pigmented forms. These flagellates have also been found to have a similar fatty acid composition to that in closely-related photosynthetic species. The abundance of DHA in photosynthetic alveolates is particularly marked in the dinoflagellates, which are a major component of aquatic phytoplankton. The synthesis of very large amounts of DHA by the non-pigmented dinoflagellate *Crypthecodinium cohnii* (Mansour et al. 1999) is an important finding. The DHA-formation pathway in *Crypthecodinium* is similar to that in the endoplasmic reticulum (ER) of *chrysophyceae* and sometimes the desaturation and elongation steps can be extended past DHA to result in very long chain PUFAs such as 28:8n-3 (Van Pelt et al. 1999). Unlike the Chromiste protists, the marine and freshwater *chlorophyceae* (Chloroplastida) are relatively deficient in EPA and DHA. This is a consequence of Δ5 desaturation and possibly Δ6 desaturation being rate limiting after the conversion of oleate to the essential 18-carbon fatty acid (Fig. 2.2) (Kajikawa et al. 2006). Interestingly, *Polytoma* species, although non-pigmented, possess the same capabilities for producing 18-carbon PUFAs (18:2n-6, 18:3n-3 and 18:4n-3) as *Chlamydomonas reinhardtii* (Erwin 1973; Vera et al. 2001). The heterotrophic euglenids (*Euglenida*) and bodonids (*Kinetoplastida*) are related groups within the Euglenozoa. Their fatty acid metabolism has been studied in considerable detail (Wallis and Browse 1999; Tripodi et al. 2005). Classic experiments using radio-labelled substrates have shown that *Euglena gracilis* has no Δ6 desaturase activity (Nichols and Appleby 1969), and this has been confirmed by recent studies involving cloning of the desaturase genes (Wallis and Browse 1999; Meyer et al. 2003). The characterization of functional Δ4 and Δ8 desaturases by these workers suggests a specific pathway (see Fig. 2.2) is present in the *Euglenida* in which 18-carbon PUFAs are elongated to 20-carbon PUFA, which then is transformed by Δ8 desaturase to produce EPA and by Δ5 desaturase to produce ARA (20:4n-6) (Wallis and Browse 1999). An important feature of *Euglenida* fatty acid metabolism is that, under heterotrophic conditions, these flagellates can produce copious amounts of EPA and DHA (Wallis and Browse 1999; Meyer et al. 2003). In protozoa belonging to the *Kinetoplastida*, PUFA biosynthesis has been studied in two parasitic genera (*Trypanosoma* and *Leishmania*). Tripodi et al. (2005) found that in *Leishmania* but not in *Trypanosoma* the functional route to 20-carbon and 22-carbon PUFAs involved Δ6, Δ5 and Δ4 desaturases. In *Trypanosoma* only the Δ4 desaturase was present. It is difficult to generalize these findings to the aquatic Bodonids especially as data on their fatty acid composition are contradictory (Zhukova and Kharlamenko 1999; Vera et al. 2001).

Cloning of fatty acid desaturases in other heterotrophic protists has been limited to Amoebozoa and ciliates (*Tetrahymena*). A bifunctional Δ12/Δ15 desaturase has been detected in *Acanthamoeba castellanii* (Sayanova et al. 2006) and because this enzyme utilizes both 16 and 18-carbon substrates, it is able to produce not only 18:2n-6 and 16:2n-4 but also 18:3n-3 and 16:3n-1 PUFAs. The Δ12/Δ15 desaturase could have been acquired by horizontal gene transfer from phototrophic microorganisms ingested by the ancestral phagocytic amoeba (Sayanova et al. 2006). The Δ9 desaturase gene in ciliates was cloned by Nakashima et al. (1996) and the expression of this gene was found to increase during adaptation of *Tetrahymena* cells to cold. In ciliates, the presence of a Δ12 desaturase is imperative for starting PUFA synthesis. Kaneshiro (1980) detected all the intermediate transformation products in the 20:4n-6 synthesis pathway in *Paramecium* and suggested that although there is functional PUFA formation in ciliates, it is directed mainly toward the n-6 fatty acids. Care is needed in generalizing this hypothesis to, for instance, all aquatic ciliates because of the very diverse range of taxa involved. From a study of algivorous ciliates, Boëchat and Adrian (2005) have suggested that species-specific differences in fatty acid composition may be a major determinant of the biochemical composition of the ciliates. Future research on lipid metabolism in HP should be directed toward elucidating PUFA biosynthetic pathways and intermediates in the most important aquatic species. Major advances can be expected from an increased application of molecular methods. For instance, the cloning and expression of desaturase genes would help identify steps involved in fatty acid formation in planktonic ciliates and non-pigmented flagellates such as *Bicosoecida*, *Choanomonada*; groups whose fatty acid compositions are unknown despite their ecological importance in marine and freshwater microbial food webs.

2.3 Variability of Heterotrophic Protist Lipid Composition

A common lament from studies on the aquatic food web is that there is little data available on lipids of freshwater or marine HP. This is surprising, considering that a number of HP are model organisms in cell membrane biology because their lipids have been of commercial interest for some time. For instance, the blocking of lipid metabolism in parasitic protists is considered a promising therapeutic target, and lipids of the Thraustochytrid are one of the most promising sources of DHA. This leads us to the conclusion that a large amount of research dealing with HP lipids has already been done. Nevertheless, in aquatic food webs the lipid composition of HP is thought to be highly variable and the lipid composition of HP appears to be less predictable than that of autotrophic protists. Diatoms (*Bacillariophyta*) are considered as to be an important source of EPA, Cryptophytes (*Cryptomonadales*) and Dinoflagellates are characterized by high levels of n-3 highly unsaturated fatty acids (HUFA), including EPA and DHA, whereas *chlorophyceae* (*Chloroplastida*) are known to contain high amounts of 16C and 18C PUFAs (Bourdier and Amblard 1987; Ahlgren et al. 1990; Sargent et al. 1995; Brett and Müller-Navarra 1997).

The variability in lipid composition in aquatic HP will arise from several sources. This includes intra-species factors which affect the lipid composition in individual species, interspecies factors which affect the relative population densities of differing protists in a very diverse population, and temporal changes to both of these in response to changes in food supply and other environmental constraints. Protozoa as eukaryotic cells contain not only different organelles (and sometimes recent endosymbionts) but also several specific typical membranous structures such as cilia, food vacuoles and oral apparatus. The membranes of these different organelles have their own lipid composition. One could argue that food web studies first require data on the whole organism rather than data dealing with intracellular variability of protozoan lipid composition. However, it should be noted that recent studies (Poerschmann et al. 2004; Boëchat 2005; Boëchat et al. 2007) have demonstrated that the PUFA concentration in protists increases when their trophic pathway shifts from heterotrophic to autotrophic. Boëchat (2005) suggested that this increase in PUFA concentration could be associated with a gain of chloroplast lipids (see also Adolf et al. 2007a).

Previous food web studies have emphasized two main factors affecting the PUFA composition of HP: their habitats (marine vs. freshwater) and their diet (bacterivorous vs. algivorous). In a review, Desvilettes et al. (1997) showed that the PUFA composition of freshwater ciliates is dominated largely by n-6 compounds. Freshwater ciliates contain mainly C_{18} PUFAs, but some genera can contain significant amounts of ARA (Kaneshiro 1980; Desvilettes et al. 1997). In contrast, the marine ciliate *Porauronema acutum* has been reported to contain high levels of n-3 HUFAs (Sul and Erwin 1997), a property it shares with other marine organisms, but which is not found in freshwater ciliates. Sul and Erwin (1997) suggested that the prevalence of n-3 HUFAs in marine organisms was not due to a fortuitous event in evolution but rather from adaptation to factors such as salinity, low temperature, and high hydrostatic pressure in the marine environment. A similar pattern exists for freshwater and marine heterotrophic flagellates. Zhukova and Kharlamenko (1999) showed that marine bacterivorous flagellates efficiently produce n-3 PUFAs, whereas freshwater bacterivorous HP are characterized by a low n-3/n-6 ratio (Vera et al. 2001; Bec et al. 2003a). The n-3/n-6 ratios are higher for marine algivorous HP (Broglio et al. 2003) than for freshwater HP (Boëchat and Adrian 2005). Dietary influence on HP lipids has been recognized for a long time (Nosawa and Thompson 1979) with recent studies clearly showing that both marine and freshwater algivorous HP exhibit higher levels of n-3 PUFAs than bacterivorous ones (Broglio et al. 2003; Bec et al. 2003b; Boëchat 2005, Fig. 2.3).

The high n-3 contents of algivorous HP have been associated with accumulation of dietary PUFAs. Such accumulation tends to be species specific as different HP species fed the same diet exhibited differences in their PUFA composition (Boëchat and Adrian 2005). When HP fed two different diets were analysed, it was concluded that the accumulation of dietary PUFA by HP is compound specific (Bec et al. 2003b; Broglio et al. 2003). Bioconversion and differential accumulation of dietary PUFAs have been hypothesized as the main factors explaining the fatty acid composition differences between HP and their diet. In some cases, an inability to extract certain PUFAs from the diet has been suggested (Broglio et al. 2003) and in

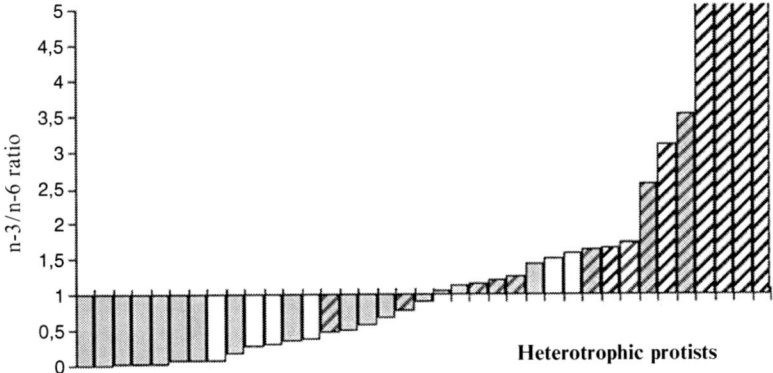

Fig. 2.3 n-3/n-6 ratio of freshwater (*grey*), marine (*white*), algivorous (*striped*), bacterivorous (*non-striped*) heterotrophic protists. Data from Desvilettes et al. (1997), Sul and Erwin (1997), Zhukova and Kharlamenko (1999), Vera et al. (2001), Bec et al. (2003a, b, 2006), Broglio et al. (2003), Boëchat (2005), Boëchat and Adrian (2005), Veloza et al. (2006)

others it has been suggested that some dietary PUFAs serve as precursors for fatty acid bioconversion processes (Bec et al. 2003b). Broglio et al. (2003) relativized the importance of diet as a determinant of PUFA composition of HP and concluded that 'individual HP species may tend to maintain a certain stochiometry in the fatty acid composition regardless of diet'. This hypothesis is consistent with recent results showing that the freshwater non-pigmented flagellate *Paraphysomonsas* sp. maintained the same fatty acid composition pattern when fed different picophytoplanktonic diets (Bec et al. 2006). Tests with diets lacking PUFA have shown that this flagellate can produce its specific PUFA profile (i.e., significant amounts of 18-carbon PUFA and of long chain HUFA such as ARA, EPA and DHA) using only de novo synthesis (Bec et al. 2006).

The fatty acid composition of HP can be altered by factors other than dietary inputs. For example, temperature, O_2, salinity, and trophic pathways are known to profoundly alter lipid composition of protists (Avery et al. 1994, 1996; Poerschmann et al. 2004). HP rapidly adjust their membrane PUFA compositions in response to environmental temperature changes (Nosawa and Thompson 1979). Increased PUFA concentrations are necessary to maintain membrane fluidity (Poerschmann et al. 2004). Because the degree of unsaturation in PUFAs is linked to membrane function and phagotrophic activity in HP, it has been suggested that adjustment in membrane lipid composition enables HP to maintain somatic growth and feeding when environmental conditions change (Avery et al. 1995).

As noted earlier, HP can alter membrane components by ingesting lipid components and/or utilizing specific biosynthetic transformations. These fatty acid bioconversion abilities are still not fully understood, but screening desaturase genes in genome databases offers one way to increase knowledge on the potential capacities of desaturases. Studies are also needed to determine the relationship between environmental factors and desaturase gene expression, and to determine effects of HP metabolism on dietary lipids.

Dietary organic compounds affect the growth and survival of cultivated protists (Vera et al. 2001). The manner which dietary factors facilitate HP to produce efficiently their optimal lipid composition pattern with regard to specific environmental constraints would probably affect drastically HP population dynamics in nature. Because membranes within HP cells have different functions, we hypothesize that their inherent ability to exploit membrane plasticity via changes in lipid composition allows them to exploit emerging or rapidly changing ecological niches eventually leading to changes in trophic function.

Finally, it should be noted that in microbial food webs, the lipid content of protists may be a factor in predator–prey interactions. For example, *Karlodinium venificum* (a toxic dinoflagellate found in coastal ecosystems) can produce anti-grazing compounds putatively named karlotoxins (Adolf et al. 2007b). The proposed mode of action of karlotoxin involves pore formation in cell membranes, and depends upon the sterol composition of target cells. It has been suggested that a predominance of 4-α methylsterols renders an organism immune to karlotoxin, whereas a predominance of des-methyl sterols such as cholesterol renders an organism susceptible to karlotoxin (Adolf et al. 2007b and references therein).

2.4 Conclusion

Heterotrophic protists are a crucial link in aquatic food chains because they repackage food particles too small (or too large) to be grazed by metazoa. Data on the efficiency of carbon incorporation by HP are scarce (Laybourn-Parry and Parry 2000). From large scale studies Landry and Calbet (2004) estimated that the gross growth efficiency (GGE) of mixed protistan communities was around 30–40% (also see Straile 1997; Rivkin and Legendre 2001), but some species of HP have a GGE of up to 70% (Caron et al. 1990).

Because the trophic transfer of HUFAs from aquatic to terrestrial food webs is important for humans beings (Arts et al. 2001), food webs studies often focus on efficient metazoan food chains. In this context, *Daphnia* is often considered as an example of trophic efficient organism. In contrast, protozoa are well less considered, as they may often repackage organic matter only for themselves. This means that organic matter can be transferred through different trophic levels of heterotrophic protists (including parasitized ones) and thus be almost entirely remineralized within these microbial food webs. However, as another fate of organic matter is to sink, rapid nutrient recycling in aquatic microbial food webs may maintain the phytoplanktonic production of essential lipid components in the photic zone (especially in a stratified oligotrophic system). Organisms such as metazoa with a long generation time then benefit indirectly from nutrients released by grazing protists (Dolan 1997). Because protozoa grow and divide as rapidly as phytoplankton, their grazing pressure is better coupled to production than is the grazing of the slower growing metazoa (Calbet and Landry 2004). Because heterotrophic protists are the first 'hungry grazers', they often are the first to find food. In marine pelagic ecosystems, the microzooplankton typically utilize 60–75% of phytoplankton production (Landry and Calbet 2004) and it appears that the highly diverse, non-pigmented protists often occupy the position of dominant primary consumers in aquatic systems. Multiple trophic transfers within microbial food webs lead to substantial carbon losses, and the efficiency of trophic transfer of primary production to metazooplankton via protists is highly dependent on the number of predatory interactions among microconsumers (Landry and Calbet 2004). In long food chains, the efficiency of trophic transfer also depends on whether the quality of food for consumption at the next trophic level is enhanced (i.e., trophic upgrading). In food chains, energy dissipates rapidly through trophic transfer processes so trophic upgrading cannot ultimately counteract carbon losses. In systems where producers are dominated by prokaryotic picoplankton, which can not substain metazoan production, the heterotrophic protists constitute a necessary biosynthetic step and provide essential lipid compounds for crustacean zooplankton growth (Bec et al. 2006). As eukaryotic cells, protists such as the metazoa contain eukaryote-specific compounds, which often make them more nutritious than organic matter from other sources (detritus, bacteria). This could explain why HP appear to be less efficient at upgrading eukaryotes (such as phytoplanktonic algae) in comparison with prokaryotes (Klein Breteler et al. 2004; Bec et al. 2006). With regard to lipids, trophic upgrading requires HP to efficiently accumulate certain specific dietary compounds and to produce others so that the dietary needs of higher trophic levels can be better satisfied. As a number of microalgae species, e.g., cryptophytes, are considered high quality food, trophic transfer via HP can be considered as not only a loss of carbon, but also a loss of essential lipid compounds. We thus regret the lack of available data on the trophic transfer efficiencies of essential lipid components in aquatic systems, especially since a recent review (Sherr and Sherr 2007) has drawn attention to the fact that in the oceans the heterotrophic dinoflagellates are the major grazers of EPA producing diatoms. It is known that the heterotrophic dinoflagellates fed diatoms contain significant amounts of EPA and that they also produce DHA (Broglio et al. 2003).

Further investigations are needed to determine the dynamics of lipids in these microbial food webs and to highlight interactions, which would better couple microbial HUFA production to metazoan grazing.

Acknowledgments We thank Professor Gilles Bourdier for having invited us (some years ago) to collaborate on his research project. We thank Diane Stoecker, Evelyn and Barry Sherr for having provided useful information and many answers. We are grateful to Dr. Keith Joblin (AgResearch Ltd., Hamilton N.Z.) for his help in improving the text from a linguistic perspective.

References

Adl, M., Simpson, A., Farmer, M.A., Andersen, R.A., Anderson, O.R., Barta, J.R., Bowser, S.S., Brugerolle, G., Fensome, R.A., Fredericq, S., James, T.Y., Karpov, S., Kugrens, P., Krug, J., Lane, C., Lewis, L., Lodge, J., Lynn, D.H., Mann, D.G., McCourt, R.M., Mendoza, L.,

Moestrup, O., Mozley-Standridge, S., Nerad, T., Shearer, C.A., Smirnov, A., Speigel, F.W., and Taylor, M.F.J.R. 2005. The new higher level classification of eukaryotes with emphasis on the taxonomy of protists. *J. Eukaryot. Microbiol.* 52:399–451.

Adolf, J.E., Place, A.R., Stoecker, D.K., and Harding Jr., L.W. 2007a. Modulation of polyunsaturated fatty acids in mixotrophic *Karlodinium Veneficum* (Dinophyceae) and its prey, *Storeatula major* (Cryptophyceae). *J. Phycol.* 43:1259–1270.

Adolf, J.E., Bachvaroff, T.R., Krupatkina, D.N., and Place, A.R. 2007b. Karlotoxin mediates grazing of *Oxyrrhis marina* on *Karlodinium veneficum* strains. *Harmful Algae* 6:400–412.

Ahlgren, G., Lundstedt, L., Brett, M.T., and Forsberg, C. 1990. Lipid composition and food quality of some freshwater phytoplankton for cladoceran zooplankters. *J. Plankton Res.* 12:809–818.

Arndt, H., Dietrich, D., Auer, B., Cleven, E., Gräfenhan, T., Weitere, M., and Mylnikov, A. 2000. Functional diversity of heterotrophic flagellates in aquatic ecosystems, pp. 240–268. *In* B.S.C. Leadbeater and J.C. Green (eds.), The flagellates unity, diversity and evolution. Taylor & Francis, London.

Arts, M.T., Ackman, R.G., and Holub, B.G. 2001. "Essential fatty acids" in aquatic ecosystems: a crucial link between diet and human health and evolution. *Can. J. Fish. Aquat. Sci.* 58:122–137.

Avery, S.V., Lloyd, D., and Harwood, J. 1994. Changes in membrane fatty acid composition and delta 12-desaturase activity during growth of *Acanthamoeba castellanii* in batch culture. *J. Eukaryot. Microbiol.* 41:396–401.

Avery, S.V., Lloyd, D., and Harwood, J. 1995. Temperature dependent changes in the plasma lipid order and the phagocytotic activity of the amoeba *Acanthamoeba castellanii* are closely correlated. *Biochem. J.* 312:811–816.

Avery, S.V., Harwood, J., Rutter, A.J., Lloyd, D., and Harwood, J. 1996. Oxygen dependent low temperature composition and delta12 desaturase induction and alteration of fatty acid composition in *Acanthamoeba castellanii* in batch culture. *Microbiology* 142:2213–2221.

Azam, F., Fenchel, T., Field, J.G., Gray, J.S., Meyer-Reil, L.A., and Thingstad, F. 1983. The ecological role of water-column microbes in the sea. *Mar. Ecol. Prog. Ser.* 10:257–263.

Bec, A., Desvilettes, C., Vera, A., Lemarchand, C., Fontvielle, D., and Bourdier, G. 2003a. Nutritional quality of a freshwater heterotrophic flagellate: trophic upgrading of its microalgal diet for *Daphnia*. *Aquat. Microb. Ecol.* 32:203–207.

Bec, A., Desvilettes, C., Vera, A., Fontvielle, D., and Bourdier, G. 2003b. Nutritional value of different food sources for the bennthic daphnidae *Simocephalus vetulus*: role of fatty acids. *Arch. Hydrobiol.* 156:145–163.

Bec, A., Martin-Creuzburg, D., and Von Elert, E. 2006. Trophic upgrading of autotrophic picoplankton by the heterotrophic flagellate *Paraphysomonas* sp. *Limnol. Oceanogr.* 51:1699–1707.

Behrouzian, B., Fauconnot, L., Daligault, F., Nugier-Chauvin, C., Patin, H., and Buist, P.H. 2001. Mechanism of fatty acid desaturation in the green alga *Chlorella vulgaris*. *Eur. J. Biochem.* 268:3545–3549.

Bettarel, Y., Sime-Ngando, T., Amblard, C., and Bouvy, M. 2005. Low consumption of virus-sized particles by heterotrophic nanoflagellates in two lakes of the French Massif Central. *Aquat. Microb. Ecol.* 39:205–209.

Bodyl, A. 2005. Do plastid-related characters support the chromalveolate hypothesis. *J. Phycol.* 41:712–719.

Boëchat, I.G. 2005. Biochemical composition of protists: dependence on diet and trophic mode and consequences for their nutritional quality. Ph.D. Thesis. Humboldt Universität zu Berlin, Berlin. 144 p.

Boëchat, I.G. and Adrian, R. 2005. Biochemical composition of algivorous freshwater ciliates: you are not what you eat. *FEMS Microbiol. Ecol.* 53:393–400.

Boëchat, I.G., Weithoff, G., Krüger, A., Gücker, B., and Adrian, R. 2007. A biochemical explanation for the success of mixotrophy in the flagellate *Ochromonas* sp.. *Limnol. Oceanogr.* 52:1624–1632.

Bourdier, G. and Amblard, C. 1987. Evolution de la composition en acides gras du phytoplancton lacustre du (lac Pavin). *Int. Revue Ges. Hydrobiol.* 11:1201–1212.

Brett, M.T. and Müller-Navarra, D.C. 1997. The role of highly unsaturated fatty acids in food web processes. *Freshw. Biol.* 38:483–499.

Broglio, E., Jonasdottir, S.H., Calbet, A., Jakobsen, H.H., and Saiz, E. 2003. Effect of heterotrophic versus autotrophic food on feeding and reproduction of the calanoid copepod *Acartia tonsa*: relationship with prey fatty acid composition. *Aquat. Microb. Ecol.* 31:267–278.

Brugerolle, G. and Müller, M. 2000. Amitochondriate flagellates, pp. 166–189. *In* B.S.C. Leadbeater and J.C. Green (eds.), The flagellates. unity, diversity and evolution. Taylor & Francis, London.

Burkholder, J.M. and Glasgow, H.B.J. 1997. *Pfiesteria piscicida* and other Pfiesteria-like dinoflagellates: behavior, impacts, and environmental controls. *Limnol. Oceanogr.* 42:1052–1075.

Calbet, A. and Landry, M.R. 2004. Phytoplankton growth, microzooplankton grazing, and carbon cycling in marine systems. *Limnol. Oceanogr.* 49:51–57.

Callieri, C. and Stockner, J.G. 2002. Freshwater autotrophic picoplankton: a review. *J. Limnol.* 61:1–14.

Caron, D.A., Goldman, J.C., and Dennett, M.R. 1990. Carbon utilization by the omnivorous flagellate *Paraphysomonas imperforata*. *Limnol. Oceanogr.* 35:192–201.

Carrias, J.-F., Quiblier-Lloberas, C., and Bourdier, G. 1998. Seasonal dynamics of free and attached heterotrophic nanoflagellates in an oligomesotrophic lake. *Freshw. Biol.* 39:91–101.

Chatton, E. 1925. *Pansporella perplexa*, amoebien à spores protégées, parasite des Daphnies. Réflexions sur la biologie et la phylogénie des Protozoaires. *Ann. Sci. Nat. Zool.* 8:5–84.

Dunstan, G.A., Volkman, J.K., Barret, S.M., Leroi, J., and Jeffrey, S.W. 1994. Essential polyunsaturated fatty acids from 14 species of diatom. *Phytochemistry* 35:155–161.

Desvilettes, C., Bourdier, G., Amblard, C., and Barth, B. 1997. Use of fatty acids for the assessment of zooplankton grazing on bacteria, protozoan and microalgae. *Freshw. Biol.* 38:629–637.

Diez, B., Pedros-Alio, C., and Massana, R. 2001. Study of genetic diversity of eukaryotic picoplankton in different oceanic regions by small-subunit rRNA gene cloning and sequencing. *Appl. Environ. Microbiol.* 67:2932–2941.

Dolan, J.R. 1997. Phosphorus and ammonia excretion by planktonic protists. *Mar. Geol.* 139:109–122.

Domergue, F., Spiekermann, P., Lerchl, J., Beckmann, C., Kilian, O., Kroth, P., Boland, W., Zähringer, U., and Heinz, E. 2003. New insight into *Phaeodactylum tricornutum* fatty acid metabolism. Cloning and functional characterization of plastidial and microsomal delta 12 fatty acid desaturases. *Plant Phycol.* 131:1648–1660.

Erwin, J.A. 1973. Lipids and biomembranes of eukaryotic microorganisms, pp. 40–143. *In* J.A. Erwin (ed.), Comparative biochemistry of fatty acids in eukaryotic microorganisms. Academic, New York.

Fauré-Fremiet, E. 1924. Contribution à la connaissance des Infusoires planctoniques. *Suppl. Bull. Biol. Fr. Bel.* 6:171.

Fogg, G.E. 1995. Some comments on picoplankton and its importance in the pelagic ecosystem. *Aquat. Microb. Ecol.* 9:33–39.

Gifford, D.J. 1991. The protozoan-metazoan trophic link in pelagic ecosystems. *J. Protozool.* 38:81–86.

Gockel, G. and Hachtel, W. 2000. Complete gene map of the plastid genome of the nonphotosynthetic euglenoid flagellate *Astasia longa*. *Protist* 151:347–351.

Haeckel, E. 1866. Generelle Morphologie der Organismen, Allgemeine Anatomie der Organismen Vol. I, Reimer, Berlin.

Hashimoto, K., Yoshizawa, A., Saito, K., Yamada, T., and Kanehisa, M. 2006. The repertoire of desaturases for unsaturated fatty acid synthesis in 397 genomes. *Genome Inform* 17:173–183.

Hessen, D. O. 1990. Carbon, nitrogen and phosphorus status in *Daphnia* at varying food conditions. *J. Plankton Res.* 12:1239–1249.

Johnson, M.D., Oldach, D., Delwiche, C.F., and Stoecker, D.K. 2007. Retention of transcriptionally active cryptophyte nuclei by the ciliate *Myrionecta rubra*. *Nature* 445:426–428.

Kajikawa, K.T., Yamato, Y., Kohzu, S., Shoji, K., Matsui, Y., Tanaka, Y., and Fukuzawa, H. 2006. A front-end desaturase from *Chlamydomonas reinhardtii* produces pinolenic and coniferonic acids by ω13 desaturation in methylotrophic yeast and tobacco. *Plant.Cell Physiol.* 47:64–73.

Kaneshiro, E.S. 1980. Positional distribution of fatty acids in the major glycerophospholipids of *Paramecium tetraurelia*. *J. Lipid Res.* 21:559–570.

Khozin-Goldberg, I., Didi-Cohen, S., Shayakhmetova, I., and Cohen, Z. 2002. Biosynthesis of EPA in the freshwater eustigmatophyte Monodus subterraneus. *J. Phycol.* 38:745–751.

Klein Breteler, W.C.M., Schogt, N., Baas, M., Schouten, S., and Kraay, G.W. 1999. Trophic upgrading of food quality by protozoans enhancing copepod growth: role of essential lipids. *Mar. Biol.* 135:191–198.

Klein Breteler, W.C.M., Koski, M., and Rampen, S. 2002. Role of essential lipids in copepod nutrition: no evidence for trophic upgrading of food quality by a marine ciliate. *Mar. Ecol. Prog. Ser.* 274:199–208.

Landry, M.R. and Calbet, A. 2004. Microzoplankton production in the oceans. *ICES J. Mar. Sci.* 61:501–507.

Laybourn-Parry, J. and Parry, J. 2000. Flagellates and the microbial loop, pp. 216–239. *In* B.S.C. Leadbeater and J.C. Green (eds.), The flagellates. unity, diversity and evolution. Taylor & Francis, London.

Lefèvre, E., Bardot, C., Noël, C., Carrias, J-F., Viscogliosi, E., Amblard, C., and SimeNgando, T. 2007. Unveiling fungal zooflagellates as members of freshwater picoeukaryotes: evidence from a molecular diversity study in a deep meromictic lake. Environ. *Microbiol.* 9:61–71.

Li, W.K.W., Subba Rao, D.V., Harrison, W.C., Smith, J.C., Cullen, J.J., Irwin, B., and Platt, T. 1983. Autotrophic picoplankton in the tropical ocean. *Science* 219:292–295.

Mansour, M.P., Volkman, J.K., Holdsworth, D.G., Jackson, A.E., and Blackburn, S.I. 1999. Very-long-chain (C28) highly unsaturated fatty acids in marine dinoflagellates. *Phytochemistry* 50:541–548.

Martin-Creuzburg, D., Bec, A., and Von Elert, E. 2005. Trophic upgrading of picocyanobacterial carbon by ciliates for nutrition of *Daphnia magna*. *Aquat. Microb. Ecol.* 41:271–280.

Martin-Creuzburg, D., Bec, A., and Von Elert, E. 2006. Supplementation with sterols improves food quality of a ciliate for *Daphnia magna*. *Protist* 157:477–486.

McManus, G.B. 1991. Flow analysis of a planktonic microbial food web model. *Mar. Microb. Food Webs* 5:145–160.

Metz, J.G., Roessler, P., Facciotti, D., Levering, C., Dittrich, F., Lassner, M., Valentine, R., Kathryn Lardizabal, K., Frederic Domergue, F., Yamada, A., Yazawa, K., Knauf, V., and John Browse, J. 2001. Production of polyunsaturated fatty acids by polyketide synthases in both prokaryotes and eukaryotes. *Science* 293:290–293.

Meyer, A., Cirpus, P., Ott, C., Schlecker, R., Zähringer, U., and Heinz, E. 2003. Biosynthesis of docosahexaenoic acid in *Euglena gracilis*: biochemical and molecular evidence for the involvement of a $\Delta 4$-fatty acyl group desaturase. *Biochemistry* 42:9779–9788.

Mignot, J-P. 1977. Etude ultrastructurale d'un flagellé du genre *Spumella* – Chrysomonadine leucoplastidié. *Protistologica* 13:219–231.

Nakashima, S., Zhao, Y., and Nozawa, Y. 1996. Molecular cloning of delta 9 fatty acid desaturase from the protozoan *Tetrahymena thermophila* and its mRNA expression during thermal membrane adaptation. *Biochem. J.* 317:29–34.

Nichols, B.W. and Appleby, R.S. 1969. The distribution and biosynthesis of arachidonic acid in algae. *Phytochemistry* 8:1907–1915.

Nozawa, Y. and Thompson, G.A. 1979. Lipids and membrane organization in *Tetrahymena*, pp.276–335. *In* M. Levandowski and S.H. Hutner (eds.), Biochemistry and Physiology of Protozoa. Academic, New York.

Park, J.S., Simpson, A.G.B., Lee, W.J., and Cho, B.C. 2007. Ultrastructure and phylogenetic placement within Heterolobosea of the previously unclassified, extremely halophilic heterotrophic flagellate *Pleurostomum flabellatum* (Ruinen 1938). *Protist* 158:397–413.

Parrish, C.C., Whiticar, M., and Puvanendran, V. 2007. Is $\omega 6$ docosapentaenoic acid an essential fatty acid during early ontogeny in marine fauna? *Limnol. Oceanogr.* 53:478–479.

Poerschmann, J., Spijkerman, E., and Langer, U. 2004. Fatty acid patterns in *Chlamydomonas* sp. as a marker for nutritional regimes and temperature under extremely acidic conditions. *Microb. Ecol.* 48:78–89.

Pomeroy, L.R. 1974. The ocean's food web, a changing paradigm. *Bioscience.* 24:499–504.

Ratledge, C. 2004. Fatty acid biosynthesis in microorganisms being used for single cell oil production. *Biochimie* 86:807–815.

Riekhof, W.R., Sears, B.B., and Benning, C. 2005. Annotation of genes involved in glycerolipid biosynthesis in *Chlamydomonas reinhardtii*: discovery of the betaine lipid synthase. *Eukaryot. Cell.* 4:242–252.
Rivkin, R.B. and Legendre, L. 2001. Biogenic carbon cycling in the upper ocean: effects of microbial respiration. *Science* 291:2398–2400.
Sanchez-Puerta, M.V., Lippmeier, J.C., Apt, K.E., and Delwiche, C.F. 2007. Plastid genes in a non-photosynthetic dinoflagellate. *Protist* 158:105–117.
Sanders, R.W. and Wickham, S.A. 1993. Planktonic protozoa and metazoa: predation, food quality and population control. *Mar. Microb. Food Webs* 7:197–223.
Sanders, R.W., Williamson, C.E., Stutsman, P.L., Moeller, R.E., Goulden, C.E., and Aoki-Goldsmith, R. 1996. Reproductive success of "herbivorous" zooplankton fed algal and non algal food resources. *Limnol. Oceanogr.* 41:1295–1305.
Sargent, J.R., Bell, M.V., and Henderson, R.J. 1995. Protists as sources of (n-3) polyunsaturated fatty acids for vertebrate development, pp. 54–64. *In* G. Brugerolle and J. P. Mignot (eds.), Protistological actualities. Proceedings of the 2nd European Conference on Protistology and the 8th European Conference on Ciliate Biology, Aubiere Cedex, France.
Sayanova, O., Haslam, R., Guschina, I., Lloyd, D., Christie, W.W., Harwood, J.L., and Napier, J.A. 2006. A bifunctional Δz12, Δ15 desaturase from *Acanthamoeba castellanii* directs the synthesis of highly unusual n-1 series unsaturated fatty acids. *J. Biol. Biochem.* 281: 36533–36541.
Scott, F.J., Davidson, A.T., and MArchant, H.J. 2001. Grazing by the antarctic sea ice ciliate *Pseudocohnolembus*. *Polar Biol.* 24:127–131.
Sekiguchi, H., Moriya, M., Nakayama, T., and Inouye, I. 2001. Vestigial chloroplasts in heterotrophic stramenopiles *Pteridomonas danica* and *Ciliophrys infusionum* (Dictyochophyceae). a*Protist*153:157–167.
Sherr, E.B. and Sherr, B.F. 1988. Role of microbes in pelagic food webs: a revised concept. *Limnol. Oceanogr.* 33:225–1227.
Sherr, E.B. and Sherr, B.F. 2002. Significance of predation by protists in aquatic microbial food webs. *Anton. Leeuw. Int. J. G.* 81:293–308.
Sherr, E.B. and Sherr, B.F. 2007. Heterotrophic dinoflagellates: a significant component of microzooplankton biomass and major grazers of diatoms in the sea. *Mar. Ecol. Prog. Ser.* 352:187–197.
Sherr, E.B., Sherr, B.F., and Paffenhöffer, G.A. 1986. Phagotrophic protozoa as food for metazoans: a 'missing' trophic link in marine pelagic food webs? Mar. *Microb. Food Webs* 1:61–80.
Sherr, B.F., Sherr, E.B., and Albright, L.J. 1987. Bacteria: link or sink? *Science* 235:88–89.
Stockner, J.G. and Antia, N.J. 1986. Algal picoplankton from marine and freshwater ecosystems: a multidisciplinary perspective. *Can. J. Fish. Aquat. Sci.* 43:2472–2503.
Stockner, J.G. and Shortreed, K.S. 1989. Algal picoplancton production and contribution to food webs in oligotrophic British Columbia lakes. *Hydrobiologia* 173:151–166.
Stoecker, D.K. 1998. Conceptual models of mixotrophy in planktonic protists and some ecological and evolutionary implications. *Eur. J. Protistol.* 34:281–290.
Stoecker, D.K. and McDowell Capuzzo, J. 1990. Predation on protozoa: its importance to zooplankton. *J. Plankton Res.* 12:891–908.
Straile, D. 1997. Gross growth efficiencies of protozoan and metazoan zooplankton and their dependence on food concentration, predator-prey weight ratio, and taxonomic group. *Limnol. Oceanogr.* 42:1375–1385.
Sul, D. and Erwin, J.A. 1997. The membrane lipids of the marine ciliated protozoan *Parauronema acutum*. *Biochim. Biophys. Acta* 1345:162–171.
Tonon, T., Harvey, D., Larson, T.R., and Graham, I.A. 2003. Identification of a very long chain polyunsaturated fatty acid Δ4-desaturase from the microalga *Pavlova lutheri*. *FEBS Lett.* 553:440–450.
Tripodi, K., Buttigliero, L., Altabe, S., and Uttaro, A. 2005. Functional characterization of front-end desaturase from trypanosomatids depicts the first PUFA biosynthetic pathway from a parasitic protozoan. *FEBS Lett.* 273:271–280.

Van Pelt, C.K., Huang, M.C., Tschanz, C.L., and Brenna, J.T. 1999. An octaene fatty acid, 4,7,10,13,16,19,22,25-octacosaoctaenoic acid (28:8n-3) found in marine oils. *J. Lipid Res.* 40:1501–1505.

Veloza, A.J., Chu, F-L.E., and Tang, K.W. 2006. Trophic modification of essential fatty acids by heterotrophic protists and its effects on the fatty acid composition of the copepod *Acartia tonsa. Mar. Biol.* 148:779–788.

Venegas-Calerón, M., Beaudoin, F., Sayanova, O., and Napier, J.A. 2007. Co-transcribed genes for long chain polyunsaturated fatty acid biosynthesis in the protozoon *Perkinsus marinus* include a plant-like FAE1 3-ketoacyl coenzyme A synthase. *Biol. Chem.* 282:2996–3003.

Vera, A., Desvilettes, C., Bec, A., and Bourdier, G. 2001. Fatty acid composition of freshwater heterotrophic flagellates: an experimental study. *Aquat. Microb. Ecol.* 25:271–279.

Volkman, J.K., Jeffrey, S.W., Nichols, P.D., Rogers, G.I., and Garland, C.D. 1989. Fatty acid and lipid composition of 10 species of microalgae used in aquaculture. *J. Exp. Mar. Biol. Ecol.* 128:219–240.

Vørs, N., Buck, K.R., Chavez, F.P., Eikrem, W., Hansen, L., Østergaard, J.B., and Thomsen, H. 1995. Nanoplankton of the equatorial Pacific with emphasis on the heterotrophic protists. *Deep-Sea Res. II* 42:585–602.

Wallis, J.G. and Browse, J. 1999. The Δ8 desaturase of *Euglena gracilis*: an alternate pathway for synthesis of 20-carbon polyunsaturated fatty acids. *Archiv. Biochem. Biophys.* 365:307–316.

Weisse, T. 1993. Dynamics of autotrophic picoplankton in marine and freshwater ecosystems. *Adv. Microb. Ecol.* 13:327–369.

Wieltschnig, C., Kirschner, A.K.T., Steitz, A., and Velimirov, B. 2001. Weak coupling between heterotrophic nanoflagellates and bacteria in a eutrophic freshwater environment. *Microb. Ecol.* 42:159–167.

Zhukova, N.V. and Kharlamenko, V.I. 1999. Sources of essential fatty acids in the marine microbial loop. *Aquat. Microb. Ecol.* 17:153–157.

Chapter 3
Ecological Significance of Sterols in Aquatic Food Webs

Dominik Martin-Creuzburg and Eric von Elert

3.1 Introduction

Sterols are indispensable for a multitude of physiological processes in all eukaryotic organisms. In most eukaryotes, sterols are synthesized de novo from low molecular weight precursors. Some invertebrates (e.g., all arthropods examined to date), however, are incapable of synthesizing sterols de novo, and therefore have to acquire sterols from their diet. Here, we aim to demonstrate that such nutritional requirements not only affect the performance of an individual in its environment but may also have major consequences for the function of aquatic ecosystems. Starting from general patterns of occurrence and biosynthesis of sterols, we next explore the physiological properties and nutritional requirements of sterols. These aspects are then integrated into a more ecological perspective. We emphasize their effects on aquatic food webs in general and on herbivorous zooplankton in particular with the major aim to outline how the interplay of physiological capabilities of individual herbivores and trophic interactions in the food web will determine the effect of low dietary provision of sterols on structure and function of aquatic ecosystems.

3.2 Occurrence of Sterols

The ability to synthesize sterols de novo is a characteristic feature of eukaryotic cells. In prokaryotes, sterols are usually absent or they are found in such small amounts that contamination from other sources cannot be excluded. However, there is evidence that at least some eubacteria are capable of synthesizing sterols de novo (e.g., *Methylococcus capsulatus*, Volkman 2003, 2005). Small amounts of sterols

---

D. Martin-Creuzburg (✉)
Limnological Institute, University of Constance, Mainaustrasse 252, 78464, Konstanz, Germany
dominik.martin-creuzburg@uni-konstanz.de

E. von Elert
Zoologisches Institut, Universität zu Koeln, Weyertal 119, 50923, Koeln, Germany
evelert@uni-koeln.de

have also been detected in some cyanobacterial strains, but the occurrence of sterols in cyanobacteria is controversial (Summons et al. 2006). The presence of a sterol biosynthetic pathway in cyanobacteria could not be confirmed by molecular data, which implies that sterols are usually absent in these prokaryotes (for critical reviews see: Volkman 2003, 2005; Summons et al. 2006).

Eukaryotes may be divided into three groups that differ in their sterol profiles: (a) the plant kingdom, which is characterized by a large set of different phytosterols, (b) the animal kingdom, where cholesterol tends to be the principle sterol, and (c) the fungal kingdom, which is generally characterized by the presence of ergosterol (Fig. 3.1). For reasons of convenience, not reflecting taxonomy, heterotrophic protists will be treated separately. Typical sterols of higher plants are sitosterol (stigmast-5-en-3β-ol), stigmasterol (24E-stigmasta-5,22-dien-3β-ol), and the two C-24 epimers campesterol (campest-5-en-3β-ol) and 22-dihydrobrassicasterol (ergost-5-en-3β-ol) (Fig. 3.2). Sterol profiles of algae are highly diverse so that a general pattern is hard to define and beyond the scope of this chapter (for reviews see: Nes and McKean 1977; Patterson 1991; Volkman et al. 1998; Volkman 2003). The sterol profile of animals is comparatively simple with characteristic high levels of cholesterol (cholest-5-en-3β-ol, Fig. 3.2; often >90% of total sterols) and only small amounts of other sterols. These minor sterols are either biosynthetic precursors or of dietary origin, in particular in herbivorous species, or are provided by symbiotic algae or other associated organisms such as fungi in the gut (Goad 1981). Ergosterol

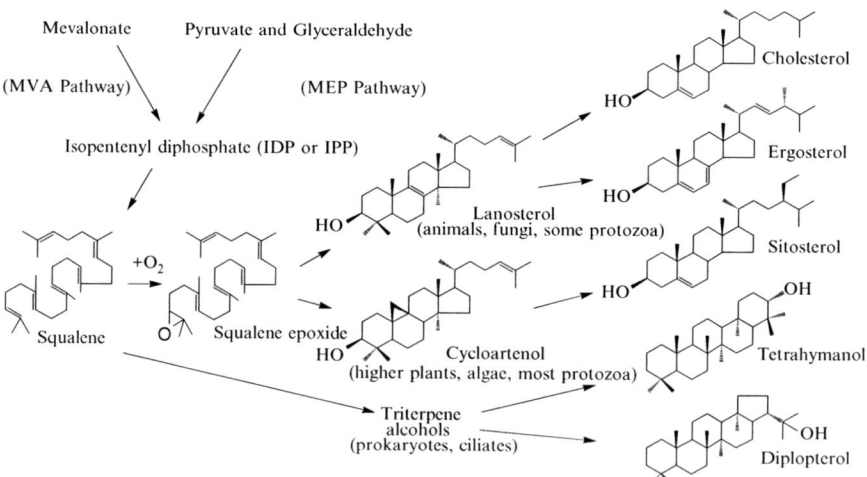

Fig. 3.1 Biosynthesis of sterols and other triterpenoid alcohols – a simplified scheme. Isopentenyl diphosphate is synthesized either via the classical mevalonate pathway (MVA) or via the more recently reported methylerythritol-phosphate pathway (MEP). Molecular oxygen is required for the cyclization of squalene via squalene epoxid. In animals, fungi, and some protozoa (e.g., dinoflagellates), the cyclization leads to lanosterol and in higher plants, algae, and most protozoa to cycloartenol, which are subsequently converted to functional products such as cholesterol, ergosterol, or sitosterol. Hopanoids (e.g., diplopterol) and the triterpenoid alcohol tetrahymanol are presumably formed nonoxidatively by direct cyclization of squalene itself (Ourisson et al. 1987)

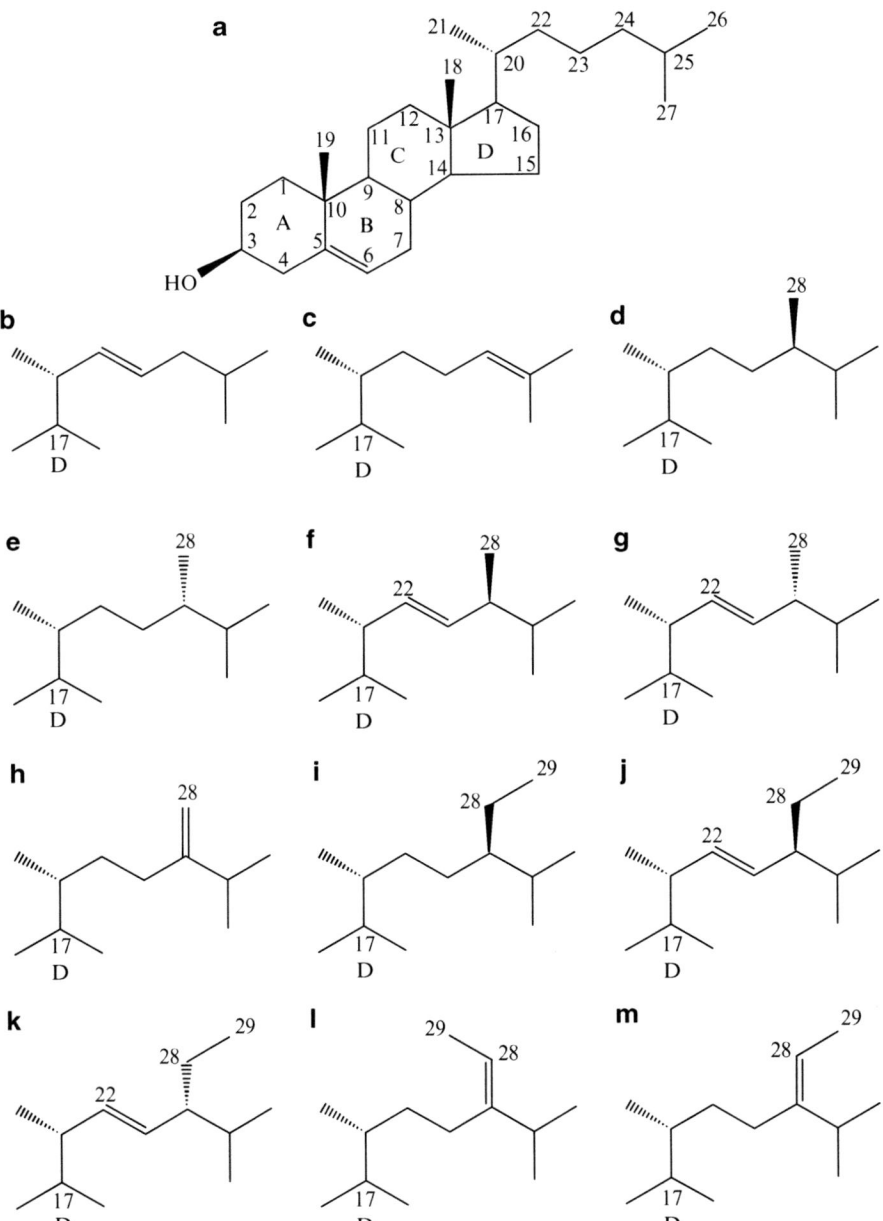

Fig. 3.2 Cholesterol (cholest-5-en-3β-ol; **a**) and other Δ⁵-sterols common in aquatic food webs (the tetracyclic sterol nucleus similar to cholesterol): 22-dehydrocholesterol ((22E)-cholesta-5,22-dien-3β-ol; **b**), desmosterol (cholesta-5,24-dien-3β-ol; **c**), campesterol (campest-5-en-3β-ol; **d**), 22-dihydrobrassicasterol (ergost-5-en-3β-ol; **e**), epibrassicasterol ((22E-campesta-5,22-dien-3β-ol; **f**), brassicasterol ((22E-ergosta-5,22-dien-3β-ol; **g**), 24-methylenecholesterol (ergosta-5,24(24)-dien-3β-ol; **h**), sitosterol (stigmast-5-en-3β-ol; **i**), stigmasterol ((24E)-stigmasta-5,22-dien-3β-ol; **j**), poriferasterol (24E-poriferasta-5,22-dien-3β-ol; **k**), fucosterol ((24E)-stigmasta-5,24(24)-dien-3β-ol; **l**), isofucosterol ((24Z)-stigmasta-5,24(24)-dien-3β-ol; **m**). The carbon atoms are numbered according to the IUPAC recommendations; *capitals* indicate the four rings (A, B, C, D) of the sterol nucleus

(22E-ergosta-5,7,22-trien-3β-ol) is the predominant sterol in fungi, it is often used as a biomarker to detect and to quantify fungi in biological samples (Gessner and Chauvet 1993). However, care has to be taken in aquatic samples since other sources of ergosterol may exist (e.g., some green algae, Thompson 1996; and some diatoms, Véron et al. 1998).

Another group of organisms that may contain sterols are the heterotrophic protists; however, data on the occurrence of sterols in heterotrophic protists are scarce. Most research has been done on ciliates, which are presumably incapable of synthesizing sterols de novo, but incorporate dietary sterols (see below). In contrast, sterol biosynthesis has been documented in heterotrophic flagellates and dinoflagellates (Williams et al. 1966; Leblond and Chapman 2002; Giner et al. 2003; Bec et al. 2006), and in amoebae (Raederstorff and Rohmer 1987).

3.3 Biosynthesis of Sterols

Sterols are synthesized from low molecular weight precursors via isopentenyl diphosphate (IDP or IPP) and squalene (Fig. 3.1). The universal C_5 building block IPP is synthesized by two different pathways: the classical Bloch–Lynen pathway, where IPP is formed from three molecules of acetyl-CoA via mevalonate (MVA pathway), and the more recently reported MEP pathway (Rohmer et al. 1993), where IPP is formed from pyruvate and glyceraldehyde via methylerythritol-phosphate (MEP). Squalene, which is synthesized from 6 IPP molecules, is oxidatively converted to oxidosqualene. Cyclization of oxidosqualene leads to either lanosterol (5α-lanosta-8,24-dien-3β-ol; animals, fungi, some protozoa) or cycloartenol (5α-cycloart-24-en-3β-ol; higher plants, algae, most protozoa), which are subsequently modified to products such as cholesterol or phytosterols (Fig. 3.1). Sterol biosynthesis is an energetically expensive process and requires molecular oxygen. For further details on the biosynthesis of sterols and the occurrence of the two different pathways (MVA and MEP) the reader is referred to Volkman (2003, 2005) and Summons et al. (2006).

3.4 Physiological Properties of Sterols

Sterols are characterized by a tetracyclic fused ring skeleton, a polar head group at position 3 (3β-OH), and an alkyl side chain at position 17 (Fig. 3.2). The large number of naturally occurring sterols is defined by the number and/or position of double bonds or by additional substituents in the sterol nucleus or in the side chain. Because of the trans-configuration of the ring skeleton, the sterol molecule is planar and rigid, except for the rather flexible side chain. Sterols are indispensable structural components of eukaryotic membranes. Their amphipathic nature enables the incorporation into phospholipid bilayers: the polar 3β-OH group oriented to the aqueous phase and the nonpolar sterol nucleus and the alkyl side chain located in

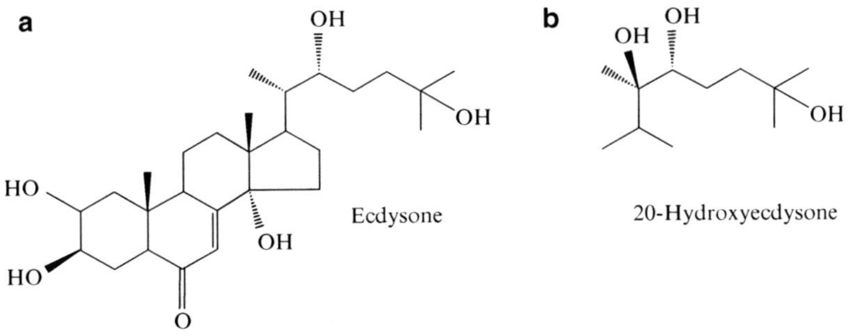

Fig. 3.3 Ecdysteroids – molting hormones of arthropods

the hydrophobic core of the phospholipid bilayer, interacting with the acyl chains of fatty acids. The sterol content is a major means by which eukaryotic cells modulate and refine membrane fluidity, permeability, and the function of various membrane proteins (Haines 2001; Ohvo-Rekilä et al. 2002; and see Chap. 10). It has been shown that the ordering capacity provided by cholesterol is of significantly greater magnitude than that of any of cholesterol's metabolic precursors (Dahl et al. 1980). Likewise, the ordering effect provided by phytosterols differs from that of cholesterol, which may limit direct substitution of cholesterol by phytosterols in animal membranes (Haines 2001).

Furthermore, sterols serve as precursors for vitamin D and for the multitude of naturally occurring steroid hormones, such as the brassinosteroids that promote growth in plants (Schaller 2003), the molt inducing ecdysteroids (Fig. 3.3) in arthropods (Grieneisen 1994), and the various steroids that are involved in the regulation of developmental processes in higher animal. Cholesterol has been found to covalently modify proteins of the "hedgehog" family, which are secreted signaling proteins that are required for developmental patterning of embryonic structures in insects, vertebrates, and other multicellular organisms (Porter et al. 1996). These examples illustrate the unique biochemical properties of sterols and their wide array of effects on biological processes.

3.5 Nutritional Requirements

3.5.1 Heterotrophic Protists

Nutritional requirements for sterols in heterotrophic protozoa are poorly investigated. The sterol composition of the freshwater heterotrophic nanoflagellate *Paraphysomonas* sp. was shown not to be affected by the sterol composition of its food source, i.e. *Paraphysomonas* sp. grown on a sterol-free food source exhibited the same sterol pattern as *Paraphysomonas* sp. grown on a sterol-containing food source, which indicates that *Paraphysomonas* sp. is capable of synthesizing sterols de novo (Bec et al. 2006). De novo biosynthesis of sterols has also been reported in some other heterotrophic flagellates and dinoflagellates (e.g. Williams et al. 1966; Klein Breteler et al. 1999; Leblond and Chapman 2002; Giner et al. 2003).

In contrast, ciliates presumably lack the ability to synthesize sterols de novo (Conner et al. 1968; Harvey and McManus 1991; Harvey et al. 1997; Klein Breteler et al. 2004; Martin-Creuzburg et al. 2005b). However, exogenously supplied sterols can be incorporated into the ciliates' cell membranes and further metabolized into various sterols (Conner et al. 1968; Harvey and McManus 1991; Harvey et al. 1997; Martin-Creuzburg et al. 2006; Boëchat et al. 2007). In the absence of exogenous sterols, many ciliates produce the pentacyclic triterpenoid alcohol tetrahymanol (gammaceran-3β-ol, Fig. 3.1) and/or hopanoids (e.g. diplopterol; hopan-22-ol, Fig. 3.1), which, in ciliates, are functionally equivalent to sterols as structural components of cell membranes (Raederstorff and Rohmer 1988; Martin-Creuzburg et al. 2006). Hence, the capability to produce tetrahymanol and/or hopanoids renders ciliates independent of a dietary sterol supply. However, the synthesis of pentacyclic triterpenoids seems to be associated with physiological costs, as suggested by the observation that the synthesis is down-regulated in the presence of dietary sterols. A few ciliates, however, require a dietary source of sterols, and this sterol requirement can be very specific; e.g. the sterol requirements of *Paramecium* are best fulfilled by adding the phytosterol stigmasterol to the growth medium but not by adding cholesterol (VanWagtendonk 1974). Other taxa may satisfy their sterol requirements by hosting symbiotic algae, which further complicates the investigation of the sterol metabolism of ciliates.

3.5.2 Invertebrates

The inability to synthesize sterols from low molecular weight precursors such as acetate or mevalonate is widespread among invertebrates. Nematodes, for instance, are incapable of synthesizing sterols de novo and require a dietary source of sterols for normal growth and development (Lozano et al. 1987; Shim et al. 2002; Merris et al. 2004). A group where biosynthesis has been demonstrated is the sponges (marine: Silva et al. 1991; freshwater: Dembitsky et al. 2003). In molluscs, reports on sterol requirements are controversial. Many gastropods have been shown to biosynthesize cholesterol, and therefore a dietary source of sterols is presumably not required (Kanazawa 2001). In contrast, experiments with marine bivalves used for aquaculture (e.g. oysters) suggest that the ability to synthesize sterols de novo is generally low or absent among bivalve species, which implies that a dietary supply of sterols is necessary for somatic growth (Voogt 1975; Trider and Castell 1980; Soudant et al. 1998). Data on sterol requirements of freshwater bivalves are scarce. Feeding experiments with radioactive-labeled acetate suggest that at least some freshwater bivalves are capable of synthesizing sterols de novo, even though the rate of incorporation of labeled acetate into sterols was found to be very low (Popov et al. 1981).

It is generally accepted that all arthropods are incapable of synthesizing sterols de novo (Goad 1981). Cholesterol is the predominant sterol found in arthropods, as in most other animals (Goad 1981). Carnivorous species are readily supplied with cholesterol, while herbivorous species cannot rely on a dietary source of cholesterol since this sterol is often hardly represented in plant material (appreciable amounts are found only in some algal classes, e.g. in eustigmatophytes and some diatoms; Véron et al. 1998). Instead, plants and algae contain several types of phytosterols that differ from cholesterol by additional substituents (e.g., methyl or ethyl groups at C-24) or by the position and/or number of double bonds in the side chain or in the sterol nucleus (Piironen et al. 2000; Moreau et al. 2002). Herbivorous arthropods can either use the sterols present in their diet directly, or they have to metabolize them to cholesterol to meet the requirements for growth and development (Svoboda and Thompson 1985). Other than a few exceptions, herbivorous insects and also the crustaceans examined to date use dietary sterols to synthesize cholesterol. Therefore, most species studied are capable of dealkylating and reducing common C-24-alkyl phytosterols, such as sitosterol or stigmasterol, to cholesterol (Grieneisen 1994; Behmer and Nes 2003). However, more than 200 different types of phytosterols have been reported in plant material (Moreau et al. 2002), and it is not surprising that not all of them are suitable as cholesterol precursors. Numerous studies have focused on dietary needs for sterols in insects and revealed that the pattern of sterol metabolism is by no means ubiquitous (Behmer and Nes 2003). In general, Δ^5 and $\Delta^{5,7}$ sterols (numbers indicate the position of the double bonds; see Fig. 3.2) meet the nutritional requirements of insects, while Δ^7 and Δ^{22} sterols are unsuitable, in at least some insect species. The ability to dealkylate C-24 methyl or ethyl sterols is widespread among insects, even though dealkylation capacities have been lost in some insect groups (Behmer and Nes 2003). It has been demonstrated that the ratio of suitable to unsuitable dietary sterols may constrain insect survival (Behmer and Elias 2000).

In crustaceans, comparable studies are scarce and mostly restricted to marine decapod crustaceans, which have received attention due to their relevance for aquaculture. As in insects, Δ^5 and $\Delta^{5,7}$ sterols are readily used by decapod crustaceans, and methyl or ethyl groups are effectively removed from the C-24 position of the side chain to form cholesterol (e.g. Teshima 1971, 1991; Grieneisen 1994). Only recently has been attention drawn to the nutritional sterol requirements of crustacean zooplankton; mostly to cladocerans of the genus *Daphnia* and to marine copepods.

In *Daphnia*, the absence of dietary sterols (accomplished by feeding a cyanobacterial diet) had serious consequences for a variety of life history traits (Martin-Creuzburg et al. 2005a). Somatic and population growth rates, the number of viable offspring, and the probability of survival were significantly reduced with the diminishing availability of sterols. Moreover, an insufficient sterol supply adversely affected the performance of the offspring, which points to strong maternal effects under sterol limitation.

In addition to low dietary sterol levels, the quality of dietary sterols may affect the assimilation of dietary carbon and thus growth and/or reproduction. In a first attempt, we have investigated to what extent structural features of sterols affect the growth and reproduction of *Daphnia galeata* by supplementing a sterol-free diet with different sterols (Martin-Creuzburg and Von Elert 2004). The results indicated that sterols containing a double bond at position Δ^5 (e.g., sitosterol, stigmasterol, ergosterol) meet the nutritional requirements of the daphnids, regardless of additional double bonds in the nucleus or in the side chain. In contrast, the Δ^7 sterol lathosterol (5α-cholest-7-en-3β-ol, Fig. 3.4) supported growth and reproduction to a significantly lower extent than cholesterol. No effect on growth of *D. galeata* was observed by supplementation with dihydrocholesterol (5α-cholestan-3β-ol), a completely saturated molecule (Δ^0) (Martin-Creuzburg and Von Elert 2004). Likewise, lanosterol, a $\Delta^{8(9),24(25)}$ sterol with additional C-4-dimethyl and C-14-methyl substituents, did not affect the growth of *D. galeata*. Growth was adversely affected by the Δ^4 sterol allocholesterol (cholest-4-en-3β-ol). These data also indicated the presence of an efficient C-24 dealkylating system in *D. galeata*.

Like daphnids, copepods lack the capacity for de novo synthesis of cholesterol, and thus require a dietary source of sterols for somatic growth and reproduction. Supplementation of a diatom diet (*Thalassiosira weissflogii*) with cholesterol has been shown to positively affect egg production rates of the marine copepods *Acartia hudsonica*, *Acartia tonsa*, and *Calanus finmarchicus* (Hassett 2004). In contrast, egg production of the copepod *Centropages hamatus* feeding on *T. weissflogii* was unaffected by cholesterol supplementation, which suggests species-specific differences in the sterol requirements or sterol storage capacities among copepod species (Hassett 2004). Interestingly, egg production rates of *A. hudsonica* were unaffected by sterol supplementation when *Thalassiosira rotula* was used as food. The sterol composition of the two *Thalassiosira* species (*T. weissflogii* and *T. rotula*)

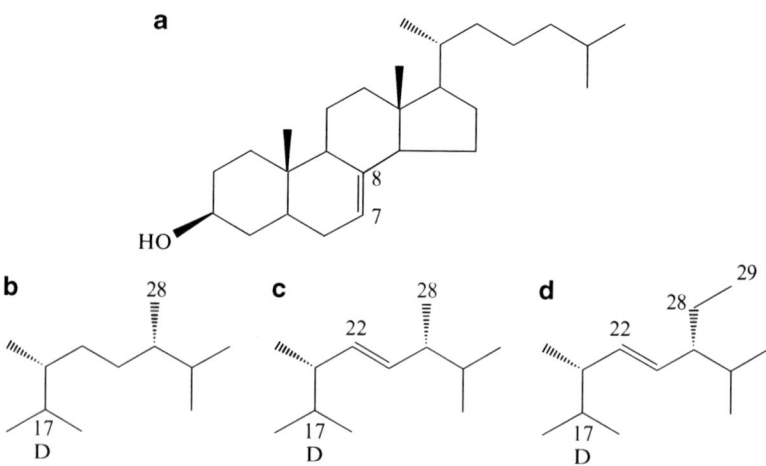

Fig. 3.4 Examples of structures of Δ^7-sterols: lathosterol (5α-cholest-7-en-3β-ol; **a**), fungisterol (5α-ergost-7-en-3β-ol; **b**), 5-dihydroergosterol ((22E)-5α-ergosta-7,22-dien-3β-ol; **c**), chondrillasterol ((24E)-5α-poriferasta-7,22-dien-3β-ol; **d**)

is very similar (Barrett et al. 1995), which makes it unlikely that the sterol composition accounted for the observed enhancement in copepod egg production. Instead, Hassett (2004) suggested that copepod egg production might have been constrained by a low concentration of dietary sterols, as the total sterol content varies considerably among and even within algal species.

Nevertheless, as described in various terrestrial insect species (Behmer and Nes 2003) and in the cladoceran *D. galeata* (Martin-Creuzburg and Von Elert 2004), structural features of dietary sterols may also affect the life history of copepods. Prahl et al. (1984) reported that in *Calanus helgolandicus* C_{28} and C_{29} sterols containing a Δ^5 or $\Delta^{5,7}$ double bond in the sterol nucleus are selectively removed from the diet (*Dunaniella primolecta*) relative to Δ^7 components, which were released unchanged as fecal lipids. The authors speculated that dietary Δ^7 sterols can be used as precursors of ecdysteroids and that the poor assimilation of these sterols provides a mechanism to avoid a haphazard production of molting hormones. Alternatively, Prahl et al. (1984) suggested that *Calanus* lacks the ability to convert Δ^7 sterols to cholesterol, and that the Δ^7 components are therefore poorly assimilated. In agreement, Klein Breteler et al. (1999) found that the ontogenetic development of the marine copepods *Temora longicornis* and *Pseudocalanus elongatus* from larvae to the adult stage was impaired by the unsuitable sterol composition of the green alga *Dunaliella* sp., which contained only traces of Δ^7 sterols and no Δ^5 or $\Delta^{5,7}$ components.

Like Δ^7 sterols, ring-saturated sterols were found to pass through the gut of copepods quantitatively, i.e., the amounts of ring-saturated sterols found in the diet and in the fecal pellets did not differ, indicating their nutritional inadequacy (Harvey et al. 1987, 1989). The ring-saturated dihydrocholesterol also did not improve the growth of *D. galeata*, when supplemented to a sterol-free diet (Martin-Creuzburg and Von Elert 2004).

Harvey et al. (1987, 1989) reported that 4-methyl and 4-desmethyl sterols (i.e., sterols with and without a methyl substituent at C-4) having a $\Delta^{8(14)}$ and $\Delta^{17(20)}$ unsaturation were readily, and, to similar degrees, removed from the diet by *Calanus helgolandicus* feeding on the dinoflagellate *Scrippsiella trochoidea*, which may indicate that a methyl group at C-4 does not prevent sterols from being assimilated and converted into cholesterol. In contrast, the presence or absence of alkyl substituents at the sterol side chain may interfere with copepod sterol nutrition. Giner et al. (2003) suggested that sterols containing an alkyl group at C-23 are refractory to the C-24-dealkylation process, with which dietary phytosterols are converted to cholesterol. Likewise, they suggested that sterols without the C-27-methyl group (27-norsterols) cannot be converted to cholesterol.

Data on sterol requirements of other zooplankton taxa, such as rotifers, are still scarce. The sterol content of the rotifer *Brachionus plicatilis* is positively correlated with the sterol content of its diet (Frolov et al. 1991). In a recent study, Boëchat and Adrian (2006) found that egg production, but not population growth, of the rotifer *Keratella quadrata* was correlated with the dietary sterol content. These findings may suggest that a dietary source of sterols is required for rotifer reproduction. To date, however, it is not known whether rotifers have a limited capacity, or even lack the ability, to synthesize sterols de novo.

3.5.3 Vertebrates

In aquatic vertebrates, sterol requirements are met by de novo synthesis and by dietary uptake in ratios depending on the amounts of sterols provided in the diet. Therefore, a dietary source of sterols is presumably not essential. In contrast, dietary phytosterols may act as endocrine and metabolic disrupters when present in high concentrations. The phytosterol sitosterol, for instance, has been shown to reduce sex steroid levels and the reproductive fitness of certain fish species in a concentration-dependent manner (e.g., MacLatchy and Van der Kraak 1995). To our knowledge, positive effects of dietary phytosterols on the performance of fish have as yet not been reported.

3.6 Ecological Implications: Competition and Succession in Zooplankton

From controlled experiments, it has been concluded that sterol limitation of somatic growth in *Daphnia magna* and *D. galeata* occurs at a sterol content of <5.4 µg/mg dietary carbon (Martin-Creuzburg et al. 2005a). Currently, the interpretation of these values in terms of the ecological relevance appears to be difficult due to the lack of data on the sterol content of eukaryotic phytoplankton. However, values as low as 5 µg/mg of carbon have been published for laboratory grown algae (Bec et al. 2006). In the field, the sterol content of algae might be even lower, as it varies significantly with environmental conditions (e.g., light and nutrient supply). It is worth emphasizing that the weight-specific sterol content of algae in general covers a wide range even within taxa (Tsitsa-Tzardis et al. 1993; Patterson et al. 1994). Thus, a limitation of *Daphnia* by a low availability of dietary sterols seems possible. Moreover, the size range of food particles ingested by the rather unselective filter feeder *Daphnia* includes bacteria, cyanobacteria, and ciliates, which all contain little or no sterols. These prey items may at least seasonally constitute a significant share of ingested carbon for *Daphnia* and thereby reduce the carbon-specific sterol content of the diet. Consequently, the dietary sterol content may strongly affect population dynamics of the herbivorous grazer *Daphnia* in nature.

In laboratory growth experiments with *D. magna* and *D. galeata*, Martin-Creuzburg et al. (2005a) showed that growth of offspring on a sterol-free diet was significantly affected by the sterol content of the mother's diet; i.e., somatic growth rates of offspring decreased with decreasing amounts of sterols in the maternal food. These findings strongly suggest that mothers allocate sterols, presumably cholesterol, to the eggs, and that this sterol content enables the offspring to buffer insufficient dietary sterol supply to some degree. The same study provides strong evidence that daphnids do not allocate a constant amount of sterol to an egg, but that less sterol is allocated under maternal sterol limitation. As a consequence, the offspring will become more sensitive to a nonsaturating sterol supply with increasing maternal sterol limitation. Thus, the negative effects of limiting dietary sterol content on population growth of *Daphnia* will increase with the persistence of a dietary sterol deficiency. Therefore, the detrimental effects of low-sterol supply on *Daphnia* in the F1 generation may be underestimated in laboratory growth assays if the maternal diet is rich in sterols. In copepods, it has been found that the dietary cholesterol content does not only affect egg production rates, but also egg viability (Crockett and Hassett 2005), which suggests that negative effects of a low sterol availability may become most pronounced at the population level.

The finding that daphnids depend on dietary sterols means that sterols are a nonsubstitutable biochemical resource. Assuming that sterols constitute a seasonally limiting resource for daphnids raises the question of the role of sterols in zooplankton competition and succession. Martin-Creuzburg et al. (2005a) looked for interspecific differences of sterol limitation and compared reaction norms of *D. magna* and *D. galeata* on a range of food with decreasing sterol content. When the individual dry mass over time was taken as a proxy for fitness (Lampert and Trubetskova 1996) both species clearly ran into increasing degrees of sterol limitation when the resource concentration (sterol) was reduced. However, the reaction norms of the two species did not cross, neither when the two species were grown on nonlimiting sterol levels (the green alga *Scenedesmus obliquus*) and in the absence of sterols (the cyanobacterium *Synechococcus elongatus*) (Fig. 3.5a) nor when growth on the cyanobacterium with and without supplementation of cholesterol was compared (Fig. 3.5b): The reaction norms obtained for *D. magna* were steeper than those for *D. galeata*; hence a lower incremental increase in dietary sterol concentration is required for

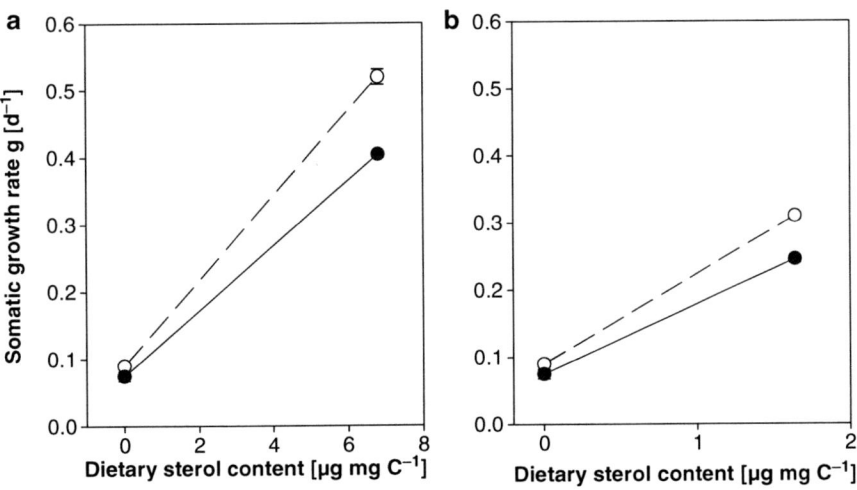

Fig. 3.5 Juvenile somatic growth rates of *Daphnia magna* and *D. galeata* fed with or without dietary sterols. Animals were grown either on (**a**) *Scenedesmus obliquus* or *Synechococcus elongatus* or (**b**) on *S. elongatus* with or without cholesterol supplementation. *Lines* represent reaction norms; *circles* are mean values ±SD. Data from Martin-Creuzburg et al. (2005a, b)

the same increase in growth in *D. magna*, which suggests a lower sterol requirement of the latter. Under all levels of sterol concentrations *D. magna* showed equal or higher juvenile growth rates than *D. galeata*, which corresponded to higher population growth rates (Martin-Creuzburg et al. 2005a) so that no effect of sterol limitation on competition could be demonstrated. Hence, as long as the extent of sterol limitation of herbivorous zooplankton in general and of different *Daphnia* species in particular in nature remains to be demonstrated, it cannot be estimated if the nonsubstitutable resource sterol affects zooplankton composition and seasonal succession.

3.7 Sterols and Carbon Transfer in Aquatic Food Webs

In aquatic food webs, the carbon transfer efficiency across the autotroph–herbivore interface is highly variable, which has far-reaching consequences for various ecosystem processes. One clear example is the low assimilation of cyanobacterial carbon by herbivorous zooplankton, which leads to a decoupling of primary and secondary production and therewith to the accumulation of cyanobacterial biomass (cyanobacterial blooms) especially in eutrophic systems. The food quality of cyanobacteria for herbivorous grazers such as *Daphnia* is determined by numerous factors (e.g., morphological and chemical properties, toxicity). However, the absence of sterols and essential fatty acids in cyanobacteria has been identified as a major food quality constraint for *Daphnia* (Müller-Navarra et al. 2000; Von Elert et al. 2003). In growth experiments with the sterol-free cyanobacterium *Synechococcus elongatus* and the sterol-containing green alga *Scenedesmus obliquus*, it was estimated that daphnids require 20–50% of the green alga in their diet to compensate for the low sterol content of a cyanobacterial food (Von Elert et al. 2003; Martin-Creuzburg et al. 2005a). Assuming that the eukaryotic phytosterols are fully available for the synthesis of cholesterol in *Daphnia*, this suggests that the food quality of natural seston is constrained by sterols if 50–80% of the biomass is prokaryotic, as is often the case with bloom-forming cyanobacteria (Oliver and Ganf 2000). It should be noted that this estimation is only valid if the phytosterol concentration of eukaryotic algae present in natural seston is similar to that of *S. obliquus*.

Like prokaryotes, ciliates presumably lack the ability to synthesize sterols de novo (see above). Recently, it was hypothesized that the comparatively poor food quality of ciliates for crustacean grazers is caused by the lack of sterols in ciliates (Martin-Creuzburg et al. 2006). This hypothesis is complicated by the finding that ciliates are able to incorporate dietary sterols into cell membranes and to metabolize them to various sterols (Conner et al. 1968; Harvey and McManus 1991; Harvey et al. 1997). In the absence of dietary sterols, ciliates produce the pentacyclic triterpenoid alcohol tetrahymanol and/or hopanoids, which are functionally equivalent to sterols as structural components of cell membranes (Conner et al. 1968; Harvey and McManus 1991; Martin-Creuzburg et al. 2005b, 2006). Hence, the occurrence of sterols in ciliates, and thereby the food quality of ciliates for higher trophic levels, might depend on the food source preyed upon by the ciliates; i.e., ciliates feeding on a sterol-free prokaryotic food are expected to be nutritionally inadequate for crustacean grazers due to the lack of sterols, whereas ciliates feeding on a sterol-containing eukaryotic food are expected to be upgraded in food quality due to the incorporation of dietary sterols. This was corroborated experimentally by using the ciliate *Colpidium campylum* (grown on bacteria) as a food for *D. magna* (Martin-Creuzburg et al. 2006). As expected, the food quality of the ciliate for *D. magna* was significantly improved by feeding the ciliate with sterol-supplemented albumin beads, which clearly indicates that the deficiency in sterols in the bacteria-fed ciliates has limited the somatic growth of the crustacean grazer. Ciliates are abundant protists in freshwater and marine ecosystems, and their significance in transferring mass and energy from the picoplankton via the microbial loop to higher trophic levels has often been recognized (Stoecker and Capuzzo 1990; Martin-Creuzburg et al. 2005b, and references therein). However, the efficiency of carbon transfer via the microbial loop might be constrained by a deficiency in sterols as long as ciliates are the predominant heterotrophic protozoa.

The carbon transfer efficiency in aquatic food webs might not only be restricted by the absence of dietary sterols, but also by the predominance of sterols which cannot be used as cholesterol precursors by the herbivorous zooplankton. For instance, the marine "red tide" dinoflagellate *Karenia brevis* is characterized by an unusual sterol profile ($\Delta^{8(14)}$-sterols methylated at C-4 or C-23, or demethylated at C-27), which is unlikely to satisfy the sterol nutritional requirements of marine invertebrates (Giner et al. 2003). The same authors proposed that the predominance of these unsuitable sterols in *Karenia brevis* may reduce predation losses, and thereby enhance the ability of the dinoflagellate to form massive blooms. Conceivably, such sterol-mediated metabolic constraints may temporarily also exist in freshwater habitats.

It should also be noted that the assimilation efficiencies of dietary phytosterols may differ significantly depending on structural features of phytosterols (see above) but also depending on food levels. Harvey et al. (1987) reported that the copepod *Calanus helgolandicus* has low assimilation efficiencies for sterols (compared with fatty acids) of the dinoflagellate *Scrippsiella trochoidea*, and that the assimilation efficiencies increase with decreasing food levels.

Detrimental effects of unsuitable dietary sterols on the performance of the consumer might not only be mediated by the inability of these sterols to serve as cholesterol precursors or by low assimilation efficiencies, but these sterols may also have adverse effects on other physiological processes. In grasshoppers, for example, the accumulation of unsuitable sterols in the body leads to high mortality rates, even if suitable sterols are present in ample concentrations (Behmer and Elias 1999). It has been suggested that the incorporation of unsuitable sterols (e.g., alkylated phytosterols) into phospholipid bilayers may result in leaky membranes because alkyl groups prevent phospholipids from packing tightly around the sterol molecules (Behmer et al. 1999b; Haines 2001). In addition, some sterols may inhibit or destroy enzymes involved in sterol dealkylation or may interfere with ecdysteroid synthesis (Giner et al. 2003). Hence, the accumulation of unsuitable dietary sterols may have severe effects on the performance of herbivores, an aspect that should be considered when analyzing carbon transfer efficiencies in aquatic food webs. Another intriguing area that has not been considered in aquatic systems is the topic of food selection. In grasshoppers, it has been demonstrated that dietary phytosterols can affect the feeding duration on a particular diet, and that unsuitable dietary phytosterols can mediate a learned aversion to the food (Champagne and Bernays 1991; Behmer et al. 1999a).

3.8 Sterol Mediated Trophic Upgrading

In the field, a sterol-depleted food source might be biochemically "upgraded" by heterotrophic protists because of their ability to produce sterols from these substrates (and other essential nutrients, see Chap. 2) thereby making them available for assimilation by higher trophic levels. For instance, Klein Breteler et al. (1999) suggested that the poor food quality of the green alga *Dunaliella* sp. for marine copepods is due to a sterol deficiency in the green alga, which contained only small amounts of Δ^7-sterols that are unsuitable to support the ontogenetic development of copepods. The authors demonstrated that the Chlorophycean food was biochemically upgraded by the heterotrophic dinoflagellate *Oxyrrhis marina* to high-quality copepod food and attributed this trophic upgrading by an intermediary protozoan to the production of Δ^5 sterols in the dinoflagellate. Likewise, sterol-mediated trophic upgrading has been documented in freshwater organisms (Bec et al. 2006). In a simplified food chain consisting of the sterol-free picocyanobacterium *Synechococcus* spp., the heterotrophic nanoflagellate *Paraphysomonas* sp., and the crustacean grazer *D. magna*, the poor food quality of the picocyanobacterial carbon was improved by the addition of sterols (and fatty acids; see Chap. 2) produced by the intermediary flagellate (Bec et al. 2006). Thus, the production of sterols by intermediary protozoans might improve carbon transfer efficiency via the microbial loop from nutritionally inadequate primary producers to metazoan grazers.

3.9 Sterols from Aquatic Sources and Their Potential Role in Human Nutrition

Humans are capable of de novo synthesis of cholesterol, which is negatively regulated by dietary cholesterol. Elevated serum cholesterol levels have been shown to increase the risk of coronary heart disease in humans, which is a leading cause of mortality in many western countries. Thus, the main target in the prevention of coronary heart disease is lowering serum cholesterol levels, either by dietary restrictions or food additives, which limit cholesterol intake (Piironen et al. 2000; Moreau et al. 2002). Phytosterols and phytostanols (and their fatty acid esters) have been shown to reduce plasma cholesterol levels by interfering with transport-mediated processes of cholesterol uptake; i.e., in presence of dietary phytosterols intestinal cholesterol absorption is reduced and more cholesterol is excreted in the feces (Trautwein et al. 2003;

Normén et al. 2004). With the growing interest in functional food, phytosterols are increasingly used as food additives, e.g., in margarine-like spreads (Moreau et al. 2002). The exploitation of sterols from aquatic microorganisms for biotechnological or nutritional purposes is largely in its infancy (Volkman 2003). However, the large variety of sterols in microalgae and the ability to mass culture these organisms may offer new biotechnological applications, e.g., for the production of steroids, nutraceuticals or food additives for human health benefits (Volkman 2003).

3.10 Perspectives

The sterol composition of numerous phytoplankton species has been investigated, mostly to assess the use of sterols as biomarkers for certain phytoplankton taxa. However, due to the large variety of naturally occurring phytosterols in eukaryotic algae, the use of sterols as biomarkers appears difficult. The sterol profile of eukaryotic algae is often dominated by $\Delta 5$ sterols alkylated at C-24 with no methyl groups at C-4 (4-desmethyl-sterols); the distributions range from the predominance of a single sterol to a mixture of ten or more sterols (Volkman 2003). Diatoms, for instance, show a great variety of sterol compositions and, although 24-methylcholesta-5,22E-dien-3β-ol (diatomsterol), and 24-methylcholesta-5,24(28)-dien-3β-ol are widely distributed, no sterol appears to be either unique or representative (Volkman et al. 1998). The sterol composition of dinoflagellates is often, but not always, dominated by 4α-methyl sterols, such as dinosterol (4α,23,24-trimethyl-5α-cholest-22E-en-3β-ol), and by sterols with a fully saturated ring system (stanols). Green algae from the Chlorophyta are often characterized by a high content of $\Delta 7$ sterols, which are not common in other eukaryotic algae (Volkman 2003). Although some general patterns have been identified, our knowledge of the sterol and composition of common phytoplankton taxa is far from being complete, in particular in freshwater habitats. Another topic that deserves further investigation is the use of sterols as trophic biomarkers in aquatic food webs. Suitable sterols, however, might be difficult to identify, as dietary sterols are often considerably metabolized at each trophic level (e.g., to cholesterol in the crustacean zooplankton).

The sterol profile of eukaryotic algae varies considerably among strains, but also within a single strain depending on the growth conditions (Ballantine et al. 1979; Wright et al. 1980). Klein Breteler et al. (2005) reported that the lipid composition (PUFA and sterols) of the diatom *Thalassiosira weissflogii* was strongly affected by nitrogen and phosphorus limitation, which, in turn, explained the observed differences in copepod growth. This example illustrates that algal growth conditions (nutrient availability, light intensity, temperature, etc.) are crucial for the interpretation of food quality experiments, and it suggests that food quality constraints caused by a stoichiometric or biochemical imbalance might be difficult to separate.

Dietary sterol requirements may vary throughout the life cycle, e.g., the impact of sterol limitation might be most severe on juvenile crustaceans, as they have higher specific growth rates than older animals, and therefore might require larger amounts of sterols to: (a) build-up and maintain cell membranes and (b) to synthesize ecdysteroids required for molting. With age, sterol requirements may gradually be reduced, as the animals increase their relative investment into reproduction (Lynch et al. 1986). Most crustaceans, however, continue to grow throughout their lifetime and hence a requirement for dietary sterols persists. It is also likely that sterols are allocated into the eggs to provide the developing embryo with sufficient amounts of sterols, which suggests that sterols are also needed for egg production (Wacker and Martin-Creuzburg 2007). Thus, sterol availability in the diet, inherent sterol biosynthetic capacity, and sterol requirements of the various zooplankton life stages have all to be determined to more fully assess the ecological role of sterols in structuring zooplankton communities.

Excess free sterols are harmful for the proper function of membranes (Leppimäki et al. 2000), therefore the dietary intake of sterols must be tightly regulated to prevent toxic effects due to over-accumulation and abnormal deposition within the body. In copepods, the cholesterol content of membranes, which contain the majority of total body cholesterol, is constant despite varying levels of dietary sterols (Crockett and Hassett 2005), suggesting sterol homeostasis. In contrast, the cholesterol content of whole-body extracts of *D. galeata* was found to decrease rapidly when they were fed a sterol-deficient diet and to increase with sterol supplementation (Martin-Creuzburg and Von Elert 2004). Sterols can be, to some extent, stored by esterifying them to long chain fatty acids (sterylesters). Therefore, this decrease in whole-body cholesterol might be attributed to the exhaustion of stored cholesterol and not to a decrease in membrane cholesterol levels. Alternatively, the observed decrease in body cholesterol levels under poor food conditions might be due to the catabolism of membranes, and it remains to be tested if low food quality or low food quantity (energy limitation) have similar effects on body cholesterol levels. In addition, more studies are needed to elucidate the ability of zooplankton to cope with varying levels of dietary sterols, i.e., to identify incipient limiting sterol levels at which somatic growth is limited by the low availability of sterols to better predict the fate of natural zooplankton populations.

The need to maintain a more or less constant level of sterols in the body implies that the absorption of sterols in the gut is strictly regulated. In vertebrates, sterol absorption is presumably a protein-mediated process, i.e. specific receptors and transporters are involved (Trautwein et al. 2003). In general, the absorption of phytosterols in the gut of vertebrates is much lower than that of cholesterol, which suggests discrimination between cholesterol and phytosterols by highly specific mechanisms (Trautwein et al. 2003). In invertebrates, sterol absorption has not been investigated in such detail. However, a selective absorption or excretion of certain sterols has also been reported (Prahl et al. 1984; Harvey et al. 1987). Further studies will have to reveal assimilation efficiencies of dietary sterols to understand sterol mediated food quality constraints on zooplankton performance.

It has already been demonstrated that structural features of dietary sterols can have pronounced effects on somatic growth and development of herbivorous grazers, in particular with regard to their suitability to serve as cholesterol precursors. However, such sterol-mediated metabolic constraints might be highly specific (as in insects, Behmer and Nes 2003) and should not be representative of the situation in invertebrate grazers in general. In addition to "free" sterols, significant amounts of sterol conjugates (steryl esters, steryl glycosides) might be present in algal food sources (Véron et al. 1998). Whether these compounds are suitable to satisfy the sterol requirements of the herbivorous grazers has not yet been investigated. Zooplankton is likely to ingest both suitable and unsuitable sterols when feeding on a mixture of eukaryotic algae. This implies that the relative proportion of suitable to unsuitable sterols may constrain proper growth and survival of the herbivorous zooplankton, as has been shown for a terrestrial insect (Behmer and Elias 2000). Thus, further studies are needed to improve our knowledge of possible metabolic constraints mediated by the multitude of free and conjugated sterols that occur in aquatic food webs.

The dietary sterol content itself or the predominance of unsuitable dietary sterols might also affect the synthesis of ecdysteroids in arthropods. In general, the prohormone ecdysone is synthesized from dietary sterols (i.e., specifically from cholesterol) and subsequently hydroxylated by target tissues to 20-hydroxyecdysone, the most active ecdysteroid in arthropods (Grieneisen 1994; Gilbert et al. 2002; Martin-Creuzburg et al. 2007; Fig. 3.3). In crustaceans, ecdysteroids are produced mainly in steroidogenic glands, the so-called Y-organs. Our knowledge of ecdysteroid biosynthesis in crustaceans others than decapods is scarce and needs to be extended to the major zooplankton taxa to assess possible effects of dietary sterols on developmental processes.

Sterols are often considered as important structural components of eukaryotic cell membranes. In ciliates, however, the triterpenoid alcohol tetrahymanol and the related hopanoids have been shown to serve as sterol surrogates, as long as a dietary source of sterols is not available; i.e. when ciliates prey on prokaryotic food sources (bacteria, picocyanobacteria; cp. Martin-Creuzburg et al. 2005b, 2006). The finding that ciliates, as intermediary grazers, improve the poor food quality of a picocyanobacterial (sterol-free) diet for subsequent use by *D. magna* implies that tetrahymanol and related compounds can be used as sterol surrogates also in crustacean tissues (Martin-Creuzburg et al. 2005b). Ederington et al. (1995) reported the assimilation of tetrahymanol in copepod tissues (mainly eggs) when ciliates were offered as food, and suggested that tetrahymanol is functionally equivalent to cholesterol in the crustacean, thereby maintaining minimal egg production. In addition to their role as structural components of cell membranes, sterols serve as precursors for many bioactive molecules. Thus, it remains to be tested whether tetrahymanol and related compounds actually improve the performance of crustaceans, possibly by supplementation of these compounds to a sterol-free diet.

3.11 Conclusions

Aquatic herbivores face a complex pattern of nutritional challenges in natural environments. Beside sterols, biochemical food quality depends on a multitude of other factors such as long chain polyunsaturated fatty acids (e.g., Von Elert 2002), amino acids (Guisande et al. 2000), and possibly vitamins (Conklin and Provasoli 1977). However, sterol limitation might be one important factor with the potential to affect the structure of aquatic food webs, at least seasonally and in certain habitats. In general, several "compensatory" mechanisms within the prey community and within the herbivorous zooplankton may lead to a less pronounced sterol limitation than would be predicted from determination of the sterol content of edible phytoplankton alone. Within the prey community, trophic upgrading within the microbial loop and synthesis of functionally equivalent compounds like tetrahymanol may reduce the impact of sterol-deficient photoautotrophs on metazoan grazers. Within the metazoan grazer community, the capability to store excess sterols and to change food selectivity may provide a means to temporarily buffer low sterol provision. However, herbivorous zooplankton differs with regard to prey size and selectivity and, though coexisting, may be limited by different resources. The relevance of the different "compensatory" mechanisms may differ substantially for different taxa, e.g., a change in food selectivity may protect calanoid copepods against sterol limitation while nonselective grazers become sterol limited. This suggested example is meant to illustrate that, counterintuitive, compensatory mechanisms may lead to even more pronounced differences of low sterol provision on different zooplankton taxa, since a particular "compensatory" mechanism may not be available to all taxa. As a result, though sterols are essential to all crustacean taxa, a low sterol provision may greatly differ in its effects on coexisting zooplankton taxa and thus affect competition among herbivorous zooplankton. Predatory zooplankton will be protected from sterol limitation due to the presence of cholesterol in its prey and will only indirectly be affected by the sterol concentration in natural seston when it runs into quantitative food limitation if the abundance of herbivorous zooplankton declines as a consequence of sterol limitation. In conclusion, results from controlled laboratory experiments strongly suggest that sterol limitation is a likely scenario for freshwater communities; one that has probably led to a variety of compensatory responses within communities and thereby affected the structure of aquatic food webs.

References

Ballantine, J.A., Lavis, A., and Morris, R.J. 1979. Sterols of the phytoplankton – effects of illumination and growth stage. Phytochemistry 18:1459–1466.
Barrett, S.M., Volkman, J.K., Dunstan, G.A., and Leroi, J.M. 1995. Sterols of 14 species of marine diatoms (bacillariophyta). J. Phycol. 31:360–369.
Bec, A., Martin-Creuzburg, D., and Von Elert, E. 2006. Trophic upgrading of autotrophic picoplankton by the heterotrophic nanoflagellate *Paraphysomonas* sp. Limnol. Oceanogr. 51:1699–1707.
Behmer, S.T., and Elias, D.O. 1999. The nutritional significance of sterol metabolic constraints in the generalist grasshopper *Schistocerca americana*. J. Insect Physiol. 45:339–348.
Behmer, S.T., and Elias, D.O. 2000. Sterol metabolic constraints as a factor contributing to the maintenance of diet mixing in grasshoppers (Orthoptera:Acrididae). Physiol. Biochem. Zool. 73:219–230.
Behmer, S.T., and Nes, W.D. 2003. Insect sterol nutrition and physiology: a global overview. Adv. Insect Physiol. 31:1–72.

Behmer, S.T., Elias, D.O., and Bernays, E.A. 1999a. Post-ingestive feedbacks and associative learning regulate the intake of unsuitable sterols in a generalist grasshopper. J. Exp. Biol. 202:739–748.
Behmer, S.T., Elias, D.O., and Grebenok, R.J. 1999b. Phytosterol metabolism and absorption in the generalist grasshopper, *Schistocerca americana* (Orthoptera: Acrididae). Arch. Insect Biochem. Physiol. 42:13–25.
Boëchat, I.G., and Adrian, R. 2006. Evidence for biochemical limitation of population growth and reproduction of the rotifer *Keratella quadrata* fed with freshwater protists. J. Plankton Res. 28:1027–1038.
Boëchat, I.G., Krüger, A., and Adrian, R. 2007. Sterol composition of freshwater algivorous ciliates does not resemble dietary composition. Microb. Ecol. 53:74–81.
Champagne, D.E., and Bernays, E.A. 1991. Phytosterol unsuitability as a factor mediating food aversion learning in the grasshopper *Schistocerca americana*. Physiol. Entomol. 16:391–400.
Conklin, D.E., and Provasoli, L. 1977. Nutritional requirements of the water flea *Moina macrocopa*. Biol. Bull. 152:337–350.
Conner, R.L., Landrey, J.R., Burns, C.H., and Mallory, F.B. 1968. Cholesterol inhibition of pentacyclic triterpenoid biosynthesis in *Tetrahymena pyriformis*. J. Protozool. 15:600–605.
Crockett, E.L., and Hassett, R.P. 2005. A cholesterol-enriched diet enhances egg production and egg viability without altering cholesterol content of biological membranes in the copepod *Acartia hudsonica*. Physiol. Biochem. Zool. 78:424–433.
Dahl, C.E., Dahl, J.S., and Bloch, K. 1980. Effect of alkyl-substituted precursors of cholesterol on artificial and natural membranes and on the viability of *Mycoplasma capricolum*. Biochemistry 19:1462–1467.
Dembitsky, V.M., Rezanka, T., and Srebnik, M. 2003. Lipid compounds of freshwater sponges: family *Spongillidae*, class *Demospongiae*. Chem. Phys. Lipids 123:117–155.
Ederington, M., McManus, G.B., and Harvey, H.R. 1995. Trophic transfer of fatty acids, sterols, and a triterpenoid alcohol between bacteria, a ciliate, and the copepod *Acartia tonsa*. Limnol. Oceanogr. 40:860–867.
Frolov, A.V., Pankov, S.L., Geradz, K.N., Pankova, S.A., and Spektrova, L.V. 1991. Influence of the biochemical composition of food on the biochemical composition of the rotifer *Brachionus plicatilis*. Aquaculture 97:181–202.
Gessner, M.O., and Chauvet, E. 1993. Ergosterol-to-biomass conversion factors for aquatic hyphomycetes. Appl. Environ. Microb. 59:502–507.
Gilbert, L.I., Rybczynski, R., and Warren, J.T. 2002. Control and biochemical nature of the ecdysteroidogenic pathway. Annu. Rev. Entomol. 47:883–916.
Giner, J.-L., Faraldos, J.A., and Boyer, G.L. 2003. Novel sterols of the toxic dinoflagellate *Karenia brevis* (Dinophyceae): a defensive function for unusual marine sterols? J. Phycol. 39:315–319.
Goad, L.J. 1981. Sterol biosynthesis and metabolism in marine invertebrates. Pure Appl. Chem. 51:837–852.
Grieneisen, M.L. 1994. Recent advances in our knowledge of ecdysteroid biosynthesis in insects and crustaceans. Insect Biochem. Mol. Biol. 24:115–132.
Guisande, C., Riverio, I., and Maneiro, I. 2000. Comparisons among the amino acid composition of females, eggs and food to determine the relative importance of the food quantity and food quality to copepod reproduction. Mar. Ecol. Prog. Ser. 202:135–142.
Haines, T.H. 2001. Do sterols reduce proton and sodium leaks through lipid bilayers? Prog. Lipid Res. 40:299–324.
Harvey, H.R., and McManus, G.B. 1991. Marine ciliates as a widespread source of tetrahymanol and hopan-3β-ol in sediments. Geochim. Cosmochim. Acta 55:3387–3390.
Harvey, H.R., Eglinton, G., O'Hara, S.C.M., and Corner, E.D.S. 1987. Biotransformation and assimilation of dietary lipids by *Calanus* feeding on a dinoflagellate. Geochim. Cosmochim. Acta 51:3031–3040.
Harvey, H.R., O'Hara, S.C.M., Eglinton, G., and Corner, E.D.S. 1989. The comparative fate of dinosterol and cholesterol in copepod feeding: implications for a conservative molecular biomarker in the marine water column. Org. Geochem. 14:635–641.

Harvey, H.R., Ederington, M.C., and McManus, G.B. 1997. Lipid composition of the marine ciliates *Pleuronema* sp. and *Fabrea salina*: shifts in response to changes in diets. J. Eukaryot. Microbiol. 44:189–193.
Hassett, R.P. 2004. Supplementation of a diatom diet with cholesterol can enhance copepod egg-production rates. Limnol. Oceanogr. 49:488–494.
Kanazawa, A. 2001. Sterols in marine invertebrates. Fish. Sci. 67:997–1007.
Klein Breteler, W.C.M., Schogt, N., Baas, M., Schouten, S., and Kraay, G.W. 1999. Trophic upgrading of food quality by protozoans enhancing copepod growth: the role of essential lipids. Mar. Biol. 135:191–198.
Klein Breteler, W.C.M., Koski, M., and Rampen, S. 2004. Role of essential lipids in copepod nutrition: no evidence for trophic upgrading of food quality by a marine ciliate. Mar. Ecol. Prog. Ser. 274:199–208.
Klein Breteler, W.C.M., Schogt, N., and Rampen, S. 2005. Effect of diatom nutrient limitation on copepod development: role of essential lipids. Mar. Ecol. Prog. Ser. 291:125–133.
Lampert, W., and Trubetskova, I. 1996. Juvenile growth rate as a measure of fitness in *Daphnia*. Funct. Ecol. 10:631–635.
Leblond, J.D., and Chapman, P.J. 2002. A survey of the sterol composition of the marine dinoflagellates *Karenia brevis*, *Karenia mikimotoi*, and *Karlodinium micrum*: distribution of sterols within other members of the class dinophyceae. J. Phycol. 38:670–682.
Leppimäki, P, Mattinen, J., and Slotte, P. 2000. Sterol-induced upregulation of phosphatidylcholine synthesis in cultured fibroblasts is affected by the double-bond position in the sterol tetracyclic ring structure. Eur. J. Biochem. 267:6385–6394.
Lozano, R., Salt, T.A., Chitwood, D.J., Lusby, W.R., and Thompson, M.J. 1987. Metabolism of sterols of varying ring unsaturation and methylation by *Caenorhabditis elegans*. Lipids 22:84–87.
Lynch, M., Weider, L.J., and Lampert, W. 1986. Measurement of the carbon balance in *Daphnia*. Limnol. Oceanogr. 31:17–33.
MacLatchy, D.L., and Van der Kraak, G. 1995. The phytoestrogen β-sitosterol alters the reproductive endocrine status of goldfish. Toxicol. Appl. Pharmacol. 134:305–312.
Martin-Creuzburg, D., and Von Elert, E. 2004. Impact of 10 dietary sterols on growth and reproduction of *Daphnia galeata*. J. Chem. Ecol. 30:483–500.
Martin-Creuzburg, D., Wacker, A., and Von Elert, E. 2005a. Life history consequences of sterol availability in the aquatic keystone species *Daphnia*. Oecologia 144:362–372.
Martin-Creuzburg, D., Bec, A., and Von Elert, E. 2005b. Trophic upgrading of picocyanobacterial carbon by ciliates for nutrition of *Daphnia magna*. Aquat. Microb. Ecol. 41:271–280.
Martin-Creuzburg, D., Bec, A., and Von Elert, E. 2006. Supplementation with sterols improves food quality of a ciliate for *Daphnia magna*. Protist 157:477–486.
Martin-Creuzburg, D., Westerlund, S.A., and Hoffmann, K.H. 2007. Ecdysteroid levels in *Daphnia magna* during a molt cycle: determination by radioimmunoassay (RIA) and liquid chromatography-mass spectrometry (LC-MS). Gen. Comp. Endocrinol. 151:66–71.
Merris, M., Kraeft, J., Tint, G.S., and Lenard, J. 2004. Long-term effects of sterol depletion in *C. elegans*: sterol content of synchronized wild-type and mutant populations. J. Lipid Res. 45:2044–2051.
Moreau, R.A., Whitaker, B.D., and Hicks, K.B. 2002. Phytosterols, phytostanols, and their conjugates in foods: structural diversity, quantitative analysis, and health-promoting uses. Prog. Lipid Res. 41:457–500.
Müller-Navarra, D.C., Brett, M., Liston, A.M., and Goldman, C.R. 2000. A highly unsaturated fatty acid predicts carbon transfer between primary producers and consumers. Nature 403:74–77.
Nes, W.R., and McKean, M.L. 1977. Biochemistry of steroids and other isopentenoids. University Park Press, Baltimore, MD.
Normén, L., Shaw, C.A., Fink, C.S., and Awad, A.B. 2004. Combination of phytosterols and omega-3 fatty acids: A potential strategy to promote cardiovascular health. Curr. Med. Chem. Cardiovasc. Hematol. Agents 2:1–12.
Ohvo-Rekilä, H., Ramstedt, B., Leppimäki, P., and Slotte, P. 2002. Cholesterol interactions with phospholipids in membranes. Prog. Lipid Res. 41:66–97.

Oliver, R.L., and Ganf, G.G. 2000. Freshwater blooms, pp. 149–194. *In* B.A. Whitton (ed.), The ecology of cyanobacteria: their diversity in time and space. Kluwer, Dordrecht.

Ourisson, G., Rohmer, M., and Poralla, K. 1987. Prokaryotic hopanoids and other polyterpenoid sterol surrogates. Ann. Rev. Microbiol. 41:301–333.

Patterson, G.W. 1991. Sterols of algae, pp. 118–157. *In* G.W. Patterson, and W.D. Nes (eds.), Physiology and biochemistry of sterols. American Oil Chemists' Society, Champaign, IL.

Patterson, G.W., Tsitsa-Tzardis, E., Wikfors, G.H., Ghosh, P., Smith, B. C., and Gladu, P.K. 1994. Sterols of eustigmatophytes. Lipids 29:661–664.

Piironen, V., Lindsay, D., Miettinen, T., Toivo, J., and Lampi, A.M. 2000. Plant sterols: biosynthesis, biological function and their importance to human nutrition. J. Sci. Food Agric. 80:939–966.

Popov, S., Stoilov, I., Marekov, N., Kovachev, G., and Andreev, S. 1981. Sterols and their biosynthesis in some freshwater bivalves. Lipids 16:663–669.

Porter, J.A., Young, K.E., and Beachy, P.A. 1996. Cholesterol modification of hedgehog signaling proteins in animal development. Science 274:255–259.

Prahl, F.G., Eglinton, G., Corner, E.D.S., O'Hara, S.C.M., and Forsberg, T.E.V. 1984. Changes in plant lipids during passage through the gut of *Calanus*. J. Mar. Biol. Ass. U. K. 64:317–334.

Raederstorff, D., and Rohmer, M. 1987. Sterol biosynthesis via cycloartenol and other biochemical features related to photosynthetic phyla in the amoebae *Naegleria lovaniensis* and *Naegleria gruberi*. Eur. J. Biochem. 164:427–434.

Raederstorff, D., and Rohmer, M. 1988. Polyterpenoids as cholesterol and tetrahymanol surrogates in the ciliate *Tetrahymena pyriformis*. Biochim. Biophys. Acta 960:190–199.

Rohmer, M., Knani, M., Simonin, P., Sutter, B., and Sahm, H. 1993. Isoprenoid biosynthesis in bacteria: a novel pathway for the early steps leading to isopentenyl diphosphate. Biochem. J. 295:517–524.

Schaller, H. 2003. The role of sterols in plant growth and development. Prog. Lipid Res. 42:163–175.

Shim, Y.-H., Chun, J.H., Lee, E.-Y., and Paik, Y.-K. 2002. Role of cholesterol in germ-line development of *Caenorhabditis elegans*. Mol. Reprod. Dev. 61:358–366.

Silva, C.J., Wünsche, L., and Djierassi, C. 1991. Biosynthetic studies of marine lipids 35. The demonstration of de novo sterol biosynthesis in sponges using radiolabeled isoprenoid precursors. Comp. Biochem. Physiol. 99B:763–773.

Soudant, P., Le Coz, J.-R., Marty, Y., Moal, J., Robert, R., and Samain, J.-F. 1998. Incorporation of microalgae sterols by scallop *Pecten maximus* (L.) larvae. Comp. Biochem. Physiol. A. 119:451–457.

Stoecker, D.K., and Capuzzo, J.M. 1990. Predation on Protozoa: its importance to zooplankton. J. Plankton Res. 12:891–908.

Summons, R.E., Bradley, A.S., Jahnke, L.L., and Waldbauer, J.R. 2006. Steroids, triterpenoids and molecular oxygen. Phil. Trans. R. Soc. B 361:951–968.

Svoboda, J.A., and Thompson, M.J. 1985. Steroids, pp. 137–175. *In* G.A. Kerkut, and L.I. Gilbert (eds.). Comprehensive insect physiology, biochemistry and pharmacology. Pergamon, New York.

Teshima, S.-I. 1971. Bioconversion of β-sitosterol and 24-methylcholesterol to cholesterol in marine crustacea. Comp. Biochem. Physiol. 39B:815–822.

Teshima, S.-I. 1991. Sterols of crustaceans, molluscs and fish, pp. 229–256. *In* G.W. Patterson, and W.D. Nes (eds.), Physiology and biochemistry of sterols. American Oil Chemists' Society, Champaign, IL.

Thompson Jr., G.A. 1996. Lipids and membrane function in green algae. Biochim. Biophys. Acta 1302:17–45.

Trautwein, E.A., Duchateau, G.S.M.J.E., Lin, Y., Mel'nikov, S., Molhuizen, H.O.F., and Ntanios, F.Y. 2003. Proposed mechanisms of cholesterol-lowering action of plant sterols. Eur. J. Lipid Sci. Technol. 105:171–185.

Trider, D.J., and Castell, J.D. 1980. Effect of dietary lipids on growth tissue composition and metabolism of the oyster (*Crassostrea virginica*). J. Nutr. 110:1303–1309.

Tsitsa-Tzardis, S.E., Patterson, G.W., Wikfors, G.H., Gladu, P.K., and Harrison, D. 1993. Sterols of *Chaetoceros* and *Skeletonema*. Lipids 28:465–467.

VanWagtendonk, W.J. 1974. Nutrition of *Paramecium*, pp. 339–376. *In* W.J. VanWagtendonk (ed.), *Paramecium*, a current survey. Elsevier, Amsterdam.

Véron, B., Dauguet, J.-C., and Billard, C. 1998. Sterolic biomarkers in marine phytoplankton. II. Free and conjugated sterols of seven species used in mariculture. J. Phycol. 34:273–279.

Volkman, J.K. 2003. Sterols in microorganisms. Appl. Microbiol. Biotech. 60:495–506.

Volkman, J.K. 2005. Sterols and other triterpenoids: source specificity and evolution of biosynthetic pathways. Org. Geochem. 36:139–159.

Volkman, J.K., Barrett, S.M., Blackburn, S.I., Mansour, M.P., Sikes, E.L., and Gelin, F. 1998. Microalgal biomarkers: a review of recent research developments. Org. Geochem. 29:1163–1179.

Von Elert, E. 2002. Determination of limiting polyunsaturated fatty acids in *Daphnia galeata* using a new method to enrich food algae with single fatty acids. Limnol. Oceanogr. 47:1764–1773.

Von Elert, E., Martin-Creuzburg, D., and Le Coz, J.R. 2003. Absence of sterols constrains carbon transfer between cyanobacteria and a freshwater herbivore (*Daphnia galeata*). Proc. R. Soc. Lond. B Bio. 270:1209–1214.

Voogt, P.A. 1975. Investigations of the capacity of synthesizing 3β-sterols in Mollusca-XIII. Biosynthesis and composition of sterols in some bivalves (*Anisomyaria*). Comp. Biochem. Physiol. 50B:499–504.

Wacker, A., and Martin-Creuzburg, D. 2007. Allocation of essential lipids in *Daphnia magna* during exposure to poor food quality. Funct. Ecol. 21:738–747.

Williams, B.L., Goodwin, T.W., and Ryley, J.F. 1966. The sterols of some protozoa. J. Protozool. 13:227–230.

Wright, D.C., Berg, L.R., and Patterson, G.W. 1980. Effect of culture conditions on the sterols and fatty acids of green algae. Phytochemistry 19:783–785.

Chapter 4
Fatty Acids and Oxylipins as Semiochemicals

Susan B. Watson, Gary Caldwell, and Georg Pohnert

4.1 Introduction

It is well established that algae and cyanobacteria are a rich source of saturated, mono and polyunsaturated fatty acids (SAFA, MUFA, PUFA), which are essential to the integrity and productivity of cells and food webs (e.g., Kainz et al. 2004; see also Chaps. 6 and 8). However, for the past 20 years, mounting evidence, from two contrasting lines of marine and freshwater studies, have shown that fatty acids (FA), FA derivatives (FADV), and FA-derived metabolites such as oxylipins may also act as semiochemicals[1] in aquatic systems. This has led to a somewhat dichotomous view of the roles of algal FA, particularly PUFA. Specific classes of PUFA are regarded as essential nutritional resources for secondary producers, particularly n-3 and n-6 PUFA, since the capability for FA desaturation at the n-3 and n-6 position is restricted to algae and plants (Cook 1996). Yet some of these same FA also have the potential to act as toxins, or to serve as precursors of toxins, repellents, attractants and possibly pheromones (Wendel and Jüttner 1996; Miralto et al. 1999; Jüttner 2005; Fink et al. 2006).

---

[1]The field of chemical ecology is developing rapidly with inevitable debate around definitions of semiochemicals. Some authors distinguish infochemicals from toxins (e.g. Dicke and Sabelis 1988), but here we adhere to the original inclusive classification by Wood (1983): "A compound that acts as a chemical messenger in intra- or interspecific communication (includes pheromones, autotoxins, allomones, kairomones, repellents and toxins" because of the multifunctional activity of many of these chemicals; which thus can no longer be partitioned under narrowly defined roles. Some metabolites have very specific activity (e.g. pheromones), while others may act along a continuum of roles or in blends (e.g. as toxins, deterrents or kairomones), depending, for example, on the organisms involved.

S.B. Watson (✉), G. Caldwell, and G. Pohnert
Aquatic Ecosystems Management Research Division, National Water Research Institute – Environment Canada, P.O. Box 5050, 878 Lakeshore Road, Burlington, ON, Canada, L7R 4A6
e-mail: sue.watson@ec.gc.ca

On the one hand, in marine systems, a long history of research established the roles of PUFA derivatives as pheromones and in grazer defense (Sects. 4.3 and 4.4). For a long time, this work was directed largely toward littoral interactions involving seaweeds (notably phaeophytes and rhodophytes). On the other hand, in freshwater systems, the nutritive value of algal PUFA in food webs has been recognized for some time, but their potential role(s) in chemical defense has been only acknowledged recently; and was similarly first convincingly elucidated in littoral communities (Jüttner 2001). Planktonic and littoral algal PUFA derivatives have also long been notorious as potent taste-odor agents in freshwater supplies, and even more extensively characterized in the food and flavor/perfume industries as sources of rancid, oily, fishy, cucumber, fruity, and floral flavors in both essential oils and spoiled lipid-rich food products (Watson et al. 2000, 2001; Watson 2003).

In this chapter, we review some of the major FA and FA-related compounds that have captured scientific interest as semiochemicals. These compounds function in a spectrum of roles that modify intra and extracellular aquatic processes. Research is identifying an increasing number of relevant structures (Chap. 13); hence our synopsis represents only a small fraction of the multidimensional processes that involve FA and FA-related chemical mediation. To date the majority of studies have focused on chemical interactions, which have important socioeconomic impacts, aquaculture applications (seaweed pheromones and grazer defense, shellfish poisoning, and tainting), harmful algal blooms (HAB) and their toxic effects on humans or commercially important species (e.g., fish and livestock) and environmental impacts (e.g., atmospheric marine sulfide production). Thus, it is likely that many other interactions and species involving these chemicals have been, and continue to be, overlooked.

To evaluate FA and oxylipins as semiochemicals, it is important to consider that their functional roles depend on their chemistry, timing, and persistence within the environment, mode(s) of delivery and the identity, proximity/susceptibility of target organism(s) (Dicke and Sabelis 1988). Energetically speaking, the best strategy is to use minimal resources to produce highly potent signals under conditions that maximize successful transfer, and where possible, to assign several roles to these signals, determined, for example, by their concentration and/or the spatial and temporal dynamics of release (cf. Boland 1995; Jüttner 2001).

The chemistry of pheromonal attraction is well characterized, particularly in terrestrial systems. On the one hand, pheromones address nonspecific individuals, and may be short-lived to minimize confounding signals from old pheromone traces. On the other hand, attractants, deterrents, allelopathic metabolites, and toxins may be produced by, and target, more than one species; they may be more persistent and/or continually synthesized, or triggered on demand. Importantly, depending on the timing, location, and level of release the same FA-derived signal might act as a toxin, deterrent, pheromone, or growth resource (e.g., Boland et al. 1983; Jüttner 2005). This wide activity spectrum highlights the difficulties of research in this field compared with, for example, the comparatively more straightforward dynamics of pheromone systems.

The widespread importance of FA and FADV from algal sources in food web energy transfer and internal/external cell signaling in littoral and planktonic communities has gained increasing scientific recognition and interest, notably following several publications describing significant deleterious effects of marine diatom PUFA-derived compounds against copepods (Miralto et al. 1999, and Sect. 4.2.5). These studies fuelled considerable debate (e.g., Ianora et al. 1999; Irigoien et al. 2002; Paffenhöfer et al. 2005; Koski et al. 2008; Jones and Flynn 2005) but, as a result of work in this area, the general concepts are now widely accepted, although much remains to be understood about these interactions.

4.2 FA and FA Esters as Toxins, Allelopathic Metabolites, and Food Web Effectors

Extensive research has characterized a diversity of specific algal toxins, which include hepatotoxins and neurotoxins (e.g., microcystins, saxitoxins, domoic acid, ciguatoxin, etc.; Cembella 2003). But a few FA such as hemolysins have also been linked to aquatic poisoning (e.g., Fu et al. 2004), and with increasing research in this area, the number of algal taxa implicated in these and other FA-related semiochemical interactions grows (Table 4.1). It is important to note, however, that most FA are not present in free form in the aquatic environment, but as covalently bound components in lipids of living tissue. Although some free SAFA and MUFA are present in intracellular pools, the more cytotoxic PUFA typically are present not in free-form in intact cells but as lipid-bound fractions (see e.g., Pohnert 2002). This thus limits their mode(s) of transfer and activity. Furthermore, once released, free polyunsaturated fatty acids may be degraded rapidly (Kieber et al. 1997). However, it is important to note that published literature often does not discriminate between toxic free FA and total FA content, which includes also lipid-bound FA that have no adverse effects. This may explain the apparent lack of evidence of direct toxicity by PUFA-rich algal taxa on freshwater cladocerans; the predominant planktonic herbivores in most freshwaters (cf. Walton and Sandgren 1995).

FA and their derivatives have been demonstrated to function as allelopathic agents in a number of algal systems. Early studies reported density-dependent inhibition of *Chlorella vulgaris* by an autoinhibitor "chlorellin"; subsequent studies implicated peroxides of MUFA and PUFA (e.g., 18:1n-9, 18:2n-6; Scutt 1964). PUFA released by senescing cells of the cyanobacterium *Phormidium tenue* have also been described as autotoxins (Yamada et al. 1994). Other studies have reported FA inhibition of chlorophyte growth, which generally increased with molecular unsaturation (McGrattan et al. 1976). These studies suggest that FA modify chloroplast membrane characteristics, inhibiting photosynthesis (Okamoto and Katoh 1977; Okamoto et al. 1977), which may, in fact, represent a more generalized effect on membrane structure. For example, Wu et al. (2006) interpreted significant K^+ release by two chlorophytes exposed to toxic FA doses as a disruption of plasma

Table 4.1 Semiochemical properties of PUAs (*HDE* 2,4-heptadienal, *DDE* 2,4-decadienal; *DDT* 2,4,7-decatrienal) and average sestonic concentration (EL). A. Species reported to produce these compounds, with average cell PUA production in parentheses where known. B. Target species where PUA bioassays show significant effects

PUA (EL; µM)	A. Producer species		B. Target species			
	RF	Species (fM PUA per cell)	RF	ACT (µM)	Mode	Target species
HDE 0.007	5	*Dinobryon divergens*, *D. cylindricum* (9–40), *Mallomonas papillosa* (82–145), *Synura petersenii*, *Uroglena americana* (37–70), *Uroglena* sp., *M. varians*, *Hydrurus foetidus*[a] (1)	4	1,000	FEC	*Temora stylifera*
			8	50	TOX	*Daphnia pulex, Hesperidiaptomus arcticus*
	7	*Asterionellopsis glacialis* (0.03), *Chaetoceros compressus* (0.9), *Fragilaria* sp. (0.1), *Guinardia deliculata* (0.2), *Melosira nummuloides* (0.6), *Skeletonema costatum* (0.1), *S. pseudocostatum* (0.2), *Thalassiosira aestivalis* (1.2), *T. anguste-lineata* (0.3), *T. minima* (0.03), *T. pacifica* (0.6), *T. rotula* (0.1–0.3)	5	0.23	T&O	
DDE 0.0001	5	*D. divergens*, *D. cylindricum* (4–8), *M. papillosa* (0.1), *S. petersenii*, cf. *Syncrypta* sp., *U. americana* (2–5), *Uroglena* sp., *H. foetidus*[a] (3.5), *Cryptomonas rostratiformis*, *Peridinium willei*	2	1.9	GI	*Thalassiosira weissflogii*
			4	132[b]	FEC	*Temora stylifera*
			4	7	EC	*Paracentrotus lividus*
			1	0.79	HCH	*Arenicola marina*
	7	*Fragilaria* sp., *Chaetoceros compressus* (0.2), *M. nummuloides* (0.2), *Skeletonema subsalsum*, *T. rotula* (0.1)[c]		15.5	FRT	*Nereis virens*
				0.13	HCH	
				33.4	FRT	*Asterias rubens*
				1.18	HCH	
				68.4	FRT	*Psammechinus miliaris*
				16.6	HCH	
				265.9	FRT	
			3	42[d]	REP	*Eudiaptomus gracilis*, *Cyclops abyssorum*, *C. vicinus*, *C. bohater*, *Cyclops* sp., *Mesocyclops leuckarti*, *Bosmina* sp., *Daphnia hyaline*, *D. galeata*
			8	14	TOX	*D. pulex, H. arcticus*
			5	0.001	T&O	

4 Fatty Acids and Oxylipins as Semiochemicals

DDT 0.0001	5	D. divergens, D. cylindricum (12–55), M. papillosa (0.3–0.5), S. petersenii, U. americana (0.4–1.8) Uroglena sp., Melosira varians, H. foetidus[a] (3.6), S. petersenii, U. americana, Uroglena sp., M. varians	4 4 8 6 3	100[b] 7[b] 12 2.5 3.5[d]	FEC T. stylifera EC Paracentrotus lividus TOX D. pulex, H. arcticus REP Daphnia magna REP Eudiaptomus gracilis, Cyclops abyssorum, C. vicinus, C. bohater, Cyclops sp., Mesocyclops leuckarti, Bosmina sp., Daphnia hyaline, D. galeata
	7	C. compressus (1.4), M. nummuloides (1.6), T. anguste-lineata (0.3), T. minima (0.03), T. rotula (0.2–3.2)[c]	5	0.007	T&O

Abbreviations used: *RF* references; *ACT* activity level at which PUA 50% effective (e.g., LC_{50}), *Mode* activity mode (*EC* egg cleavage inhibitor; *FEC* reduced fecundity; *GI* growth inhibitor; *HCH* reduced hatching success, *FRT* reduced fertilization, *REP* repellent; values represent dose at 50% inhibition). *T&O* = taste-odor source – activity level denotes odor Threshold Level where detectable to humans

References: 1 (Caldwell et al. 2002); 2 (Casotti et al. 2005); 3 (Jüttner 2005); 4 (Miralto et al. 1999); 5 (Watson 2003; Watson et al. 2001); 6 (Watson et al. 2007); 7 (Wichard et al. 2005b); 8 This chapter (Fig. 4.8)

[a] Units in mM PUA (g wet wt)$^{-1}$
[b] LC_{50}
[c] Strain specific production
[d] RC_{50} values (repellent concentration causing 50% reduction of crustacean density in the assay vial) (Jüttner 2005)

membrane permeability and integrity, with inhibitory or toxic effects. This mechanism may also account for reports of allelopathic activity manifested as "blistering" of cell plasma membranes where the allelogen has not been characterized (cf. Rengefors and Legrande 2001).

PUFA liberation during algal cell damage has been identified as an effective grazer defense mechanism by Jüttner (2001). He suggested that PUFA are highly suited to serve as a wound activated grazer defense. As integral cellular components (e.g., in membranes and galactolipids), they only require additional lipases for liberation. Since lipids and lipases are ingested in particulate form, efficient transfer and maximum herbivore exposure are guaranteed. Jüttner (2001) identified EPA and other PUFA (18:3n-3, 18:3n-6) as key toxic agents in interactions between freshwater periphytic diatoms and the grazer *Thamnocephalus platyurus*, with an LC_{50} for EPA of 34 µM. Such free PUFA concentrations clearly far exceed those occurring in natural seston.

Numerous studies have implicated algal FA in semiochemical food web interactions, largely in a toxigenic capacity. In marine systems, algal taxa releasing free FA have been implicated in mass mortality events, although in many cases the toxin(s) have not been identified. Findlay and Patil (1984) identified the phytylester of eicosapentaenoic acid (EPA), phytyl eicosa-5,8,11,14,17-pentaenoate as the main antimicrobial compound produced by *Navicula delognei* f. *elliptica*, along with hexadeca-6,9,12,15-tetranoic acid, octadeca-6,9,12,15-tetraenoic acid, and hexadeca-6,9,12-trienoic acid. Murakami et al. (1989) reported that hexadeca-4,7,10,13-tetraenoic acid from *Pediastrum* was highly embryotoxic against echinoderms, but showed no antimicrobial activity. Octadecapentenoic acid from *Gymnodinium* cf. *mikimotoi* was observed to inhibit embryonic cleavage and elicit abnormal larval development in sea urchins (Sellem et al. 2000). Lipid-rich flagellates such as *Gyrodinium aureolum*, *Gymnodinium mikimotoi*, *Chrysochromulina polylepis*, *Heterosigma akashiwo*, *Prymnesium parvum*, and the colonial chlorophyte *Pediastrum* sp. produce hemolysins composed of FA and galactolipids, which damage gill epithelial tissue in fish and shellfish (Shilo 1981; Taylor 1985; Murakami et al. 1989; Yasumoto et al. 1990; Sellem et al. 2000). The fact that reproduction requires high levels of EPA, docosahexenoic acid (DHA), and arachidonic acid (ARA) (Harrison 1990; Jeckel et al. 1989), but that, on the other hand, several studies report adverse effects of FA on reproductive systems (Rukmini 1990; Bell et al. 1996) needs additional experimental verification with a parallel monitoring of lipase activities on lipids and free FA.

In freshwaters, FA-related toxic events are less well known, but studies have implicated flagellates, cyanobacteria, and chlorophytes (cf. Hansen et al. 1994; Ikawa et al. 1996; Watson 2003; Chiang et al. 2004). Several identified and unidentified PUFA (notably (γ-linolenic) produced by planktonic cyanobacteria (*Anabaena* sp., *Microcystis aeruginosa*, *Oscillatoria agardhii*), haptophytes (*Chrysochromulina parva*), and chrysophytes (*Poterioochromonas malhamensis*, *Uroglena* cf. *volvox*)) have been reported as toxic to grazers, insect larvae, and fish (Kamiya et al. 1979; Hansen et al. 1994; Leeper and Porter 1995; Bury et al. 1998; Reinikainen et al. 2001).

4.3 Oxylipins in Aquatic Chemical Ecology

As products of PUFA oxidative metabolism, oxylipins represent a suite of important bioactive molecules in terrestrial and aquatic systems with a wide range of semiochemical functions (Gerwick 1994; Stanley-Samuelson 1994; Feussner and Wasternack 2002; Pohnert and Boland 2002). Eicosanoids, for example, are metabolites derived from polyunsaturated C_{20} FA, and comprise three major groups: (1) prostaglandins and thromboxanes, (2) epoxyeicosatri- and tetraenoic acids, and, (3) lipoxygenase (LOX) products including hydroperoxy-, and hydroxy acids, lipoxins, and leukotrienes. EPA is widespread among aquatic organisms (Chuecas and Riley 1969; Clare and Walker 1986; Yongmanitchai and Ward 1991) and thus provides an ample precursor supply for eicosanoid synthesis. The functioning of eicosanoids in aquatic chemical ecology is wide-ranging, with particular relevance to the reproductive biology and ecology of many organisms. Selected examples involving LOX products include reinitiation of meiosis in prophase arrested starfish ooyctes (Meijer et al. 1986), regulation of spawning behavior of scallops (Matsutani and Nomura 1987), prevention of polyspermy during fertilization of sea urchins (Schuel et al. 1985), and the regulation of body pattern and regeneration in *Hydra vulgaris* (Di Marzo et al. 1993). In the following sections, we will focus on two groups of LOX-derived oxylipins with common biosynthetic and regulatory pathways, which play important roles in aquatic food web signaling; polyunsaturated aldehydes (PUA), and volatile unsaturated hydrocarbons.

4.3.1 Polyunsaturated Aldehydes

PUA (e.g., 2,4,7-decatrienal, 2,4-decadienal, 2,4-nonadienal, 2,6-nonadienal, 2,4-octadienal, 2,4-heptadienal; Fig. 4.1) account for much of the semiochemical activity attributed to oxylipins in both terrestrial and aquatic food webs. PUA are

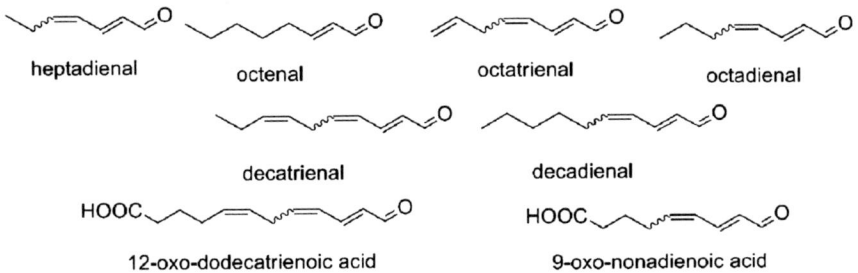

Fig. 4.1 Major C_7–C_{10} unsaturated and polyunsaturated aldehydes (PUA) produced by LOX-mediated transformation of polyunsaturated fatty acids (PUFA); also shown are the C_9 and C_{12}-oxo acids produced during this process

produced by a range of lipid-rich aquatic taxa. Their bioactivity and mode(s) of signaling are highly dependent on their induction, biosynthesis, and interspecific transfer. Recently, there has been increased focus on PUA among aquatic ecologists (cf. Paffenhöfer et al. 2005 and references therein), because of their importance in aquatic signaling, particularly in relation to a proposed mechanism for bottom-up regulation of planktonic food webs (Miralto et al. 1999; Ianora et al. 2004). PUA are not produced by intact cells, but production is rapidly activated by cell damage, an observation that motivated much research activity to elucidate PUA biosynthesis. This process initially yields (*E,Z*) PUA, which, once released into the environment, spontaneously isomerizes to the more stable (*E,E*) forms, which differ in molecular properties (e.g., volatility, odor characteristics, and strength). Since it has not been well established if the isomers also differ significantly in food-web bioactivity, this process has relevance both to the drinking water and food industries as well as bioassays using direct immersion assays (vs. internal exposure routes), which base their protocols, analytical, and sensory standards on the commercially available (*E,E*) form (Adolph et al. 2003; Watson 2003).

4.3.2 PUA and Volatile Hydrocarbon Biosynthesis

In diatoms, the formation of PUA is dependent upon the availability of free precursor FA. Intact diatom cells contain mostly free SAFA, with free PUFA concentrations only increasing following cell damage. EPA was shown to be the precursor for the production of 2,4,7-decatrienal and 2,4-heptadienal (Fig. 4.1) by the diatom *Thalassiosira rotula* (Pohnert 2000; d'Ippolito et al. 2004). However, this FA can be ruled out as the precursor of octadienal and octatrienal, since their unusual double bond position cannot be explained with a transformation of any n-3-fatty acid, and instead, 16:3 n-4 and 16:4 n-1, respectively, have been identified as the precursors (d'Ippolito et al. 2003; Pohnert et al. 2004). Wound-activated PUA formation is under the control of strong lipolytic activity, which acts directly after algal tissue disruption. Initially, phospholipids were identified as the primary sources for EPA, and enzyme inhibitor studies with *T. rotula* suggested that phospholipase A2 activity triggered this release (Pohnert 2002). However, further studies demonstrated that two of the precursor PUFA (16:3 n-4 and 16:4 n-1) are not found in phospholipids but are instead liberated from galactolipids – which are also a major source of EPA, indicating that lipolytic activity acts on both polar lipid pools to release the immediate precursors for PUA-biosynthesis (d'Ippolito et al. 2004; Cutignano et al. 2006). The current model suggests that upon tissue decompartmentalization, chloroplastic monogalactosyldiacylglycerols and phospholipids are exposed to lipolytic activity, thereby releasing elevated amounts of PUFA (Fig. 4.2).

Early investigations of the freshwater diatom *Melosira* sp. by Wendel and Jüttner (1996) showed that 2,4,7-decatrienal production requires molecular oxygen (O_2). This indicated that PUFA transformation to PUA in diatoms could involve LOX activity, as observed for the production of short chain aldehydes by higher

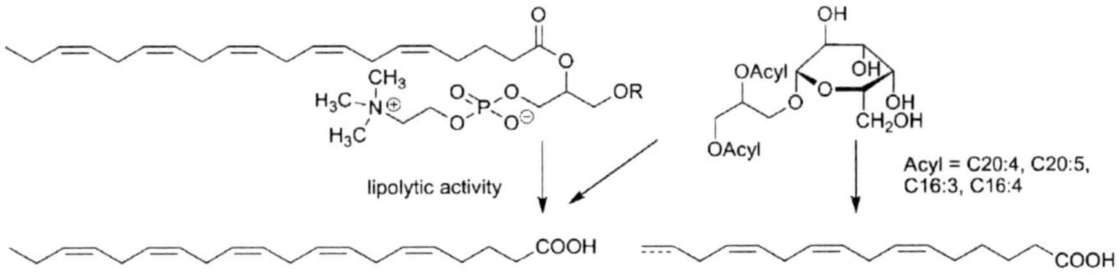

4 Fatty Acids and Oxylipins as Semiochemicals

Fig. 4.2 Decompartmentalization of chloroplast-derived monogalactosyldiacylglycerol and phospholipids by lipolytic activity, releasing PUFA, the immediate precursors for PUA-biosynthesis to feed the downstream activities of lipoxygenases

plants (Feussner and Wasternack 2002). This hypothesis was supported by the observation that known LOX inhibitors suppressed PUA production in *Melosira* sp. (Wendel and Jüttner 1996). LOX catalyze the introduction of O_2 into a skipped 1,5-diene PUFA segment and generate a reactive hydroperoxide. Later work was unable to detect reactive hydroperoxides even using the most sensitive analyses, indicating a rapid downstream transformation. Hombeck et al. (1999) provided the first direct evidence of the role of LOX in PUA production, using suspensions of the freshwater periphytic diatom *Gomphonema parvulum*, wounded (via sonication to simulate cell damage) in the presence of excess glutathione and glutathione peroxidase. The glutathione/glutathione peroxidase system quickly reduced the intermediate hydroperoxides produced by the diatom and allowed isolation and characterization of the trapped 9-hydroxyeicosapentaenoic acid. This showed that *G. parvulum* uses a 9S-LOX to activate PUFA for downstream metabolism (Fig. 4.3). The involvement of LOX in the transformation of PUFA is further supported by the observation that the marine diatom *Stephanopyxis turris* accepts externally added hydroperoxyeicosatetraenoic acid as a substrate for the production of a 12-oxo acid (Wichard and Pohnert 2006).

Work by d'Ippolito and coworkers has shown that extracts of the wounded diatom *T. rotula* contain a series of additional metabolites known to be derived from LOX metabolism. In accordance with a LOX-mediated transformation of C_{16} and C_{20}-PUFA by this diatom, a number of oxygenated long-chain FA could be detected (Fig. 4.4). In particular, the occurrence of 11R-hydroxyeicosapentaenoic and 9S-hydroxyhexadecatetraenoic acids point toward 9-LOX and 11-LOX that could also be involved in the biosynthesis of C_8-PUA and C_{10}-PUA (d'Ippolito et al. 2005, 2006).

The further transformation of the intermediate hydroperoxy fatty acids might result either in the formation of PUA or in the release of unsaturated hydrocarbons. These latter compounds have been detected as early as 1979 in investigations of freshwater plankton and biofilms. Seasonal peaks of several intensely "seawater-smelling" C_8- and C_{11}–compounds, which are structurally identical to known brown algal metabolites (fucoserratene, dictyopterenes, and hormosirene; Jüttner and Müller 1979; Jüttner 1984; Jüttner et al. 1986; Wendel and Jüttner 1996; Jüttner and Dürst 1997), are well characterized as pheromones in marine phaeophytes such as *Fucus serratus* and *Hormosira banksii* (Pohnert and Boland 2002). Discrete seasonal peaks of ectocarpene and dictyopterenes have been observed to coincide with

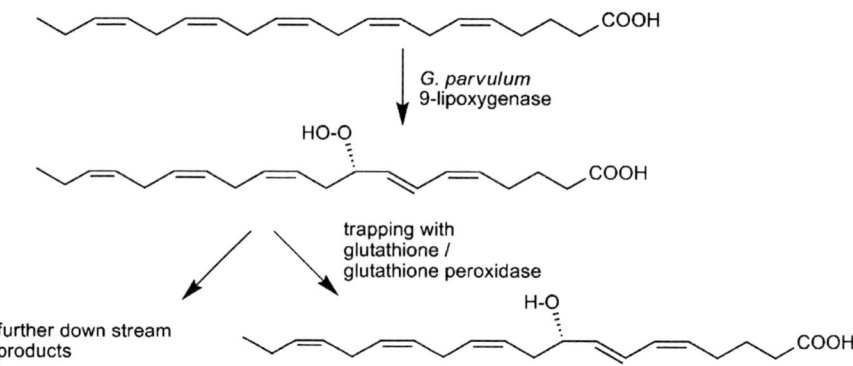

Fig. 4.3 Biosynthetic pathways deduced via trapping experiments, using cell suspensions of the freshwater diatom *Gomphonema parvulum* wounded in the presence of excess glutathione and glutathione reductase

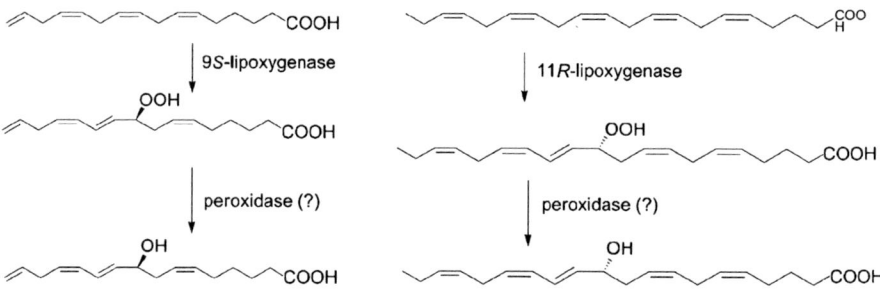

Fig. 4.4 Lipoxygenase-mediated transformation of C_{16} and C_{20}-fatty acids by the marine diatom *Thalassiosira rotula*

successional events among major epilithic diatoms (Jüttner 1992, 2001; Jüttner and Dürst 1997). However, as yet, there have been no direct demonstrations of their biological activity among freshwater taxa.

In general, it can be stated that three different enzymatic mechanisms of fatty acid hydroperoxide cleavage are observed:

1. *Simultaneous production of hydrocarbons and PUA*: in the freshwater diatoms *G. parvulum* and *A. formosa* hydroperoxide lyases act by abstracting a proton concomitant with the cleavage of the C–C bond besides the hydroperoxide (Fig. 4.5) (Pohnert and Boland 1996; Hombeck and Boland 1998; Hombeck et al. 1999). As a second fragment of the cleavage of EPA in *A. formosa*, a C_{12}-oxo acid is detected along with the C_8-hydrocarbon fucoserratene (Fig. 4.5). In close analogy, hormosirene production in *G. parvulum* occurs concomitant with the release of a C_9-oxo acid – identified as a selective grazer deterrent in pheophytes (Schnitzler et al. 2001).

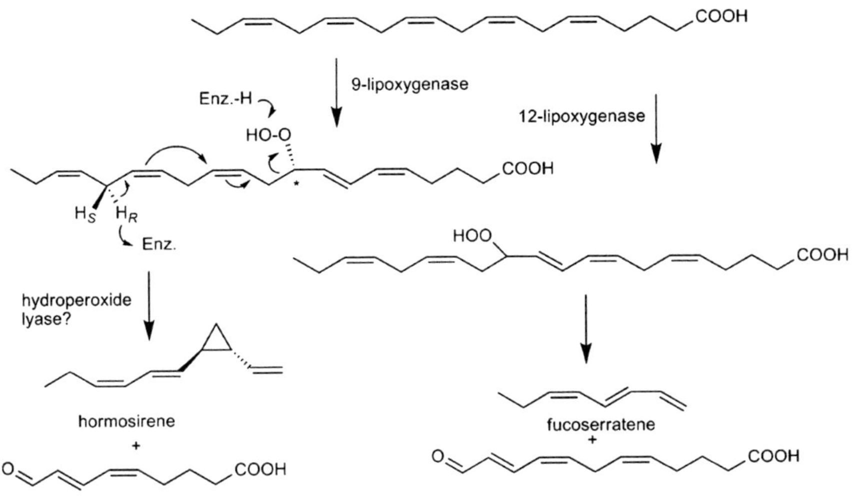

Fig. 4.5 Simultaneous production of hydrocarbons and PUA in the fresh water diatoms *Gomphonema parvulum* and *Asterionella formosa*

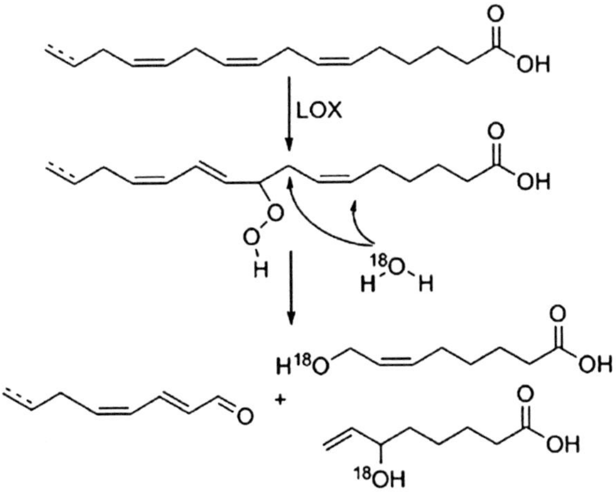

Fig. 4.6 Simultaneous production of alcohols and PUA in the marine diatom *Thalassiosira rotula* where the fatty acid hydroperoxide cleavage is assisted by the nucleophilic addition of water

2. *Simultaneous production of alcohols and PUA*: the biosynthesis of octadienal, octatrienal, and decatrienal in the diatom *T. rotula* does not occur in a similar fashion to that described in (1) above, since the corresponding hydrocarbons as second fragments are not detected. Instead, FA hydroperoxide cleavage seems to be assisted by the nucleophilic addition of water (Fig. 4.6) (Barofsky and Pohnert 2007).

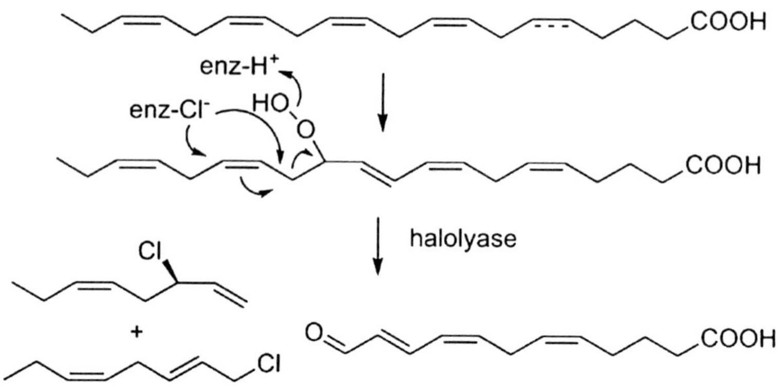

Fig. 4.7 Proposed pathway for the production of halogenated hydrocarbons and PUA in the diatom *Stephanopyxis turris* mediated by the enzyme haloyase, which uses a different mechanism to that used in the formation of alcohols shown in Fig. 4.6

3. *Simultaneous production of halogenated hydrocarbons and PUA*: a unique enzyme termed haloyase is responsible for PUA formation in the diatom *Stephanopyxis turris* (Fig. 4.7) (Wichard and Pohnert 2006). This mechanism might be seen as a variation of that required for the formation of alcohols (Fig. 4.6). Instead of the enzyme-assisted nucleophilic attack of water, a halogenide-like chloride or bromide is presumably used by the enzyme to compensate for the electronic requirements during the carbon–carbon bond cleavage (Wichard and Pohnert 2006). All three pathways of hydroperoxy FA-transformations in diatoms have attracted much attention, since they follow pathways that are not observed in the related oxylipin production of higher plants and mosses (Wichard et al. 2005a; Matsui 2006).

4.3.3 Semiochemistry of Oxylipins

Bioactivity of oxylipins in aquatic systems has been reported in allelopathic and grazer defense contexts, thereby potentially affecting aquatic food webs at a number of trophic levels over a range of spatial and temporal scales. The grazer defense hypothesis is not without its critics due to inconsistencies between laboratory and field studies, and, in particular, between marine and freshwater studies. PUA have also been linked to allelopathic relationships in planktonic systems. Casotti et al. (2005) and Ribalet et al. (2007b) describe growth inhibition and cellular damage in a range of marine phytoplankton induced by PUA. In an elegant study, Vardi et al. (2006) showed that diatoms could detect the presence of PUA and initiate cell death via a nitric oxide pathway. This may represent chemical control of diatom bloom dynamics.

As with many other metabolic processes, PUA production is strongly modified by growth and environment, resulting in variable production even by individual species (and possibly also explaining discrepant reports of toxicity). For example, mesocosm populations of *Uroglena americana* showed a marked change in PUA chemistry over the growth cycle, with a predominance of 2,4,7-decatrienal production during initial growth changing to the less toxic 2,4-heptadienal toward the stationary stage (Watson et al. 2000). Similarly, Wee et al. (1994) reported variance of in vitro production of 2,6-nonadienal within and among different *Synura* species. Only strains of *S. petersenii* were reported to produce this PUA, and this occurred largely toward the stationary phase. Observations of variable oxylipin production were also made in field surveys of marine plankton. While a spring bloom of *T. rotula* in the English Channel produced predominantly C_{10} PUA, the same species released C_7 PUA during autumn (Wichard et al. 2008). Resource limitation also appears to modify algal PUA production. Watson and Satchwill (2003) reported significant differences in per capita yield and relative production of C_7 and C_{10} PUA by chrysophytes (*U. americana, D. cylindricum, M. papillosa*) under different light, P and Fe regimes, suggesting changes in either PUFA composition or enzyme activity. Similar age and resource-dependant results have been obtained for the marine diatom *Skeletonema marinoi* (Ribalet et al. 2007a).

The use of aldehydes as defensive compounds in terrestrial systems and by macroalgae is well known (Cavill and Hinterberger 1960; Kubo et al. 1976; Camazine et al. 1983; Morimoto et al. 2002; Schnitzler et al. 2001). PUA are often semiochemically active in several different ways, for example antimicrobial, fertilization inhibiting, embryotoxic, sperm motility inhibiting, larval toxic, grazer toxic, and feeding deterrent (Caldwell et al. 2002, 2004a, 2005; Poulet et al. 2003, 2007; Tosti et al. 2003; Adolph et al. 2004; Lewis et al. 2004). Bioactivity is directed at cellular and molecular targets such as microtubule and microfilament stability, DNA replication, and the activity of the G2-M promoting complex cyclin B-Cdk1, which are all crucial during embryonic development (Romano et al. 2003; Hansen et al. 2004a). PUA toxic to developmental stages of a number of aquatic invertebrate species, including copepods, sea urchins, polychaetes, and ascidians. Reported symptoms of PUA exposure include gamete infertility and dysfunction, fertilization failure, and abnormal morphological development, indicating that the timing and/or level of exposure may result in subtle, yet deleterious departures from normal reproductive patterns (e.g., Miralto et al. 1999; Ianora et al. 2004; Adolph et al. 2004; Poulet et al. 2007). Given the importance of oxylipins in reproduction and development, exposure of invertebrates to altered environmental titers at particular ontogenetic stages may therefore disrupt highly sensitive endocrine processes.

It is important to note that PUA production is not ubiquitous, even among closely related taxa. In a comprehensive survey of 51 planktonic diatom species (71 strains, mostly marine), Wichard et al. (2005b) found that only ~36% produced PUA (two of which were freshwater strains; *Fragilaria* sp. and *Skeletonema subsalum*), while per capita production varied over four orders of magnitude among species and strains. Thus, diatom blooms per se cannot be assumed toxic, which may account for some of the reported differences in their detectable effects on grazers (cf. Irigoien et al. 2002; Koski et al. 2008). Other factors such as the nutritional status of the prey, and access to mixed or alternative prey resources can also mitigate the negative effects of these blooms on grazer populations (Jones and Flynn 2005; Barreiro et al. 2007). Two recent studies demonstrate that even under standard experimental conditions certain copepods are not affected by PUA. Dutz et al. (2008) observed that egg hatching rates of the copepod *Temora longicornis* decreased after 4 days in four treatments with different diatoms, irrespective of the diatom PUA production. In assays with the same copepod, Wichard et al. (2007) showed that external additions of EPA to diatom-rich diets resulted in increased egg hatching success, despite the fact that this FA treatment also caused significantly higher PUA production. These authors attributed the observed higher performance of the copepods to an increased availability of the essential FA that is otherwise depleted by PUA production. But recent field and laboratory studies with *Calanus helgolandicus* collected in the English Channel also show no impact of diatom PUA (Wichard et al. 2008). Fontana et al. (2007) demonstrated that PUA production is not mandatory to induce reproductive failure in grazers. Indeed, other LOX derived products of diatoms may also be responsible for grazer defense, even in non-PUA producing strains. The authors observed the production of reactive oxygenated metabolites in the marine diatoms *Skeletonema marinoi*, *Chaetoceros socialis*, and *C. affinis*. The short-lived compounds were found to be significantly more toxic than PUA, and therefore may represent an unanticipated additional level of LOX-mediated grazer defense. The extent to which additional compounds that are structurally related to PUA, selective feeding behavior, or active detoxification by copepods lead to such strong variability has yet to be determined.

PUA toxicity toward marine planktonic grazers (copepods) is, in general, manifested as reduced fecundity and hatching success, not as reduced adult survivorship (e.g., Miralto et al. 1999). Similar bioassays with freshwater crustaceans have also failed to detect consistent effects either on adult survivorship and growth (Carotenuto et al. 2005; Watson unpublished data). Nevertheless, with exposure to high concentrations of C_6–C_{10} PUA, clonal adult *Daphnia pulex* and field populations of *Hesperodiaptomus arcticus* gave LD_{50} values similar to those reported for marine copepods, decreasing with carbon number and saturation (further influenced by double bond placement; Fig. 4.8). Carotenuto et al. (2005) report reduced embryonic development of late clutches in *Daphnia pulicaria* with a diet of PUA producers or direct immersion in a C_{10} PUA (decadienal)-containing medium, but question whether this has any ecological significance given that population parameters (i.e., life-time fecundity and instantaneous rate of population growth) were not affected by PUA. Nevertheless, the reproductive lifespan of some daphnids may overlap successive biomass peaks of several PUA-producing species (e.g., chrysophytes diatoms; Sandgren 1988) and the potential effects of this exposure on grazer communities need elucidation with carefully controlled laboratory and field-based studies. For example, our preliminary trials with *D. pulex* showed increased ephippial production with direct immersion to C_6 and C_{10} PUA (Watson and Caldwell unpublished data), a known adaptive response to environmental stress (Alekseev and Lampert 2001).

4 Fatty Acids and Oxylipins as Semiochemicals

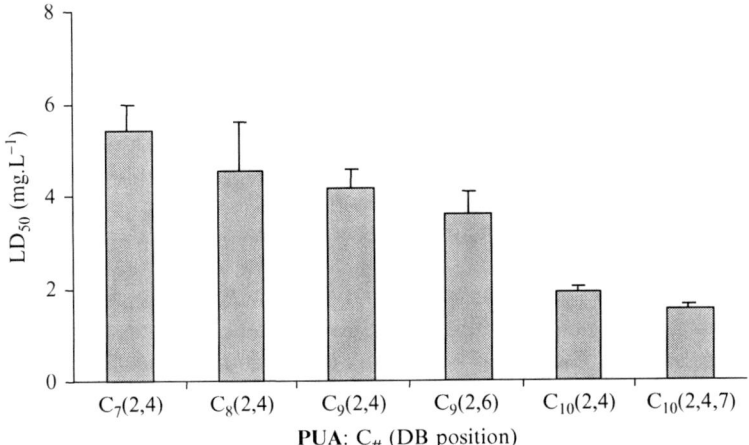

Fig. 4.8 Adult survival ($T = 24$ h) of *Daphnia pulex* (clonal individuals, raised on standard food) with exposure to PUA of increasing carbon number and varying double bond (DB) placement and number (C_7: heptadienal, C_8: octadienal, C_9: nonadienal, C_{10}: decadienal, C_{10}: decatrienal)

There are several likely explanations for these apparent contradictions. Daphnids normally filter-feed on a mix of particles and seston that do not approach the PUFA levels represented in diatom-dominated biofilms. Fecundity in aquatic grazers often has a seasonal pattern that may be affected by factors such as food availability and quality, regulating somatic growth and reproductive fitness (Bottrell et al. 1976; Ban 1994; LaMontagne and McCauley 2001). The development of newly spawned copepod eggs is affected by the maternal diet (Guisande et al. 1999; Laabir et al. 1999), yet algal diets that are regarded as beneficial for growth may not support fecundity or egg viability (Chen and Folt 1993; Poulet et al. 1994; Ianora et al. 1999). Early studies of the effects of diatoms and dinoflagellates on recruitment in copepods suggested that reproductive anomalies could be explained by the differing nutritional quality of the algal taxa, with some diatoms considered nutritionally insufficient (Støttrup and Jensen 1990; Kleppel 1993; Jónasdóttir and Kiørboe 1996).

It is difficult to predict the true role of these reactive compounds in nature. Most bioassays inevitably impose conditions that may not be representative of natural environments. With current analytical techniques, it is often not possible to determine local semiochemical concentrations or tissue exposure levels. The complex mode of release and delivery of most PUFA-related compounds and their instability once released do not allow standard quantification (cf. Caldwell et al. 2004b). Nevertheless, in situ trapping with derivatization agents and subsequent quantification is feasible for certain structures, such as unsaturated aldehydes and ketones (Wichard et al. 2005c). A major criticism often leveled at the inhibitory diatom hypothesis is that in vitro studies of this mechanism have applied concentrations of cells, cell extracts, or PUA in excess of natural levels. However, this is not the case with every study – using copepods or other invertebrates; several have applied algal cell densities in the order of 10^5–10^7 cells L^{-1}; well within concentrations seen during blooms (e.g., see Chaudron et al. 1996; Carotenuto et al. 2005; Casotti et al. 2005; Caldwell et al. 2005). Furthermore, Poulet et al. (1994) hypothesized that inhibitory compounds are accumulated in oocytes following diatom feeding, thus exceeding the ambient threshold level required for inhibition. A cumulative toxin loading would therefore permit intoxication even at naturally occurring diatom densities.

Can inhibitory FA-derived compounds similarly be stored and transferred from the female digestive system to vitellogenic oocytes, and is this a realistic explanation for observed patterns of hatching failure? Circumstantial evidence comes from the time lag between initiation of a diatom diet and the production of nonviable eggs. There is a lag phase of 24–72 h during which time viable eggs are spawned, followed by progressive reduction in oocyte viability. Additionally, Laabir et al. (1995) demonstrated that by reducing diatom cell densities from 10^5 to 10^4 cells mL^{-1}, the time required to reach total egg inhibition was increased from 7 to 12 days. The copepod gut epithelium is in close contact with the ovary, so a diffusion mechanism may explain the transfer of inhibitory compounds from food to eggs. In order for this accumulative mechanism to take place, the active metabolites must, by necessity, traverse lipid membranes. The degree of passive incorporation will be governed by the characteristic partition coefficient between water and the lipid membranes. Evidence of *trans*-lipid PUA movement was reported for bacterial cells by Trombetta et al. (2002) using model phosphatidylcholine liposomes. Therefore, gonadal accumulation is feasible in theory, despite the reactivity of PUA resulting in significant losses before reaching the target organs. PUA will likely form DNA adducts within animal tissues (Loureiro et al. 2000), the significance of this within aquatic grazers is unknown but may be related to observed patterns of apoptosis in gonadal, embryonic, and larval tissue (Romano et al. 2003; Poulet et al. 2003, 2007). At the present time, this remains purely theoretical, since it is extremely difficult to demonstrate, and a number of protocol issues need to be addressed before this mechanism can be systematically examined (cf. Caldwell et al. 2004b).

Controversial discussions, partially contradicting evidence and complex indirect actions of oxylipins make clear that there is no universal activity of these metabolites (e.g., Dutz et al. 2008, Wichard et al. 2008; Jones and Flynn 2005). Water is not a homogenous medium, but varies in time and space in its physical and chemical structure and characteristics at both micro and macroscale levels around cells, from the boundary layer out. Future progress requires the development of bioassay designs, which can measure changes in processes and compounds along these different spatial and temporal scales around cells and surrounding environs to resolve the true role of PUA in plankton interactions. Despite these caveats, the above-mentioned studies, and other increasing evidence, suggests that under certain environmental conditions, algal diets that produce PUA can be detrimental and influence plankton interactions, potentially with a multitude of mechanisms and consequences.

4.4 Grazer Counter-Defense: Behavioral Avoidance, Adaptive Strategies, and Detoxification

Copepods have developed chemically-mediated behavioral responses in relation to food acquisition and/or avoidance (Poulet and Marsot 1978), for example, prey rejection and feeding cessation can be initiated by the presence of particular toxic dinoflagellate species (Huntley 1982; Huntley et al. 1986). There is also evidence from freshwater systems that PUA may act as grazer deterrents by modifying swimming behavior. Jüttner (2005) observed that mixed zooplankton assemblages in "U" tubes (species of *Cyclops*, *Bosmina*, *Daphnia*, and *Eudiaptomus*) exhibited significant movement away from both a PUA mix liberated from diatoms, and controlled applications of 2,4,7-decatrienal ($P < 0.001$, relative to controls). Similarly, Watson et al. (2007) observed significant avoidance behavior to 2,4,7-decatrienal by *Daphnia magna* ($P < 0.0001$) using an inline behavioral monitoring system. Such a mechanism may result in avoidance of PUA-liberating blooms; for example during senescence or initial grazer attack. Carotenuto and Lampert (2004) observed no indication of selective grazing by *D. pulicaria* on different algal food sources when feeding on oxylipin producers and nonproducers, indicating that such avoidance behavior does not include altered filtering rates.

Recent work by Fink et al. (2006) demonstrated that grazers can develop selective foraging behavior in response to specific mixes of FA-derivatives in algal prey (termed "foraging kairomones"). These authors found that the freshwater herbivorous gastropod *Radix ovata* is attracted to a particular mix of C_5 and C_7 oxylipins released by damaged cells of the filamentous chlorophyte *Ulothrix fimbriata* (1-penten-3-one, 1-penten-3-ol, (*Z*)-2-pentenal, (*E*)-2-pentenal and 2,4-heptadienal), demonstrating the potential for synergistic effects among FA-related semiochemicals; a complex and difficult phenomenon to characterize.

Miralto et al. (1999) raise the question why copepods have not evolved mechanisms to protect their embryos from PUA and characterized the inhibitory effect of diatoms on copepod reproduction as "insidious." The question is intriguing, since there is a remarkable degree of variation in the susceptibility of grazers both between and within species to phycotoxins. A good example is the response of the bivalve mollusc *Mya arenaria* to exposure to the toxigenic dinoflagellate *Alexandrium tamarensis*. *M. arenaria* not previously exposed to *A. tamarensis* show higher mortality, impaired burrowing capacity, lower respiration, reduced food clearance rates, and tenfold lower toxin accumulation compared with individuals from sites historically exposed to the toxins. Resistance is due to a single nucleotide mutation in the outer pore loop of Domain II of the Na^+ channel pore, replacing glutamic acid with aspartic acid (Bricelj et al. 2005). In experiments with *Daphnia*, it has been shown that genotypes exposed to cyanobacterial toxins can increase their toxin tolerance in <10 years (Hairston et al. 2001). If copepods are unable to mitigate PUA toxicity, this suggests that either the evolution of PUA in aquatic ecosystems is relatively recent, or that other factors such as fitness benefits from the PUFA-rich diets can alleviate negative impacts; however, as of yet no convincing evidence has been produced. Chaudron et al. (1996) surmised that at low diatom densities, marine copepods may possess a limited capacity to break down diatom inhibitors by digestive enzymes in the gut. Poulet et al. (1994) suggested that grazers may synchronize egg production cycles to coincide with periods of low diatom production. However, field data from freshwater prairie systems shows the highest abundance of copepod nauplii in May–June, suggesting that *eggs* were produced in spring during the typical spring diatom peak (M.T. Arts, personal communication).

It can be argued, however, that if an allele coding for a detoxification mechanism has evolved based on the Malthusian parameter, and the model described by Charnov (1997), such an allele should improve reproductive fitness and thus spread rapidly through the population until it became the norm. Factors that would promote this include improved fecundity, growth, and survival (Olive 1992). Yet it would appear that for copepods in particular, if a PUA detoxification pathway exists, it is not highly conserved. This disparity with Charnov's model adds further ambiguity of the "paradox of diatom–copepod interactions" (cf. Ban et al. 1997).

4.5 Algal Producers

A variety of freshwater and marine algal species have already been identified as important sources of oxylipins (Table 4.1A). However, since the large majority of taxa are as yet uncharacterized, this likely represents a small fraction of the actual producers. In marine systems, the bioactivity of oxylipins was not recognized until the late 1990s, when attention was then focused on planktonic diatoms as major sources (Miralto et al. 1999; Wichard et al. 2005b), although recent work now shows that other marine taxa, such as the prymnesiophyte *Phaeocystis*, also produce these metabolites (Hansen et al. 2004b). In comparison, PUA and many of their major algal producers were known as early as 1981 in freshwaters, largely because of their impacts on drinking water taste-odor (Jüttner 1981; Watson et al. 2000; Watson 2003). In these waterbodies, flagellates are the more common planktonic sources of these compounds, notably synurophytes and mixotrophic chrysophytes (e.g., *Dinobryon cylindricum, D. divergens, Uroglena americana, U. volvox, Mallomonas papillosa, Synura petersenii, S. uvella, Chrysosphaerella longispina*). These taxa are common in oligo-mesotrophic lentic systems (lakes, small ponds), and frequently bloom in systems that have been restored from eutrophication. Population outbreaks of these species can produce significant levels of fishy-smelling C_7, C_9, and C_{10} PUA in lakes with high natural or anthropogenic organic loading, sufficient to impair drinking water supplies (Watson et al. 2000, 2001). Other freshwater planktonic producers include diatoms, dinoflagellates, cryptophytes, and chlorophytes (*Fragilaria crotonensis, Aulacoseira islandica, Peridinium willei, Cryptomonas rostratiformis, Ulothrix fimbriata*) (e.g., Jüttner 1995; Watson 2003). As noted earlier, many freshwater littoral algal communities are also a rich source of PUA, notably diatom biofilms and the fetid-smelling thallic chrysophyte *Hydrurus foetidus*, which forms dense beds in oligotrophic high mountain streams (Watson et al. 2000, unpublished data; Jüttner 2001).

In freshwaters, diatoms have been identified as the primary sources of the C_8- and C_{11}-hydrocarbons discussed above. These are primarily periphytic species (e.g., *Amphora venetia*, *Gomphonema parvulum*) and the single planktonic taxon *Asterionella formosa*. A few marine planktonic diatom producers have also been reported (e.g., *Skeletonema costatum*, *Lithodesmium undulatum*; Derenbach and Pesando 1986).

4.6 Other FA-Related Semiochemicals

It is highly probable that the above interactions represent a small fraction of the FA-mediated semiochemical interactions occurring in aquatic systems. Many FA and their derivatives have either not been examined in this light, or are yet to be identified. Nevertheless, the list of newly discovered compounds continues to grow (cf Chap. 13). For example, many algal taxa produce FA derivatives, which demonstrate nonspecific bioactivity. Early studies reported that *n*-hexane and unsaturated hydrocarbons act as nonspecific attractants for "+" gametes of *Chlamydomonas* (Tsubo 1961). Similarly, butyl alcohol, methyl butenyl ether, diethyl ether, methyl amyl ketone, *n*-propyl acetate, isobutyl acetate, methyl and ethyl propionates, and methyl *n*-butyrate were reported as male gamete chemo-attractants in marine phaeophytes (*Ectocarpus*, *Fucus* and *Ascophyllum*; e.g., Cook and Elvidge 1951). Given their nonspecific action, however, it is unlikely that these compounds are released as pheromones. Chlorophytes and other algal taxa produce other lipid derivatives, some known for their grassy odors (e.g., (*E*)-2-butenal, (*Z*)-3-hexanol, (*Z*)-, and (*E*)-2-hexenal, hexanal, 2-heptanone, 2-octanone, 2-nonanone). Their semiochemical activity in aquatic systems is, to our knowledge, yet untested, but they are known to be highly bioactive among terrestrial organisms as chemo-attractants, stimulants, and deterrents (Bradow and Connick 1990; Dittberner et al. 1995).

Among freshwater chrysophytes and synurophytes, sexual reproduction is reported as density-dependent, and triggered by unidentified compounds (Sandgren 1988). PUFA derivatives are released at cell lysis and both their chemistry and yields vary with environment (see Sect. 4.3.1) but no clear connection to reproductive regulation of these metabolites is yet available. As noted, the freshwater diatom *Asterionella formosa* produces 1,3,5-octatriene (fucoserratene), which is known to be a pheromone among marine Fucaceae. However, sexual reproduction is rarely, if ever, observed in this and many other planktonic diatoms (Mann 1999), and aquatic levels of 1,3,5-octatriene show a correspondence with *Asterionella* population growth and decline (Jüttner et al. 1986). Today we know that the production of this compound is initiated upon wounding of the cells, but again, no clear evidence for a function in cell–cell communication is available (Pohnert 2000). As discussed above 1,3,5-octatriene may simply represent a side product formed during the synthesis of a bioactive moiety, 12-oxododeca-5,8,10-trienoic acid (Fig. 4.5).

4.7 Summary and Conclusions

The significance of FA and oxylipins in semiochemical interactions influencing food web structure and function has been demonstrated in several cases. There is compellingly evidence that the chemical ecology of FA and their metabolites has a major influence on intra and intercellular processes at all levels of the food web. There has been rapid development in this field over the past decade, with the identification of key functions of free FA (notably PUFA), the elucidation of biochemical pathways and bioactive oxylipins, and the characterization of the transfer modes and effects of these metabolites on target organisms. Nevertheless, our knowledge is, as yet, very limited, and there is a considerable need for future careful and critical research. For example, relatively few species have been investigated as producers and target organisms; numerous lipid-rich taxa and grazers remain uncharacterized. Little is known about the genetic regulation of these FA-mediated interactions. We have yet to explain the reasons for induced or innate differences among taxa or strains in production and/or susceptibility, and why/how these are affected by environment and cell energetics/growth cycles. Counter defense and avoidance mechanisms are poorly understood, and likely far more widespread that we suspect. Importantly, since PUFA are essential food components for many herbivores, interactive/counteractive effects of semiochemical and nutritional qualities of algal diets may play significant roles in grazer reproductive strategies and success. Indeed, initial results indicate that even PUA resistant copepods can suffer from FA depletion from the rapid transformation of PUFA to PUA in diatom diets (Wichard et al. 2007).

FA-mediated chemical interactions represent an exciting field, which offers considerable challenges. We have presented clear evidence that these interactions are widespread across freshwater and marine systems, where they likely play significant but as yet unmeasured roles in food web interactions. A greater insight into these (and other) chemical interactions may account for some of the unexplained variability in aquatic communities and food webs, and also provide important chemical (and in some cases, olfactory) clues to their context sensitive function. This is particularly important as global and/or local anthropogenic activities can have strong impacts on both pelagic and benthic aquatic communities. Significant shifts in the structure and integrity of aquatic food webs are occurring along with the widespread degradation of many surface waters, changes in physical and chemical aquatic regimes, the introduction of exotic biota and the loss of keystone species. Such changes will also likely have a significant affect on the scope and nature of FA-mediated and other chemical signaling. Further advances in the field will require systematic collaboration among physicists, chemists, biologists, ecologists, and modelers, along with other related disciplines such as medicine and the food and water industries where there is already a tremendous body of information.

References

Adolph, S., Poulet, S.A., and Pohnert, G. 2003. Synthesis and biological activity of $\alpha,\beta,\gamma,\delta$-unsaturated aldehydes from diatoms. Tetrahedron 59:3003–3008.

Adolph, S., Bach, S., Blondel, M., Cueff, A., Moreau, M., Pohnert, G., Poulet, S., Wichard, T., and Zuccaro, A. 2004. Cytotoxicity of diatom-derived oxylipins in organisms belonging to different phyla. J. Exp. Biol. 207:2935–2946.

Alekseev, V. and Lampert, W. 2001. Maternal control of resting-egg productoin in *Daphnia*. Nature 414:899–900.

Ban, S.H. 1994. Effect of temperature and food concentration on post-embryonic development, egg production and adult body size of the calanoid copepod *Eurytemora affinis*. J. Plankton Res. 16:721–735.

Ban, S., Burns, C., Castel, J., Chaudron, Y., Christou, E., Escribano, R., Umani, S.F., Gasparini, S., Ruiz, F.G., Hoffmeyer, M., Ianora, A., Kang, H.K., Laabir, M., Lacoste, A., Miralto, A., Ning, X., Poulet, S., Rodriguez, V., Runge, J., Shi, J., Starr, M., Uye, S., and Wang, Y. 1997. The paradox of diatom-copepod interactions. Mar. Ecol. Prog. Ser. 157:287–293.

Barofsky, A., and Pohnert, G. 2007. Biosynthesis of polyunsaturated short chain aldehydes in the diatom *Thalassiosira rotula*. Org. Lett. 9:1017–1020.

Barreiro, A., Guisande, C., Maneiro, I., Vergara, A.R., Riveiro, I., and Iglesias, P. 2007. Zooplankton interactions with toxic phytoplankton: some implications for food web studies and algal defense strategies of feeding selectivity behaviour, toxin dilution and phytoplankton population diversity. Acta Oecol. 32:279–290.

Bell, M.V., Dick, J.R., Thrush, M., and Navarro, J.C. 1996. Decreased 20:4n-6/20:5n-3 ratio in sperm from cultured sea bass, *Dicentrarchus labrax*, broodstock compared with wild fish. Aquaculture 144:189–199.

Boland, W. 1995. The chemistry of gamete attraction: chemical structures, biosynthesis, and abiotic degradation of algal pheromones. Proc. Natl Acad. Sci. U. S. A. 92:37–43.

Boland, W., Marner, F., Jaenicke, L., Mueller, D., and Foelster, E. 1983. Comparative receptor study in gamete chemotaxis of the seaweeds *Ectocarpus siliculosus* and *Cutleria multifida*: an approach to interspecific communication of algal gametes. Eur. J. Biochem. 134:97–104.

Bottrell, H.H., Duncan, A., Gliwics, Z.M., Grygierek, E., and Herzig, A. 1976. A review of some problems in zooplankton production. Norw. J. Zool. 24:419–456.

Bradow, J., and Connick, J.W. 1990. Volatile seed germination inhibitors from plant residues. J. Chem. Ecol. 16:645–667.

Bricelj, V.M., Connell, L., Konoki, K., MacQuarrie, S.P., Scheuer, T., Catterall, W.A., and Trainer, V.L. 2005. Sodium channel mutation leading to saxitoxin resistance in clams increases risk of PSP. Nature 434:763–767.

Bury, N.R., Codd, G.A., Bonga, S.E.W., and Flik, G. 1998. Fatty acids from the cyanobacterium *Microcystis aeruginosa* with potent inhibitory effects on fish gill Na^+/K^+-ATPase activity. J. Exp. Biol. 201:81–89.

Caldwell, G.S., Olive, P.J.W., and Bentley, M.G. 2002. Inhibition of embryonic development and fertilization in broadcast spawning marine invertebrates by water soluble diatom extracts and the diatom toxin 2-trans, 4-trans decadienal. Aquat. Toxicol. 60:123–137.

Caldwell, G.S., Bentley, M.G., and Olive, P.J.W. 2004a. First evidence of sperm motility inhibition by the diatom aldehyde 2E, 4E-decadienal. Mar. Ecol. Prog. Ser. 273:97–108.

Caldwell, G.S., Watson, S.B., and Bentley, M.G. 2004b. How to assess toxin ingestion and postingestion partitioning in zooplankton? J. Plankton Res. 26:1369–1377.

Caldwell, G.S., Lewis, C., Olive, P.J.W., and Bentley, M.G. 2005. Exposure to 2,4-decadienal negatively impacts upon marine invertebrate larval fitness. Mar. Environ. Res. 59:405–417.

Camazine, S.M., Resch, J.F., Eisner, T., and Meinwald, J. 1983. Mushroom chemical defense: pungent sesquiterpenoid dialdehyde antifeedant to opossum. J. Chem. Ecol. 9:1439–1447.

Carotenuto, Y., and Lampert, W. 2004. Ingestion and incorporation of freshwater diatoms by *Daphnia pulicaria*: do morphology and oxylipin production matter? J. Plankton Res. 26:563–569.

Carotenuto, Y., Wichard, T., Pohnert, G., and Lampert, W. 2005. Life-history responses of *Daphnia pulicaria* to diets containing freshwater diatoms: effects of nutritional quality versus polyunsaturated aldehydes. Limnol. Oceanogr. 50:449–454.

Casotti, R., Mazza, S., Brunet, C., Vantrepotte, V., Ianora, A., and Miralto, A. 2005. Growth inhibition and toxicity of the diatom aldehyde 2-trans, 4-trans-decadienal on *Thalassiosira weissflogii* (Bacillariophyceae). J. Phycol. 41:7–20.

Cavill, G.W.K., and Hinterberger, H. 1960. The chemistry of ants IV. Terpenoid constituents of some *Dolichoderus* and *Iridomyrmex* species. Aust. J. Chem. 13:514–519.

Cembella, A.D. 2003. Chemical ecology of eukaryotic microalgae in marine ecosystems. Phycologia 42:420–447.

Charnov, E.L. 1997. Trade-off-invariant rules for evolutionary stable life histories. Nature 387:3993–3994.

Chaudron, Y., Poulet, S.A., Laabir, M., Ianora, A., and Miralto, A. 1996. Is hatching success of copepod eggs diatom density-dependent? Mar. Ecol. Prog. Ser. 144:185–193.

Chen, W.H., and Folt, C.L. 1993. Measures of food quality as demographic predictors in freshwater copepods. J. Plankton Res. 15:1247–1261.

Chiang, I.Z., Huang, W.Y., and Wu, J.T. 2004. Allelochemicals of *Botryococcus braunii* (Chlorophyceae). J. Phycol. 40:474–480.

Chuecas, L., and Riley, J.P. 1969. Component fatty acids of the total lipids of some marine phytoplankton. J. Mar. Biol. Ass. U. K. 49:97–116.

Clare, A.S., and Walker, G. 1986. Further studies on the control of the hatching process in *Balanus balanoides* (L). J. Exp. Mar. Biol. Ecol. 97:295–304.

Cook, H.W. 1996. Fatty acid desaturation and chain elongation in eukaryotes, pp. 129–152. *In* D.E. Vance and J.E. Vance (eds.), Biochemistry of Lipids, Lipoproteins and Membranes. Elsevier, Amsterdam.

Cook, A., and Elvidge, J. 1951. Fertilization in the Fucaceae: investigations of the nature of the chemotactic substance produced by eggs of *Fucus serratus* and *F.vesiculosus*. Proc. R. Soc. Lond. Ser. B 138:97–114.

Cutignano, A., d'Ippolito, G., Romano, G., Lamari, N., Cimino, G., Febbraio, F., Nucci, R., and Fontana, A. 2006. Chloroplastic glycolipids fuel aldehyde biosynthesis in the marine diatom *Thalassiosira rotula* Chem. BioChem. 7:450–456.

Derenbach, J.B., and Pesando, D. 1986. Investigations into a small fraction of volatile hydrocarbons. iii. Two diatom cultures produce ectocarpene, a pheromone of brown algae. Mar. Chem. 19:337–342.

Dicke, M., and Sabelis, M.W. 1988. Infochemical terminology: based on cost-benefit analysis rather than origin of compounds? Funct. Ecol. 2:131–139.

Di Marzo, V., De Petrocellis, L., Gianfrani, C., and Cimino, G. 1993. Biosynthesis, structure and biological activity of hydroxyeicosatetraenoic acids in *Hydra vulgaris*. Biochem. J. 295:23–29.

Dittberner, U., Eisenbrand, G., and Zankl, H. 1995. Genotoxic effects of the α,β-unsaturated aldehydes 2-trans-butenal, 2-trans-hexenal and 2-trans, 6-cis-nonadienal. Mutat. Res. 335:259–265.

Dutz, J., Koski, M., and Jonasdottir, S.H. 2008. Copepod reproduction is unaffected by diatom aldehydes or lipid composition. Limnol. Oceanogr. 53:225–235.

Feussner, I., and Wasternack, C. 2002. The lipoxygenase pathway. Ann. Rev. Plant Biol. 53:275–297.

Findlay, J.A., and Patil, A.D. 1984. Antibacterial constituents of the diatom *Navicula delognei*. J. Nat. Prod. 47:815–818.

Fink, P., von Elert, E., and Jüttner, F. 2006. Volatile foraging kairomones in the littoral zone: attraction of an herbivorous freshwater gastropod to algal odors. J. Chem. Ecol. 32:1867–1831.

Fontana, A., d'Ippolito, G., Cutignano, A., Romano, G., Lamari, N., Gallucci, A.M., Cimino, G., Miralto, A., and Ianora, A. 2007. LOX-induced Lipid peroxidation mechanism responsible for the detrimental effect of marine diatoms on zooplankton grazers. Chembiochem: A European Journal of Chemical Biology 8:1810–1818.

Fu, M., Koulman, A., Van Rijssel, M., Luetzen, A., De Boer, M.K., Tyl, M.R., and Liebezeit, G. 2004. Chemical characterisation of three haemolytic compounds from the microalgal species *Fibrocapsa japonica* (Raphidophyceae). Toxicon 43:355–363.

Gerwick, W.H. 1994. Structure and biosynthesis of marine algal oxylipins. Biochim. Biophys. Acta 1211:243–255

Guisande, C., Maneiro, I., and Riveiro, I. 1999. Homeostasis in the essential amino acid composition of the marine copepod *Euterina acutifrons*. Limnol. Oceanogr. 44:691–696.

Hairston, N.G., Holtmeier, C.L., Lampert, W., Weider, L.J., Post, D.M., Fischer, J.M., Caceres, C.E., Fox, J.A., and Gaedke, U. 2001. Natural selection for grazer resistance to toxic cyanobacteria: evolution of phenotypic plasticity? Evolution 55:2203–2214.

Hansen, L.R., Kristiansen, J., and Rasmussen, J.V. 1994. Potential toxicity of the freshwater *Chrysochromulina* species *C. parva* (Prymnesiophyceae). Hydrobiologia 287:157–159.

Hansen, E., Even, Y., and Genevière, A.M. 2004a. The $\alpha,\beta,\gamma,\delta$-unsaturated aldehyde 2-trans-4-trans-decadienal disturbs DNA replication and mitotic events in early sea urchin embryos. Toxicol. Sci. 81:190–197.

Hansen, E., Ernstsen, A., and Eilertsen, H.C. 2004b. Isolation and characterisation of a cytotoxic polyunsaturated aldehyde from the marine phytoplankter *Phaeocystis pouchetii* (Hariot) Lagerheim. Toxicology 199:207–217.

Harrison, K.E. 1990. The role of nutrition in maturation, reproduction and embryonic development of decapod crustaceans: a review. J. Shellfish Res. 9:1–28.

Hombeck, M., and Boland, W. 1998. Biosynthesis of the algal pheromone fucoserratene by the freshwater diatom *Asterionella formosa* (Bacillariophyceae). Tetrahedron 54:11033–11042.

Hombeck, M., Pohnert, G., and Boland, W. 1999. Biosynthesis of dictyopterene A: stereoselectivity of a lipoxygenase/hydroperoxide lyase from *Gomphonema parvulum* (Bacillariophyceae). Chem. Commun. 3:243–244.

Huntley, M.E. 1982. Yellow water in La Jolla Bay, California, July 1980. II. Suppression of zooplankton grazing. J. Exp. Mar. Biol. Ecol. 63:81–91.

Huntley, M., Sykes, P., Rohan, S., and Marin, V. 1986. Chemically mediated rejection of dinoflagellate prey by the copepods *Calanus pacificus* and *Paracalanus parvus* – mechanism, occurrence and significance. Mar. Ecol. Prog. Ser. 28:105–120.

Ianora, A., Miralto, A., and Poulet, S.A. 1999. Are diatoms good or toxic for copepods? Reply to comment by Jónasdottir et al. Mar. Ecol. Prog. Ser. 177:305–308.

Ianora, A., Miralto, A., Poulet, S.A., Carotenuto, Y., Buttino, I., Romano, G., Casotti, R., Pohnert, G., Wichard, T., Colucci-D'Amato, L., Terrazzano, G., and Smetacek, V. 2004. Aldehyde suppression of copepod recruitment in blooms of a ubiquitous planktonic diatom. Nature 429:403–407.

Ikawa, M., Haney, J.F., and Sasner, J.J. 1996. Inhibition of *Chlorella* growth by the lipids of cyanobacterium *Microcystis aeruginosa*. Hydrobiologia 331:167–170.

d'Ippolito, G., Romano, G., Caruso, T., Spinella, A., Cimino, G., and Fontana, A. 2003. Production of octadienal in the marine diatom *Skeletonema costatum*. Organ. Lett. 5:885–887.

d'Ippolito, G., Tucci, S., Cutignano, A., Romano, G., Cimino, G., Miralto, A., and Fontana, A. 2004. The role of complex lipids in the synthesis of bioactive aldehydes of the marine diatom *Skeletonema costatum*. Biochim Biophys. Acta 1686:100–107.

d'Ippolito, G., Cutignano, A., Briante, R., Febbraio, F., Cimino, G., and Fontana, A. 2005. New C-16 fatty-acid-based oxylipin pathway in the marine diatom *Thalassiosira rotula*. Org. Biomol. Chem. 3:4065–4070.

d'Ippolito, G., Cutignano, A., Tucci, S., Romano, G., Cimino, G., and Fontana, A. 2006. Biosynthetic intermediates and stereochemical aspects of aldehyde biosynthesis in the marine diatom *Thalassiosira rotula*. Phytochemistry 67:314–322.

Irigoien, X., Harris, R.P., Verheye, H.M., Joly, P., Runge, J., Starr, M., Pond, D., Campbell, R., Shreeve, R., Ward, P., Smith, A.N., Dam, H.G., Peterson, W., Tirelli, V., Koski, M., Smith, T., Harbour, D., and Davidson, R. 2002. Copepod hatching success in marine ecosystems with high diatom concentrations. Nature 419:387–389.

Jeckel, W.H., de Moreno, J.E.A., Oreno, V., and Moreno, V.J. 1989. Biochemical composition, lipid classes and fatty acids in the male reproductive system of the shrimp *Pleoticus muelleri* Bate. Comp. Biochem. Physiol. B 93:807–811.

Jónasdóttir, S.H., and Kiørboe, T. 1996. Copepod recruitment and food composition: do diatoms affect hatching success? Mar. Biol. 125:743–750.

Jones, R.H., and Flynn, K.J. 2005. Nutritional status and diet composition affect the value of diatoms as copepod prey. Science 307:1457–1459.

Jüttner, F. 1981. Detection of lipid degradation products in the water of a reservoir during a bloom of *Synura uvella*. Appl. Environ. Microbiol. 41:100–106.

Jüttner, F. 1984. Dynamics of the volatile organic substances associated with cyanobacteria and algae in a eutrophic shallow lake. Appl. Environ. Microbiol. 47:815–820.

Jüttner, F. 1992. Flavour compounds in weakly polluted rivers as a means to differentiate pollution sources. Water Sci. Technol. 25:155–164.

Jüttner, F. 1995. Physiology and biochemistry of odourous compounds from freshwater cyanobacteria and algae. Water Sci. Technol. 31:69–78.

Jüttner, F. 2001. Liberation of 5,8,11,14,17-eicosapentaenoic acid and other polyunsaturated fatty acids from lipids as a grazer defense reaction in epilithic diatom biofilms. J. Phycol. 37:744–755.

Jüttner, F. 2005. Evidence that polyunsaturated aldehydes of diatoms are repellents for pelagic crustacean grazers. Aquatic Ecol. 39:271–282.

Jüttner, F., and Müller, H. 1979. Excretion of octadiene and octatrienes by a freshwater diatom. Naturwissenschaften 66:363–364.

Jüttner, F., and Dürst, U. 1997. High lipoxygenase activities in epilithic biofilms of diatoms. Archiv. Hydrobiol. 138:451–463.

Jüttner, F., Hoflacher, B., and Wurster, K. 1986. Seasonal analysis of volatile organic biogenic substances in freshwater phytoplankton populationsdominated by *Dinobryon Microcystis* and *Aphanizomenon*. J. Phycol. 22:169–175.

Kainz, M., Arts, M.T., and Mazumder, A. 2004. Essential fatty acids in the planktonic food web and their ecological role for higher trophic levels. Limnol. Oceanogr. 49:1784–1793.

Kamiya, H., Naka, K., and Hashimoto, K. 1979. Ichthyotoxicity of a flagellate *Uroglena volvox*. Bull. Jpn Soc. Sci. Fish. 45:129.

Kieber, R., Hydro, L., and Seaton, P. 1997. Photooxidation of triglycerides and fatty acids in seawater: implication toward the formation of marine humic substances. Limnol. Oceanogr. 42:1454–1462.

Kleppel, G.S. 1993. On the diets of calanoid copepods. Mar. Ecol. Prog. Ser. 99:183–195.

Koski, M., Wichard, T., and Jónasdóttir, S. 2008. "Good" and "bad" diatoms: development, growth and juvenile mortality of the copepod *Temora longicornis* on diatom diets. Mar. Biol. 154:719–734.

Kubo, I., Lee, Y.-W., Pettei, M., Pilkiewicz, F., and Nakanishi, K. 1976. Potent army worm antifeedants from the east African Warburgia plants. Chem. Commun. 24:1013–1014.

Laabir, M., Poulet, S.A., Ianora, A., Miralto, A., and Cueff, A. 1995. Reproductive response of *Calanus helgolandicus*. II. *In situ* inhibition of embryonic development. Mar. Ecol. Prog. Ser. 129:97–105.

Laabir, M., Poulet, S.A., Cueff, A., and Ianora, A. 1999. Effect of diet on levels of amino acids during embryonic and naupliar development of the copepod *Calanus helgolandicus*. Mar. Biol. 134:89–98.

LaMontagne, J.M., and McCauley, E. 2001. Maternal effects in *Daphnia*: what mothers are telling their offspring and do they listen? Ecol. Lett. 4:64–71.

Leeper, D.A., and Porter, K.G. 1995. Toxicity of the mixotrophic chrysophyte *Poterioochromonas malhamensis* to the cladoceran *Daphnia ambigua*. Arch. Hydrobiol. 134:207–222.

Lewis, C., Caldwell, G.S., Bentley, M.G., and Olive, P.J.W. 2004. Effects of a bioactive diatom-derived aldehyde on developmental stability in *Nereis virens* (Sars) larvae: an analysis using fluctuating asymmetry. J. Exp. Mar. Biol. Ecol. 304:1–16.

Lincoln, J.A., Turner, J.T., Bates, S.S., Leger, C., and Gauthier, D.A. 2001. Feeding, egg production, and *egg* hatching success of the copepods *Acartia tonsa* and *Temora longicornis* on diets of the toxic diatom *Pseudo-nitzschia multiseries* and the non-toxic diatom *Pseudo-nitzschia pungens*. Hydrobiologia 453:107–120.

Loureiro, A.P.M., Di Masco, P., Gomes, O.F., and Medeiros, M.H.G. 2000. trans,trans-2,4-Decadienal-induced 1,*N*-2-etheno-2'-deoxyguanosine adduct formation. Chem. Res. Toxicol. 13:601–609.

Mann, D. 1999. The species concept in diatoms. Phycologia 38:437–495.

Matsui, K. 2006. Green leaf volatiles: hydroperoxide lyase pathway of oxylipin metabolism. Curr. Opin. Plant Biol. 9:274–280.

Matsutani, T., and Nomura, T. 1987. *In vitro* effects of serotonin and prostaglandins on release of eggs from the ovary of the scallop, *Patinopecten yessoensis*. Gen. Comp. Endocrinol. 67:111–118.

McGrattan, D., Sullivan, J., and Ikawa, M. 1976. Inhibition of *Chlorella* (Chlorophyceae) growth by fatty acids, using the paper disk method. J. Phycol. 12:129–131.

Meijer, L., Brash, A., Bryant, R., Ng, K., Maclouf, J., and Sprecher, H. 1986. Stereospecific induction of starfish oocyte maturation by (8r)-hydroxyeicosatetraenoic acid. J. Biol. Chem. 261:7040–7047.

Miralto, A., Barone, G., Romano, G., Poulet, S., Ianora, A., Russo, G., Buttino, I., Mazzarella, G., Laabir, M., Cabrini, M., and Giacobbe, M. 1999. The insidious effect of diatoms on copepod reproduction. Nature 402:173–176.

Morimoto, M., Tanimoto, K., Sakatani, A., and Komai, K. 2002. Antifeedant activity of an anthraquinone aldehyde in *Galium aparine* L. against *Spodoptera litura*. Phytochemistry 60:163–166.

Murakami, M., Makebe, K., Yamaguchi, K., and Konosu, S. 1989. Cytotoxic polyunsaturated fatty acid from *Pediastrum*. Phytochemistry 28:625–626.

Okamoto, T., and Katoh, S. 1977. Linolenic acid binding by chloroplasts. Plant Cell Physiol. 18:539–550.

Okamoto, T., Katoh, S., and Murakuami, S. 1977. Effects of linolenic acid on spinach chloroplast structure. Plant Cell Physiol. 18:551–560.

Olive, P.J.W. 1992. The adaptive significance of seasonal reproduction in marine invertebrates: the importance of distinguishing between models. Invertebr. Reprod. Dev. 22:165–174.

Paffenhöfer, G.A., Ianora, A., Miralto, A., Turner, J., Kleppel, G.S., Ribera d'Alcalà, M., Casotti, R., Caldwell, G., Pohnert, G., Fontana, A., Müller-Navarra, D., Jónasdóttir, S., Armbrust, V., Båmstedt, U., Ban, S., Bentley, M.G., Boersma, M., Bundy, M., Buttino, I., Calbet, A., Carlotti, F., Carotenuto, Y., d'Ippolito, G., Frost, B., Guisande, C., Lampert, W., Lee, R., Mazza, S., Mazzocchi, M., Nejstgaard, J.C., Poulet, S.A., Romano, G., Smetacek, V., Uye, S., Wakeham, S., Watson, S., and Wichard, T. 2005. Colloquium on diatom–copepod interactions. 2005. Mar. Ecol. Prog. Ser. 286:293–305.

Pohnert, G. 2000. Wound-activated chemical defense in unicellular planktonic algae. Angew. Chem. Int. Ed. Engl. 39:4352–4354.

Pohnert, G. 2002. Phospholipase A2 activity triggers the wound-activated chemical defense in the diatom *Thalassiosira rotula*. Plant Physiol. 129:103–111.

Pohnert, G. 2005. Diatom/copepod interactions in plankton: the indirect chemical defense of unicellular algae. Chem. BioChem. 6:946–959.

Pohnert, G., and Boland, W. 1996. Biosynthesis of the algal pheromone hormosirene by the freshwater diatom *Gomphonema parvulum* (Bacillariophyceae). Tetrahedron 52:10073–10082.

Pohnert, G., and Boland, W. 2002. The oxylipin chemistry of attraction and defense in brown algae and diatoms. Nat. Prod. Rep. 19:108–122.

Pohnert, G., Adolph, S., and Wichard, T. 2004. Short synthesis of labeled and unlabeled 6Z,9Z,12Z,15-hexadecatetraenoic acid as metabolic probes for biosynthetic studies on diatoms. Chem. Phys. Lipids 131:159–166.

Poulet, S.A., and Marsot, P. 1978. Chemosensory grazing by calanoid copepods (Arthropoda: Crustacea). Science 200:1403–1405.

Poulet, S.A., Ianora, A., Miralto, A., and Meijer, L. 1994. Do diatoms arrest embryonic development in copepods? Mar. Ecol. Prog. Ser. 111:79–86.

Poulet, S.A., de Forges, M.R., Cueff, A., and Lennon, J.F. 2003. Double-labelling methods used to diagnose apoptotic and necrotic cell degradations in copepod nauplii. Mar. Biol. 143:889–895.

Poulet, S.A., Cueff, A., Wichard, T., Marchetti, J., Dancie, C., and Pohnert, G. 2007. Influence of diatoms on copepod reproduction. III. Consequences of abnormal oocyte maturation on reproductive factors in *Calanus helgolandicus*. Mar. Biol. 152:415–428.

Reinikainen, M., Meriluoto, J.A.O., Spoof, L., and Harada, K. 2001. The toxicities of a polyunsaturated fatty acid and a microcystin to *Daphnia magna*. Exp. Toxicol. 16:444–448.

Rengefors, K., and Legrande, C. 2001. Toxicity in *Peridinium aciculiferum* – an adaptive strategy to outcompete other winter phytoplankton? Limnol. Oceanogr. 46:1990–1997.

Ribalet, F., Wichard, T., Pohnert, G., Ianora, A., Miralto, A., and Casotti, R. 2007a. Age and nutrient limitation enhance polyunsaturated aldehyde production in marine diatoms. Phytochemistry 68:2059–2067.

Ribalet, F., Berges, J.A., Ianora, A., Casotti, R. 2007b. Growth inhibition of cultured marine phytoplankton by toxic algal-derived polyunsaturated aldehydes. Aquat. Toxicol. 85:219–227.

Romano, G., Russo, G.L., Buttino, I., Ianora, A., and Miralto, A. 2003. A marine diatom-derived aldehyde induces apoptosis in copepod and sea urchin embryos. J. Exp. Biol. 2006:3487–3494.

Rukmini, C. 1990. Reproductive toxicology and nutritional studies on manhua oil (*Madhuca latifolia*). Food Chem. Toxicol. 28:601–605.

Sandgren C. 1988. The ecology of chrysophyte flagellates: their growth and perenniation strategies as freshwater phytoplankton, pp. 99–105. In Sandgren C. (ed.), Growth and Reproductive Strategies of Freshwater Phytoplankton. Cambridge University Press, Cambridge.

Schnitzler, I., Pohnert, G., Hay, M., and Boland, W. 2001. Chemical defense of brown algae (*Dictyopteris* spp.) against the herbivorous amphipod *Ampithoe longimana*. Oecologia 126:515–521.

Schuel, H., Moss, R., and Schuel, R. 1985. Induction of polyspermic fertilization in sea urchins by the leukotriene antagonist FPL-55712 and the 5-lipoxygenase inhibitor BW755C. Gamete Res. 11:41–50.

Scutt, J. 1964. Autoinhibition production by *Chorella vulgaris*. Am. J. Bot. 51:581–584.

Sellem, F., Pesando, D., Bodennec, G., El Abed, A., and Girard, J.P. 2000. Toxic effects of *Gymnodinium* cf. *mikimotoi* unsaturated fatty acids to gametes and embryos of the seaurchin *Paracentrotus lividus*. Water Res. 34:550–556.

Shilo, M. 1981. The toxic principles of *Prymnesium parvum*, pp. 37–47. In W.W. Carmichael (ed.), Algal Toxins and Health. Plenum, New York.

Stanley-Samuelson, D.W. 1994. The biological significance of prostaglandins and related eicosanoids in invertebrates. Am. Zool. 34:589–598.

Støttrup, J.G., and Jensen, J. 1990. Influence of algal diet on feeding and *egg* production of the calanoid copepod *Acartia tonsa* Dana. J. Exp. Mar. Biol. Ecol. 141:87–105.

Taylor, F.J.R. 1985. The taxonomy and relationships of red tide flagellates, pp. 11–26. *In* D. Anderson, A. White, and D. Baden (eds.), 3rd International Conference on Toxic Dinoflagellates. Elsevier, New York.

Tosti, E., Romano, G., Buttino, I., Cuomo, A., Ianora, A., and Miralto, A. 2003. Bioactive aldehydes from diatoms block the fertilization current in ascidian oocytes. Mol. Reprod. Dev. 66:72–80.

Trombetta, D., Saija, A., Bisignano, G., Arena, S., Caruso, S., Mazzanti, G., Uccella, N., and Castelli, F. 2002. Study on the mechanisms of the antibacterial action of some plant alpha,beta-unsaturated aldehydes. Lett. Appl. Microbiol. 35:285–290.

Tsubo, Y. 1961. Chemotaxis and sexual behavior in *Chlamydomonas*. J. Protozool. 8:114–121.

Vardi, A., Formiggini, F., Casotti, R., Martino, A., Ribalet, F., Miralto, A., and Bowler, C. 2006. A stress surveillance system based on calcium and nitric oxide in marine diatoms. PLOS Biol. 4:411–419.

Walton, W., and Sandgren, C. 1995. The influence of zooplankton herbivory on the biogeography of chrysophyte algae, pp. 269–302. In C. Sandgren, J. Smol, and J. Kristiansen (eds.), Chrysophyte Algae. Cambridge University Press, Cambridge.

Watson, S.B. 2003. Cyanobacterial and eukaryotic algal odour compounds: signals or by-products? A review of their biological activity. Phycologia 42:332–350.

Watson, S.B., and Satchwill, T. 2003. Chrysophyte odour production: the impact of resources at the cell and population levels. Phycologia 42:393–405.

Watson, S.B., Satchwill, T., and McCauley, E. 2000. Drinking water taste and odour: a chrysophyte perspective. Nova Hedwigia 122:119–146.

Watson, S.B., Satchwill, T., and McCauley, E. 2001. Under-ice blooms and source-water odour in a nutrient-poor reservoir: biological, ecological and applied perspectives. Freshwater Biol. 46:1–15.

Watson, S.B., Jüttner, F., and Köster, O. 2007. *Daphnia* behavioural responses to taste and odour compounds: ecological significance and application as an inline treatment plant monitoring tool. Water Sci. Technol. 55:23–31.

Wee, J.L., Harris, S.A., Smith, J.P., Dionigi, C.P., and Millie, D.F. 1994. Production of the taste/odour compound, trans-2, cis-6-nonadienal within the Synurophyceae. J. Appl. Phycol. 6:365–369.

Wendel, T., and Jüttner, F. 1996. Lipoxygenase-mediated formation of hydrocarbons and unsaturated aldehydes in freshwater diatoms. Phytochemistry 41:1445–1449.

White, A.W. 1981. Marine zooplankton can accumulate and retain dinoflagellate toxins and cause fish kills. Limnol. Oceanogr. 26:103–109.

Wichard, T., and Pohnert, G. 2006. Formation of halogenated medium chain hydrocarbons by a lipoxygenase/hydroperoxide halolyase-mediated transformation in planktonic microalgae. J. Am. Chem. Soc. 128:7114–7115.

Wichard, T., Gobel, C., Feussner, I., and Pohnert, G. 2005a. Unprecedented lipoxygenase/hydroperoxide lyase pathways in the moss *Physcomitrella patens*. Angew. Chem. Int. Ed. Engl. 44:158–161.

Wichard, T., Poulet, S., Halsband-Lenk, C., Albaina, A., Harris, R., Liu, D., and Pohnert, G. 2005b. Survey of the chemical defense potential of diatoms: screening of fifty one species for $\alpha,\beta,\lambda,\delta$ unsaturated aldehydes. J. Chem. Ecol. 31:949–958.

Wichard, T., Poulet, S.A., and Pohnert, G. 2005c. Determination and quantification of alpha,beta,gamma,delta-unsaturated aldehydes as pentafluorobenzyl-oxime derivates in diatom cultures and natural phytoplankton populations: application in marine field studies. J. Chromatogr. B: Anal. Technol. Biomed. Life Sci. 814:155–161.

Wichard, T., Gerecht, A., Boersma, M., Poulet, S.A., Wiltshire, K., and Pohnert, G. 2007. Lipid and fatty acid composition of diatoms revisited: rapid wound-activated change of food quality parameters influences herbivorous copepod reproductive success. Chembiochem: A European Journal of Chemical Biology 8:1146–1153.

Wichard, T., Poulet, S.A., Boulesteix, A.L., Ledoux, J.B., Lebreton, B., Marchetti, J., and Pohnert, G. 2008. Influence of diatoms on copepod reproduction. II. Uncorrelated effects of diatom-derived α,β,χ,δ-unsaturated aldehydes and polyunsaturated fatty acids on *Calanus helgolandicus* in the field. Prog. Oceanogr. 77:30–44.

Wood, W.F. 1983. Chemical ecology: chemical communication in nature. J. Chem. Ed. 60:531–539.

Wu, J.T., Chiang, Y.R., Huang, W.Y., and Jane, W.N. 2006. Cytotoxic effects of free fatty acids on phytoplankton algae and cyanobacteria. Aquat. Toxicol. 80:338–345.

Yamada, N., Murakami, N., Kawamura, N., and Sakakibara, J. 1994. Mechanism of an early lysis by fatty acids from axenic *Phormidium tenue* (musty odor-producing cyanobacterium) and its growth prolongation by bacteria. Biol. Pharm. Bull. 17:1277–1281.

Yasumoto, T., Underdal, B., Aune, T., Hormazabal, V., Skulberg, O.M., and Oshima, Y. 1990. Screening for hemolytic and ichthyotoxic components of *Chrysochromulina polylepis* and *Gyrodinium aureolum* from Norwegian coastal waters, pp. 436–440. *In* E. Graneli, B. Sundström, L. Edler, and D.M. Anderson (eds.), 4th International Conference on Toxic Marine Phytoplankton. Elsevier, New York.

Yongmanitchai, W., and Ward, O.P. 1991. Screening of algae for potential alternative sources of eicosapentaenoic acid. Phytochemistry 31:2963–2967.

Chapter 5
Integrating Lipids and Contaminants in Aquatic Ecology and Ecotoxicology

Martin J. Kainz and Aaron T. Fisk

5.1 Introduction

Heterotrophic organisms in marine and freshwater food webs ingest a wide range of essential and xenobiotic compounds. *Essential* compounds are physiologically required by consumers, yet cannot be synthesized de novo, or cannot be synthesized in quantities sufficient to meet an organism's need for somatic growth, reproduction, and survival (*see* Goulden and Place 1990, for daphnids; Tocher 2003, for teleost fishes). For example, some polyunsaturated fatty acids (PUFA) and trace elements such as zinc (Zn), iron (Fe), calcium (Ca) are considered essential, and if inadequate amounts are available in the diet, the health and fitness of an organism can be reduced. *Xenobiotic* compounds have no physiological value for organisms, but can be accumulated by consumers and can be toxic in cases were concentrations are sufficiently high (Watson et al. – Chap. 4). Xenobiotic compounds include many of the classic contaminants, such as PCBs, DDT, and mercury (Hg), and more recently recognized contaminants, such as estradiol, and can also be accumulated from non-dietary sources. It should be noted that essential compounds can also be toxic if concentrations are high enough or if they are converted to other molecules. For example, it has been suggested that PUFA in diatoms can be converted to unsaturated aldehydes, which reduce egg hatching rates in marine herbivorous copepods (Miralto et al. 1999).

Lipids are recognized to be amongst the most important nutritional factors that affect the fitness of aquatic organisms, supplying energy and essential compounds for general metabolic functioning, somatic growth and reproduction (Müller-Navarra et al. 2000), and enhanced immunocompetency (Kiron et al. 1995). Among lipid classes, storage lipids, such as triacylglycerols, serve as high-energy sources, whereas structural lipids, such as phospholipids, are essential building blocks for cell membranes. Essential lipids are of particular nutritional importance for aquatic consumers as they support physiological development and health of organisms and, in a larger

---

M.J. Kainz (✉) and A.T. Fisk
WasserKluster Lunz – Biologische Station, Dr. Carl Kupelwieser Promenade 5,
A-3293 Lunz am See, Austria
e-mail: martin.kainz@donau-uni.ac.at

sense, strengthen the nutritional status of aquatic food webs. Although de novo synthesis of omega-3 (n-3) PUFA in aquatic ecosystems is generally restricted to algae, it has been reported that some heterotrophic nanoflagellates (Bec et al. 2006; and see Devilettes and Bec – Chap. 2) and protozoans (Klein Breteler et al. 1999) have the enzymatic ability to produce n-3 PUFA, as well as sterols, from their respective precursors. Thus, such key organisms from lower trophic levels can be involved in *trophically upgrading* their food for their own physiological benefit as well as, inadvertently, for the benefit of consumers at higher trophic levels (Fig. 5.1). In aquatic consumers, such conversion may occur when there is conditional dietary need for long-chain polyunsaturates (Cunnane 2003). It is, however, important to note that biosynthesis of fatty acids (FA) is greater in organisms at the base of the aquatic food chain, whereas higher trophic organisms such as zooplankton and fish are not likely to biosynthesize highly unsaturated fatty acids (HUFA) de novo to any significant extent (Tocher 2003; Chap. 9). Thus, the *trophic transfer* of essential

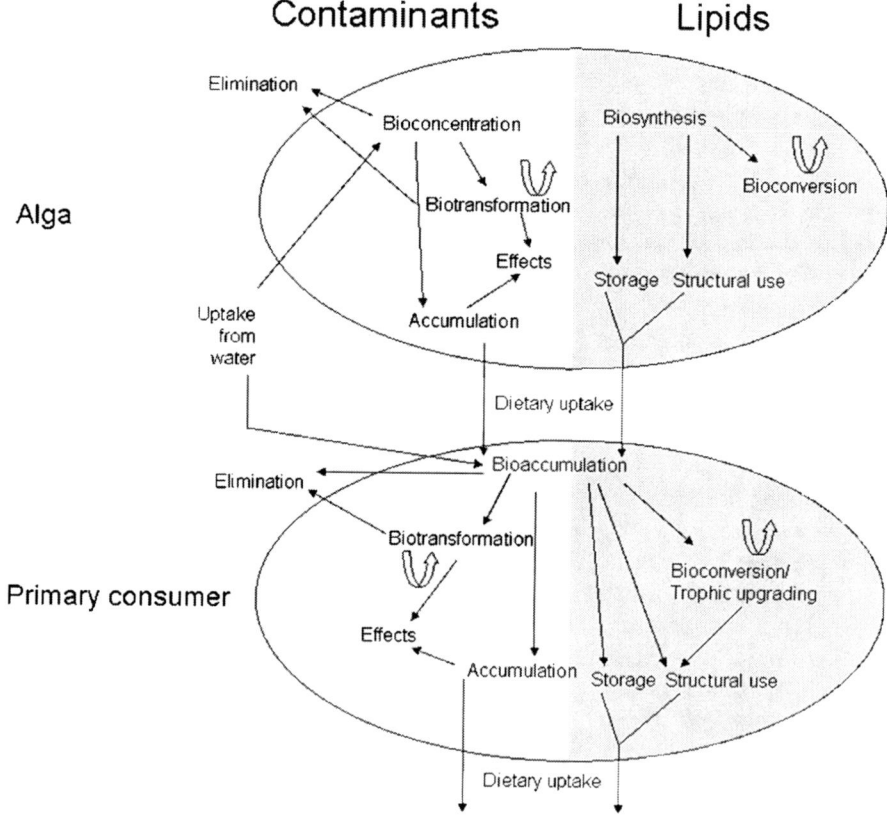

Fig. 5.1 Schematic of the relevant processes controlling the movement of lipids and contaminants through aquatic food webs. The relative importance of various processes and mechanisms can vary widely with the contaminant or lipid and organisms in question. Processes and pathways in the primary consumer are similar in secondary and higher level consumers

lipids from primary producers to upper trophic levels is critical for the health of higher trophic level organisms. Trophic transfer is also commonly used when discussing contaminants, and refers to the movement of a contaminant from resource to consumer (Borgå et al. 2004). The presence of organisms capable of trophically upgrading FA within the food web is often crucial for supplying a diet rich in nutritionally beneficial PUFA and sterols and changes to food web structure and components may alter this important process.

Xenobiotic chemicals, which we will also refer to as contaminants, in the diet of animals come from a variety of sources, with those of anthropogenic origin that bioaccumulate in aquatic organisms of particular concern and the focus of this chapter. Contaminants that are a concern for aquatic food webs include: metals and metalloids (such as Al, Ar, Cd, Cr, Cu, Ni, Pb, Hg, Se); organic contaminants including polycyclic aromatic hydrocarbons (PAH), polychlorinated biphenyls (PCB), polybrominated biphenyls (PBB), chlorinated phenols, pesticides, polychlorinated dibenzodioxins (PCDD) and dibenzofurans (PCDF), and other industrial chemicals (e.g., perflourinated alkanes); and organometallic compounds such as the potentially powerful neurotoxin methyl mercury (MeHg). Contaminants that bioaccumulate have the potential to counteract the mostly favorable physiological effects of essential dietary nutrients, particularly at higher trophic levels, and eventually in humans, because their dietary pathways, or trophic transfer, may follow similar pathways as those of lipids (Newman 1998; for essential fatty acids, see Kainz et al. 2006, 2008).

Most lipid ecologists have focused on increasing our understanding of lipid synthesis and distribution (Hagen and Auel 2001; Graeve et al. 2005; Guschina and Harwood 2006), the trophic relationships of lipids (Dalsgaard et al. 2003; Iverson et al. 2004; Kainz et al. 2004), and the physiological effects of different lipid classes and their constituent FA on aquatic organisms (Müller-Navarra et al. 2000; Tocher 2003; Martin-Creuzburg and von Elert 2004). Similarly, most aquatic ecotoxicologists focus their research on how contaminants move through aquatic food webs (Campfens and MacKay 1997; Russell et al. 1999; Borgå et al. 2004) and how they affect the physiological performance of organisms (Wang and Fisher 1999; Tanabe 2002; Scott and Sloman 2004). Since most FA and contaminants are trophically transferred through aquatic food webs, but potentially have very different effects and relevance, there is a need to understand how the fate of these two chemical groups varies across food webs and amongst different aquatic ecosystems. It is, therefore, important to investigate sources, biological uptake, biotransformation, physiological implications, accumulation and elimination of both lipids and contaminants within aquatic food webs. For example, changes in food webs associated with increasing stress from climate change, invasive species, and habitat destruction may result in significant changes in contaminant and lipid dynamics within ecosystems (Kelly et al. 2006). Furthermore, differences in FA and contaminant trophic transfer in food webs have the potential to provide novel insights on ecological function and the influence and effects of stress. Thus, the main objectives of this chapter are to evaluate and contrast trophic transfer of dietary lipids and contaminants in aquatic food webs. In doing so we will examine their dynamics in aquatic ecosystems and demonstrate their potential for assessing and predicting aquatic food web health, structure, and function.

5.2 Trophic Transfer of Contaminants in Aquatic Food Webs

The bioaccumulation and trophic transfer of contaminants is well studied and has been summarized in a number of recent reviews (e.g., Morel et al. 1998; Gobas and Morrison 2000, Borgå et al. 2004). Here, we summarize some major factors that influence contaminant bioaccumulation that are relevant to the goals of this review.

A key point regarding the accumulation of contaminants is that they can be accumulated by aquatic organisms, (a) directly from the water, called *bioconcentration* (the net process by which the chemical concentration in an aquatic organism achieves a level that exceeds that in the water, as a result of chemical uptake through chemical exposure in water) or (b) via food and water, defined as *bioaccumulation* (the net process by which the chemical concentration in an aquatic organism achieves a level that exceeds that in the water, as a result of chemical uptake through all possible chemical exposure, i.e., water or food, and elimination from all possible routes) (Gobas and Morrison 2000). This differs from dietary lipids, which can only be acquired from the diet. While lipids can by synthesized by aquatic organisms, contaminants cannot, although some can be modified via biotransformation (e.g., Konwick et al. 2006). The relative importance of food or water as a source of contaminant accumulation is highly dependent on the contaminant and organism in question, and for metals/elements the characteristics of the water, and is discussed below. The accumulation of organic contaminants and elements/metals are very different and are dealt with separately.

5.2.1 Trophic Transfer of Organic Contaminants

Most organic contaminants (OCs), such as PCBs and pesticides, are hydrophobic, and this drives their accumulation by aquatic organisms (Mackay 1982). Accumulation of OCs is generally considered to be a passive process, driven by diffusion and differences in fugacities of a chemical when in different matrices (e.g., water and aquatic organisms; Mackay and Paterson 1981). The fugacity of a chemical can be defined as a molecule's urge to escape or flee a system and is based on the differences in chemical potentials of a contaminant between matrices (Mackay and Paterson 1981). Contaminants will flow from high to low fugacity, with fugacity based on a combination of the properties of the contaminant and matrix. For example, at equal concentrations in water and lipid, contaminants that are hydrophobic (e.g., PCBs) will have a much higher fugacity in the water and thus will diffuse from water to lipids until the fugacities in each matrix are equal. Concentrations of PCBs can be as much as 7 orders of magnitude greater in the lipids than in the water when fugacities are equal. Even when OCs are not in equilibrium between matrices, which is the most common case, their concentrations are in general much higher in aquatic organisms than water, which is consistent with their hydrophobic and lipophilic characteristics.

Passive diffusion of contaminants into organisms can also occur in the gastrointestinal (GI) tract of an organism, although it is somewhat different than accumulation from water. As lipids are broken down in the GI tract, they form micelles that diffuse across the intestinal wall (Gobas et al. 1993; Kelly et al. 2004). Hydrophobic contaminants in the stomach contents are often transported along with the micelles or may diffuse into the cell walls directly (Burreau et al. 1997). Regardless of the mechanism, the movement of contaminants from the GI contents to the organism is, for the vast majority of contaminants of concern, a passive process.

The potential for bioaccumulation increases with increasing levels of hydrophobicity and slower elimination rates. Increasing hydrophobicity eventually results in diet being a much more important exposure route for heterotrophic aquatic organisms when compared with water (Thomann 1981). For phytoplankton, bioconcentration is the only mechanism of contaminant accumulation. In bioaccumulation studies, the hydrophobicity of OCs is most often evaluated using the log octanol–water partition coefficient (log K_{ow}). This coefficient measures how hydrophilic ("water loving") or hydrophobic ("water fearing") a chemical is using octanol as a surrogate for lipids (Finizio et al. 1997). Differences in K_{ow} between OCs are generally due to changes in water solubility; most OCs are highly octanol soluble, and hence highly lipid soluble, but differ substantially in their water solubility (Mackay et al. 2000). However, log K_{ow} and water solubility correlations are rarely 1:1 and can vary substantially among groups of contaminants (Schwarzenbach et al. 2003). Log K_{ow} also provides a fairly accurate quantitative prediction of bioconcentration; relationships between log K_{ow} and log bioconcentration factors (BCF; [organism]$_{lipid}$/[water]) for recalcitrant compounds are generally 1:1 (Mackay 1982; Fox et al. 1994; Finizio et al. 1997; Fig. 5.2a). Log K_{ow} values have also been used to evaluate other parameters such as biomagnifications factors (BMF = [predator]$_{lipid}$/[prey]$_{lipid}$) (Fisk et al. 1998, 2001) (Fig. 5.2b). The log K_{ow} of most OCs range from 3 to 8 (Mackay et al. 2000), and OCs with a value of ≥5 are accumulated almost completely via the diet (Thomann 1981).

Once an OC is assimilated by an organism, its fate, or bioaccumulation potential, is determined largely by a combination of its hydrophobicity and susceptibility to biotransformation. OCs can be eliminated either via diffusion, which is passive and controlled by the physico-chemical properties of the chemical, or via biotransformation by the organism, which is active and varies with the organism and environmental conditions (particularly temperature; Buckman et al. 2007). The more hydrophobic an OC is, the longer it is retained (i.e., longer half-life) and the greater its bioaccumulation (Fisk et al. 1998), but if the chemical is biotransformed it will be eliminated more quickly (i.e., short half life) and will have lower bioaccumulation (Fisk et al. 2000). However, the metabolite of the contaminant may itself bioaccumulate; sometimes to a greater extent than the parent compound (Konwick et al. 2006). These concepts are demonstrated by examining the bioaccumulation of PAHs and PCBs. Both of these contaminant groups have similarly high hydrophobicities (log K_{ow} range from 5 to 8) (Mackay et al. 2000). PCBs are highly bioaccumulated by fish because most congeners cannot be biotransformed (Kwon et al. 2006; Wong et al. 2004); however, PAHs do not achieve high concentrations in fish because they are readily biotransformed (D'Adamo et al. 1997). PAHs are also readily bioaccumulated

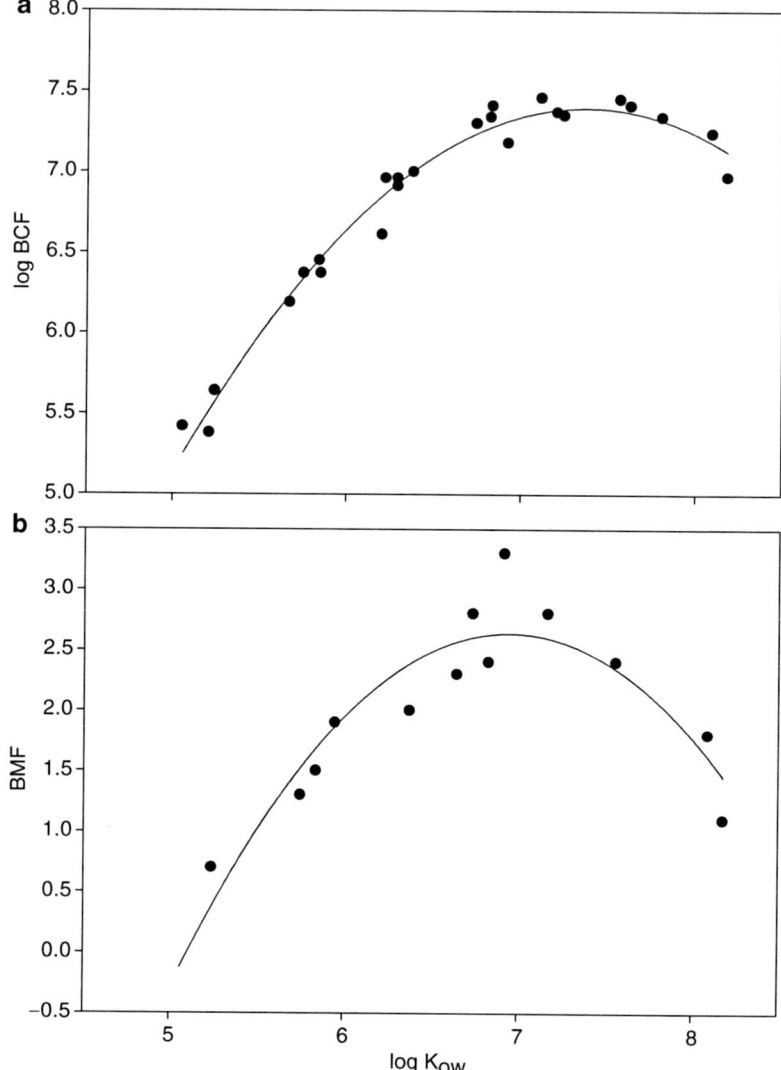

Fig. 5.2 Relationships between log K_{ow} and log BCF (**a**; log BCF = −14.5 + 5.9 * log K_{ow} − 0.4 log K_{ow}^2, $r^2 = 0.99$, $p < 0.0001$) and BMF (**b**; BMF = −34.8 + 10.0 * log K_{ow} − 0.8 log K_{ow}^2, $r^2 = 0.91$, $p < 0.001$) from laboratory studies. BCF data is from Fox et al. (1994) for zebrafish (*Brachydanio rerio*), BMF data from Fisk et al. (1998) for juvenile rainbow trout (*Oncorhyncus mykiss*) and K_{ow} from Hawker and Connel (1988)

by many classes of invertebrates, notably mussels, because they lack the enzymes that biotransform PAHs (D'Adamo et al. 1997). Note, although not accumulated, PAHs and their metabolites can, nonetheless, influence the health of fish (see e.g., Incardona et al. 2006).

Finally, if a chemical is sufficiently hydrophobic (log K_{ow} > ~4; Fisk et al. 2001) and recalcitrant (i.e., cannot be biotransformed), it will have a tendency to biomagnify through food webs. *Biomagnification* is the increase in chemical concentration with each trophic level transformation in the food web (Gobas and Morrison 2000), resulting in the highest concentrations in the upper trophic levels (Fisk et al. 2001). It should be stressed that even if a contaminant does not biomagnify, dietary exposure may still be the most important exposure route for aquatic organisms (Borgå et al. 2004).

5.2.2 Factors that Influence Organic Contaminant Bioaccumulation

A number of biological factors influence bioaccumulation and trophic transfer of OCs, and detailed reviews on this subject have been previously published (e.g., Borgå et al. 2004). These are briefly summarized here because many of these factors will also influence the trophic transfer of individual lipid compounds, although in different ways, and assessments of lipid and contaminant dynamics need to account for the fact that differences between food webs or organisms within the food webs may result in variation in contaminant and lipid behavior.

Most OCs of interest in aquatic food webs are highly lipid soluble, and water insoluble, and significantly influenced by the organism's lipid dynamics. In fact, comparative studies on OC levels in aquatic biota often account for the influence of lipids by either (a) lipid normalizing the concentration ([OC]/lipid content) (Thomann 1989), (b) using total lipid content as a covariate in statistical models (Hebert and Kennleyside 1995), or (c) using the residuals of the regression of OC levels on lipids for further analysis (Hebert and Kennleyside 1995, Hop et al. 2002). All of these approaches assume that the accumulation of OCs is linearly correlated with the organism's total lipid content, although in phytoplankton organic carbon can be used (see e.g., Swackhamer and Skoglund 1993). However, the types of lipids in the organism are rarely considered, even though this can influence contaminant uptake dynamics. For example, marine zooplankton store energy as dense wax esters in addition to triacylglycerols (Hagen and Auel 2001; Scott et al. 2002). In general, linear relationships between total lipid content and OC concentrations can be found within a population, although this can vary seasonally (Greenfield et al. 2005). Most parameters that quantify lipophilic OC bioaccumulation (e.g., BMFs) or trophic transfer (food web studies, see below) use lipid normalization to remove the influence of variable lipid contents (e.g., Fisk et al. 2001). This is particularly important in food web studies as total lipid concentrations can vary substantially between organisms at different trophic levels (Kidd et al. 1998).

For OCs that bioaccumulate and biomagnify, concentrations generally increase with body size and age in aquatic invertebrates and fish (Fisk et al. 2003; McIntyre and Beauchamp 2007). Age and size of fish are often highly correlated within a fish population (Johnston et al. 2002) but can often vary among populations, ecosystems, or temporally, even for the same species. The relationship between age or size and OC concentration will also vary with the contaminant. Very hydrophobic OCs (log K_{ow} > 6) are generally found to increase throughout a fish's life; that is, they may never achieve equilibrium between the fish and its environment (Paterson et al. 2006). Whereas, more moderately hydrophobic OCs (log K_{ow} < 6) may reach an equilibrium between fish and the environment and not increase in concentration once a specific age or size is reached (Paterson et al. 2006). Hidden within the age or size and OC relationships, but rarely acknowledged, is the influence of growth rate (*but see* Trudel and Rasmussen 2006, Paterson et al. 2006). The high growth rates commonly seen in the early spring and summer periods in temperate aquatic systems will generally decrease the observed OC concentrations per unit biomass (Paterson et al. 2006). During the late summer, autumn, and winter, when feeding decreases, growth rates decline or even stop and OC concentrations will increase because body mass decreases and slow OC elimination rates lag behind (Paterson et al. 2007).

Temperature and reproduction may also have an important influence on observed OC concentrations in aquatic organisms. As temperatures decrease, elimination rates of OCs in aquatic organisms decrease (Buckman et al. 2007). Decreasing temperature will also reduce feeding and growth rates, as discussed above, resulting in lower exposure to contaminants. Thus, contaminant dynamics can vary among systems that have different climatic regimes, such as tropical (Kidd et al. 2001) or arctic (Kidd et al. 1998). Reproduction by female aquatic organisms provides an opportunity to eliminate OCs through lipid rich eggs (Fisk and Johnston 1998), although the influence of this is much less important in fish (Johnston et al. 2002) than in mammals, where lactation provides an efficient means of eliminating lipophilic OCs (Fisk et al. 2001).

Another important factor that influences observed OC concentrations in aquatic organisms is trophic position (Rasmussen et al. 1990; McIntyre and Beauchamp 2007). Since many OCs biomagnify, higher trophic level organisms will have greater OC concentrations. The influence of trophic position is often more important than body size, age, or reproductive state (Borgå et al. 2004), although trophic position and body size are highly correlated in aquatic ecosystems.

5.2.3 Trophic Transfer of Elements/Metals

The bioaccumulation of elements/metals[1] by aquatic organisms is in several regards more complicated than OCs and involves the interaction of chemically, physically,

---

[1] The term elements is considered more appropriate than the more commonly used terms metals or heavy metals because modern analytical methods provide data for a wide suite of elements, which include both metallic and non-metallic members (Duffus 2002).

and biologically mediated processes. Elements enter aquatic environments through both natural and anthropogenic sources, and human activity can result in high levels of elements in aquatic environments. Elements differ from most OCs in that they occur naturally in the environment and are classified as either essential (e.g., copper, zinc, manganese), because they are necessary to an organism for life, or nonessential (e.g., arsenic, cadmium, mercury), which may be present in an organism, but serve no known positive biological role. Essential elements are regulated by organisms to specific internal levels, although information on these ranges for most elements and aquatic organisms is limited. It has, however, been suggested that levels in a species should not vary widely among systems (McMeans et al. 2007). Nonessential elements can sometimes behave like essential elements if regulated through the same processes, but are generally assumed to be regulated less efficiently than essential elements (Kraemer et al. 2005) and can thus reflect local levels and vary spatially within the same species (McMeans et al. 2007).

Elements are accumulated by aquatic organisms from either the surrounding water or from food with the relative importance of these exposure routes varying amongst the different elements, and even the forms of the element. Elements can occur as ions (cations or anions) or as complexes (with inorganic ligands, chelates with organic ligands, or sorbed onto particle surfaces) with each different form referred to as an element species. The species of an element that is observed in an aquatic ecosystem will vary with the physio-chemical characteristics of that ecosystem (Gundersen et al. 2001), in particular pH and redox state, and thus element bioaccumulation can vary widely among ecosystems.

Most elements do not biomagnify (Newman 1998), and if differences in metal concentrations among animal species are accounted for, generally do not show patterns related to food web structure. For example, Cd and Ag concentrations in marine copepods were 3× higher than in ingestible (>0.2 μm), suspended particles, whereas Co and Se concentrations were 4× lower in copepods than in suspended particles (Fisher et al. 2000). In fact, many elements have been found to be *biodiluted*, i.e., decreased in concentration with increasing trophic level (Campbell et al. 2005a). This results in greater accumulation in lower trophic level organisms (e.g., some invertebrates) due to a combination of greater physiological need (essential elements), greater accumulation from water (surface-to-volume ratio) and/or poorer elimination capacity (Fig. 5.3). The exception is Hg, which has been widely shown to biomagnify in aquatic food webs (Kainz et al. 2006; Campbell et al. 2005a; Cabana and Rasmussen 1994) and can reach concentrations in some biota that warrant concern for both the wildlife and humans who consume them (Fisk et al. 2003) (Fig. 5.3). The biomagnification of Hg is driven by the species MeHg, which bonds with sulphur-containing amino acids. This differs from other OCs in that hydrophobicity and lipid content do not explain the behavior of MeHg in food webs. Cesium, zinc, thalium, and rubidium have also been shown, in some but not in all studies, to biomagnify (Campbell et al. 2005a,b; Dietz et al. 1996), although to a much lower extent than mercury.

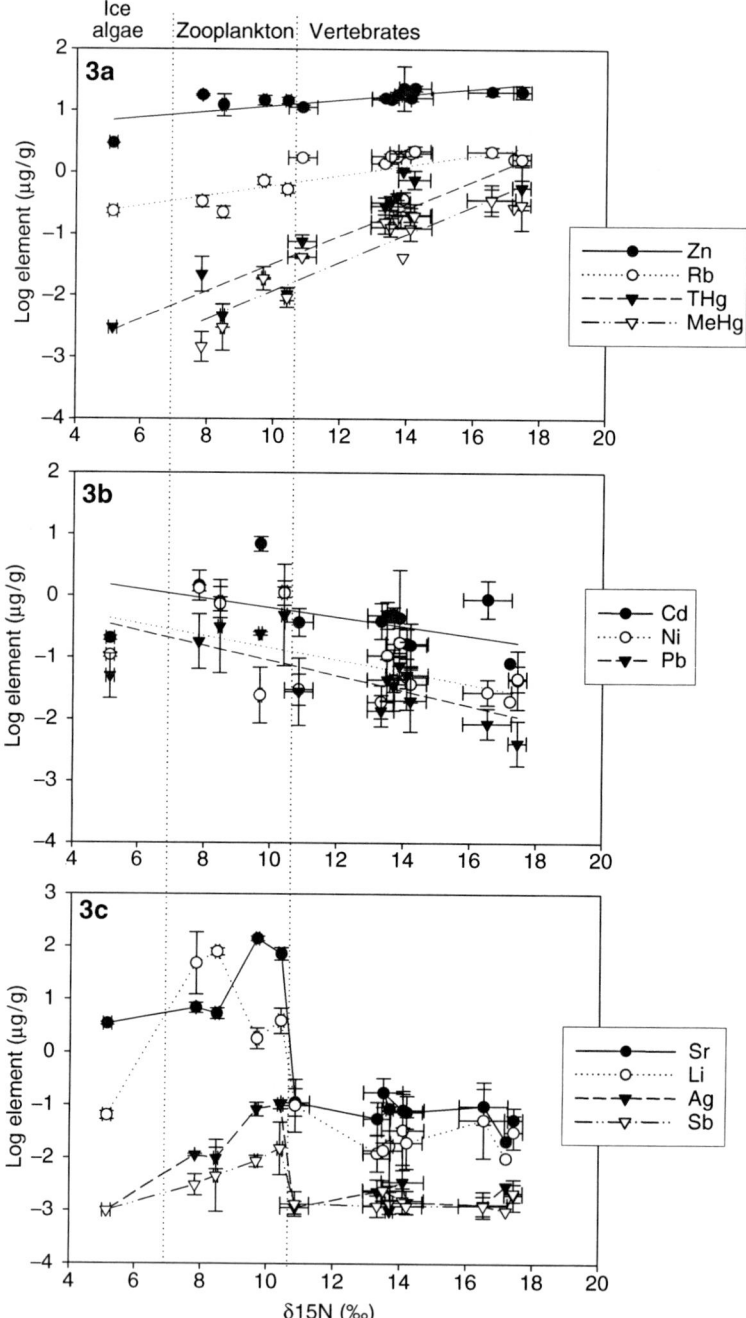

Fig. 5.3 Mean ± S.D. of log-transformed elements versus $\delta^{15}N$ and regressions of various metals versus $\delta^{15}N$ values of individual biota (from Campbell et al. 2005a). The vertical dotted lines indicate which data points represent ice algae, zooplankton and vertebrates (seabirds, fish and seals). Part (**a**) shows the metals that are biomagnifying through the food web with regression lines; (**b**) indicates the metals that are biodiluting through the food web with regression lines; while (**c**) indicates selected metals that are higher in zooplankton than in vertebrates by at least an order of magnitude

5.3 Trophic Transfer of Lipids

In addition to containing dietary energy for organisms at higher trophic levels, lipids provide structural (i.e., phospholipids and sterols) and storage (i.e., triacylglycerols and wax esters) functions, both of which are important when evaluating the ecological condition of food webs and ecotoxicological influence of contaminants on them. Although lipids are ecotoxicologically important for determining the extent to which lipophilic OCs are bioaccumulated in aquatic organisms (see above), there is physiological evidence that lipids, including PUFA and sterols (see Martin-Creuzberg and von Elert – Chap. 3), provide a nutritionally stabilizing function for aquatic organisms. However, the retention of dietary lipids is not driven directly, as the case with contaminants, by the biochemical properties of the lipids, but rather by the organism's specific lipid requirement and/or its ability to directly gain physiological benefits. For example, the cyanobacterium *Spirulina platensis* and the eukaryotic microalgae *Chlorella vulgaris* and *Botryococcus braunii* increase their relative content of unsaturated FA with decreasing temperature ("FA plasticity of algae"; e.g., Sushchik et al. 2003), indicating that algal thermoadaptation can directly affect the quality of FA transferred to consumers. When feeding on the eicosapentaenoic acid (20:5n-3; EPA)-containing marine diatom *Thalassiosira weissflogii* the copepod *Acartia tonsa* had 10× higher egg production rate than when feeding on the EPA-impoverished ciliate *Pleuronema* spp. (Ederington et al. 1995). Moreover, concentrations the n-6 PUFA arachidonic acid (20:4n-6; ARA) and the EPA increased 10.8× and 4.2×, respectively, from lacustrine seston to macrozooplankton (Kainz et al. 2004), which lends support to the argument that these PUFA are physiologically important for zooplankton.

However, not all "essential fatty acids" (see Parrish – Chap. 13) are retained similarly in aquatic organisms. For example, while docosahexaenoic acid (22:6n-3; DHA) is the most highly retained PUFA in most freshwater fish (Ahlgren et al. 1994), there is a significant difference in DHA concentration between copepods and cladocerans (e.g., Persson and Vrede 2006). Bioaccumulation patterns of essential fatty acid concentrations between macrozooplankton (>500 μm) and planktivorous rainbow trout (*Oncorhynchus mykiss*) showed that linoleic acid (LIN; 18:2n-6), α-linolenic acid (ALA; 18:3n-3), ARA, and EPA were 31%, 60%, 29%, and 65% lower, respectively, in dorsal muscle tissues of rainbow trout than in cladocera-dominated macrozooplankton prey (Kainz et al. 2006). Furthermore, while concentrations of bacteria-derived FA, including odd-numbered saturated and branched-chain FA (Kaneda 1991), increased 5.8× from seston to macrozooplankton (Kainz and Mazumder 2005), the ecological significance of such bacterial FA retention is still unclear. These divergent FA concentration patterns suggest that the transfer of nutritionally important lipids is dependent on the organism's physiological requirements as well as the taxonomic structure of aquatic food webs and displays conceptual differences to bioaccumulation patterns of contaminants. Such conceptual differences are explained by the organism's ability to selectively retain specific nutrients, but its inability to exert control over bioaccumulation of OC.

5.4 Concurrent Flow of Lipids and Contaminants

As detailed earlier, lipids and contaminants are accumulated by aquatic organisms and transfer through trophic levels. The major difference between contaminants, particularly OCs, and lipids, is that contaminants are passively bioaccumulated and concentrations are not regulated, whereas physiologically required lipids are selectively regulated. These differences have a direct bearing on the dynamics of lipophilic OCs and are of particular importance when examining the concurrent flow of lipids and weakly lipophilic contaminants. This is because the ecotoxicological fate of weakly lipophilic contaminants, such as MeHg, is not related to selective lipid metabolism in aquatic organisms, a feature that may result in higher bioaccumulation of some contaminants than lipids (Kainz et al. 2006). To increase our understanding of flow dynamics of lipids and contaminants and to subsequently evaluate the nutritional value of aquatic food web components, it is necessary to investigate concurrent pathways of dietary chemicals (nutrients and contaminants) in natural communities under realistic exposure conditions.

5.4.1 Lipids as Chemical Tracers in Ecotoxicology

When diet is the major conveyor of contaminants to aquatic consumers, ecotoxicologists often use tracers to indicate dietary sources of these contaminants. For example, stable isotopes of naturally occurring elements (Broman et al. 1992) and specific contaminants of concern (e.g., stable isotopes of Hg; Orihel et al. 2006) are applied to quantify bioaccumulation of contaminants to specific trophic levels within the aquatic food web. The application of stable isotopes, $\delta^{15}N$ as an indicator of consumer trophic position (Campbell et al. 2005a; Cabana and Rasmussen 1994) and $\delta^{13}C$ as an indicator of the dietary source (Campbell et al. 2000), in ecotoxicology is widespread (Borgå et al. 2004). As some essential fatty acids bioaccumulate along aquatic food webs, they have also been used as an index of MeHg bioaccumulation with increasing trophic position of zooplankton (Kainz et al. 2006). These authors found that MeHg concentrations were significantly correlated with ARA ($R^2 = 0.80$) and EPA concentrations ($R^2 = 0.65$). In a study on herring gull trophodynamics from sites across the Laurentian Great Lakes, Hebert et al. (2006) showed that egg EPA concentrations and n-3/n-6 FA ratios correlated significantly with egg $\delta^{15}N$ values (and contaminant levels; Hebert, pers. comm.) providing further information on how food web structure influences lipid dynamics in aquatic ecosystems.

Fatty acids are useful as source-specific biomarkers because it is often possible to quantify algal, bacteria, and allochthonous-derived FA compounds. Napolitano (1999) described PUFA in plankton as markers to assess algal-derived FA and odd-saturated and branched-chain FA as bacterial biomarkers. In zooplankton, it is assumed that most FA are largely dietary in origin and can thus be used as diet indicators (see Brett et al. – Chap. 6). Hence, Kainz et al. (2002) suggested that measuring the contributions of algal, bacterial, and allochthonous matter contributions to zooplankton diets by looking at group specific FA provides more detailed information on how the bioaccumulation of MeHg in zooplankton is related to the retention of specific diet sources. It was shown that MeHg concentrations in freshwater zooplankton were significantly ($p < 0.01$) associated with concentrations of bacterial ($R^2 = 0.50$) and, to a lesser degree, with algal ($R^2 = 0.35$) FA (Kainz and Mazumder 2005).

In an effort to shed more light on transport and chemical reactivity of PAH concentrations in a river estuary, Countway et al. (2003) found that PAH, except perylene, were correlated with allochthonous sterols (i.e., campesterol, stigmasterol, and β-sitosterol) during the fall/winter sampling and concluded that specific sterols play an important role in the fate and transport of PAHs. Correlations between the saturated long-chain lignoceric acid (24:0), used as allochthonous organic matter indicator (Sun and Wakeham 1994), and MeHg concentrations in zooplankton (Kainz et al. 2002) and lake sediments (Kainz et al. 2003) were not significantly associated, indicating that MeHg and this allochthonous organic matter indicator follow different uptake pathways. Although such source-specific FA biomarkers provide more detailed information than bulk organic matter analyses (e.g., total organic carbon concentrations, atomic C/N ratios, and $\delta^{13}C$), it is critical to know how such lipid compounds are retained and bioconverted in aquatic organisms when using them as biomarkers for ecotoxicological studies.

5.4.2 Implications for Aquatic Food Web Health

The biochemical composition and concentrations of lipids and contaminants largely determine the nutritional value of aquatic food organisms for their consumers. Most FA, PUFA in particular, in marine and freshwater systems are primarily of autochthonous origin, where primary producers at the base of the food web supply FA to organisms at higher trophic levels (see Gladyshev et al. – Chap. 8). Metals, however, can be taken up by algae from the ambient water and subsequently passed on to consumers. Because aquatic consumers can take up metals directly from the water (Pickhardt and Fisher 2007; Wang and Fisher 1998) as well as from their food, there are more entry routes for metals into consumers than for lipids. Although it has not yet been clearly demonstrated, metals, such as MeHg, may negatively affect pathways of lipid synthesis in algae. This could occur because MeHg accumulates in the cell cytoplasm of algae (Mason et al. 1996) where enzymes required for FA synthesis are located (Rangan and Smith 2004).

Algae play a central role in determining the health of aquatic food webs as they provide varying levels of both dietary nutrients and contaminants to upper trophic levels. Algal lipid composition varies among taxa (Guschina and Harwood – Chap. 1; Viso and Marty 1993; Volkman et al. 1998; Pereira et al. 2004) and with productivity of aquatic systems; Müller-Navarra et al. (2004) demonstrated that dietary supply of n-3 PUFA declined with increasing lake productivity (i.e., n-3 PUFA-poor cyanobacteria) at the plant–animal interface. Moreover, access to algal-derived FA may be constrained by food limitation (DeMott et al. 2001), taxa size-related FA composition, as well as by the presence of toxic algae (Jüttner 2005). Dietary lipid retention strongly depends on the physiological requirements of consumers for individual FA. For zooplankton, there is laboratory and field evidence that cladocerans are clearly impoverished in DHA (Persson and Vrede 2006), whereas copepods (Evjemo et al. 2003) and rotifers (Parrish et al. 2007) are generally DHA-enriched. The availability of dietary n-3 PUFA is important for the somatic growth of marine (Copeman et al. 2002), freshwater (Engstrom-Ost et al. 2005), and anadromous fish (Sargent et al. 1999), and DHA is the most highly retained PUFA in a variety of freshwater fish (Ahlgren et al. 1994). Therefore, the taxonomic composition of the lower aquatic food web clearly affects the dietary supply of FA to higher consumers.

Bioaccumulation patterns of metals are strongly determined by metal-specific predator–prey enrichment factors and appear to be independent of lipid bioaccumulation processes (see Sect. 5.2.3). Phytoplankton classes clearly differ in their ability to synthesize and retain various FA (Sushchik et al. 2003; Guschina and Harwood – Chap. 1), sterols and other lipids (Volkman et al. 1998), whereas Pickhardt and Fisher (2007) reported no appreciable differences of MeHg uptake by three eukaryotic algal groups. Differences in specific lipid synthesis but similar contaminant uptake patterns by different algae species have important implications for their consumers. This is evident for zooplankton taxa that differ in their ability to retain dietary FA (Persson and Vrede 2006), but do not differ in their ability to bioaccumulate contaminants such as MeHg (Kainz et al. 2006). Such discordant concentration patterns of essential nutrients and contaminants in zooplankton confirm that food web structure is functionally important for understanding the quantity and quality of lipids as well as contaminants flowing to organisms at higher trophic levels.

5.4.3 Lipid Composition and Contaminants

The ecotoxicological role of lipids in aquatic organisms is important, but the influence of lipids on hydrophilic and hydrophobic contaminants is different. As presented above, the term "lipophilic" is applied for chemicals with log K_{ow} values ≥ 5 (Borgå et al. 2004). The K_{ow} value is a measure of total lipid solubility and does not refer to any specific lipid class or compounds. Although the term lipid solubility in the contaminant literature is generally not related to specific lipid classes or compounds, the biochemical composition of lipids in aquatic organisms is, nonetheless, important for the bioaccumulation of lipophilic contaminants. For example, it has been reported that chlorobenzenes preferentially bind to storage lipids in African catfish (*Clarias gariepinus*; van Wezel and Opperhuizen 1995) and, total PCB concentrations in ribbed mussels (*Geukensia demissa*) correlated best with triacylglycerol concentrations (Bergen et al. 2001). Moreover, although it is still unclear whether PCBs at the concentrations found in aquatic ecosystems cause physiological damage to biota, phospholipids (structural lipids) in gonads and muscles of the eastern oyster (*Crassostrea virginica*) were shown to decrease following PCB exposure (Chu et al. 2003). Such findings show that bioaccumulation patterns of

Table 5.1 Ecological and ecotoxicological relevance of lipids and contaminants for aquatic food webs (with references where appropriate)

	Lipids	Contaminants
Source	Fatty acid synthesis	Geological
	SAFA – plants[a], animals[b]	Anthropogenic
	MUFA – plants[a], animals[b]	Biological (e.g., methylation)[c]
	PUFA – plants[a]	
Uptake	Dietary[d]	Dietary and via water[e]
Physiological relevance	Structural lipids[f]	Benign if concentrations are below toxicity threshold
	Support/enhance somatic growth[g–i]	
	Support/enhance reproduction[g]	Toxic if concentrations are above toxicity threshold, effects include:
	Storage lipids	Bind to lipid classes (PCBs – storage lipids[j])
		Reduce reproduction[k]
Bioaccumulation	Organism dependent[l]	Organism[m] (e.g., lipid content, size, age, etc.), trophic position[n] and environment dependent (e.g., temperature)[o]
Trophic status of aquatic ecosystem	Decreasing nutritional quality of fatty acids with eutrophication[p]	Decreasing bioaccumulation with eutrophication[q]

[a]Guschina and Harwood 2006;
[b]Tocher 2003;
[c]St. Louis et al. 2004;
[d]Brett et al. 2006;
[e]Wang and Fisher 1998;
[f]E.g., temperature adaptation, Dey et al. 1993;
[g]Martin-Creuzburg and von Elert 2004;
[h]Ravet et al. 2003;
[i]von Elert 2002;
[j]For ribbed mussles (*Geukensia demissa*): Bergen et al. 2001;
[k]Rohr and Crumrine 2005;
[l]Persson and Vrede 2006;
[m]Kainz et al. 2006;
[n]Cabana and Rasmussen 1994;
[o]Borgå et al. 2004;
[p]Müller-Navarra et al. 2004;
[q]For methyl mercury: Pickhardt et al. 2002

lipophilic OCs are linked to selective lipid class dynamics, but also indicate the scientific need to further identify how contaminants affect the production and bioconversion of essential lipids in aquatic organisms (Table 5.1).

5.5 Conclusions

Lipids play a major role in the accumulation of lipophilic contaminants in aquatic organisms. To further "eco"-toxicological understanding of lipids and contaminants in aquatic food webs, future research will need to investigate concurrent flows of contaminants and lipids, at concentrations relevant to aquatic ecosystems, and their effects on proper physiological functioning from cell to whole organism levels. Such approaches require detailed identification of the spatial positioning of contaminants and lipid compounds within cells, tissues, and organs. Because recent research has strongly advanced our knowledge on how some PUFA and sterols positively affect the physiological development of aquatic organisms (Martin-Creuzberg and von Elert – Chap. 3; Ahlgren et al. – Chap. 7), it is equally important to understand how the concurrent presence of contaminants affect aquatic organisms and eventually food webs.

From a management perspective, it is clear that the maintenance, protection, and where possible, improvement in the flow of essential lipids along aquatic food webs is desirable. It is also clear that management strategies should simultaneously strive to limit the bioaccumulation of contaminants. Integrating our knowledge of contaminant bioaccumulation and the beneficial effects of certain dietary lipids is necessary and promising; for example, increased dietary supply of PUFA, in particular EPA, as a clear somatic growth-enhancing nutrient for daphnids (Müller-Navarra et al. 2000, von Elert 2002), may result in lower dietary contaminant concentrations per unit biomass of zooplankton and possibly organisms at higher trophic levels. Such approaches, driven by diet quality rather than quantity, could alter our interpretation of the phenomenon of "growth dilution" of contaminants that has thus far been attributed to high algal biomass (Pickhardt et al. 2002). It is furthermore expected that increased dietary access to PUFA increases the health of aquatic ecosystems because organisms would be able to enhance their immuno-competency (see Arts and Kohler – Chap. 10). Ultimately, the goal is to identify, maintain, protect, and, where possible, improve those food web structures (including food–fish interactions in aquaculture) that provide an optimal supply of essential lipids with limited contamination to aquatic organisms and eventually humans.

References

Ahlgren, G., Blomqvist, P., Boberg, M., and Gustafsson, I.B. 1994. Fatty acid content of the dorsal muscle – an indicator of fat quality in freshwater fish. J. Fish Biol. 45:131–157.

Bec, A., Martin-Creuzburg, D., and von Elert, E. 2006. Trophic upgrading of autotrophic picoplankton by the heterotrophic nanoflagellate *Paraphysomonas* sp. Limnol. Oceanogr. 51:1699–1707.

Bergen, B.J., Nelson, W.G., Quinn, J.G., and Jayaraman, S. 2001. Relationships among total lipid, lipid classes, and polychlorinated biphenyl concentrations in two indigenous populations of ribbed mussels (*Geukensia demissa*) over an annual cycle. Environ. Toxicol. Chem. 20:575–581.

Borgå, K., Fisk, A.T., Hoekstra, P.F., and Muir, D.C.G. 2004. Biological and chemical factors of importance in the bioaccumulation and trophic transfer of persistent organochlorine contaminants in arctic marine food webs. Environ. Toxicol. Chem. 23:2367–2385.

Brett, M. T., Müller-Navarra, D. C., Ballantyne, A. P., Ravet, J. L., and Goldman, C. R. 2006. *Daphnia* fatty acid composition reflects that of their diet. Limnol. Oceanogr. 51:2428–2437.

Broman, D., Näf, C., Rolff, C., Zebuhr, Y., Fry, B., and Hobbie, J. 1992. Using ratios of stable nitrogen isotopes to estimate bioaccumulation and flux of polychlorinated dibenzo-*p*-dioxins (PCDDs) and dibenzofurans (PCDFs) in two food chains from the northern Baltic. Environ. Toxicol. Chem. 11:331–345.

Buckman, A.H., Brown, S.B., Small, J.M., Muir, D.C.G., Parrott, J.L., Solomon, K.R., and Fisk, A.T. 2007. The role of temperature and enzyme induction in the biotransformation of PCBs and bioformation of OH-PCBs by rainbow trout (*Oncorhynchus mykiss*). Environ. Sci. Technol. 41:3856–3863.

Burreau, S., Axelman, J., Broman, D., and Jakobsson, E. 1997. Dietary uptake in pike (*Esox lucius*) of some polychlorinated biphenyls, polychlorinated naphthalenes and polybrominated diphenyl ethers administered in natural diet. Environ. Toxicol. Chem. 16:2508–2513.

Cabana, G., and Rasmussen, J.B. 1994. Modelling food chain structure and contaminant bioaccumulation using stable nitrogen isotopes. Nature 372:255–257.

Campbell, L.M., Schindler, D.W., Muir, D.C.G., Donald, D.B., and Kidd, K.A. 2000. Organochlorine transfer in the food web of subalpine Bow Lake, Banff National Park. Can. J. Fish. Aquat. Sci. 57:1258–1269.

Campbell, L.M., Norstrom, R.J., Hobson, K.A., Muir, D.C.G., Backus, S., and Fisk, A.T. 2005a. Mercury and other trace elements in a pelagic Arctic marine food web (Northwater Polynya, Baffin Bay). Sci. Total Environ. 351:247–263.

Campbell, L.M., Fisk, A.T., Wang, X., Köck, G., and Muir, D.C.G. 2005b. Evidence of biomagnification of rubidium in aquatic and marine food webs. Can. J. Fish. Aquat. Sci. 62:1161–1167.

Campfens, J., and MacKay, D. 1997. Fugacity-based model of PCB bioaccumulation in complex aquatic food webs. Environ. Sci. Technol. 31:577–583.

Chu, F.L.E., Soudant, P., and Hall, R.C. 2003. Relationship between PCB accumulation and reproductive output in conditioned oysters *Crassostrea virginica* fed a contaminated algal diet. Aquat. Toxicol. 65:293–307.

Copeman, L.A., Parrish, C.C., Brown, J.A., and Harel, M. 2002. Effects of docosahexaenoic, eicosapentaenoic, and arachidonic acids on the early growth, survival, lipid composition and pigmentation of yellowtail flounder (*Limanda ferruginea*): a live food enrichment experiment. Aquaculture 210:285–304.

Countway, R.E., Dickhut, R.M., and Canuel, E.A. 2003. Polycyclic aromatic hydrocarbon (PAH) distributions and associations with organic matter in surface waters of the York River, VA Estuary. Org. Geochem. 34:209–224.

Cunnane, S.C. 2003. Problems with essential fatty acids: time for a new paradigm? Prog. Lipid Res. 42:544–568.

D'Adamo, R., Pelosi, S., Trotta, P., and Sansone, G. 1997. Bioaccumulation and biomagnification of polycyclic aromatic hydrocarbons in aquatic organisms. Mar. Chem. 56:45–49.

Dalsgaard, J., St. John, M., Kattner, G., Müller-Navarra, D.C., and Hagen, W. 2003. Fatty acid trophic markers in the pelagic marine environment, pp. 225–340. In A.J. Southward, P.A Tyler, C.M. Young, C.M. Fuiman and L.A. [eds.], Advances in marine biology. Elsevier, Amsterdam.

DeMott, W.R., Gulati, R.D., and Van Donk, E. 2001. *Daphnia* food limitation in three hypereutrophic Dutch lakes: Evidence for exclusion of large-bodied species by interfering filaments of cyanobacteria. Limnol. Oceanogr. 46:2054–2060.

Dey, I., Buda, C., Wiik, T., Halver, J.E., and Farkas, T. 1993. Molecular and structural composition of phospholipid-membranes in livers of marine and freshwater fish in relation to temperature. Proc. Natl. Acad. Sci. USA 90:7498–7502.

Dietz, R., Riget, F., and Johansen, P. 1996. Lead, cadmium, mercury and selenium in Greenland marine animals. Sci. Total Environ. 186:67–93.

Duffus, J.H. 2002. "Heavy metals"-a meaningless term? IUPAC Technical Report. Pure Appl. Chem. 74:793–807.

Ederington, M.C., McManus, G.B., and Harvey, H.R. 1995. Trophic transfer of fatty acids, sterols, and a triterpenoid alcohol between bacteria, a ciliate, and the copepod *Acartia tonsa*. Limnol. Oceanogr. 40:860–867.

Engstrom-Ost, J., Lehtiniemi, M., Jonasdottir, S.H., and Viitasalo, M. 2005. Growth of pike larvae (*Esox lucius*) under different conditions of food quality and salinity. Ecol. Freshw. Fish 14:385–393.

Evjemo, J.O., Reitan, K.I., and Olsen, Y. 2003. Copepods as live food organisms in the larval rearing of halibut larvae (*Hippoglossus hippoglossus* L.) with special emphasis on the nutritional value. Aquaculture 227:191–210.

Finizio, A., Vighi, M., and Sandroni, D. 1997. Determination of N-octanol/water partition coefficient (K_{ow}) of pesticide critical review and comparison of methods. Chemosphere 34:131–161.

Fisher, N.S., Stupakoff, I., Sanudo-Wilhelmy, S., Wang, W.X., Teyssie, J.L., Fowler, S.W., and Crusius, J. 2000. Trace metals in marine copepods: a field test of a bioaccumulation model coupled to laboratory uptake kinetics data. Mar. Ecol. Prog. Ser. 194:211–218.

Fisk, A.T., and Johnston, T.A. 1998. Maternal transfer of organochlorines to eggs of walleye (*Stizostedion vitreum*) in Lake Manitoba and western Lake Superior. J. Great Lakes Res. 24:917–928.

Fisk, A.T., Norstrom, R.J., Cymbalisty, C.D., and Muir, D.C.G. 1998. Dietary accumulation and depuration of hydrophobic organochlorines: Bioaccumulation parameters and their relationship with the octanol/water partition coefficient. Environ. Toxicol. Chem. 17:951–961.

Fisk, A.T., Tomy, G.T., Cymbalisty, C.D., and Muir, D.C.G. 2000. Dietary accumulation and QSARs for depuration and biotransformation of short (C10), medium (C14) and long (C18) carbon chain polychlorinated alkanes by juvenile rainbow trout (*Oncorhynchus mykiss*). Environ. Toxicol. Chem. 19:1508–1516.

Fisk, A.T, Hobson, K.A, and Norstrom, R.J. 2001. Influence of chemical and biological factors on trophic transfer of persistent organic pollutants in the Northwater Polynya food web. Environ. Sci. Technol. 35:732–738.

Fisk, A.T., Hoekstra, P.F., Gagnon, J-M., Norstrom, R.J., Hobson, K.A., Kwan, M., and Muir, D.C.G. 2003. Biological characteristics influencing organochlorine contaminants in Arctic marine invertebrates. Mar. Ecol. Prog. Ser. 262:201–214.

Fox K., Zauke, G.-P., and Butte, W. 1994. Kinetics of bioconcentration and clearance of 28 polychlorinated biphenyl congeners in zebrafish (*Brachydanio rerio*). Ecotox. Environ. Safety 28:99–109.

Gobas, F.A.P.C., McCorouodale, J.R., and Haffner, G.D. 1993. Intestinal-absorption and biomagnification of organochlorines. Environ. Toxicol. Chem. 12:567–576.

Gobas, F.A.P.C., and Morrison, H.A. 2000. Bioconcentration and biomagnification in the aquatic environment, pp. 189–231. In R.S. Boethling and D. Mackay [eds.], Handbook of property estimation methods for chemicals: environmental and health sciences. Lewis, Boca Raton.

Goulden, C.E., and Place, A.R. 1990. Fatty acid synthesis and accumulation rates in daphniids. J. Exp. Zool. 256:168–178.

Graeve, M., Albers, C., and Kattner, G. 2005. Assimilation and biosynthesis of lipids in Arctic Calanus species based on feeding experiments with a 13C labelled diatom. J. Exp. Mar. Biol. Ecol. 317:109–125.

Greenfield, B.K., Davis, J.A., Fairey, R., Roberts, C., Crane, D., and Ichikawa, G. 2005. Seasonal, interannual, and long-term variation in sport fish contamination, San Francisco Bay. Sci. Total Environ. 336:25–43.

Gundersen, P., Olsvik, P.A., and Steinnes, E. 2001. Variations in heavy metal concentrations and speciation in two mining-polluted streams in central Norway. Environ. Toxicol. Chem. 20:978–984.

Guschina, I.A., and Harwood, J.L. 2006. Lipids and lipid metabolism in eukaryotic algae. J. Lipid Res. 45:160–186.

Hagen, W., and Auel, H. 2001. Seasonal adaptations and the role of lipids in oceanic zooplankton. Zool.-Anal. Comp. Syst. 104:313–326.

Hawker, D.W.; Connell, D.W. 1988. Octanol-water partition coefficients of polychlorinated biphenyl congeners. Environ. Sci. Technol. 22:382–387.

Hebert, C.E., and Keenleyside, K.A. 1995. To normalize or not to normalize? Fat is the question. Environ. Toxicol. Chem. 14:801–807.

Hebert, C.E., Arts, M.T., and Weseloh, D.V.C. 2006. Ecological tracers can quantify food web structure and change. Environ. Sci. Technol. 40:5618–5623.

Hop, H., Borgå, K., Gabrielsen, G.W., Kleivane, L.K., and Skaare, J.U. 2002. Food web magnification of persistent organic pollutants in poikilotherms and homeotherms from the Barents Sea. Environ. Sci. Technol. 36:2589–2597.
Incardona, J.P., Day, H.L., Collier, T.K., and Scholz, N.L. 2006. Developmental toxicity of 4-ring polycyclic aromatic hydrocarbons in zebrafish is differentially dependent on AH receptor isoforms and hepatic cytochrome P4501A metabolism. Toxicol. Appl. Pharmacol. 217:308–321.
Iverson, S.J., Field, C., Bowen, W.D., and Blanchard, W. 2004. Quantitative fatty acid signature analysis: a new method of estimating predator diets. Ecol. Monogr. 74:211–235.
Johnston, T.A., Fisk, A.T., Whittle, D.M., and Muir, D.C.G. 2002. Variation in organochlorine bioaccumulation by a predatory fish; gender, geography, and data analysis methods. Environ. Sci. Technol. 36:4238–4244.
Jüttner, F. 2005. Evidence that polyunsaturated aldehydes of diatoms are repellents for pelagic crustacean grazers. Aquat. Ecol. 39:271–282.
Kainz, M., and Mazumder, A. 2005. Effect of algal and bacterial diet on methyl mercury concentrations in zooplankton. Environ. Sci. Technol. 39:1666–1672.
Kainz, M., Lucotte, M., and Parrish, C.C. 2002. Methyl mercury in zooplankton – the role of size, habitat and food quality. Can. J. Fish. Aquat. Sci. 59:1606–1615.
Kainz, M., Lucotte, M., and Parrish, C.C. 2003. Relationships between organic matter composition and methyl mercury content of offshore and carbon-rich littoral sediments in an oligotrophic lake. Can. J. Fish. Aquat. Sci. 60:888–896.
Kainz, M., Arts, M.T., and Mazumder, A. 2004. Essential fatty acids within the planktonic food web and its ecological role for higher trophic levels. Limnol. Oceanogr. 49:1784–1793.
Kainz, M., Telmer, K., and Mazumder, A. 2006. Bioaccumulation patterns of methyl mercury and essential fatty acids in the planktonic food web and fish. Sci. Total Environ. 368:271–282.
Kainz, M., Arts, M.T., and Mazumder, A. 2008. Essential versus potentially toxic dietary substances a seasonal assessment of essential fatty acids and methyl mercury concentrations in the planktonic food web. Env. Poll.; 155:262–270.
Kaneda, T. 1991. Iso- and anteiso-fatty acids in bacteria – biosynthesis, function, and taxonomic significance. Microbiol. Rev. 55:288–302.
Kelly, B.C., Gobas, F.A.P.C., and McLachlan, M.S. 2004. Intestinal absorption and biomagnification of organic contaminants in fish, wildlife, and humans. Environ. Toxicol. Chem. 23:2324–2336.
Kelly, E.N., Schindler, D.W., St. Louis, V.L., Donald, D.B., and Vlaclicka, K.E. 2006. Forest fire increases mercury accumulation by fishes via food web restructuring and increased mercury inputs. Proc. Natl. Acad. Sci. USA 103:19380–19385.
Kidd, K.A., Schindler, D.W., Hesslein, R.H., Ross, B.J., Koczanski, K., Stephens, G.R., and Muir, D.C.G. 1998. Effects of trophic position and lipid on organochlorine concentrations in fishes from subarctic lakes in Yukon Territory. Can. J. Fish. Aquat. Sci. 55:869–881.
Kidd, K.A., Bootsma, H.A., Hesslein, R.H, Muir, D.C.G, and Hecky, R.E. 2001. Biomagnification of DDT through the benthic and pelagic food webs of Lake Malawi, East Africa: Importance of trophic level and carbon source. Environ. Sci. Technol. 35:14–20.
Kiron, V., Fukuda, H., Takeuchi, T., and Watanabe, T. 1995. Essential fatty acid nutrition and defense mechanisms in rainbow trout *Oncorhynchus mykiss*. Comp. Biochem. Physiol. A Physiol. 111:361–367.
Klein Breteler, W.C.M., Schogt, N., Baas, M., Schouten, S., and Kraay, G.W. 1999. Trophic upgrading of food quality by protozoans enhancing copepod growth: role of essential lipids. Mar. Biol. 135:191–198.
Konwick, B.J., A.W. Garrison, M.C. Black, J.K. Avants and A.T. Fisk. 2006. Bioaccumulation, biotransformation, and metabolite formation of fipronil and chiral legacy pesticides in rainbow trout. Environ. Sci. Technol. 40:2930–2936.
Kraemer, L.D., Campbell, P.G.C., and Hare, L. 2005. Dynamics of Cd, Cu and Zn accumulation in organs and sub-cellular fractions in field transplanted juvenile yellow perch (*Perca flavescens*). Environ. Pollut. 138:324–337.
Kwon, T.D., Fisher, S.W., Kim, G.W., Hwang, H., and Kim, J.E. 2006. Trophic transfer and biotransformation of polychlorinated biphenyls in zebra mussel, round goby, and smallmouth bass in Lake Erie, USA. Environ. Toxicol. Chem. 25:1068–1078.

Mackay, D. 1982. Correlation of bioconcentration factors. Environ. Sci. Technol. 16:274–278.
Mackay, D., and Paterson, S. 1981. Calculating fugacity. Environ. Sci. Technol. 15:1006–1013.
Mackay, D., Shiu, W.-Y., and Ma, K.C. 2000. Physical-chemical properties and environmental fate handbook on CD. CRC Press, Boca Raton.
Martin-Creuzburg, D., and von Elert, E. 2004. Impact of 10 dietary sterols on growth and reproduction of *Daphnia galeata*. J. Chem. Ecol. 30:483–500.
Mason, R.P., Reinfelder, J.R., and Morel, F.M.M. 1996. Uptake, toxicity, and trophic transfer of mercury in a coastal diatom. Environ. Sci. Technol. 30:1835–1845.
McIntyre, J.K., and Beauchamp, DA. 2007. Age and trophic position dominate bioaccumulation of mercury and organochlorines in the food web of Lake Washington. Sci. Total Environ. 372:571–584.
McMeans, B.C., Borgå, K., Bechtol, W.R., Higginbotham, D., and Fisk, A.T. 2007. Essential and non-essential element concentrations in two sleeper shark species collected in arctic waters. Environ. Pollut. 148:281–290.
Miralto, A., Barone, G., Romano, G., Poulet, S. A., Ianora, A., Russo, G. L., Buttino, I., Mazzarella, G., Laabir, M., Cabrinik, M., Giacobbe, M. G. 1999. The insidious effect of diatoms on copepod reproduction. Nature 402:173–176.
Morel, F.M.M., Kraepiel, A.M.L., and Amyot, M. 1998. The chemical cycle and bioaccumulation of mercury. Ann. Rev. Ecol. Syst. 29:543–566.
Müller-Navarra, D.C., Brett, M.T., Park, S., Chandra, S., Ballantyne, A.P., Zorita, E., Goldman, C. R. 2004. Unsaturated fatty acid content in seston and tropho-dynamic coupling in lakes. Nature 427:69–72.
Müller-Navarra, D.C., Brett, M.T., Liston, A.M., and Goldman, C.R. 2000. A highly unsaturated fatty acid predicts carbon transfer between primary producers and consumers. Nature 403:74–77.
Napolitano, G.E. 1999. Fatty acids as trophic and chemical markers in freshwater ecosystems, pp. 21–44. In M.T. Arts, and B.C wainman (eds.), Lipids in freshwater ecosystems. Springer, New York.
Newman, M.C. 1998. Fundamentals of Ecotoxicology. Ann Arbor Press, Chelsea, MI, p. 402.
Orihel, D. M., Paterson, M.J., Gilmour, C.C., Bodaly, R.A., Blanchfield, P.J., Hintelmann, H., Harris, R. C., Rudd, J. W. M. 2006. Effect of loading rate on the fate of mercury in littoral mesocosms. Environ. Sci. Technol. 40:5992–6000.
Parrish, C.C, Whiticar, M., and Puvanendran, V. 2007. Is omega 6 docosapentaenoic acid an essential fatty acid during early ontogeny in marine fauna? Limnol. Oceanogr. 52:476–479.
Paterson, G., Drouillard, K.G, and Haffner, G.D. 2006. An evaluation of stable nitrogen isotopes and polychlorinated biphenyls as bioenergetic tracers in aquatic systems. Can. J. Fish. Aquat. Sci. 63:628–641.
Paterson, G., Drouillard, K.G., and Haffner, G.D. 2007. PCB elimination by yellow perch (*Perca flavescens*) during an annual temperature cycle. Environ. Sci. Technol. 41:824–829.
Pereira, S.L., Leonard, A.E., Huang, Y.S., Chuang, L.T., and Mukerji, P. 2004. Identification of two novel microalgal enzymes involved in the conversion of the omega 3-fatty acid, eicosapentaenoic acid, into docosahexaenoic acid. Biochem. J. 384:357–366.
Persson, J., and Vrede, T. 2006. Polyunsaturated fatty acids in zooplankton: variation due to taxonomy and trophic position. Freshw. Biol. 51:887–900.
Pickhardt, P.C., and Fisher, N.S. 2007. Accumulation of inorganic and methylmercury by freshwater phytoplankton in two contrasting water bodies. Environ. Sci. Technol. 41:125–131.
Pickhardt, P.C., Folt, C.L., Chen, C.Y., Klaue, B., and Blum, J.D. 2002. Algal blooms reduce the uptake of toxic methylmercury in freshwater food webs. Proc. Natl. Acad. Sci. USA 99:4419–4423.
Rangan, V.S., and Smith, S. 2004. Fatty acid synthesis in eukaryotes, pp. 151–179. In D.E. Vance and J.E. Vance [eds.], Biochemistry of lipids, lipoporteins and membranes. Elsevier, Amsterdam.
Rasmussen, J.B., Rowan, D.J., Lean, D.R.S., and Carey, J.H. 1990. Food chain structure in Ontario lakes determines PCB levels in lake trout (*Salvelinus namaycush*) and other pelagic fish. Can. J. Fish. Aquat. Sci. 47:2030–2038.
Ravet, J.L., Brett, M.T., and Müller-Navarra, D.C. 2003. A test of the role of polyunsaturated fatty acids in phytoplankton food quality for *Daphnia* using liposome supplementation. Limnol. Oceanogr. 48:1938–1947.

Rohr, J.R., and Crumrine, P.W. 2005. Effects of an herbicide and an insecticide on pond community structure and processes. Ecol. Appl. 15:1135–1147.

Russell, R.W., Gobas, F., and Haffner, G.D. 1999. Role of chemical and ecological factors in trophic transfer of organic chemicals in aquatic food webs. Environ. Toxicol. Chem. 18:1250–1257.

Sargent, J.R., McEvoy, L., Estevez, A., Bell, G., Bell, M., Henderson, J., Tocher, D. 1999. Lipid nutrition of marine fish during early development: current status and future directions *Aquaculture*. 179:217–229.

Schwarzenbach, R.P., Gschwend, P.M., and Imboden, D.M., 2003. Environmental Organic Chemistry 2nd Edition, wiley-Interscience.

Scott, C.L., Kwasniewski, S., Falk-Petersen, and S., Sargent, R.J. 2002. Species differences, origins and functions of fatty alcohols and fatty acids in the wax esters and phospholipids of *Calanus hyperboreus, C. glacialis* and *C. finmarchicus* from arctic waters. Mar. Ecol. Prog. Ser. 235:127–134.

Scott, G.R., and Sloman, K.A. 2004. The effects of environmental pollutants on complex fish behaviour: integrating behavioural and physiological indicators of toxicity. Aquat. Toxicol. 68:369–392.

St. Louis, V.L., Rudd, J.W.M., Kelly, C.A., Bodaly, R.A., Paterson, M.J., Beaty, K.G., Hesslein, R.H., Heyes, A., and Majewski, A.R. 2004. The rise and fall of mercury methylation in an experimental reservoir. Environ. Sci. Technol. 38:1348–1358.

Sun, M.-Y., and Wakeham, S.G. 1994. Molecular evidence for degradation and preservation of organic matter in the anoxic Black Sea basin. Geochim. Cosmochim. Acta 58:3395–3406.

Sushchik, N.N., Kalacheva, G.S., Zhila, N.O., Gladyshev, M.I., and Volova, T.G. 2003. A temperature dependence of the intra- and extracellular fatty-acid composition of green algae and *cyanobacterium*. Russ. J. Plant Physiol. 50:374–380.

Swackhamer, D.L., and Skoglund, R.S. 1993. Bioaccumulation of PCBs by algae: kinetics versus equilibrium. Environ. Toxicol. Chem. 12:831–838.

Tanabe, S. 2002. Contamination and toxic effects of persistent endocrine disrupters in marine mammals and birds. Mar. Poll. Bull. 45:69–77.

Thomann, R.V. 1981. Equilibrium model of fate of microcontaminants in diverse aquatic food chains. Can. J. Fish. Aquat. Sci. 38:280–296.

Thomann, R.V. 1989. Bioaccumulation model of organic-chemical distribution in aquatic food-chains. Environ. Sci. Tehcnol. 23: 699–707.

Tocher, D.R. 2003. Metabolism and functions of lipids and fatty acids in teleost fish. Rev. Fish. Sci. 11:107–184.

Trudel, M., and Rasmussen, J.B. 2006. Bioenergetics and mercury dynamics in fish: a modelling perspective. Can. J. Fish. Aquat. Sci. 63:1890–1902.

van Wezel, A.P, and Opperhuizen, A. 1995. Thermodynamics of partitioning of a series of chlorobenzenes to fish storage lipids, in comparison to partitioning to phospholipids. Chemosphere 31:3605–3615.

Viso, A.C., and Marty, J.C. 1993. Fatty acids from 28 marine microalgae. Phytochem. 34:1521–1533.

Volkman, J.K., Barrett, S.M., Blackburn, S.I., Mansour, M.P., Sikes, E.L., and Gelin, F. 1998. Microalgal biomarkers: A review of recent research developments. Org. Geochem. 29:1163–1179.

von Elert, E. 2002. Determination of limiting polyunsaturated fatty acids in Daphnia galeata using a new method to enrich food algae with single fatty acids. Limnol. Oceanogr. 47:1764–1773.

Wang, W.X., and Fisher, N.S. 1998. Accumulation of trace elements in a marine copepod. Limnol. Oceanogr. 43:273–283.

Wang, W.X., and Fisher, N.S. 1999. Assimilation efficiencies of chemical contaminants in aquatic invertebrates: A synthesis. Environ. Toxicol. Chem. 18:2034–2045.

Wong, C.S., Mabury, S.A., Whittle, D.M., Backus, S.M., Teixeira, C., DeVault, D.S., Bronte, C.R., and Muir, D.C.G. 2004. Organochlorine compounds in Lake Superior: Chiral polychlorinated biphenyls and biotransformation in the aquatic food web. Environ. Sci. Technol. 38:84–92.

Chapter 6
Crustacean Zooplankton Fatty Acid Composition

Michael T. Brett, Dörthe C. Müller-Navarra, and Jonas Persson

6.1 Introduction

Fatty acids (FA) are among the most important molecules transferred across the plant–animal interface in aquatic food webs. Particular classes of FA, such as the n-3 highly unsaturated fatty acids (HUFA), are important somatic growth limiting compounds for herbivorous zooplankton (Müller-Navarra 1995a; Müller-Navarra et al. 2000; Ravet et al. 2003). These molecules are also critical for the growth, disease resistance, and general well being of juvenile fish (Adams 1999; Olsen 1999; Sargent et al. 1999). Thus, knowing how nutritionally important FA are conveyed through food webs has important implications for understanding economically important fisheries. A very substantial literature shows these same molecules have a wide range of positive impacts on human health (Simopoulos 1999; Arts et al. 2001). Specific FA may also help interpret trophic relations in aquatic systems (Dalsgaard et al. 2003), as the group specific FA composition of primary producers varies greatly (Volkman et al. 1989; Ahlgren et al. 1992). Therefore, it is important to understand how much the FA composition of zooplankton is determined by taxonomic affiliation, changed by diet, and modified by starvation or temperature. It is also essential to know whether zooplankton maintain a semiconstant FA profile relative to their diets or, alternatively, bioconvert some FA into other FA molecules. This review will summarize the published information on how these factors regulate the FA composition of freshwater and marine zooplankton.

M.T. Brett (✉), D.C. Muller-Navarra, and J. Persson
Department of Civil and Environmental Engineering, University of Washington, Seattle, Washington, USA
e-mail: mtbrett@u.washington.edu

6.1.1 Historical Context

The analysis of zooplankton FA started with Lovern (1935), who compared FA in the marine calanoid copepod *Calanus finmarchicus* and the freshwater zooplankters *Cyclops strenuous*, *Daphnia galeata*, and *Diaptomus gracilis* with the FA of fish caught from the same environments. Lovern observed that the FA composition of these zooplankton was quite similar to the lipids contained in "typical fish-oil" and concluded this indicated fish deposit dietary lipids into their tissue largely unchanged. Subsequently, Ackman and Eaton (1966) showed the most prevalent FA in the euphausiid *Meganyctiphanes norvegica* affected the FA composition of the fin whale in the North Atlantic. Variation in zooplankton FA composition on a seasonal basis was first explored by the pioneering research of Tibor Farkas (Csengeri and Halver 2006). When examining zooplankton samples collected from Lake Balaton, Hungary, in 1958, Farkas observed that zooplankton lipids always had lower melting points than the ambient water temperatures (Csengeri and Halver 2006). He also noted the proportions of eicosapentaenoic acid (20:5n-3; EPA) and especially docosahexaenoic acid (22:6n-3; DHA) in zooplankton lipids increased with decreasing temperatures (Farkas and Herodek 1964). He was the first to note that cladocerans nearly exclusively accumulate EPA whereas copepods predominately accumulate DHA (Farkas 1979). In several laboratory studies (Farkas 1979; Farkas et al. 1984), Farkas suggested copepods could readily adjust their n-3 HUFA and especially DHA content in response to cold stress, whereas the results he obtained for *Daphnia magna* suggested *Daphnia* only had a minimal capacity to adjust HUFA composition in response to temperature. Farkas explained these results within a homeoviscous[1] adaptation context and suggested that these differences were due to varying over-wintering strategies. He suggested that cladocerans as a group were primarily active when water temperatures exceeded 10°C, and over-wintered as inactive resting eggs. Farkas concluded that because cladocerans did not modify their DHA content in response to cold stress, they were unable to maintain lipid melting points below ambient winter water temperatures and therefore could not over-winter in an active life stage. In contrast, copepods readily increased their DHA content when exposed to lower temperatures and many species over-wintered in an active stage.

The first published studies of environmental impacts on the FA of marine zooplankton (Lewis 1969; Jeffries 1970) followed an approach similar to Farkas and focused on seasonal affects (changing water temperatures and phytoplankton community composition) on *Acartia* spp. FA. During the winter and spring, the phytoplankton at Jeffries' field site (Narragansett Bay, Rhode Island) was dominated by diatoms, and during the summer and fall, it was dominated by flagellates. Jeffries noted that during winter and spring, *Acartia* had higher monounsaturated fatty acid (MUFA) contents, and during summer and fall, they had higher saturated fatty acid (SAFA)

---

[1] Homeoviscous response refers to the modification of membrane lipid composition to maintain similar physical properties across a range of water temperatures.

contents. Paradoxically, he reported that *Acartia* accumulated more DHA during the warmer summer/fall months. This latter result could be because dinoflagellates (which often have high DHA content) were prevalent in the summer during his study.

One of the most pivotal studies of marine zooplankton FA was Lee et al.'s (1971) study of dietary influences on the accumulation and composition of wax esters. Wax esters are neutral storage lipids that are the dominant lipid class in polar/north temperate and deep-living calanoid copepods. These storage lipids play a critical role in the life history of copepods in these regions because they are dependent on brief, but intense, vernal phytoplankton blooms. On the basis of *Calanus helgolandicus* feeding experiments with three diatoms and one dinoflagellate, Lee et al. (1971) observed the wax ester and triacylglyceride (TAG) FA composition of this copepod closely matched the FA composition of their diets. They also noted the correspondence between diet and copepod FA increased when food concentrations were higher. In contrast to the results observed for TAG and wax esters, Lee and colleagues reported the FA composition of the structural phospholipids (PL) was not dependent on diet. Many studies have subsequently examined lipid accumulation in marine zooplankton; particularly for zooplankton from polar regions (Kattner and Hagen – Chap. 11).

6.1.2 *Emphasis in the Marine and Freshwater Literature*

The emphasis in the marine and freshwater dietary vs. zooplankton FA literature has been different for several reasons. Many early studies with both marine and freshwater zooplankton were focused on the nutritional needs of aquatic organisms as this affected their nutritional value as food for aquaculture fish (Provasoli and D'Agostino 1969). Also as previously noted, the marine literature was also quite focused on storage lipids and marine researchers were the first to realize the potential utility of FA as trophic markers (Graeve et al. 1994; reviewed by Dalsgaard et al. 2003).

The earliest freshwater field studies, e.g. Farkas (1964), focused on temperature impacts on zooplankton FA composition as this related to "cold adaptation". Subsequently, the importance of essential FA for zooplankton nutrition in nature was investigated (Ahlgren et al. 1990; Müller-Navarra 1995a; Jónasdóttir et al. 1995). Most recent freshwater studies looking at zooplankton FA composition (e.g., Persson and Vrede 2006; Brett et al. 2006; Müller-Navarra 2006) have focused on the somatic growth regulating properties of HUFA for zooplankton and fish and have therefore emphasized the trophic transfer of polyunsaturated fatty acids (PUFA) and HUFA[2] with an eye toward the availability of these molecules for upper trophic levels.

---

[2] In this chapter, we will use PUFA to refer to 16 and 18 carbon chain (C_{16} and C_{18}) FA with two or more double bonds and HUFA to represent the subset of C_{20} and C_{22} PUFA.

6.2 Zooplankton Taxonomic Differences in Fatty Acid Composition

Much is known about the FA dynamics of copepods from north temperate and polar marine systems (Dalsgaard et al. 2003, Kattner and Hagen – Chap. 11). Marine copepods are particularly rich in lipids (i.e., 37 ± 19% of dry mass), and these are strongly dominated by wax esters (56 ± 32% of lipids) and TAG (13 ± 18% of lipids), which serve as storage lipids (reviewed by Lee et al. 2006). Wax esters comprise a particularly important class of lipids, especially for polar, temperate, upwelling or deep water copepods, which are exposed to short but intense phytoplankton blooms and have adapted by developing an ability to accumulate pronounced seasonal lipid stores (Lee et al. 2006). Wax esters may also be important to seasonally diapausing copepods because the thermal expansion and compressibility of these molecules allows copepods to remain neutrally buoyant at great depth (Lee et al. 2006). Tropical epipelagic zooplankton species do not deposit storage lipids because they encounter much weaker seasonal food pulses and have higher metabolic rates. Instead, marine copepods in tropical regions rapidly utilize available food for growth and reproduction (Kattner and Hagen – Chap. 11).

Wax ester synthesis has been particularly well studied for marine *Calanus* spp. copepods. The following pattern can be concluded from the literature (e.g., Dalsgaard et al. 2003): In contrast to the FA moiety, the fatty alcohol component (mostly 20:1n-9, 22:1n-11, and 22:1n-9; Hagen et al. 1993) of wax esters is synthesized by copepods from the related FA, which can then be used as markers for fish copepod consumption (e.g. Sargent and Henderson 1986). Fatty alcohols can also be synthesized de novo from dietary carbohydrates and proteins (Lee et al. 2006). A recent ^{13}C labeling experiment (Graeve et al. 2005) concluded that the abundant MUFA 20:1n-9 and 22:1n-11 and corresponding fatty alcohols in three species of Arctic *Calanus* were most likely synthesized de novo from nonlipid dietary sources. In contrast, structural FA such as EPA and DHA were taken up directly from the diet and highly retained in the body (Graeve et al. 2005). Kattner and Hagen (Chap. 11) compared the wax esters and phospholipids (PL) of four calanoid copepods and found the FA of the wax ester fraction was composed of 61 ± 24% MUFA, especially 16:1n-7, 18:1n-9, 20:1n-9, and 22:1n-11. Fatty alcohols of the 20:1n-9 and 22:1n-11 moieties are also important components of wax esters. In contrast, the FA of the PL of these copepods only had 10 ± 4% MUFA, and was instead dominated by DHA (36 ± 6%), EPA (18 ± 2%) and the SAFA 16:0 (25 ± 3%). Scott et al. (2002) reported nearly identical results for the same copepod species.

Persson and Vrede (2006) found that freshwater zooplankton could be separated into groups based on their PUFA and HUFA composition. These authors found copepods contained a large fraction of DHA while cladocerans were rich in EPA and arachidonic acid (20:4n-6; ARA). Persson and Vrede (2006) also found that herbivorous zooplankton contained more HUFA than did seston, and that carnivores contained more HUFA than herbivores. Similar differences between *Daphnia* spp. and various copepod species have been noted previously (e.g. Farkas 1970; Ballantyne et al. 2003). The compilation of ARA, EPA, and DHA content in wild caught zooplankton in Table 6.1 shows that these conclusions also hold for a wider dataset. The proportion of ARA was similar in cladocerans and copepods, but cladocerans have relatively high proportions of EPA compared with the copepods. Copepods have high proportions of DHA, while DHA is nearly absent in cladocerans. The greater relative content of EPA and ARA in cladocerans compared with copepods may be related to the cladocerans' higher potential for reproduction (Persson and Vrede 2006; Smyntek et al. 2008). The relationship between EPA and growth and reproductive capacity is speculative, however, and the physiological functions of EPA and ARA in crustaceans remain to be clarified.

Recently, Scott et al. (2002), Persson and Vrede (2006) suggested that the high DHA content of copepods might be due to a more highly developed nervous system compared with other zooplankton. Copepods have highly developed prey attack and predator avoidance strategies, which allow them to respond to stimuli within milliseconds (Lenz et al. 2000). They also have abundant chemoreceptors on their antennae and mouth-parts, which allow them to taste food and track mates, see references in Persson and Vrede (2006). Lenz et al. (2000) noted that some calanoid copepods have thick myelin sheaths covering axons in their nervous system, which allow them to achieve exceptionally quick nerve impulse response times. Similar to other nervous tissues, DHA may be critical for the proper functioning of myelin and associated neural tissues. As Scott et al. (2002) concluded "the possibility that [DHA] has special properties in copepods relating to their mobility and migrations rather than to adaptation to low temperatures is worthy of future research". In contrast, Smyntek et al. (2008) suggested the high DHA content of copepods was an adaptation for over-wintering in an active life stage, as previously hypothesized by Farkas (1979).

The carnivorous cladoceran *Bythotrephes longimanus* and the carnivorous calanoid copepods *Epischura nevadensis* and *Heterocope* spp. have considerably higher proportions of PUFA than do herbivorous zooplankton. *B. longimanus* contains 22% EPA while the mean for the filter feeding cladocerans is 14% and *E. nevadensis* and *Heterocope* spp. contain 18% and 21% DHA, respectively, while omnivorous calanoids average 13% DHA. In a study analyzing zooplankton from Lake Tahoe, Müller-Navarra (unpublished data) found that *E. nevadensis* had a higher absolute n-3 PUFA content, and especially DHA, than the herbivore *Diaptomus tyrelli*, but lower than what was found in *Mysis relicta*. The higher relative PUFA proportions in carnivorous zooplankton might be a direct result of the fact that the food they consume (i.e., rotifers and crustacean zooplankton) is richer in PUFA than the seston diets of filter feeding cladocerans and the seston/micro-zooplankton diets of omnivorous calanoids. Since these differences in food have been present on an evolutionary time scale, it can be hypothesized that they have adapted to the high PUFA intake and that they may now be completely dependent on direct dietary sources of ARA, EPA, and DHA to meet their physiological demands. In this regard, it is worth noting several strictly carnivorous fish species such as northern pike (*Esox lucius*) have very limited abilities to convert LIN to ARA, and ALA to EPA and DHA (Henderson et al. 1995). Similarly, we speculate that carnivorous zooplankton may be dependent on a high intake of ARA, EPA, and DHA to meet their physiological demands.

The total FA composition of the major zooplankton groups for which substantial FA data exist (i.e., freshwater cladocerans and copepods, marine calanoid copepods

Table 6.1 Mean zooplankton fatty acid composition (as a percent of total FA) by FA functional groups

Group Trophic mode System	Cladoceran Herbivorous Freshwater	Cladoceran Carnivorous Freshwater	Calanoid copepod Omnivorous Freshwater	Calanoid copepod Carnivorous Freshwater	Cyclopoid copepods Omni.-Carni. Freshwater	Calanoid copepod Omnivorous Marine	Mysids Carnivorous FW & Marine	Euphausia superba Omnivorous Marine
n	13	6	9	3	4	11	4	8
SAFA	34.1 ± 7.2	34.6 ± 6.2	33.6 ± 5.9	34.6 ± 4.2	28.4 ± 10.6	25.5 ± 14.3	27.6 ± 3.5	32.9 ± 4.9
MUFA	23.5 ± 5.5	18.7 ± 5.1	13.2 ± 4.8	11.6 ± 1.4	17.6 ± 10.4	34.2 ± 18.4	33.1 ± 7.4	27.8 ± 4.0
LIN	6.2 ± 1.6	5.3 ± 0.8	4.7 ± 1.5	4.8 ± 0.9	5.2 ± 1.5	2.5 ± 2.2	3.4 ± 1.9	2.6 ± 0.2
ALA + SDA	14.4 ± 5.8	8.1 ± 1.3	13.1 ± 5.3	11.8 ± 1.0	13.1 ± 3.7	8.3 ± 5.6	4.6 ± 4.3	5.5 ± 3.8
ARA	5.2 ± 2.1	8.9 ± 1.0	3.8 ± 2.2	4.4 ± 0.6	3.8 ± 1.0	0.3 ± 0.6	2.2 ± 1.8	0.8 ± 0.4
EPA	14.7 ± 3.9	22.1 ± 2.3	13.0 ± 6.0	11.0 ± 1.7	10.9 ± 3.4	14.4 ± 4.2	16.5 ± 2.2	19.4 ± 5.9
22:2n-6	0.2 ± 0.3	0.3 ± 0.4	1.2 ± 1.3	1.8 ± 1.9	0.8 ± 0.7	0.0 ± 0.0	0.2 ± 0.4	0.0 ± 0.0
DHA	1.7 ± 1.5	2.0 ± 0.9	17.6 ± 9.1	20.1 ± 1.8	20.2 ± 7.3	14.8 ± 7.6	12.3 ± 3.4	11.0 ± 5.0
n-3:n-6 ratio	3.0 ± 0.7	2.3 ± 0.2	5.2 ± 2.8	4.2 ± 0.3	4.7 ± 1.8	18.1 ± 9.6	6.7 ± 4.6	9.8 ± 2.6

The values presented are average percent of total FA ± 1 SD. The freshwater zooplankton fatty acids data were obtained from Hessen and Leu (2006), Persson and Vrede (2006), Smyntek et al. (2008), M.T. Arts (unpublished data), M.T. Brett (unpublished data), C.W. Burns (unpublished data), and D.C. Müller-Navarra (unpublished data). The marine copepod data was taken from Peters et al. (2006), Veloza et al. (2006) and Kattner and Hagen (Chap. 11). The euphausiid FA data was obtained from Cripps et al. (1999), Hagen et al. (2001), Stübing et al. (2003) and Schmidt et al. (2006). Freshwater mysid FA data were obtained from D.C. Müller-Navarra (unpublished data) and M.T. Arts (unpublished data), and marine mysid data were obtained from Richoux et al. (2005)

and euphausiids) indicates considerable differences amongst these groups (Table 6.1). Freshwater cladocerans were notable for having much lower DHA (at ≈2%) than the other zooplankton groups. Cladocerans also had the lowest n-3:n-6 ratios (i.e., 2.4–3.0). Carnivorous freshwater cladocerans could be distinguished from herbivorous cladocerans by a higher ARA and EPA content, and lower ALA + SDA content and n3:n6 ratios (Table 6.1). Compared with other zooplankton, carnivorous cladocerans had particularly high proportions of ARA, i.e., 8.9 ± 1.0% (±1 SD) and low n-3:n-6 ratios, 2.3 ± 0.2. Freshwater calanoid and cyclopoid copepods had the highest proportion DHA (≈20%) and intermediate n-3:n-6 ratios (i.e. 4–6). In general, freshwater cladocerans and copepods had twice as much n-6 and n-3 PUFA, and 10× as much ARA, as did marine copepods and euphausiids. In contrast, marine zooplankton averaged twice as much MUFA and had much higher n-3:n-6 ratios (i.e. 10–20). The FA composition of marine omnivorous copepods differed from that of freshwater omnivorous copepods specifically, and from all freshwater copepods more generally, in their much higher MUFA content and n-3:n-6 ratios, and their much lower ARA and lower LIN and ALA + SDA content.

The zooplankton FA composition data summarized above (n = 58) was analyzed using discriminant function analysis (DFA; see Fig. 6.1). This DFA correctly classified 66% of the samples according to their major group (i.e., herbivorous

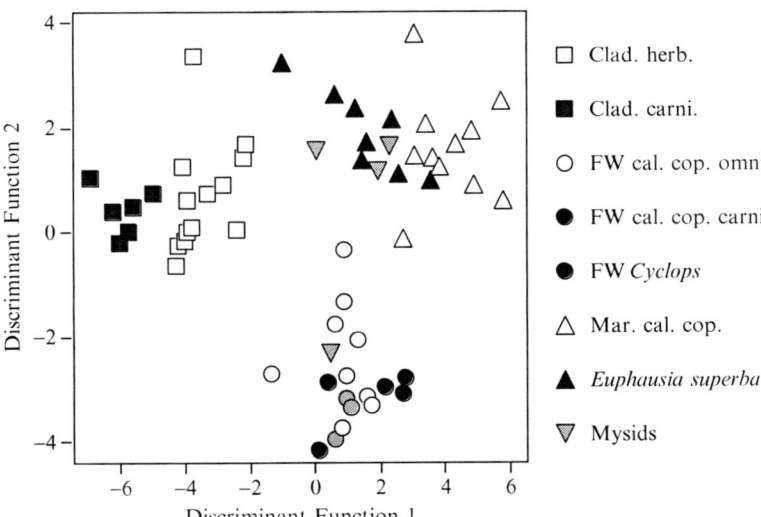

Fig. 6.1 A bivariate plot of the results of a discriminant function analysis of zooplankton fatty acid composition data presented in Table 6.1 (n = 58). The first axis explained 67.5% of the variability. This axis was positively correlated with the log(n-3:n-6) ratio and DHA and negatively correlated with ARA and LIN. The second axis explained 22.6% of the variability, and was positively associated with MUFA and EPA. *Clad. herb.* herbivorous cladocerans, *Clad. carni.* carnivorous cladocerans, *FW cal. cop. omni* freshwater omnivorous calanoid copepods, *FW Cyclops* freshwater cyclopoid copepods, *Mar. cal. cop.* marine calanoid copepods, and *Mysids* marine and freshwater mysids

cladocerans, carnivorous cladocerans, omnivorous freshwater calanoid copepods, etc.) using a "leave-one-out" algorithm. The large majority of the misclassification errors were within the freshwater copepod and marine and freshwater mysid groups. Overall, this DFA explained 98% of the variability using 3 axes, with the first axis explaining 67.5% of the variability. This axis was strongly positively correlated with the log(n-3:n-6) ratio and moderately positively correlated with DHA. The first axis was also strongly negatively correlated with ARA and moderately negatively correlated with LIN. The second axis explained 22.6% of the variability, and was positively associated with MUFA and EPA. This plot shows freshwater cladocerans and copepods formed distinct clusters, and the marine zooplankton formed a third distinct cluster. Within these groups, carnivorous and herbivorous cladocerans could be readily distinguished and marine copepods and euphausiids were mostly separated. Freshwater and marine mysids were poorly classified and tended to be confused with *Euphausia superba*. The results of this DFA strongly support Persson and Vrede's (2006) hypothesis that carnivorous cladocerans can be distinguished from herbivorous cladocerans. However, this DFA does not support their hypothesis that freshwater carnivorous copepods can be distinguished from freshwater omnivorous copepods.

6.3 Phytoplankton Fatty Acid Composition as Food for Zooplankton

The dominant phytoplankton available to herbivorous zooplankton in freshwater and marine planktonic systems differ greatly in their FA composition (Volkman et al. 1989; Ahlgren et al. 1992) (Table 6.2). When comparing the FA composition of freshwater and marine phytoplankton, a few differences are quite apparent. These dissimilarities may be adaptations to the respective environment of the algae. However, differences in experimental focus and/or methods cannot be excluded. For example, the marine literature reports considerably more EPA and DHA in chlorophytes than does the freshwater literature. On average these two FA comprise 4.8% and 1.0% of marine chlorophyte FA, respectively, but these FA are often not detected in freshwater chlorophytes (Table 6.2). The higher n-3 HUFA content of marine chlorophytes may be real or it may be due to the fact that most surveys of marine phytoplankton FA composition are geared toward identifying taxa with potential value as mariculture food stocks, and therefore HUFA-rich chlorophytes may be overly represented. This "aquaculture bias" does not exist in the freshwater phytoplankton literature. Probably because of this mariculture emphasis, and because of the fact that diatoms are quite important in marine systems, there are also far more observations of diatom FA composition for marine than for freshwater taxa. Conversely, there are far more observations of cyanophyte FA composition for freshwater than for marine taxa. This is probably because cyanobacteria are more prevalent in freshwater systems. In addition, they have a low n-3 FA content and are therefore of less interest to mariculturalists. There is also a very substantial

Table 6.2 Mean phytoplankton fatty acid composition (as a percent of total FA) by FA functional groups

Group system	Chlorophytes freshwater	Cryptophytes freshwater	Diatoms freshwater	Cyanophytes freshwater	Chlorophytes marine	Cryptophytes marine	Diatoms marine	Isochrysis galbana marine
n	11	9	6	9	10	11	14	8
SAFA	32.5 ± 9.5	28.4 ± 9.8	23.8 ± 11.0	58.6 ± 18.5	29.1 ± 10.4	23.3 ± 9.5	25.8 ± 6.3	31.2 ± 12.3
MUFA	27.3 ± 12.5	9.9 ± 5.1	40.3 ± 12.8	24.8 ± 16.6	15.3 ± 4.3	9.4 ± 3.6	24.4 ± 5.6	22.7 ± 5.5
C_{16} PUFA	0.0 ± 0.0	0.1 ± 0.4	9.1 ± 5.8	0.0 ± 0.0	17.8 ± 4.9	0.8 ± 1.5	18.5 ± 7.7	1.3 ± 1.3
LIN	14.4 ± 5.6	3.3 ± 2.4	2.0 ± 1.8	7.2 ± 6.5	7.7 ± 4.9	5.8 ± 6.4	1.8 ± 1.1	6.4 ± 2.0
ALA + SDA	25.5 ± 9.7	39.7 ± 10.4	2.9 ± 3.1	7.0 ± 10.2	24.3 ± 10.2	43.5 ± 12.7	2.4 ± 1.8	22.4 ± 7.9
ARA	0.2 ± 0.3	0.1 ± 0.2	2.3 ± 1.7	1.0 ± 2.4	0.9 ± 0.8	0.8 ± 1.0	1.9 ± 2.1	0.1 ± 0.1
EPA	0.1 ± 0.2	15.1 ± 6.1	16.9 ± 8.2	0.6 ± 1.2	4.0 ± 2.4	9.5 ± 3.1	22.0 ± 5.5	1.4 ± 0.8
22:2n-6	0.0 ± 0.0	0.6 ± 1.1	0.1 ± 0.2	0.1 ± 0.1	0.0 ± 0.0	0.4 ± 0.7	0.3 ± 0.4	1.2 ± 1.9
DHA	0.0 ± 0.0	2.9 ± 1.8	2.5 ± 3.0	0.7 ± 2.1	0.9 ± 1.4	6.5 ± 2.2	2.9 ± 1.7	13.3 ± 8.3
n-3:n-6	1.9 ± 0.9	16.8 ± 8.2	7.6 ± 4.8	1.0 ± 1.0	4.5 ± 2.4	22.4 ± 20.8	11.6 ± 8.9	5.1 ± 1.7

The values presented are average percent ±1 SD. The freshwater chlorophyte, cryptophyte and cyanobacteria FA data summarized in this table was taken from Ahlgren et al. (1992), Brett et al. (2006) and C.W. Burns (unpublished data). The freshwater diatom data was taken from Müller-Navarra (1995b), Desvilettes et al. (1997), Gatenby et al. (2003), Müller-Navarra (2006) and Caramujo et al. (2008). The marine chlorophyte data was taken from Volkman et al. (1989), Renaud et al. (1999), and Lourenco et al. (2002). The marine cryptophyte data was taken from Renaud et al. (2002), Broglio et al. (2003), Dunstan et al. (2005), Veloza et al. (2006), and Tremblay et al. (2007). The marine diatom FA data was taken from Dunstan et al. (1994). The Isochrysis galbana FA data were taken from Volkman et al. (1989), Reitan et al. (1994), Nanton and Castell (1998), Renaud et al. (1999), Lourenco et al. (2002), Renaud et al. (2002), Wacker et al. (2002), and Patil et al. (2007)

literature reporting variation in the FA composition of the marine prymnesiophyte *Isochrysis galbana* because this species is the most important phytoplankton food source for aquaculture. Another striking difference between the marine and freshwater literature is that most marine studies report results for a wide range of 16 carbon chain (C_{16}) PUFA and many freshwater studies do not report these FA. In fact, some of the common FA standards used in freshwater studies (e.g. the Supelco® 37-FAME Standard [47885U]) do not contain these FA, making their identification in actual samples problematic. From the marine literature, it is clear C_{16} PUFA are very prevalent in diatoms and chlorophytes, but they do not appear to be common in cryptophytes (but see Müller-Navarra 2006), cyanophytes or the prymnesiophyte *I. galbana*.

Amongst freshwater phytoplankton, chlorophytes are notable for having a high proportion of C_{18} n-6 FA, in particular linoleic acid (18:2n-6; LIN). Freshwater chlorophytes also tend to have very little C_{20} and C_{22} n-3 and n-6 FA. Marine chlorophytes have similar FA composition, except they have, on average, about half as much MUFA and C_{18} n-6 FA, more EPA (4.8% vs. 0.1%), more DHA (1.0% vs. 0%) and a clearly higher n-3:n-6 ratio than freshwater chlorophytes (Table 6.2). Much of the difference between marine and freshwater chlorophyte FA composition may be due to the fact that FA composition data have been reported for a wide variety of freshwater chlorophytes, whereas the marine literature seems to be focused on species with potential aquaculture value (and hence tend to have a high n-3 HUFA content).

Freshwater and marine cryptophytes have a low MUFA content, very high and roughly equal proportions of the C_{18} n-3 FA α-linolenic acid (18:3n-3; ALA) and stearidonic acid (18:4n-3; SDA), high EPA and moderately high DHA content, and a very high n-3:n-6 ratio (i.e., ≈17:1 and 22:1, respectively). Marine cryptophytes have about one third less EPA and twice as much DHA as do freshwater cryptophytes. In general, diatoms have the highest MUFA content, low proportions of both n-3 and n-6 C_{18} FA, high EPA and ARA content and moderately high DHA. Diatoms also have considerable amounts of C_{16} MUFA and PUFA, which are a characteristic of this group. Few studies have reported the FA composition of marine cyanobacteria, but freshwater cyanobacteria are characterized by having a very high SAFA content, very little n-3 FA in general, and a particularly low n-3:n-6 ratio. The marine flagellate *I. galbana* has nearly the global average FA composition for phytoplankton, except it has little EPA and exceptionally high DHA content (Table 6.2). Its high DHA content and the ease with which it can be grown are the reasons this species is very widely used in aquaculture.

The FA composition of the freshwater and marine phytoplankton summarized above ($n = 74$) was analyzed using DFA (Fig. 6.2). This DFA correctly classified 91% of the samples according to their major group (i.e., diatom, chlorophyte, cryptophyte, cyanophyte, and *Isochrysis*) using a "leave-one-out" algorithm. The DFA correctly classified 100% of the diatom and *Isochrysis* monocultures and 95% of the cryptophyte monocultures. Two marine chlorophytes were misclassified

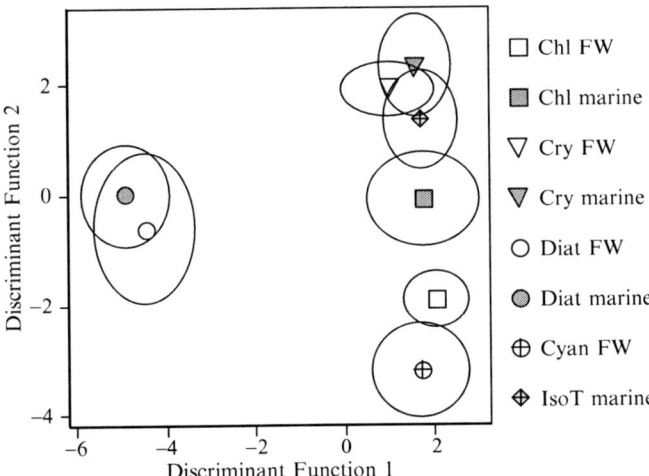

Fig. 6.2 A bivariate plot of the results of a discriminant function analysis of phytoplankton fatty acid composition for the phytoplankton data presented in Table 6.2 ($n = 78$). The ovoids around the phytoplankton group centroids represent the area delineated by ±1 SD on the X and Y axes. The first axis of this DFA explained 61.4% of the overall variation in these data and was correlated positively with C_{18} n-3 and C_{18} n-6 FA and negatively with EPA, arachidonic acid (20:4n-6; ARA) and MUFA. The second axis explained 22.0% of the variation and was positively associated with the n-3:n-6 ratio, DHA and C_{18} n-3 FA and negatively associated with SAFA. The third axis (not shown) explained an additional 10.3% of the overall variation and was positively associated with DHA. *Chl* chlorophytes, *Cry* cryptophytes, *Diat* diatoms, *Cyan* cyanophytes, *IsoT Isochrysis galbana*, and *FW* freshwater taxa. This analysis is based on data obtained from the sources described in Table 6.2

as cryptophytes and two freshwater cyanophytes (both *Oscillatoria*) were misclassified as chlorophytes. The first axis of this DFA explained 61.4% of the overall variation and was correlated positively with C_{18} n-3 and C_{18} n-6 FA and negatively with EPA, ARA, and MUFA. This axis distinguished freshwater and marine diatoms from all other phytoplankton. The second axis explained 22.0% of the variation and was positively associated with the n-3:n-6 ratio, DHA and C_{18} n-3 FA and negatively associated with SAFA. This axis clustered freshwater and marine cryptophytes with *Isochrysis*, and diatoms with marine chlorophytes. These two clusters where easily distinguished from each other, as well as from freshwater chlorophytes and cyanophytes, which were distinct from each other and all other groups. The third axis (not shown) explained an additional 10.3% of the overall variation and was positively associated with DHA. This DF strongly distinguished *Isochrysis* from all other taxonomic groups. The marine and freshwater diatom and cryptophytes clusters were nearly indistinguishable. In contrast, the marine chlorophyte cluster was most similar to the cryptophyte cluster and the freshwater chlorophyte cluster was most similar to the cyanobacteria cluster, consistent with an over-representation of n-3 HUFA rich taxa in the marine literature.

6.4 Dietary Impacts on Zooplankton Fatty Acid Composition

6.4.1 Freshwater Zooplankton: Laboratory Studies

Several studies have examined the impact of algal diets on crustacean FA composition and found a great similarity between the consumer's FA pattern and that of their diet (see e.g. Lewis 1969 for marine amphipods; Bourdier and Amblard 1989 for *Acanthodiaptomus denticornis*; Elendt 1990 for *Daphnia magna*). The FA composition of the neutral lipid fraction (mainly TAGs) is especially affected by dietary FA (Langdon and Waldock 1981; Parrish et al. 1995). Some FA could even be traced across several trophic levels, from phytoplankton via zooplankton to fish larvae (Fraser et al. 1989).

Elendt (1990) was one of the first to study dietary impacts on daphnid FA composition using artificial supplements. D'Abramo and Sheen (1993) found that the FA composition in the freshwater prawn (*Macrobrachium rosenbergii*) whole body tissue reflects that of purified artificial diets. However, concentrations of SAFA and MUFA seemed to change in relation to additions of PUFA. ALA, EPA, and ARA were conserved in the polar lipid fraction of the tissue even when these FA were not provided with the diet. In contrast, n-3 PUFA decreased in the neutral lipids when not provided in the diet whereas n-6 PUFA remained unchanged or increased. They suggest further that n-6 and n-3 PUFA have different metabolic and nutritional functions (*see* Ahlgren et al. – Chap. 7). Using HUFA enriched supplements, Weers et al. (1997) showed that when *Daphnia galeata* were fed combinations of the alga *Chlamydomonas reinhardtii* and emulsions with varying DHA to EPA ratios, the DHA content of *D. galeata* increased with the emulsion DHA:EPA ratio, but even at the highest DHA/EPA ratio tested ($\approx$4:1) *D. galeata* still contained 4× more EPA. These authors suggested this indicated *D. galeata* were retro-converting much of the DHA to EPA. Weers et al. (1997) also showed that *D. galeata* that consumed *Cryptomonas* spp. contained 3× more SDA and 25× more EPA than *D. galeata* that consumed *Scenedesmus* spp. These findings are supported by recent research which has also shown that phytoplankton FA composition has pronounced impacts on the FA profiles of *Daphnia* spp. (Brett et al. 2006; Müller-Navarra 2006). Most FA groups (i.e., SAFA, MUFA, C_{18} n-6, etc.) show moderate correlations ($r^2 = 0.40$–0.68) between the diet and *Daphnia* FA. However, EPA and, even more so, the sum of EPA and DHA show a particularly strong correlation between diet and *Daphnia* FA composition ($r^2 \approx 0.85$). Differences between these studies might suggest some differences in the FA accumulation patterns for different *Daphnia* species. For example, Brett et al. (2006) studied dietary impacts on the FA composition of a clone of *D. pulex* isolated from a lake in California and found diet and somatic FA were most strongly correlated for ARA and EPA. In contrast, Burns et al. (unpublished data) studied a clone of *D. carinata* isolated in New Zealand and observed the best correlations for MUFA, C_{18} n-3s, EPA + DHA, and the n-3:n-6 ratio.

Despite the strong dietary impacts on *Daphnia* FA, they tend to accumulate less SAFA, more MUFA, and especially more ARA than what is found in their diets.

Also, when consuming diets that contain DHA, *Daphnia* tend to accumulate far less of this FA than what is present in their food. However, the differences in *Daphnia* FA composition when consuming different phytoplankton monoculture diets is pronounced. For example, *Daphnia* that consumed cryptophytes had on average 16 ± 4% (±1 SD) EPA in their FA pool, whereas *Daphnia* that consumed chlorophytes averaged only 1 ± 1% EPA.

As previously noted, the major phytoplankton groups have distinct FA profiles by which they can be readily separated using discriminant function analysis (Fig. 6.2). We used the freshwater phytoplankton FA data depicted in Fig. 6.2 and the FA composition of *Daphnia* fed monoculture phytoplankton diets (Brett et al. 2006; Müller-Navarra 2006; Müller-Navarra et al. unpublished data, Burns et al. unpublished data) to graphically demonstrate the strong impact of dietary FA on *Daphnia* FA composition (Fig. 6.3). In this DFA, the phytoplankton and *Daphnia* samples were treated as a single class, and 94.6% of these samples were correctly classified to phytoplankton group (e.g., chlorophyte), or to *Daphnia* eating phytoplankton from that group, using the "leave-one-out" algorithm. The first axis of this DFA explained 54.7% of the overall variation in these data and was strongly negatively

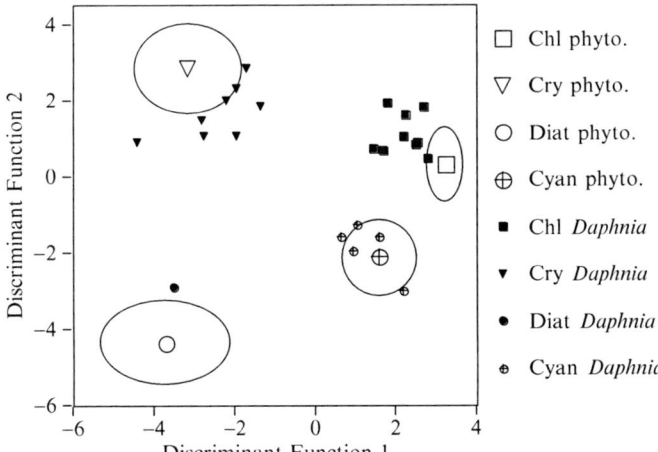

Fig. 6.3 A bivariate plot of the results of a discriminant function analysis of phytoplankton fatty acid composition and the FA composition of *Daphnia* consuming these phytoplankton. The ovoids around the phytoplankton group centroids represent the area delineated by ±1 SD on the X and Y axes. The large open symbols represent the phytoplankton group centroids. The small filled symbols represent the individual *Daphnia*-phytoplankton monoculture treatments. The first axis of this DFA explained 54.7% of the overall variation in these data and was very strongly negatively correlated with EPA, moderately negative correlated with DHA and the n-3:n-6 ratio and moderately positively correlated with C_{18} n-6 FA. The second axis explained 36.1% of the variation and was strongly positively associated with C_{18} n-3 FA, and moderately negatively associated with MUFA. A third axis (not shown) explained 9.1% of the variability and was moderately correlated with SAFA. *Chl* chlorophytes, *Cry* cryptophytes, *Diat* diatoms, *Cyan* cyanophytes, and *phyto.* phytoplankton. This figure is based on data from Brett et al. (2006), Müller-Navarra (2006), C.W. Burns (unpublished data) and D.C. Müller-Navarra (unpublished data)

correlated with EPA, moderately negatively correlated with DHA, and the n-3:n-6 ratio and moderately positively correlated with LIN. The second axis explained 36.1% of the variation and was strongly positively associated with C_{18} n-3 FA, and moderately negatively associated with MUFA. A third axis (not shown) explained 9.1% of the variability and was moderately correlated with SAFA. The results in Fig. 6.3 show diet FA composition had a distinct impact on *Daphnia* FA composition. However, this figure also shows that in 22 out of 23 cases the FA of the *Daphnia* was slightly more inclined toward a general central tendency relative to their diets. These data demonstrate that diet has a dominating impact on *Daphnia* FA composition but that, irrespective of diet, *Daphnia* retain some internally consistent features of their FA profiles.

The unpublished data of Burns et al. showed the cladoceran *Ceriodaphnia dubia* had similar responses to dietary FA as those noted above for *Daphnia* spp. *C. dubia* accumulated more ARA relative to their diets, and even when their diets were rich in DHA they accumulated very little of this FA. The SAFA, MUFA, and EPA content and n-3:n-6 ratio of *C. dubia* was moderately correlated with their monoculture diets, whereas the C_{18} n-3 and LIN where strongly correlated ($r^2 \approx 0.85$).

Less is known about dietary impacts on freshwater copepods, but Bourdier and Amblard (1989) explored this subject for the calanoid *Acanthodiaptommus denticornis*, Burns and colleagues (unpublished data) studied the calanoid *Boeckella* spp., and Caramujo et al. (2008) studied the harpacticoid *Attheyella trispinosa*. After feeding with chlorophyte, cyanophyte, and diatom monocultures, Bourdier and Amblard (1989) noted the neutral lipid composition of *A. denticornis* was closely linked to that of their diets. These authors also indicated that diet did not affect the FA composition of the structural polar lipids. These different dietary responses for neutral and polar lipids are well characterized for marine copepods (e.g., Lee et al. 1971). Burns et al. (unpublished data) noted that when fed chlorophyte, cyanophyte, and cryptophyte monocultures, the calanoid *Boeckella* spp. accumulated significantly more ARA, EPA, and DHA than its diet. The MUFA and LIN content and n-3:n-6 ratio of *Boeckella* was moderately correlated ($r^2 = 0.52$–0.62) with their diets, while SAFA and LIN were strongly correlated ($r^2 \approx 0.90$). Caramujo et al. (2008) fed diatom and cyanobacteria monocultures to *A. trispinosa* and found the neutral lipid 16:1n-7, LIN and EPA content of this harpacticoid was clearly influenced by diet. These authors also observed the FA of polar lipids was not influenced by diet.

6.4.2 Freshwater Zooplankton: Field Studies

Several recent studies have examined the FA composition of freshwater zooplankton collected from the field. Ballantyne et al. (2003) noted that cladocerans collected from Lake Washington tend to accumulate EPA, whereas copepods accumulated both EPA and especially DHA. Kainz et al. (2004) examined the accumulation of essential fatty acids (EFA) in different zooplankton size classes for a series of lakes on Vancouver Island, British Columbia. These authors found that all zooplankton size classes accumulated 2–4× more EFA than the seston, and that copepod-dominated "meso-zooplankton" (200–500 μm) tended to accumulate DHA, while cladoceran-dominated "macro-zooplankton" (>500 μm) tended to accumulate EPA. As suggested in their study, and more clearly shown in subsequent studies, size per se is not a rational basis for examining differences in zooplankton FA composition because very large differences exist in the EPA and DHA accumulation patterns between cladocerans and copepods, which are unrelated to size. That is, small herbivorous cladocerans have much more similar FA profiles to large herbivorous cladocerans than they do to small-sized copepods (Persson and Vrede 2006).

Persson and Vrede (2006) noted that zooplankton from oligotrophic alpine lakes in Sweden were greatly enriched with PUFA and HUFA relative to seston. Persson and Vrede (2006) also found that the FA composition of zooplankton was unrelated to that of the seston, but was related to zooplankton taxonomic affiliation and trophic mode. Similary, Smyntek et al. (2008) noted the FA profiles of the major freshwater zooplankton groups (i.e., cladocerans and copepods) differed systematically in the large lake systems they sampled, but within individual zooplankton taxa FA profiles appeared to be independent of the seston's FA composition. These results were similar to those of Müller-Navarra (2006), who found pronounced food dependency of *Daphnia's* FA composition when fed cultured algae but much weaker patterns for natural diets. In the field, significant relationships between the FA composition of seston and zooplankton were recorded for 18:4n-3 and LIN for gravid daphnids and in *Eudiaptomus* spp. for ARA, DHA (gravid animals), and ALA (animals without eggs) (Müller-Navarra 2006). As Müller-Navarra (2006) noted, the weaker relationships between seston and zooplankton FA composition in the field may be because there was less variation in the FA composition of the seston than there is for the phytoplankton monocultures utilized as food in laboratory studies. Persson and Vrede (2006) also noted that the seston FA composition varied little in the suite of relatively similar lakes they sampled making it more difficult to detect dietary impacts.

In contrast to the field studies mentioned earlier, Ravet et al. (2009) observed strong relations between seston and zooplankton FA composition in the mesotrophic Lake Washington. Lake Washington has dramatic shifts in phytoplankton biomass and community composition (Arhonditsis et al. 2003) that make it particularly amendable to studies of natural seston impacts on zooplankton FA composition. Overall, Ravet et al. (2009) found quite similar results for *Leptodiaptomus ashlandi* feeding on natural seston in Lake Washington compared with Burns et al.'s results for *Boeckella* spp. feeding on phytoplankton monocultures in the lab. Lake Washington *L. ashlandi* had a significantly lower proportion MUFA and LIN and significantly more C_{18} n-3, ARA and DHA and a higher n-3:n-6 ratio than did the seston collected on the same dates. This pattern was particularly pronounced for DHA, which was, on average, 4× more prevalent in *L. ashlandi* than in the seston. The seston's SAFA and n-3 HUFA content was moderately correlated with that of *L. ashlandi* ($r^2 \approx 0.77$). Although the sample sizes were much smaller for *Cyclops bicuspidatus thomasi* ($n = 5$), *Epischura nevadensis* ($n = 5$), and *Daphnia* spp. ($n = 4$), than for *L. ashlandi* ($n = 15$), these zooplankton also showed evidence of seston impacts on their FA composition. Similar to *L. ashlandi*, *Cyclops* had significantly less LIN and significantly more C_{18} n-3 and DHA and a higher n-3:n-6 ratio than the seston. SAFA, DHA, and the n-3:n-6 ratio had the strongest relations with diet for *Cyclops* ($r^2 \approx 0.90$). Probably because this copepod is predominantly predaceous, the FA composition of *E. nevadensis* was not correlated with any of the main FA functional groups in seston. However, *E. nevadensis* had 38% less SAFA, 2× more C_{18} n-3 and 21× as much DHA as the seston from the same dates. Lake Washington *Daphnia* had less SAFA, and more C_{18} n-3 FA, ARA, and EPA than their diets and the MUFA, C_{18} n-3 FA and EPA content of *Daphnia* was correlated with that of the seston.

6.4.3 Marine Calanoid Copepods

It is well established that the FA composition of the storage lipid fraction in marine copepods is influenced by their diet, and it is generally believed that dietary FA are incorporated, unmodified, into these lipids (Lee et al. 1971). These authors also noted the total lipid content of copepods was correlated with phytoplankton concentrations, as was the strength of the association between the FA composition of the diet and storage lipids. Lee et al. (1971) also reported that the FA composition of the structural phospholipids was not affected by diet. In another classic study, Graeve et al. (1994) showed that the FA profile of the boreal herbivorous copepod *Calanus finmarchicus* could be changed from a presumptive dinoflagellate dominated (as indicated by a high 18:4n-3 content) to a diatom dominated profile (i.e., high 16:1n-7 content) by feeding wild collected *C. finmarchicus* a diatom monoculture diet for 42 days. Similarly, these authors were able to switch the FA composition of *C. hyperboreus* from diatom-like to dinoflagellate-like by feeding this copepod a dinoflagellate diet for 47 days. Since these studies, many marine copepod field studies have assumed the FA 16:1n-7 and EPA represent diatom consumption and 18:4n-3 and DHA represent dinoflagellate consumption (Kattner et al. 1994; Scott et al. 2002). It has also been suggested that C_{14} and C_{16} SAFA and the MUFA 18:1n-9 are trophic markers for omnivorous feeding on ciliates and 18:1n-7 indicates bacterial consumption (Stevens et al. 2004; Peters et al. 2006). Recently, Peters et al. (2006) used a FA trophic marker approach to infer that the glacial relict copepod *Pseudocalanus acuspes* exhibited an opportunistic feeding strategy in the Baltic Sea. The FA profiles of *P. acuspes* indicated that their diet was dominated by ciliates, diatoms, dinoflagellates, and cyanobacteria depending on the time of year. In contrast, the FA patterns of carnivorous and omnivorous copepods cannot be as easily linked to diet as is the case for herbivorous copepods from temperate to polar regions. It should be noted that there is considerable overlap in the FA composition of the major phytoplankton groups, so caution should be exercised when attributing consumer FA to particular dietary sources based on individual FA. For example, LIN is prevalent in both cyanobacteria and chlorophytes, whereas cryptophytes share a high EPA and DHA content with diatoms and a high ALA and 18:4n-3 content with chlorophytes.

6.4.4 Harpacticoid Copepods

When looking at dietary impacts on the FA composition of the marine harpacticoid *Tisbe holothuriae,* Norsker and Støttrup (1994) reported this copepod accumulated between 27 and 50% n-3 HUFA when consuming diets containing 1–13% HUFA. These authors concluded that *T. holothuriae* was able to synthesize n-3 HUFA from ALA at high rates. However, despite this bioconversion capacity *T. holothuriae* achieved considerably higher nauplii production when consuming HUFA-rich diets. Similarly, Nanton and Castell (1998) found *Tisbe* sp. had a high n-3 HUFA content (i.e., 19–41% of total FA) regardless of the HUFA content of baker's yeast and phytoplankton diets (which had n-3 HUFA content varying between 1 and 36%). Furthermore, these authors found the DHA/EPA ratio (a larval fish nutritional index) of these copepods varied between 2.6:1 and 3.3:1 despite the fact that this ratio in their diets ranged between 0.1:1 and 12:1. Nanton and Castell (1998) concluded *Tisbe* had a high capacity to convert ALA to EPA and DHA, which they suggested was an adaptation to the fact that *Tisbe* occupies detritus-rich benthic habitats where n-3 HUFA might be scarce.

6.4.5 Artemia spp

Considerable research effort has been devoted to understanding how the FA composition of aquaculture food organisms like *Artemia* spp. is affected by diet and supplements. The vast majority of the research on *Artemia* FA concerns short-term supplementation (i.e. <24 h) designed to boost the EPA and DHA content in these naturally HUFA deficient crustaceans (Palmtag et al. 2006). *Artemia* are, in some regards, ideal food sources for aquaculture because their nauplii are of suitable size for a wide range of first feeding larval fish and these nauplii do not themselves require food. Unfortunately, *Artemia* normally have very low EFA content, particularly DHA. Furthermore, when starved *Artemia* readily catabolize EFA(Coutteau and Mourente 1997). One of the few studies that extended *Artemia* diet studies beyond 24 h found that when fed algae containing ARA and EPA, *Artemia* readily accumulated these FA; however, *Artemia* accumulated very little DHA irrespective of diet (Vismara et al. 2003).

6.4.6 Euphausiids (krill)

Krill are one of the most important zooplankton groups in terms of understanding the global trophic transfer of EFA from primary producers to upper trophic levels. Because of this, considerable effort has been directed at untangling the environmental factors that exert the greatest influence on euphausiid FA composition, particularly for the Antarctic species *Euphausia superba*. The FA of krill larval stages are dominated by EPA, DHA and 16:0. In contrast, in the adult stages with larger TAG stores, the FA 14:0, 16:0, and 18:1n-9 are dominant (Hagen et al. 2001). Lipid concentrations are the highest in gravid females. However, during the spawning season females may lose half of their total lipids because they typically spawn multiple times (e.g., Hagen et al. 2001). Because of these ontogenetic variations accompanied with the accumulation of very large lipid stores, it is particularly challenging to assess dietary impacts on euphausiid FA composition (Stübing et al. 2003). For example, *E. superba* store up sufficient lipid reserves to last through 6 months of near starvation conditions. Because of these reserves and experimental constraints (i.e., most feeding experiments by necessity last <2 months), it is difficult to impart strong dietary signals in the lipid composition of adult or juvenile euphausiids. Furthermore, some researchers have observed declining adult body mass during experiments indicating the krill were not feeding efficiently (Stübing et al. 2003), and it is generally thought that dietary impacts on zooplankton FA composition will be most evident when they are actively accumulating lipids.

In contrast to the studies above, which suggest krill maintain FA profiles that are somewhat independent of their diets, Falk-Petersen et al. (2000) used FA trophic markers to deduce the trophic levels of various polar euphasiids. These authors concluded that high 16:1n-7, 18:1n-7, SDA and EPA composition indicated herbivory. Specifically 16:1n-7 and EPA are indicators of diatom consumption and SDA and DHA are indicators of dinoflagellate consumption. High 18:1n-9 content and/or a high 18:1n-9/18:1n-7 ratio were suggested to be indicators of carnivory. Furthermore, a high 20:1n-9 and 22:1n-11 content was suggested as indicating carnivory on calanoid copepods specifically. On the basis of these assumptions, Falk-Petersen et al. (2000) inferred the euphausiids *Thysanoessa inermis* and *Euphausia crystallorophias* are herbivores, *T. rashii*, *T. macrura*, and *E. superba* are omnivores (with switching between phytoplankton and zooplankton diets during the year), and *T. longicaudata* and *Meganyctiphanes norvegia* are carnivores that primarily feed on the copepod *Calanus*. These inferences depend on an assumption that these euphausiids metabolize and bioconvert FA in similar ways. However, it is well established that many marine copepods can bioconvert C_{18} MUFA to 20:1n-9 and 22:1n-11 (Lee et al. 2006), and if any euphausiids had this capacity (but clearly some, such as *E. superba*, do not) they could be misclassified as copepod predators according to Falk-Petersen et al.'s scheme.

Stübing et al. (2003) showed that the FA composition of larval krill is very clearly modified by their diet. Larval *E. superba*, which start out with far less lipids than adults, also gained weight during these experiments and this weight gain was primarily due to lipids. In contrast to the sometimes ambiguous results obtained in dietary studies with adult krill FA, Stübing observed very clear trends when comparing the FA composition of krill diets to those of "fecal strings." These comparisons showed larval *E. superba* preferentially assimilated EPA, 16:1n-7, 16:4n-1, and DHA, with the pattern being particularly strong for EPA. Because of this, the residual FA in the fecal strings were relatively enriched with 16:0, 18:0 and 18:1n-9.

However, these results do not prove the preferentially assimilated FA were also accumulated, as the HUFA could have been catabolized. This type of question could be more easily resolved if ^{13}C-labeled FA were employed in feeding experiments as Graeve et al. (2005) did for marine copepods. More recently, Pond et al. (2005) showed that short-term inter-moult growth responses in krill were positively correlated with the concentration of diatom lipid biomarkers (i.e., 16:4n-1 and EPA) and the flagellate biomarker SDA. Furthermore, when feeding adult krill an exclusive diet of copepods, Cripps and Atkinson (2000) demonstrated clear changes in *E. superba* FA in 16 day feeding trails. In these experiments, the krill's EPA + DHA content increased from 20 to 45% and the sum of SAFA and MUFA declined from 41 to 27%.

6.4.7 *The FATM Approach Applied to Zooplankton*

As previously noted, there is great interest in using FA as trophic markers (Iverson – Chap. 12), particularly for studies of zooplankton feeding ecology (Dalsgaard et al. 2003). The FATM approach is a semiquantitative approach to reveal consumption of items with distinctive FA signatures. For more quantitative information (e.g., food web, carbon mass-balance calculations) information is needed on how the various FA are selectively metabolized as they are conveyed through aquatic food webs. Recently, a quantitative experimental approach was advanced that used ^{13}C-labeled phytoplankton to examine FA turnover times in marine copepods (e.g., Graeve et al. 2005). Knowing FA turnover times is particularly important (especially during short term experiments) as the FATM concept assumes equilibrium conditions in the system. Recent marine studies have tried to improve applications of the FATM approach. Instead of single FA and FA ratios, the whole FA pattern is analyzed by means of multivariate statistical methods (e.g., Quantitative FA Signature Approach, QFASA; see also Iverson – Chap. 12).

It is also possible to combine the FATM or QFASA approach with other food web tracer methods such as analyses of stable isotopes, sterols, gut pigments, gut content (including genetic markers), and lipophilic or even non-lipophilic anthropogenic substances (e.g., new and legacy contaminants) (Kainz and Fisk – Chap. 5). Studies have shown that lipid production in the zooplankton and seasonal variability for zooplankton dietary sources increases uncertainty within diet-zooplankton FA patterns. The ecological relevance of zooplankton species, their feeding strategies, and life history traits (incl. biosynthetic pathways and specific lipid composition requirements, differential catabolism of specific lipids, etc.) still needs to be better understood to allow us to more accurately interpret patterns observed in field studies before the full potential of the QFASA approach is realized (Iverson – Chap. 12). However, despite these caveats this approach has the potential to greatly improve our understanding of food web mass transfer, especially for the nutritionally critical lipids.

6.4.8 Inconsistencies Between Taxa and Laboratory and Field Studies

The published research on dietary impacts on zooplankton FA composition shows a wide range of responses from laboratory studies of *Daphnia* fed phytoplankton monocultures, where strong dietary impacts are observed, to field studies of adult euphausiids, where dietary responses are much more muted. Several field studies of freshwater zooplankton have failed to observe dietary impacts even for those taxa which show clear responses in laboratory studies. Some of these differences might be due to the fact that some zooplankton (e.g., *Daphnia*) are very fast growing and relatively lean so it only takes a short period of time to replace their lipid reserves with new dietary lipids (see Sect. 6.6). At the other extreme, marine copepods and especially euphausiids grow much more slowly and build up much larger lipid reserves over a period of months. Thus, recently acquired lipids are diluted into a much larger pool of previously stored lipids. In this case, it may only be possible to discern clear dietary signals in experiments lasting several months or more, which can be problematic for marine zooplankton that might be difficult to rear in the laboratory. It is noteworthy that studies looking at FA accumulation in larval euphausiids, which grow faster and have much smaller lipid reserves, show clearer evidence of dietary impacts on FA composition (Stübing et al. 2003).

It should also be noted that in oligotrophic freshwater systems zooplankton may often be food quantity limited (Persson et al. 2007) and for that reason zooplankton collected from these systems may have accumulated less storage lipids – which are more readily influenced by diet. In contrast, many laboratory studies utilize relatively high food concentrations >1 mg C (l^{-1}), which makes it more likely that the zooplankton will acquire new lipids. In addition, because most laboratory studies employ phytoplankton monocultures, and natural phytoplankton assemblages are rarely dominated by one group, the differences in the dietary FA profiles are much greater in laboratory than field studies. Finally, zooplankton probably also vary in their tendency to modify the FA profiles of stored lipids relative to their diets. Knowing how the major zooplankton groups differ in this regard is an important and as of yet unresolved research question.

6.5 Homeostatic Fatty Acid Composition Responses

Even amongst the zooplankton taxa that have FA profiles strongly influenced by diet, clear evidence of quasi-homeostatic[3] responses to dietary FA availability is evident. A quasi-homeostatic EFA content could be observed for freshwater

---

[3] A homeostatic response refers to a generally fixed elemental or biochemical composition in a consumer despite considerable variation in their diet. A "quasi-homeostatic" response indicates some (but much less) variation in the elemental or biochemical composition of a consumer compared to their diet.

zooplankton, although variability for individual FA was considerably higher than for phosphorus in daphnids (Müller-Navarra 2006). For example, when feeding *Daphnia galeata*, three algal cultures (*Scenedesmus obliquus, Cryptomonas erosa, Nitzschia palea*), for which EPA concentrations varied by a factor of 20×, EPA only varied by a factor of 2 in *D. galeata* (Müller-Navarra 2006). In Schöhsee, *Daphnia* spp. and *Eudiaptomus* spp. had higher and much less variable PUFA composition than the seston, especially for LIN and n-3 HUFA. (Müller-Navarra 2006). In Lake Washington, Ravet et al. (2009) observed the SAFA content of the freshwater calanoid copepod *Leptodiaptomus ashlandi* was clearly correlated with that of the seston over the course of a yearly sampling cycle ($r^2 = 0.75$) (see Fig. 6.4a). However, *L. ashlandi* had more SAFA than the seston during the spring diatom bloom in Lake Washington, i.e., 35 ± 1 (±1 SD) vs. 27 ± 1, respectively, and considerably less SAFA than the seston during the summer stratified period, i.e., 49 ± 6 vs. 66 ± 8, respectively. Since these samples were collected from the field, seasonal fluctuations in water temperatures as they affect the homeoviscous response should have also influenced the FA composition of *L. ashlandi*. However, in this case the trends observed (i.e., more SAFA than the seston when the water temperatures were low and less SAFA than the seston during the warm summer period) were more consistent with a quasi-homeostatic than a homeoviscous response.

In a laboratory study, Burns et al. (unpublished data) showed that the proportion of ALA plus SDA in the diet correlated strongly with the proportion of these FA in *Ceriodaphnia dubia* (see Fig. 6.4b). However, when *C. dubia* were fed cyanobacteria that had very little C_{18} n-3 FA, they still had ≈15% of these FA in their FA pool.

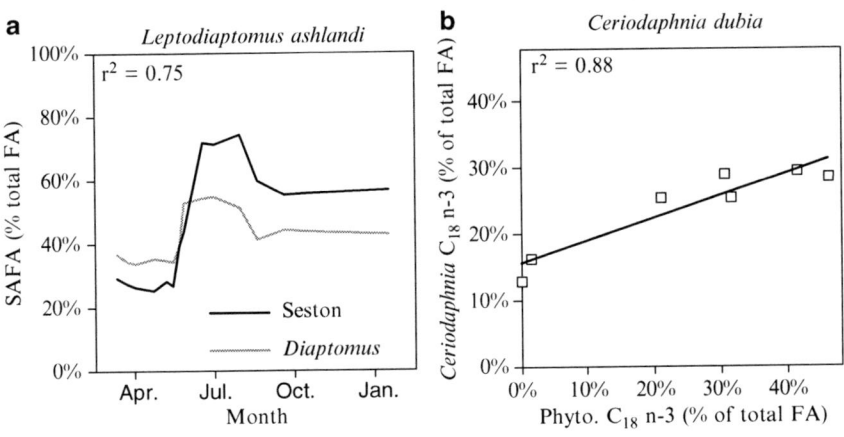

Fig. 6.4 Quasi-homeostatic responses in zooplankton-dietary FA composition. (**a**) This shows the relationship between the FA composition of natural seston and the calanoid copepod *Leptodiaptomus ashlandi* collected over an annual cycle in Lake Washington, USA (Ravet et al. 2009). (**b**) This shows the relationship between the FA composition of phytoplankton monocultures and the cladoceran *Ceriodaphnia dubia* obtained in a laboratory experiment (Burns et al. unpublished data)

Conversely, when *C. dubia* consumed cryptophytes that had ≈46% C_{18} n-3s, they contained ≈30% of these FA. Burns et al. (unpublished data) obtained similar results for *Daphnia carinata* and the copepod *Boeckella* spp. In both cases, the quasi-homeostatic responses primarily manifested themselves at the low range of dietary ALA and SDA availability where *D. carinata* and *Boeckella* contained substantially more C_{18} n-3 FA than their diets.

Daphnia spp. that consume SAFA-rich cyanobacteria tend to have only about half as much of these FA as their diet. In contrast, when *Daphnia* consume MUFA poor cryptophytes, they accumulate about twice the proportion of these FA as in their diets (Brett et al. 2006; Müller-Navarra 2006). This quasi-homeostatic response was particularly clear for *Daphnia* n-3:n-6 ratios, which averaged 16 ± 8 (±1 SD) in cryptophytes and 7 ± 5 in *Daphnia* that consume cryptophytes. Cyanobacteria had an average n-3:n-6 ratio of 0.3 and *Daphnia* that consumed cyanobacteria had an average n-3:n-6 ratio of 1.0. Similarly, Nanton and Castell (1998) observed the DHA/EPA ratio of the marine harpacticoid copepod *Tisbe* spp. only varied between 2.6 and 3.3 despite the fact that this same ratio in their phytoplankton and baker's yeast diets ranged between 0.1 and 12. It should also be noted that zooplankton that have systematically different FA composition than their diets also represent a form of homeostatsis.

6.6 Fatty Acid Turnover Times in Freshwater and Marine Zooplankton

When applying the FATM approach to zooplankton studies or merely trying to infer the impact of diet on zooplankton FA composition, it is important to know how long it takes for zooplankton to turn over their FA. That is, does the observed FA profile in a zooplankter reflect the food they consumed in the last 24 h or the last 2 months? In fact, much of the differences in zooplankton FA composition responses to dietary FA (see Sect. 6.4) may be due to widely varying FA turnover times (and experimental conditions employed) for the major zooplankton groups summarized. In its simplest sense, the FA turnover question can be conceptualized as a simple dilution model (see Jobling 2004, for an example with a fish context). According to a dilution model, the turnover time for particular FA should be most strongly influenced by the initial pool of a particular FA in the zooplankton tissues, normalized according to accrual of new FA. Even when lipid reserves are not being actively accumulated, FA can be replaced as older molecules of a particular FA are replaced by newer ones. The dilution of preexisting FA with new dietary FA should be simpler to conceptualize and easier to measure for neutral storage FA than for polar FA which have important structural roles and should therefore vary less with diet (Jobling 2004).

In experimental systems, where very fast growing zooplankton such as *Daphnia* are used and the experiments are initiated with <24 h old neonates, this is easy to address because new biomass is accrued so rapidly that nearly all of the maternal lipids initially present will be rapidly diluted by new growth. For example, a *Daphnia* growing at a rate of 0.4 $(d)^{-1}$ will increase its mass by a factor of 10 in 6 d. In this case, one could assume that nearly all of the observed FA were accrued during the experiment. However, in natural systems, even fast growing zooplankton like *Daphnia* often grow well below their optima. In these cases, the FA accumulated in the daphnids may have accrued over one or more weeks. This could make it more difficult to detect unequivocal signals of dietary impacts on the FA composition of wild collected zooplankton. This question is even more challenging for zooplankton such as marine copepods and especially euphausiids, which accumulate very large lipid reserves over long periods. In this case, accumulation of new FA on a daily basis is minor relative to the lipids previously stored. This and the fact that many marine zooplankton are difficult to rear in a laboratory setting can make it more challenging to detect dietary signals.

To date only one study has addressed the zooplankton FA turnover question in a systematic fashion. Graeve et al. (2005) used ^{13}C labeled diets to determine how long it took three different species of Arctic *Calanus* to turn over their FA and fatty alcohol pools. These authors concluded the marine copepod *C. hyperboreus* exchanged nearly all of its original lipid pool after 11 d. In contrast, *C. finmarchicus* and *C. glacialis* exchanged 22% and 45% of their lipids, respectively, after 14 days, even though they were not actually growing. Since lipid turnover rates will depend on growth rates, FA turnover should also vary with zooplankter life stage, food availability, and water temperature.

6.7 Zooplankton Reproductive Investment in Fatty Acids

Müller-Navarra (2006) compared the FA composition of *Daphnia* spp. somatic tissues and subitaneous eggs to that of monoculture chlorophyte, cryptophyte, and diatom diets. She demonstrated that the LIN, ALA, SDA, and EPA composition and n-3:n-6 ratios of *Daphnia* somatic tissues were strongly correlated with those of their diets. Müller-Navarra's results also showed that *D. galeata* tended to have a higher absolute and relative content of ARA and less DHA than their diets, both in somatic tissue and eggs. Compared with the respective somatic tissues, subitaneous eggs had substantially higher total FA, SAFA, n-3 PUFA and n-6 PUFA content and higher n-3:n-6 ratios. Overall the FA composition of both somatic tissues and subitaneous eggs was very strongly influenced by diet when fed monocultures. In a similar study, Wacker and Martin-Creuzburg (2007) compared the FA composition of *Daphnia magna* somatic tissues and subitaneous eggs to those of high and low food quality diets. They found *D. magna* eggs contained significantly more SAFA, MUFA, and n-6 and n-3 PUFA than somatic tissues irrespective of diet quality. Among the n-6 PUFA, ARA was significantly enriched in eggs but LIN was not. Müller-Navarra (2006) found that ALA, SDA, and EPA were 2–3× enriched in eggs over somatic tissues. Wacker and Martin-Creuzburg (2007) found that EPA was the most strongly enriched, being 2.4× higher in subitaneous eggs than in somatic tissues. These authors suggested that *Daphnia* enrich their subitaneous eggs with EPA as a buffer against low food quality cyanobacteria blooms.

Abrusán et al. (2007) compared the FA composition of *Daphnia pulicaria* subitaneous and resting eggs when they consumed *Scenedesmus obliquus*. [Unfortunately, these authors did not present FA composition data for the somatic tissues of the *Daphnia* they used]. These authors noted that *D. pulicaria* hatched from resting eggs had higher growth and egg production rates than *D. pulicaria* from subitaneous eggs, so it is likely *D. pulicaria* invest more essential biochemicals in resting eggs. Consistent with this expectation, Abrusán et al. (2007) found that when *D. pulicaria* consumed EPA deficient *S. obliquus*, their starvation induced resting eggs contained 3.5× higher amounts of total FA and these resting eggs contained a substantially lower proportion of the SAFA 16:0 and the MUFA 16:1n-9. These resting eggs also had higher proportions of the C_{18} PUFA LIN, and in particular ALA and SDA, and a higher n-3:n-6 ratio. Furthermore, resting eggs contained approximately 0.7 ± 0.5% EPA, but this FA was barely detected in the subitaneous eggs of *D. pulicaria* fed *S. obliquus*. Overall, both n-3 and n-6 PUFA and HUFA accounted for 52% of total FA in the resting eggs and only 27% in the subitaneous eggs. Therefore, these studies show *Daphnia* invest heavily in the total FA content of both subitaneous and resting eggs and that these eggs were preferentially enriched with LIN and ARA and especially n-3 FA, although the specific EFA enriched (i.e., LIN or ARA, ALA or EPA) seems to vary from one case to the other.

The results obtained by Ederington et al. (1995) for the marine copepod *Acartia tonsa* contrast notably in several regards from the *Daphnia* results presented above. These authors' data showed that when *Acartia* consumed diatoms both their somatic tissues and their eggs were clearly enriched with EPA, whereas the somatic tissues and eggs of *A. tonsa* consuming the bacterivorous ciliate *Pleuronema* (which had 18:1n-11 as its dominant FA) were enriched with 18:1n-11. However, irrespective of diet both the somatic tissues and eggs of *A. tonsa* were dramatically enriched with the SAFA 16:0 and 18:0. Remarkably, Ederington et al.'s (1995) results suggest that the eggs of *A. tonsa* fed ciliates were almost completely devoid of n-3 and n-6 FA, and were almost entirely composed of 18:0 (50%), 16:0 (30%), and 18:1n-11 (12%) with other SAFA and MUFA making up the balance of FA in these eggs. While these authors did report that ciliate consumption was associated with dramatically reduced egg production, they also reported that 95% of the eggs from copepods fed ciliates were viable, which they suggested meant EFA were not essential for egg hatching. This latter result is paradoxical and should be validated in future studies.

Overall, little is known about the PUFA requirements for subitaneous and diapause eggs in both marine and freshwater copepods. In copepods, variability of FA independent of diets may be due to different requirements of the different developmental stages, with differences for somatic growth and egg production. During egg production, conversion of storage lipids and/or dietary lipids to PL takes place. These PL are then transported to the gonads where they become part of the egg yolk (lipovitelin production). In cladocerans, these fractions seem to be especially rich in PUFA as *Daphnia* eggs have several fold higher PUFA and especially HUFA contents than the somatic tissue (Müller-Navarra 2006). Thus, one can hypothesize that egg production in copepods may also be HUFA intensive. However, the FA composition of freshwater and marine copepod eggs warrants further research.

6.8 Temperature Impacts on Zooplankton Fatty Acids: The Homeoviscous Response

Although the first studies to examine the FA composition of zooplankton were specifically directed at examining temperature impacts on zooplankton FA (Farkas & Herodek 1964), few studies have focused on this topic since this pioneering research. Farkas and colleagues initially noted that those freshwater zooplankton that over-winter in an active life stage (i.e., copepods) were able to strongly modify their FA composition in response to temperature variation (Farkas 1979; Farkas et al. 1984). Copepods exposed to cold stress greatly increased the proportion of DHA and decreased the proportion of the SAFA 18:0 in their FA pools. It is now well known that nearly all poikilotherms adapt to cold stress, by increasing the proportions of PUFA, and in particular HUFA, in their membrane lipids (Hazel 1995). It was also suggested that zooplankton that do not over-winter in an active phase (i.e., cladocerans) do not have the capacity to modify their lipid composition in response to cold challenges and therefore over-winter as resting-eggs (Farkas 1979; Farkas et al. 1984). These authors also stressed the particular importance of DHA for zooplankton cold-water adaptation.

Several of Farkas assertions were challenged in a recent study by Schlechtriem et al. (2006), as well as by Arts et al.'s (1992) observation that some *Daphnia* are able to over-winter in ice-covered lakes. While supporting Farkas' fundamental observation that zooplankton adapt to cold stress by increasing the proportion of HUFA in their lipids, Schlechtriem et al. challenged the assertion that *Daphnia* cannot over-winter in an active phase and presented very clear evidence that *Daphnia* can increase their n-3 HUFA composition in response to cold stress. Schlechtriem et al. (2006) showed *Daphnia pulex* fed *Ankistrodesmus falcatus* at 11°C had 4× as much EPA (12.7% vs. 3.1%) compared with *Daphnia* grown at 22°C. Furthermore, cold-adapted *Daphnia* had 4× more 16:1n-7 (13.0% vs. 3.3%) and only half as much LIN (6.8% vs. 14.3%) and ALA (11.3% vs. 22.4%) as warm adapted *Daphnia* (see Fig. 6.5). Cold-adapted *Daphnia* also did not accumulate DHA, which is consistent with the broader observation that cladocerans as a group do not accumulate this FA. This taxa-specific observation also contradicts Farkas' assertion that DHA plays an essential role in membrane adaptation for zooplankton.

Kattner and Hagen (Chap. 11) also noted the high EPA and DHA content of tropical marine copepods calls into question the membrane fluidity hypothesis, and suggested these molecules may play important roles in membrane structure and function besides maintaining fluidity. Nanton and Castell (1999) reported the FA composition responses of the benthic marine harpacticoid copepods *Amonardia* and *Tisbe* to 6, 15, and 20°C temperature treatments. Unfortunately, this study produced rather equivocal results because the copepod FA profiles were generally more similar for the 6 and 20°C treatments than for the 15°C treatment. Specifically, both copepods had substantially lower DHA content when cultured at 15°C (Nanton and Castell 1999).

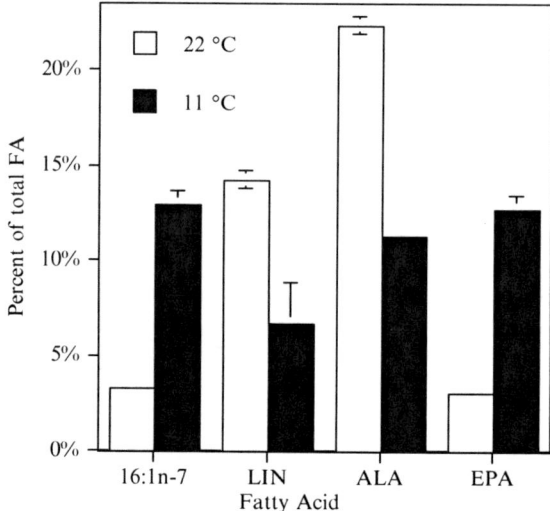

Fig. 6.5 The percent of total fatty acids for those FA in *Daphnia pulex* that showed a clear response (i.e. 16:1n-7, LIN, ALA and EPA) to high (22°C) and low (11°C) temperature treatments. All other FA showed much smaller responses to the temperature treatments. The results presented are treatment means ±1 SD. This figure is based on the data reported in Schlechtriem et al. (2006)

6.9 Starvation Impacts on Zooplankton Fatty Acids

Several studies have attempted to quantify starvation impacts on zooplankton FA composition, with Schlechtriem et al. (2006) presenting the clearest results. These authors cultured *Daphnia pulex* on the chlorophyte *Ankistrodesmus falcatus* and then starved individuals at 22 and 11°C. The daphnids starved at 22°C died within three days, during which time their FA composition did not change markedly. However, many of the daphnids starved at 11°C survived until day six during which time their FA composition changed dramatically. In particular, the mass of SAFA, MUFA, and ALA per individual declined markedly, whereas LIN, ARA, and EPA were conserved. This study could serve as a model for other studies attempting to assess starvation impacts on zooplankton FA composition.

6.10 Unanswered Questions

1. Why do freshwater zooplankton accumulate ARA? A number of studies have shown pronounced ARA accumulation by a variety of zooplankters, but so far no study has demonstrated a clear physiological justification (e.g., enhanced growth or reproduction) for ARA accumulation.

2. Are differences in the homeoviscous response the reason why cladocerans are less likely to over-winter in an active phase than copepods? Farkas and colleagues originally hypothesized that cladocerans were forced to over-winter as resting eggs because they were unable to modify the HUFA content of their lipids. In contrast, copepods were observed to dramatically increase their DHA content in response to cold stress. However, recent results challenge this hypothesis for daphnids and suggest *Daphnia* are able to exist in an active phase even at very low temperatures. Do the storage lipids and greater starvation tolerance of copepods also play an important role in over-wintering as suggested by research on marine copepods and krill?
3. Is the FA composition of polar lipids unaffected by diet for all zooplankton? Many studies have mentioned structural lipids are relatively unaffected by diet, whereas the composition of neutral lipids should be highly responsive to dietary lipid accumulation. Is this generally the case or are there exceptions to this "rule"? [See Stübing et al. (2003) and Hagen et al. (1996) for contradictory results in certain euphausiids].
4. How much endogenous FA production occurs in zooplankton? Many studies have cited Goulden and Place's (1990) observation that only 2% of *Daphnia* FA are produced de novo when consuming *Ankistrodesmus falcatus*. However, the generality of this result has not been tested for different *Daphnia* species or different diets, much less for other freshwater zooplankton. In contrast to *Daphnia*, it is well known that polar copepods synthesize C_{20} and C_{22} MUFA from dietary carbohydrates and proteins (Lee et al. 2006). Are marine zooplankton better able to synthesize FA than freshwater zooplankton, or is this difference between cladocerans and copepods independent of habitat?
5. To what extent are zooplankton capable of converting one EFA molecule to another; for example, LIN to ARA or ALA, and SDA to EPA or DHA? If zooplankton can make these conversions, what are the energetic costs of these conversions and how do these costs vary from one group to another (e.g., between cladocerans and copepods in freshwater and copepods and euphausids in marine systems)? Are carnivorous zooplankton less able to bioconvert C_{18} PUFA to C_{20} and C_{22} HUFA than herbivorous zooplankton? Knowing how these conversion capacities vary among zooplankton taxa will allow us to better understand to what extent zooplankton are able to "upgrade" the FA content of the food they consume.
6. Which zooplankton groups have FA profiles that are the most strongly influenced by diet and which have FA profiles which are the most "fixed" and why? The research conducted to date indicates cladocerans such as *Daphnia* have quite plastic FA profiles that are strongly influenced by diet, whereas some zooplankton have less obvious responses to the FA composition of their diets. Are these differences due to the greater role (and slower turnover) of storage lipids in the latter group? Or alternatively: Are some zooplankton more inclined to modify the composition of the dietary lipids they accumulate?
7. Why do cladocerans tend to accumulate relatively more EPA whereas copepods accumulate DHA? Various hypotheses have been offered to explain these patterns.

For example, the hypothesis that DHA plays an important role in conveying nerve impulses. However, so far none of these hypotheses have been rigorously tested.

8. What role do the "congener" molecules of ARA, EPA, and DHA (e.g., 20:3n-6, 20:3n-3, and 22:5n-3, respectively) play in the nutritional ecology and FA metabolism of marine and freshwater zooplankton? While most studies only report results for these HUFA, it is clear that congeners of these molecules are commonly encountered and may in some cases actually be more prevalent than the classically considered forms of HUFA. At this time, virtually nothing is known about how these molecules are utilized by zooplankton.

6.11 Conclusions

The various crustacean zooplankton groups found in the ocean and freshwater lakes have markedly different FA composition. Marine copepods and euphausids in temperate and polar regions store large pools of lipids as wax esters and to a lesser extent TAG composed predominantly of MUFA. Freshwater copepods accumulate EPA and especially DHA, while freshwater cladocerans preferentially accumulate EPA. Both freshwater copepods and cladocerans accumulate ARA compared with their diets. In general, nearly all zooplankton accumulate less SAFA, and a higher portion n-3 PUFA and especially n-3 and n-6 HUFA relative to their available diets. Diet has a very strong impact on the FA composition of some zooplankton (in particular *Daphnia* spp.), a moderate impact on some copepods, and only a small impact on the FA composition of zooplankton such as detritivorous harpacticoid copepods and adult euphausids. However, even the zooplankton taxa with the most plastic FA composition show clear homeostatic FA responses to diets with widely varying FA composition. Most zooplankton exposed to cold-stress show a clear homeoviscous response, with greatly increased accumulation of n-3 HUFA. When starved, zooplankton preferentially retain n-3 and n-6 HUFA. It has been shown that *Daphnia* heavily invest both C_{18} n-3 PUFA and EPA in their eggs. As outlined in this review, there remain many challenging research questions pertaining to zooplankton FA composition to occupy aquatic ecologists during the next decade.

References

Abrusán, G., Fink, P., and Lampert, W. 2007. Biochemical limitation of resting egg production in *Daphnia*. Limnol. Oceangr. 52:1724–1728.

Ackman, R.G. and Eaton, C.A. 1966. Lipids of the fin whale (*Baluenoptera physalus*) from North Atlantic waters. III. Occurrence of eicosenoic and docosenoic fatty acids in the zooplankter *Meganyctiphanes norvegica* (M. Sars) and their effect on whale oil composition. Can. J. Biochem. 44:1561–1566.

Adams, S.M. 1999. Ecological role of lipids in the health and success of fish populations, pp.132–160. In M.T. Arts and B.C. Wainman [eds.], Lipids in Freshwater Ecosystems. Springer, New York.

Ahlgren, G., Lundstedt, L., Brett, M.T., and Forsberg, C. 1990. Lipid composition and food quality of some freshwater phytoplankton for cladoceran zooplankters. J. Plankton Res. 12:809–818.

Ahlgren, G., Gustafsson, I.-B., and Boberg, M. 1992. Fatty acid content and chemical composition of freshwater microalgae. J. Phycol. 28:37–50.

Arhonditsis, G.B., Brett, M.T., and Frodge, J. 2003. Environmental control and limnological impacts of a large recurrent spring bloom in Lake Washington, USA. Env. Manag. 31:603–618.

Arts, M.T., Ackman, R.G., and Holub, B.J. 2001. "Essential fatty acids" in aquatic ecosystems: a crucial link between diet and human health and evolution. Can J. Fish. Aq. Sci. 58:122–137.

Arts, M.T., Evans, M.S., and Robarts, R.D. 1992. Seasonal patterns of total and energy reserve lipids of dominant zooplanktonic crustaceans from a hypereutrophic lake. Oecologia 90:560–571.

Ballantyne, A.P., Brett, M.T., and Schindler, D.E. 2003. The importance of dietary phosphorus and highly unsaturated fatty acids for sockeye (*Oncorhynchus nerka*) growth in Lake Washington – a bioenergetics approach. Can J. Fish. Aq. Sci. 60:12–22.

Bourdier, G., and Amblard, C. 1989. Lipids in *Acanthodiaptomus denticornis* during starvation and fed on three different algae. J. Plankton Res. 11:1201–1212.

Brett, M.T., Müller-Navarra, D.C., Ballantyne, A.P., and Ravet, J.L. 2006. *Daphnia* fatty acid composition reflects that of their diet. Limnol. Oceanogr. 51:2428–2437.

Broglio, E., Jonasdóttir, S.H., Calbet, A., Jakobsen, H.H., Saiz, E. 2003. Effect of heterotrophic versus autotrophic food on feeding and reproduction of the calanoid copepod *Acartia tonsa*: relationship with prey fatty acid composition. Aquat. Microb. Ecol. 31:267–278.

Caramujo, M.-J., Boschker, H.T.S., and Admiraal, W. 2008. Fatty acid profiles of algae mark the development and composition of harpacticoid copepods. Freshw. Biol. 53:77–90.

Coutteau, P., and Mourente, G. 1997. Lipid classes and their content of n-3 highly unsaturated fatty acids (HUFA) in *Artemia franciscana* after hatching, HUFA-enrichment and subsequent starvation. Mar. Biol. 130:81–91.

Cripps, G.C., and Atkinson, A. 2000. Fatty acid composition as an indicator of carnivory in Antarctic krill, *Euphausia superba*. Can. J. Fish. Aq. Sci. 57:31–37. Suppl. 3.

Cripps, G.C., Watkins, J.L., Hill, H.J., and Atkinson, A. 1999. Fatty acid content of Antarctic krill *Euphausia superba* at South Georgia related to regional populations and variations in diet. Mar. Ecol. Prog. Ser. 181:177–188.

Csengeri, I., and Halver, J.E. 2006. Tibor Farkas 1929-2003. A biographical memoir. National Academy of Sciences, Washington, D.C. 27 pgs.

D'Abramo, L.R., and Sheen, S.-S. 1993. Polyunsaturated fatty acid nutrition in juvenile freshwater prawn *Macrobrachium rosenbergii*. Aquaculture 115:63–86.

Dalsgaard, J., St. John, M., Kattner, G., Müller-Navarra, D.C., and Hagen, W. 2003. Fatty acid trophic markers in the pelagic marine food environment. Adv. Mar. Biol. 46:226–340.

Desvilettes, C., Bourdier, G., Amblard, C., and Barth, B. 1997. Use of fatty acids for the assessment of zooplankton grazing on bacteria, protozoans and microalgae. Freshw. Biol. 38:629–637.

Dunstan, G.A., Volkman, J.K., Barrett, S.M., Leroi, J.M., and Jeffrey, S.W. 1994. Essential polyunsaturated fatty-acids from 14 species of diatom (Bacillariophyceae). Phytochemistry 35:155–161.

Dunstan, G.A., Brown, M.R., and Volkman, J.K. 2005. Cryptophyceae and Rhodophyceae; chemotaxonomy, phylogeny, and application. Phytochemistry 66:2557–2570.

Ederington, M.C., McManus, G.B., and Harvey, H.R. 1995. Trophic transfer of fatty-acids, sterols, and a triterpeniod alcohol between bacteria, a ciliate, and the copepod *Acartia tonsa*. Limnol. Oceanogr. 40:860–867.

Elendt, B.-P. 1990. Nutritional quality of a microencapsulated diet for *Daphnia magna*. Effects on reproduction, fatty acid composition, and midgut ultrastructure. Arch. Hydrobiol. 118:461–475.

Falk-Petersen, S., Hagen, W., Kattner, G., Clarke, A., and Sargent, J.R. 2000. Lipids, trophic relationships, and biodiversity in Arctic and Antarctic krill. Can. J. Fish. Aq Sci. 57:178–191. Suppl. 3.

Farkas, T. 1970. Fats in fresh water crustaceans. Acta Biol. Acad. Sci. Hung. 21:225–233.

Farkas, T. 1979. Adaptation of fatty-acid compositions to temperature-study on planktonic crustaceans. Comp. Biochem. Physiol. 64B:71–76.

Farkas, T., and Herodek, S. 1964. Effect of environmental temperature on fatty acid composition of crustacean plankton. J. Lipid Res. 5:369–373.

Farkas, T., Nemecz, G., and Csengeri, I. 1984. Differential response of lipid-metabolism and membrane physical state by an actively and passively over wintering planktonic crustacean. Lipids 19:436–442.

Fraser, A.J., Sargent, J.R., Gamble, J.C., and Seaton, D.D. 1989. Formation and transfer of fatty acids in an enclosed marine food chain comprising phytoplankton, zooplankton and herring (*Clupea harengus* L.) larvae. Mar. Chem. 27:1–18.

Gatenby, C.M., Orcutt, D.M., Kreeger, D.A., Parker, B.C., Jones, V.A., and Neves, R.J. 2003. Biochemical composition of three algal species proposed as food for captive freshwater mussels. J. Appl. Phycol. 15:1–11.

Goulden, C. E., and Place, A. R. 1990. Fatty acid synthesis and accumulation rates in daphnids. J. Exp. Zool. 256:168–178.

Graeve, M., Kattner, G., and Hagen, W. 1994. Diet-induced changes in the fatty acid composition of Arctic herbivorous copepods: experimental evidence of trophic markers. J. Exp. Mar. Biol. Ecol. 182:97–110.

Graeve, M., Albers, C., and Kattner, G. 2005. Assimilation and biosynthesis of lipids in Arctic *Calanus* species based on feeding experiments with a ^{13}C labelled diatom. J. Exp. Mar. Biol. Ecol. 317:109–125.

Hagen, W., Kattner, G., and Graeve, M. 1993. *Calanoides acutus* and *Calanus propinquus*, Antarctic copepods with different lipid storage modes via wax esters or triacylglycerols. Mar. Ecol. Prog. Ser. 97:135–142.

Hagen, W., van Vleet, E.S., and Kattner, G. 1996. Seasonal lipid storage as overwintering strategy of Antarctic krill. Mar. Ecol. Progr. Ser. 134:85–89.

Hagen, W., Kattner, G., Terbruggen, A., and Van Vleet, E.S. 2001. Lipid metabolism of the Antarctic krill *Euphausia superba* and its ecological implications. Mar. Biol. 139:95–104.

Hazel, J.R. 1995. Thermal adaptation in biological-membranes- is homeoviscous adaptation the explanation. Ann. Rev. Physiol. 57:19–42.

Henderson, R.J., Park, M.T., and Sargent, J.R. 1995. The desaturation and elongation of ^{14}C-labelled polyunsaturated fatty acids by pike (*Esox lucius* L.) *in vivo*. Fish Physiol. Biochem. 14:223–235.

Hessen, D.O., and Leu, E. 2006. Trophic transfer and trophic modification of fatty acids in high Arctic lakes. Freshw. Biol. 51:1987–1998.

Jeffries, H.P. 1970. Seasonal composition of temperate plankton communities: fatty acids. Limnol. Oceanogr. 15:419–426.

Jobling, M. 2004. Are modifications in tissue fatty acid profiles following a change in diet the result of dilution? Test of a simple dilution model. Aquaculture 232:551–562.

Jónasdóttir, S.H., Fields, D., and Pantoja, S. 1995. Copepod egg production in Long Island Sound USA, as a function of the chemical composition of seston. Mar. Ecol. Progr. Ser. 119:87–98.

Kainz, M., Arts, M.T., and Mazumder, A. 2004. Essential fatty acids in the planktonic food web and their ecological role for higher trophic levels. Limnol. Oceanogr. 49:1784–1793.

Kattner, G., Graeve, M., and Hagen, W. 1994. Ontogenetic and seasonal changes in lipid and fatty acid/alcohol compositions of the dominant Antarctic copepods *Calanus propinquus*, *Calanoides acutus* and *Rhincalanus gigas*. Mar. Biol. 118:637–644.

Langdon, C.J., and Waldock, M.J. 1981. The effect of algal and artificial diets on the growth and fatty acid composition of *Crassostrea gigas* spat. J. Mar. Boil. Ass. U.K. 61:431–448.

Lee, R.F., Nevenzel, J.C., and Paffenhofer, G.-A. 1971. Importance of wax esters and other lipids in the marine food chain: phytoplankton and copepods. Mar. Biol. 9:99–108.

Lee, R.F., Hagen, W., and Kattner, G. 2006. Lipid storage in marine zooplankton. Mar. Ecol. Prog. Ser. 307:273–306.

Lenz, P.H., Hartline, D.K. and Davis, A.D. 2000. The need for speed. I. Fast reactions and myelinated axons in copepods. J. Comp. Phys. A 186:337–345.

Lewis, R.W. 1969. The fatty acid composition of arctic marine phytoplankton and zooplankton with special reference to minor acids. Limnol. Oceanogr. 14:35–40.

Lourenco, S.O., Barbarino, E., Mancini-Filho, J., Schinke, K.P., and Aidar, E. 2002. Effects of different nitrogen sources on the growth and biochemical profile of 10 marine microalgae in batch culture: an evaluation for aquaculture. Phycologia 41:158–168.
Lovern, J.A. 1935. The fats of some plankton crustacea. Biochem. J. 29:847–849.
Müller-Navarra, D.C. 1995a. Evidence that a highly unsaturated fatty acid limits *Daphnia* growth in nature. Arch. Hydrobiol. 132:297–307.
Müller-Navarra, D.C. 1995b. Biochemical versus mineral limitation in *Daphnia*. Limnol. Oceanogr. 40:1209–1214.
Müller-Navarra, D.C. 2006. The nutritional importance of polyunsaturated fatty acids and their use as trophic markers for herbivorous zooplankton: Does it contradict? Arch. Hydrobiol. 167:501–513.
Müller-Navarra, D.C., Brett, M.T., Liston, A., and Goldman, C.R. 2000. A highly-unsaturated fatty acid predicts biomass transfer between primary producers and consumers. Nature 403:74–77.
Nanton, D.A., and Castell, J.D. 1998. The effects of dietary fatty acids on the fatty acid composition of the harpacticoid copepod, *Tisbe* sp, for use as a live food for marine fish larvae. Aquaculture 163:251–261.
Nanton, D.A., and Castell, J.D. 1999. The effects of temperature and dietary fatty acids on the fatty acid composition of harpacticoid copepods, for use as a live food for marine fish larvae. Aquaculture 175:167–181.
Norsker, N.H., and Støttrup, J.G. 1994. The importance of dietary HUFA for fecundity and HUFA content in the harpacticoid, *Tisbe holothuriae* Humes. Aquaculture 125:155–166.
Olsen, Y. 1999. Lipids and essential fatty acids in aquatic food webs: what can freshwater ecologists learn from mariculture?, pp. 161–202. In M.T. Arts and B.C. Wainman [eds.], Lipids in Freshwater Ecosystems. Springer, New York.
Palmtag, M.R., Faulk, C.K., and Holt, G.J. 2006. Highly unsaturated fatty acid composition of rotifers (*Brachionus plicatilis*) and *Artemia* fed various enrichments. J. World Aquaculture Soc. 37:126–131.
Parrish, C.C., McKenzie, C.H., MacDonald, B.A., and Hatfield, E.A. 1995. Seasonal studies of seston lipids in relation to microplankton species composition and scallop growth in South Broad Cove, Newfoundland. Mar. Ecol. Progr. Ser. 129:151–164.
Patil, V., Kallqvist, T., Olsen, E., Vogt, G., and Gislerod, H.R. 2007. Fatty acid composition of 12 microalgae for possible use in aquaculture feed. Aquaculture Int. 15:1–9.
Persson, J., and Vrede, T. 2006. Polyunsaturated fatty acids in zooplankton: variation due to taxonomy and trophic position. Freshw. Biol. 51:887–900.
Persson, J., Brett, M.T., Vrede, T., and Ravet, J.L. 2007. Food quantity and quality regulation of trophic transfer between primary producers and a keystone grazer (*Daphnia*) in pelagic freshwater food webs. Oikos 116:1152–1163.
Peters, J., Renz, J., van Beusekom, J., Boersma, M., and Hagen, W. 2006. Trophodynamics and seasonal cycle of the copepod *Pseudocalanus acuspes* in the Central Baltic Sea (Bornholm Basin): evidence from lipid composition. Mar. Biol. 149:1417–1429.
Pond, D.W., Atkinson, A., Shreeve, R.S., Tarling, G., and Ward, P. 2005. Diatom fatty acid biomarkers indicate recent growth rates in Antarctic krill. Limnol. Oceanogr. 50:732–736.
Provasoli, L., and D'Agostino, A. 1969. Development of artificial media for *Artemia salina*. Biol. Bull. 136:434–453.
Ravet, J.L., Brett, M.T., and Arhonditsis, G.B. 2009. The effects of seston lipids on zooplankton fatty acid composition in Lake Washington. Ecology (in press).
Ravet, J.L., Brett, M.T., and Müller-Navarra, D.C. 2003. A test of the role of polyunsaturated fatty acids in phytoplankton food quality for *Daphnia* using liposome supplementation. Limnol. Oceanogr. 48:1938–1947.
Reitan, K.I., Rainuzzo, J.R., and Olsen, Y. 1994. Effect of nutrient limitation on fatty-acid and lipid-content of marine microalgae. J. Phycol. 30:972–979.
Renaud, S.M., Thinh, L.V., and Parry, D.L. 1999. The gross chemical composition and fatty acid composition of 18 species of tropical Australian microalgae for possible use in mariculture. Aquaculture 170:147–159.

Renaud, S.M., Thinh, L.V., Lambrinidis, G., and Parry, D.L. 2002. Effect of temperature on growth, chemical composition and fatty acid composition of tropical Australian microalgae grown in batch cultures. Aquaculture 211:1–4.

Richoux, N.B., Deibel, D., Raymond, J., Thompson, R.J., and Parrish, C.C. 2005. Seasonal and developmental variation in the fatty acid composition of *Mysis mixta* (Mysidacea) and *Acanthostepheia malmgreni* (Amphipoda) from the hyperbenthos of a cold-ocean environment (Conception Bay, Newfoundland). J. Plankton Res. 27:719–733.

Sargent, J.R., McEvoy, L., Estevez, A., Bell, G., Bell, M., Henderson, J., and Tocher, D. 1999. Lipid nutrition of marine fish during early development: current status and future directions. Aquaculture 179:217–229.

Sargent, J.R., and Henderson, R.J., 1986. Lipids, pp. 59–108. In: E.D.S. Corner and S. O'Hara [eds.], Biological Chemistry of Marine Copepods, Oxford University Press, Oxford.

Schlechtriem, C., Arts, M.T., and Zellmer, I.D. 2006. Effect of temperature on the fatty acid composition and temporal trajectories of fatty acids in fasting *Daphnia pulex* (crustacea, cladocera) Lipids 41:397–400.

Schmidt, K., Atkinson, A., Petzke, K.J., Voss, M., and Pond, D.W. 2006. Protozoans as a food source for Antarctic krill, *Euphausia superba*: Complementary insights from stomach content, fatty acids, and stable isotopes. Limnol. Oceanogr. 51:2409–2427.

Scott, C.L., Kwasniewski, S., Falk-Petersen, S., and Sargent, J.R. 2002. Species differences, origins and functions of fatty alcohols and fatty acids in the wax esters and phospholipids of *Calanus hyperboreus*, *C. glacialis* and *C. finmarchicus* from Arctic waters. Mar. Ecol. Prog. Ser. 235:127–134.

Simopoulos, A.P. 1999. Essential fatty acids in health and chronic disease. Am. J. Clin. Nutr. 70:560S–569S.

Smyntek, P.M., Teece, M.A., Schulz, K.L., and Storch, A.J. 2008. Taxonomic differences in the essential fatty acid composition of groups of freshwater zooplankton relate to reproductive demands and generation time. Freshw. Biol. 53:1768–1782.

Stevens, C.J., Deibel, D., and Parrish, C.C. 2004. Copepod omnivory in the North Water Polynya (Baffin Bay) during autumn: spatial patterns in lipid composition. Deep-Sea Res. Part I 51:1637–1658.

Stübing, D., Hagen, W., and Schmidt, K. 2003. On the use of lipid biomarkers in marine food web analyses: An experimental case study on the Antarctic krill, *Euphausia superba*. Limnol. Oceanogr. 48:1685–1700.

Tremblay, R., Cartier, S., Miner, P., Pernet, F., Quere, C., Moal, J., Muzellec, M.L., Mazuret, M., and Samain, J.F. 2007. Effect of *Rhodomonas salina* addition to a standard hatchery diet during the early ontogeny of the scallop *Pecten maximus*. Aquaculture 26:410–418.

Veloza, A.J., Chu, F.L., and Tang, K.W. 2006. Trophic modification of essential fatty acids by heterotrophic protists and its effects on the fatty acid composition of the copepod *Acartia tonsa*. Mar. Biol. 48:779–788.

Vismara, R., Vestri, S., Barsanti, L., and Gualtieri, P. 2003. Diet-induced variations in fatty acid content and composition of two on-grown stages of *Artemia salina*. J. Appl. Phycol. 15:477–483.

Volkman, J.K., Jeffrey, S.W., Nichols, P.D., Rogers, G.I., and Garland, C.D. 1989. Fatty-acid and lipid-composition of 10 sepcies of microalgae used in mariculture. J. Exp. Mar. Biol. Ecol. 128:219–240.

Wacker, A., and Martin-Creuzburg, D. 2007. Allocation of essential lipids in *Daphnia magna* during exposure to poor food quality. Funct. Ecol. 21:738–747.

Wacker, A., Becher, P., and von Elert, E., 2002. Food quality effects of unsaturated fatty acids on larvae of the zebra mussel *Dreissena polymorpha*. Limnol. Oceanogr. 47:1242–1248.

Weers, P.M.M., Siewertsen, K., and Gulati, R.D. 1997. Is the fatty acid composition of *Daphnia galeata* determined by the fatty acid composition of the ingested diet? Freshw. Biol. 38:731–738.

Chapter 7
Fatty Acid Ratios in Freshwater Fish, Zooplankton and Zoobenthos – Are There Specific Optima?

Gunnel Ahlgren, Tobias Vrede, and Willem Goedkoop

7.1 Introduction

Two groups of polyunsaturated fatty acids (PUFA), termed omega-3 and omega-6 in food (or here as n-3 and n-6 PUFA, respectively), are essential for all vertebrates and probably also for nearly all invertebrates. The absolute concentrations of the different PUFA are important, as is an appropriate balance between the two. The optimal ratio of n-3/n-6 is not known for most organisms but is anticipated to be more or less species-specific (Sargent et al. 1995). The three most important PUFA in vertebrates are eicosapentaenoic acid (EPA, 20:5n-3), docosahexaenoic acid (DHA, 22:6n-3) and arachidonic acid (ARA, 20:4n-6). Both EPA and ARA are precursors for biologically active eicosanoids that are vital components of cell membranes and play many dynamic roles in mediating and controlling a wide array of cellular activities (Crawford et al. 1989; Harrison 1990; Henderson et al. 1996; see Chap. 9). Since n-3 and n-6 PUFA cannot be synthesized de novo by most metazoans, they must be included in the diet, either as EPA, DHA and ARA, or as their precursors, such as α-linolenic acid (ALA, 18:3n-3, precursor of EPA and DHA) and linoleic acid (LIN, 18:2n-6, precursor of ARA) (Bell et al. 1986; Sargent et al. 1995). Both ALA and LIN are produced in the thylacoid membranes of algae and plants with chlorophyll (Sargent at al. 1987). The same set of enzymes, Δ6 and Δ5 desaturases, are required for the elongation and desaturation of ALA and LIN to produce EPA and ARA (Bell et al. 1986; Sargent et al. 1995; see Chap. 9). This results in competition between the two fatty acid families for these enzymes. During the conversion of EPA to DHA, Δ6 desaturase is also needed resulting in additional competition for this enzyme (Voss et al. 1991; Sargent et al. 1995). Consequently, both the absolute concentrations and the dietary proportions between n-3 and n-6 PUFA are important (Sargent et al. 1997). If one fatty acid is greatly in excess of the other, suppression of the desaturation process of the less abundant FA will take place.

G. Ahlgren (✉), T. Vrede, and W. Goedkoop
Department of Ecology and Evolution (Limnology), Uppsala University, PO Box 573, Uppsala, Sweden, SE-751 23
e-mail: Gunnel.Ahlgren@ebc.uu.se

In mammals, both DHA and ARA have been determined to be limiting factors in the evolution of the brain. Early *Homo sapiens* were associated with aquatic environments that provided them access to fish and shellfish rich in the major lipids (especially DHA) found in mammalian brain tissue (Broadhurst et al. 1998; Crawford et al. 1999). In contrast, the high ARA content in human grey matter probably originated from the seeds of higher plants. In addition, there is a remarkable similarity in DHA/ARA ratios with an average of 2 ± 0.2 (standard deviation, SD) in all mammalian brain phospholipids including man (Crawford et al. 1976). In contrast, there are great differences in mammalian liver DHA/ARA ratios, with graminivores (both non-ruminants and grass-eating ruminants) having a ratio of 0.2 ± 0.2, but small mammals having a ratio ≤1. In carnivores the liver DHA/ARA ratio is higher than in omnivorous humans, 1 and 0.5, respectively (Crawford et al. 1976). However, the dolphin, a marine mammal, has a DHA/ARA ratio of 4 (Crawford et al. 1976). The general rule that in aquatic food webs the n-3/n-6 ratios are >1 and in terrestrial food webs <1 (Crawford and Marsh 1989; Olsen 1999) seems to hold also for DHA/ARA ratios. Koussoroplis et al. (2008) showed a gradual decrease in adipose tissue DHA/LIN ratios (0.46–0.01) for six semi-aquatic mammals when their dependence on terrestrial food increased. A similar decrease is also seen in the DHA/ARA ratio (2.0–0.1).

The concentrations of DHA in vertebrates are especially high in neural tissues such as brain, retina, and auditory and olfactory nerves. In comparison, ARA and EPA are preferentially incorporated into structural lipids in the phosphoglycerides of cell membranes. In addition, ARA and EPA are important precursors of the hormone-like eicosanoids that modulate and regulate many functions in cells, e.g. blood flow, glandular secretions and smooth-muscle contraction (Crawford et al. 1989). For example, there is a delicate balance between vasodilation and vasoconstriction of blood vessels in fish to help maintain proper regulation of blood flow that is probably maintained by an optimum ratio between n-3 and n-6 PUFA. When mammals evolved, they required n-6 fatty acids for the evolution of the placenta and mammary glands. Hence, ARA has been hypothesized to be the 'missing biochemical link' (Crawford et al. 1999).

Knowledge that the long-chained n-3 PUFA in fish are beneficial to human health (e.g. Arts et al. 2001; see Chap. 14) has led aquatic food web researchers to focus mainly on n-3 fatty acids. However, the importance of ARA has been emphasized in two reviews concerning the optimal of dietary PUFA compositions for marine larval fish (Sargent et al. 1997; 1999). Both fish and mammals have similar eicosanoid precursors for prostanoid and leukotriene production (Bell et al. 1986; Sargent 1995). Fish eggs are exceedingly rich in DHA, probably because of its important role in the formation of neural cellular membranes; however, small quantities of ARA are also required in fish eggs for eicosanoid formation (Sargent 1995). For example, ARA is important in adult fish for the formation of the eicosanoids that regulate egg shedding via synchronization of ovulation and spawning thereby preventing over-ripening of eggs (Sargent et al. 1997). In addition, ARA seems to be important to the general survival of juvenile turbot, whereas DHA improves their somatic growth (Castell et al. 1994). The high ARA content of turbot gills and kidneys, and sea bass sperm, suggests ARA has a specific physiological role in these tissues (Castell et al. 1994; Bell et al. 1996). The presence or absence of ARA in the food of seabream larvae had no effect on their size or growth; however, ARA-enriched diets (DHA/ARA = 5.6) improved survival during stress challenges compared to DHA-enriched diets (DHA/ARA = 30) (Koven et al. 2001). Sargent (1995) suggested that the optimum n-3/n-6 ratio for marine brood-stock diets and fish eggs lies between 5:1 and 10:1.

Studies of fatty acid metabolism in zooplankton and benthic animals are just commencing; thus, the physiological role of PUFA in these groups is incomplete. However, invertebrates were evolutionary forerunners of vertebrates, resulting in the basic biochemistry being similar throughout the entire animal kingdom, for example, amino acids in proteins and mechanisms for energy production (Crawford and Marsh 1989; Crawford et al. 1999). It is therefore highly likely that the function of PUFA is similar in both invertebrates and vertebrates. Thus far, EPA and DHA have been highlighted in feeding experiments with zooplankters (e.g. Elendt 1990; Brett and Müller-Navarra 1997; Ravet et al. 2003), while the potential importance of n-6 fatty acids and especially ARA has gone unrecognized. In fact, only a few feeding experiments with invertebrates have included manipulations of ARA (e.g. Von Elert and Wolffrom 2001; Von Elert 2002; Becker and Boersma 2005). It is highly likely that the n-6 PUFA are also important in invertebrates and that the balance between n-3 and n-6 PUFA plays an important role. The n-3/n-6 ratio is suggested as a putative marker of the amount of autochthonous versus allochthonous matter and was positively correlated with chlorophyll *a* content in the food sources for invertebrates (Torres-Ruiz et al. 2007). The DHA/ARA ratio is likely of importance in strictly carnivorous animals which as a group lack the Δ5 desaturase enzyme required to convert ALA to DHA and LIN to ARA (Crawford et al. 1989). However, the ability of invertebrates to elongate and desaturate C_{18} PUFA into biologically active EPA, DHA and ARA may vary, both amongst and within species during different life stages.

The PUFA requirements, when culturing commercially important crustacean decapods (shrimps, mussels, oyster and crayfish), have been extensively covered in the aquaculture literature (e.g. Harrison 1990; Xu et al. 1993). Harrison (1990) stressed that research is needed to determine not only the optimal levels of certain fatty acids but also the correct proportions of dietary n-3 and n-6 PUFA. In feeding experiments with the Chinese prawn (*Penaeus chinensis*) Xu et al. (1993) found that additions of ARA resulted in the same moulting frequency as additions of DHA, leading these authors to conclude ARA is involved in the moulting process. It is also worth mentioning that eicosanoids produced from ARA were found to be involved in egg-laying processes in several terrestrial insect species (Stanley-Samuelson 1994).

The primary objective of this chapter is to review the research on n-3 and n-6 PUFA, focusing on n-3/n-6 and DHA/ARA ratios, and to discuss the physiological importance and ecological implications of these ratios in freshwater fish and the invertebrates that commonly comprise fish prey. The second objective is to identify knowledge gaps regarding the role of PUFA in freshwater food webs. Comparisons of wild versus cultured species and their food are made with the presumption that 'natural is best' (Sargent 1995). An argument in support of this statement is that species can be expected to adapt by evolving their biochemical requirements to best suit the food resources that are available in their environment. If environments are stable over a longer time period, other, less well-adapted, species will eventually be out-competed. The following questions are addressed in this review:

- Are there systematic differences in the n-3/n-6 and DHA/ARA ratios of herbivorous and carnivorous fish, and are there ontogenetic differences within species, as found in terrestrial animals?
- Are there differences in the PUFA ratios of wild versus cultured fishes?
- What are the characteristic n-3/n-6 and DHA/ARA ratios of wild freshwater invertebrates?
- Are there specific optima in n-3/n-6 ratios in aquatic invertebrates?

7.2 Methods

The data in Tables 7.1–7.7 (located at the end of this chapter) were obtained by calculating n-3/n-6 and DHA/ARA ratios using: (a) values obtained from raw data files provided by the authors (Persson and Vrede 2006; Kainz et al. 2006), (b) from published tables, or (c) estimated from published figures. It should be noted that these ratios are independent of the units of the original FA data (e.g. μg L^{-1}, mg g^{-1} dry weight, mg g^{-1} C). Differences amongst groups were tested by one- or two-way analysis of variance (ANOVA) and Tukey-Kramer Honestly Significant Difference (HSD) post hoc tests with $\alpha = 0.05$. Prior to the statistical analyses, ratios were $\log_{10}$-transformed to meet the normality assumption required by ANOVA.

7.3 n-3/n-6 and DHA/ARA Ratios in Freshwater Metazoans

7.3.1 Freshwater Fish

7.3.1.1 Wild Fish

The n-3/n-6 and DHA/ARA ratios of the dorsal muscle tissue of 130 freshwater fish samples, representing 15 different species collected from 20 different, mostly small Swedish freshwaters, were on average 2.8 ± 1.1 and 3.1 ± 1.7, respectively (Ahlgren et al. 1994, 1996, 1999). A significant correlation was not found between PUFA ratios and fork length for the whole data set or by fork length within three functional feeding groups, herbivorous–omnivorous, carnivorous–benthivorous and carnivorous–piscivorous. This suggests these ratios do not change during ontogenetic development (Fig. 7.1). However, both n-3/n-6 and DHA/ARA ratios differed significantly between the functional feeding groups (one-way ANOVA, $p < 0.0001$),

7 Fatty Acid Ratios in Freshwater Fish

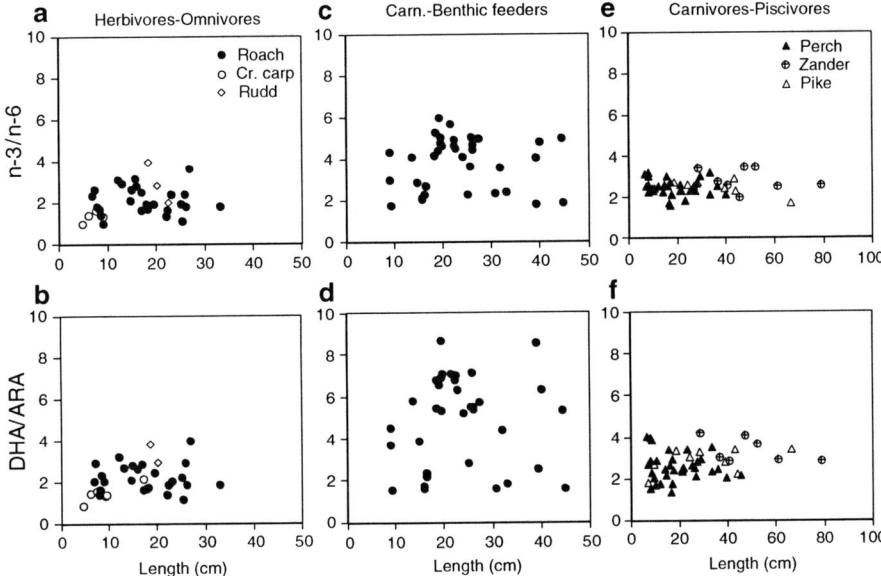

Fig. 7.1 PUFA ratios vs. fork length in: (**a, b**) Three species of herbivorous–omnivorous fish: Crucian carp (*Carassus carassus*), Roach (*Rutilus rutilus*), Rudd (*Scardinius erythrophthalmus*). Average and standard deviation (SD) of n-3/n-6 and DHA/ARA ratios are 2.07 ± 0.71 and 2.17 ± 0.72, respectively ($n = 33$), (**c, d**) Nine species of carnivorous–benthic feeders: Bleak (*Alburnus alburnus*), Bream (*Abramis brama*), Burbot (*Lota lota*), Grayling (*Thymallus thymallus*), Ide (*Leuciscus idus*), Ruffe (*Gymnocephalus cernuus*), Tench (*Tinca tinca*), White bream (*Blicca bjoerkna*), Whitefish (*Coregonus* sp,). Average and SD of n-3/n-6 and DHA/ARA ratios are 3.77 ± 1.26 and 4.76 ± 2.16, respectively ($n = 40$), (**e, f**) Three species of carnivorous–piscivorous fish: Perch (*Perca fluviatilis*), Zander (*Lucioperca lucioperca*), Pike (*Esox lucius*). Average and SD of n-3/n-6 and DHA/ARA ratios are 2.59 ± 0.69 and 2.73 ± 0.71, respectively ($n = 57$)

and post hoc comparisons showed that all groups were significantly different from each other. The ratios were highest in carnivorous–benthivorous fish (3.7 and 4.6, respectively), intermediate in carnivorous–piscivorous fish (2.5 and 2.6) and lowest in herbivorous–omnivorous fish (2.0 and 2.1). When comparing n-3/n-6 and DHA/ARA ratios in the herbivorous–omnivorous roach (*Rutilus rutilus*) and the carnivorous perch (*Perca fluviatilis*) from a nutrient-rich versus a nutrient-poor lake, there was a significant difference in n-3/n-6 ratio (two-way ANOVA, $p = 0.0002$) but not in DHA/ARA ratio (two-way ANOVA, $p = 0.06$) (Fig. 7.2). The n-3/n-6 ratios were higher in perch than in roach ($p = 0.0005$), and higher in eutrophic Lake Valloxen than in oligotrophic Lake Siggeforasjön ($p = 0.02$). Interestingly, the lake-time-species interaction was also significant ($p = 0.005$). At a length of 10–15 cm a shift in food quality is noticeable in both species. However, the new food increased in both ratios for perch and decreased for roach, resulting in large differences in FA quality between the two species (Fig. 7.2). Thus, the herbivorous–omnivorous roach was more flexible concerning food items, and the carnivorous perch fed on food of similar fatty acid composition in the two lakes.

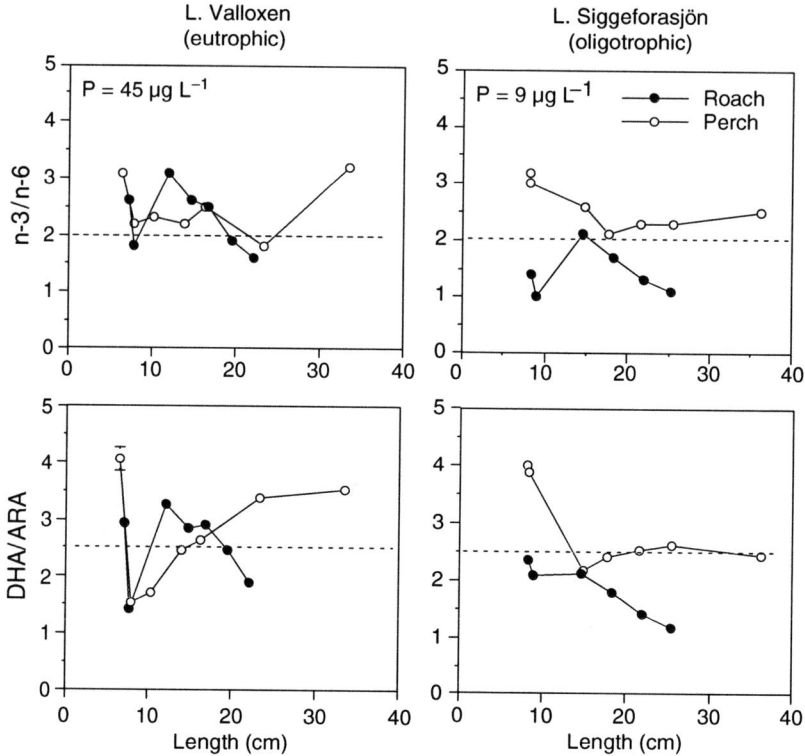

Fig. 7.2 PUFA ratios vs. length of roach and perch in two different lake types. The points represent 2–10 fish mixed in the samples in the eutrophic lake and 1–5 fish mixed in the oligotrophic lake (the smaller the fish, the higher the number mixed). The broken lines make the comparison easier between the two lakes. P, total phosphorous concentration

A similar pattern was found in fish from five tropical lakes in Ethiopia (Zenebe et al. 1998a; b). Great variation was observed in the n-3/n-6 and DHA/ARA ratios of herbivorous tilapia (*Oreochromis niloticus*) compared to ratios for the carnivorous catfish (*Clarias gariepinus*) among lakes. Significant differences could also be seen in both ratios between these two fish species taken from the two most productive lakes (lakes Chamo and Haiq; Fig. 7.3). Thus, the supply and the quality of food are probably the main mechanisms controlling PUFA content in herbivorous–omnivorous fish, whereas species identity is more important for the PUFA pattern in carnivorous fish. Comparative t-tests between the n-3/n-6 ratios in the Swedish (temperate) and the Ethiopian (tropical) freshwater fish showed no significant differences (Zenebe et al. 1998a).

7.3.1.2 Feeding Experiments with Fish

Studies of marine fish by Navas et al. (1993) and Thrush et al. (1993) (see also Sargent et al. 1997) were among the first to indicate that ARA is important to egg

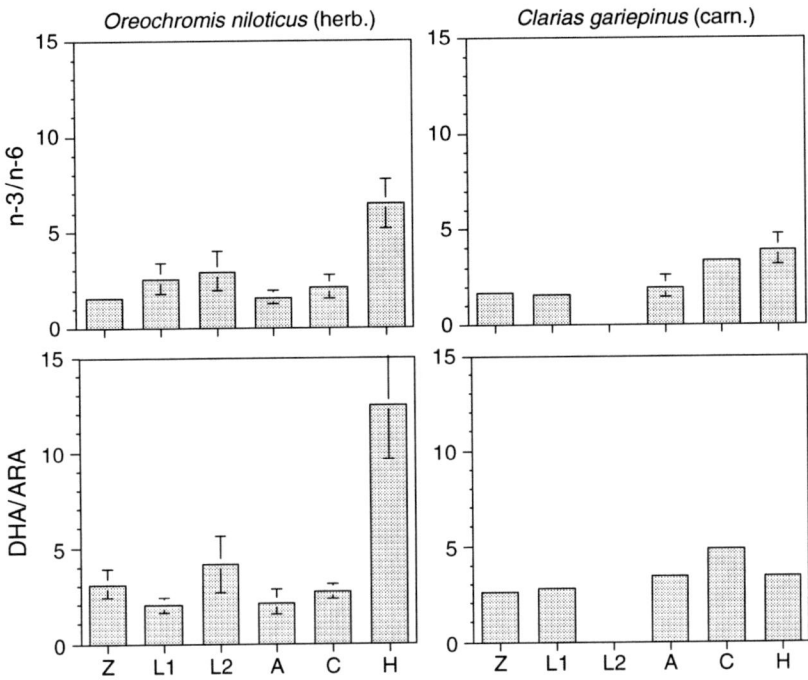

Fig. 7.3 PUFA ratios in five lakes of two tropical fish species, the herbivorous *Oreochromis niloticus* and the carnivorous *Clarias gariepinus*. Bars indicate standard deviation within each lake ($n = 3$): Ziway (Z), Langano (L), Awassa (A), Chamo (C) and Haiq (H). Average and SD of n-3/n-6 and DHA/ARA ratios are for *Oreochromis* 2.9 ± 1.9 and 4.4 ± 3.5, respectively, and for *Clarias* 2.5 ± 1.1 and 3.4 ± 0.9, respectively

quality in fish. In feeding experiments with sea bass (*Dicentrarchus labrax*), the fish in a high ARA treatment (EPA/ARA = 1.5:1) produced eggs with higher survival and more successful hatching than did fish in a low ARA treatment (EPA/ARA = 15:1). In contrast, larvae of yellowtail flounder (*Limanda ferruginea*) grew best on DHA-enriched rotifers (DHA/ARA = 25) (Copeman et al. 2002). The control diet, consisting of rotifers fed olive oil and no additions of PUFA (mainly 18:1n-9, DHA/ARA = 2.5), resulted in a body weight which was only one-fourth that of the DHA-enriched treatments. Despite the other differences observed, pigmentation results were equal in the two dietary regimes. However, the percentage of complete eye migration in the small flounder was 75% for the fish fed the DHA + EPA-enriched rotifer diet (DHA/ARA = 11), but only 46–47% for fish larvae fed either DHA- or ARA-enriched rotifers (DHA/ARA = 25 and 3, respectively). Thus, the optimum DHA/ARA ratios vary depending on which criterion is used. Surprisingly, when fed the control diets the flounders were enriched with DHA by a factor 4.1 and with ARA by a factor 4.0 (Copeman et al. 2002). ARA was concentrated in all diets except the DHA + ARA diet, whereas DHA was concentrated only in the control diet. Apparently, the flounder larvae required ARA nearly as much as DHA.

Yang and Dick (1994) carried out a series of LIN and ALA addition feeding experiments with freshwater populations of the Arctic charr (*Salvelinus alpinus*) and rainbow trout (*Oncorhynchus mykiss*). Both species were able to convert C_{18} PUFA to DHA and ARA, but rainbow trout were somewhat more efficient than Arctic charr (Table 7.1). However, both species grew best on commercial marine fish-based diets that contained both DHA and ARA. Two piranha species, the herbivorous *Mylossoma aureum* and the carnivorous *Serrasalmus natterie*, were fed oat flakes and *Chironomus* larvae, respectively (Henderson et al. 1996) and their tissues strongly mirrored the n-3/n-6 ratios of their diets (Table 7.1). The herbivore contained no DHA but contained small amounts of ARA, whereas the carnivore contained DHA that was probably produced from EPA in the *Chironomus* larvae. Liver samples from both species contained the two PUFA with a DHA/ARA ratio of 0.29 and 1.8, respectively. In feeding experiments with sea bass, even changes in the DHA/ARA ratios of the fish sperm were seen (Bell et al. 1996). Commercial pellets and fish meal derived from *Boops boops* had DHA/ARA ratios of 15 and 5, respectively, resulting in ratios of 28 and 13, respectively, in phosphatidylserine (the greatest fraction of phospholipids) in sperm. In a feeding experiment with Tilapia, somatic growth was only seen when the fish were fed cichlid pellets and *Microcystis* colonies (Zenebe et al. 2003). DHA decreased and ARA increased in the feed series from pellets to unfed, showing a decrease in the DHA/ARA ratio in the fish (Table 7.1). The phytoplankton diets, *Microcystis* (single-celled), *Scenedesmus*, and helical filaments of *Arthrospira*, were apparently not sufficiently nutritious to support growth in juvenile *Tilapia*.

7.3.1.3 Wild Versus Cultured Fish

Several studies have shown various species of cultured fish differ in lipid composition from their wild counterparts (e.g. Otwell and Richards 1981/1982; Van Vliet and Katan 1990; Bell et al. 1994). These studies have shown the n-3/n-6 ratio was twofold to threefold higher in cultured fish compared to wild fish and ARA was very low in cultured fish, giving a threefold to fivefold higher DHA/ARA ratio than in wild fish. Comparison between wild and cultured grayling (*Thymallus thymallus*) also showed great differences in the fatty acid patterns and contents (Ahlgren et al. 1999). In cultured fish, the mean n-3/n-6 ratio was doubled and the DHA/ARA ratio was 8X higher than in wild fish (Table 7.2). The fatty acid pattern in pellets used as food (EWOS ST-40, based on marine fish meal) for cultured grayling differed substantially from that of natural food (various invertebrates), DHA/ARA 24 and 0.3, respectively.

Data from wild salmon show average n-3/n-6 and DHA/ARA ratios of 5 and 3.5, respectively, compared with ratios of ≥10 for both n-3/n-6 and DHA/ARA in farmed stocks (Pickova et al. 1999) (Table 7.2). In two other salmonids, Arctic charr and rainbow trout, the n-3/n-6 ratios were doubled and the DHA/ARA ratios were 4X higher in cultured fish compared to wild fish (Yang and Dick 1994). In

Atlantic salmon (*Salmo salar*) from Norwegian rivers, DHA/ARA ratios can be much higher in both wild and reared fish, 25 and 31, respectively (Olsen 1999). The river salmon was starved and much leaner than the farmed fish with a lipid content of 3–4 and 14–23% of fresh weight, respectively. The percentage of individual fatty acids was similar, showing that reduction during starvation was nearly proportional. The starved fish had probably been ocean going and thus eating the same food of marine origin as the farmed fish. Bell et al. (1994) reported commercial Atlantic salmon pellets had a DHA/ARA ratio of 15, whereas the same ratio in wild salmon food was only 0.24. This is a more characteristic difference between marine and freshwater prey. The sometimes great differences in the fatty acid ratios between wild and cultured fish reveal the commercial feed are often far from natural food. Farmed fish often develop dermal lesions, such as fin erosion, as well as high fry mortality. One reason for this might be unnaturally high DHA concentrations in the feed. We agree with Bell et al. (1994), who suggested that commercial feed should resemble natural food. It is also not advisable, as is commonly done with freshwater salmonids, to feed typically freshwater fish feed of marine origin.

7.3.2 Freshwater Zooplankton

7.3.2.1 Wild Zooplankton

The PUFA content of crustacean zooplankton appears to be affected both by species identity and the availability of dietary PUFA. It is known that wild cladocerans contain large amounts of EPA, smaller amounts of ARA and minute amounts of DHA, whereas copepods contain mostly DHA, with relatively less EPA and ARA (Persson and Vrede 2006; see Chap. 6). Persson and Vrede (2006) suggested that the high concentrations of DHA in copepods could be a prerequisite for a more highly developed nervous system compared to cladocerans. Copepods have abundant chemoreceptors and mechanoreceptors on their antennae and mouth appendices (Yen et al. 1992) that are essential for their abilities to feed selectively (DeMott 1986), track mates (Weissburg et al. 1998) and detect and escape predators. In addition to these phylogenetic differences, dietary supply affects the PUFA content of zooplankton. For example, the PUFA content of predatory zooplankton, which fed on a PUFA-rich diet, was higher than that of filter-feeding zooplankton with a diet low in PUFA (Persson and Vrede 2006). Likewise, *Daphnia* PUFA content and fatty acid ratios varied depending on the PUFA content of their diet (Brett et al. 2006; see Chap. 6).

Despite the variation in PUFA content of zooplankton due to differences in dietary PUFA availability and feeding modes, n-3/n-6 ratios are surprisingly stable within wild cladocerans and copepods (Higgs et al. 1995; Persson and Vrede 2006; Hessen and Leu 2006; Müller-Navarra 2006). Filter-feeding cladocerans had an average n-3/n-6 ratio of 3.0 and the carnivorous cladoceran *Bythotrephes* had a ratio of 2.4, which is within the range of variation observed among the filter-feeding cladoceran taxa (Table 7.3).

The copepods had an average n-3/n-6 ratio of 4.1 mainly due to their high DHA content, which is significantly higher than the n-3/n-6 ratio of cladocerans (one-way ANOVA, $p = 0.01$). There is no apparent difference in n-3/n-6 ratios between predominantly herbivorous diaptomid copepods and carnivorous copepods in the genera *Heterocope* and *Cyclops* (Table 7.3). The biochemical difference between cladocerans and copepods (again due primarily to differences in DHA concentrations) is even greater when DHA/ARA ratios are compared, generally <1 and 6, respectively (Table 7.3). Compared to seston, copepods have both high DHA content and the presence of n-6 docosapentaenoic acid (n-6DPA, 22:5n-6), whereas seston have low DHA content and little or no n-6DPA; both of which suggests that copepods have a great capacity to convert not only ALA and LIN to EPA and ARA but also EPA and ARA to longer-chain PUFA (Persson and Vrede 2006). A copepod diet with high levels of LIN, ALA and/or ARA could lead to a low EPA to DHA conversion rate because of competition for the Δ6 enzyme.

In marine copepods, even higher n-3/n-6 and DHA/ARA ratios have been found than in freshwaters. For example, polar copepods have relatively stable n-3/n-6 ratios of 11 ± 2.5 and variable DHA/ARA ratios of 18 ± 9.6 (Dalsgaard et al. 2003). This freshwater and marine difference merely reflects the differences in the composition of the available seston between these systems. High ratios were also found in mesocosm studies in the southern Baltic Sea (Ahlgren et al. 2005), where high n-3/n-6 ratios in phytoplankton resulted in an abnormally high DHA/ARA ratio in copepods (≥40) compared to data from the Norwegian Sea (≈20). The very high ratios indicated an imbalance in fatty acid composition at the base of the food webs that might have negative implications for copepods as well as for organisms at higher trophic levels in the Baltic.

7.3.2.2 Feeding Experiments with Crustaceans Fed on Chlorophytes

It has been shown both in plankton food web studies and in feeding experiments with *Daphnia* spp. that ARA and EPA are preferentially accumulated and that the n-3/n-6 ratio is correlated with that of the diet (Kainz et al. 2004; Brett et al. 2006; Hessen and Leu 2006). High accumulation of ARA from the diet suggests that *Daphnia* are able to convert C_{18} n-6 PUFA to ARA and that ARA is important for certain physiological functions in these animals. However, many feeding experiments with different daphnid species have documented highly variable outcomes in terms of somatic growth and egg production rates. Several explanations for this high variability are possible: use of varying algal and cyanobacteria species or strains as feed; varying types and amounts of PUFA additions; species and/or population specific differences in the capacity to convert ALA to EPA and LIN to ARA; running very short-term experiments (sometimes only 2–4 days); and use of PUFA additions that are inappropriate for the species being examined. All of these factors might lead to an imbalance between n-3 and n-6 PUFA. The C and total FA content of the experimental diets were usually constant in the separate experiments, i.e. only the type of FA was varied, so the outcome should solely depend on the PUFA quality. *Daphnia galeata* and *D. magna* are the most commonly used species, but *D. pulex*, *D. pulicaria* and *Bosmina freyi* have also been used in some experiments.

In the first feeding experiment in Table 7.4, *D. galeata* was fed the green alga *Scenedesmus quadricauda* which was grown under four different nitrogen regimes: 100% and 10% NO_3-N; and 100% and 10% NH_4-N (Ahlgren et al. 2000). *Daphnia* growth was considerably depressed in both N-limited diets and ceased altogether after 4–6 days, which resulted in them gradually dying off by the end of the 13-day experiment. However, after 6 days, the number of neonates was 3X higher in the 10% NO_3-N than in the 10% NH_4-N treatment. Consequently, if the experiment had been stopped on days 4–6, quite different results would have been presented. The only explanation, given by the authors, for the negative growth in both the N-limited cultures during the second part of the experiment was that the n-3/n-6 ratio was <2 in the N-limited cultured algae compared to ~3 in non-limited cultures.

In another experiment, *D. galeata* fed N-limited *Chlamydomonas* showed the same growth rate and produced the same numbers of neonates as when fed non-N-limited *Chlamydomonas* (Weers and Gulati 1997b). However, P-limited *Chlamydomonas* reduced the specific growth rate and number of individuals. The n-3/n-6 ratio of the P-limited alga was as low as 1.2. The next experiment showed that additions of EPA increased growth rates by 13% and 24% and more young were produced when the alga *Scenedesmus acutus* was added (Weers and Gulati 1997a). However, in a different feeding experiment, ALA additions to *S. obliquus* gave the same increase in growth rate as EPA (Von Elert 2002). Differences in the n-3/n-6 ratios between the two *Scenedesmus* species used as feed for *D. galeata* might explain the different outcome; the n-3/n-6 ratio was 12 in the *S. acutus* used by Weers and Gulati (1997a) and 2.9 in *S. obliquus* control and additions of ALA or EPA gave moderate high ratios of 5.6 (Von Elert 2002). The high ratios used in Weers and Gulati (1997a) were probably not appropriate for *D. galeata* because the highest observed specific growth rate was much lower than in the Von Elert (2002) study.

D. magna grew best and produced most eggs when the n-3/n-6 ratio was close to 5 (Elendt 1990; Boersma 2000). Thus, these experiments with *D. galeata* and *D. magna* show the same apparent capacity to convert ALA to EPA for both species at the necessary rate as long as the n-3/n-6 ratio is ≥3 for *D. galeata* and ≥5 for *D. magna*.

The last experiment in Table 7.4 shows the results of a feeding experiment with *Bosmina freyi* fed either *S. acutus* or Ohio River seston (Acharya et al. 2005). After 6 days, there was no difference in either growth rate or fecundity between the treatments. However, after 22 days both somatic growth rate and fecundity were higher in the *Scenedesmus* treatment. This study and that of Ahlgren et al. (2000) reveal that critical information may not come to light in short-term experiments. This is particularly true regarding aspects of food quality, such as in PUFA feeding studies, where the experimental duration should allow for measurements of egg production and hatching success. An exciting challenge is to discover why Acharya et al.

(2005) observed higher fecundity for *Bosmina* on a green alga diet than a seston diet. One approach might be to check the animal's PUFA ratio. Unfortunately, FA analyses of the treated *Bosmina* were not available. However, *Bosmina* from oligotrophic sub-alpine lakes are very similar to other cladocerans regarding their n-3/n-6 and DHA/ARA ratios (Persson and Vrede 2006). In addition to the suggestion by Acharya et al. (2005) that the high content of ALA or n-3 FA in *Scenedesmus* may be the answer, we suggest that the higher content of n-6 FA, i.e. LIN and ARA in the alga compared to the river seston, may result in more favourable FA ratios. An n-3/n-6 ratio of 7 in seston means that there was a relative lack of n-6. The mean n-6 content was 0.7 mg g^{-1} C in river seston and 11 mg g^{-1} C in *Scenedesmus*.

7.3.2.3 Feeding Experiments with Crustaceans Fed Cyanophytes

Several feeding experiments with different outcomes have also been performed with cyanobacteria used as food for zooplankton (Table 7.5). In an experiment with *D. galeata* fed *Synechococcus elongatus* (n-3/n-6 = 3.7), additions of fish oils (high in EPA) increased the growth rate moderately (DeMott and Müller-Navarra 1997). However, the same numbers of eggs were produced as when fed the control alga *Scenedesmus acutus*. In contrast, the next experiment with *D. galeata* fed the cyanobacteria *Synechococcus elongatus*, a strain that lacks both sterols and PUFA, gave low growth rate and additions of PUFA did not affect *Daphnia* growth rate compared to *Synechococcus* alone (Von Elert and Wolffrom 2001; Von Elert et al. 2003). Additions of sterols produced somewhat higher growth rate, whereas additions of both sterols and PUFA produced a further slight increase in growth rate. However, this growth rate was still ~30% lower than that of control treatments with *Scenedesmus*. Note that for all of these additions to *Synechococcus*, the n-3/n-6 ratios of the diets were either ≤1.5 or >12, whereas in *Scenedesmus* the ratio was 2.5. When the cyanobacteria *Anabaena variabilis* was fed to *D. galeata*, additions of sterols and ALA + EPA gave the same high growth rate as *Scenedesmus obliquus* (Von Elert et al. 2003). A possible explanation for why *D. galeata* responded with a higher growth rate to *Anabaena* with additions than to *Synechococcus* with additions is that *Anabaena* contained both n-3 and n-6 FA, whereas the strain of *Synechococcus* used contained neither n-3 nor n-6 PUFA.

Daphnia magna fed plain *Synechococcus* (n-3/n-6 = 3.7) showed a growth rate of 60% compared to the control alga *Scenedesmus acutus*, but the number of eggs was only 12.5% of control values (DeMott and Müller-Navarra 1997). Additions of fish oils to *Synechococcus* increased the growth rate to 80% and egg production to about half that when fed *Scenedesmus*. This result differs from that of *D. galeata* where additions of fish oils gave a 60% growth rate but the same egg production as the control.

Daphnia pulicaria, in contrast to *D. magna* but similar to *D. galeata,* produced no eggs when fish oils were added to *Synechococcus* (DeMott and Müller-Navarra 1997). Thus, the same strain of the feed alga *Synechococcus* (n-3/n-6 = 3.7) was used for the three species of daphnids: *D. galeata, D. magna* and *D. pulicaria*. The results showed that, of the three species, *D. magna* made the best use of the low-quality cyanobacteria food.

Several algal species and the cyanobacterium *Microcystis* sp. were tested for their appropriateness as food for *D. pulex* (Table 7.5; Lürling and Van Donk 1997). *D. pulex* showed moderate growth and was able to produce some offsprings when fed *Microcystis* that had an n-3/n-6 ratio of 1.6. However, two flagellates (i.e. *Cryptomonas* and *Chlamydomonas*) gave about the same high growth rate even though only *Cryptomonas* contained EPA. Surprisingly, *D. pulex* produced more offspring until the third brood when fed *Chlamydomonas* than when fed *Cryptomonas* with n-3/n-6 ratios of 6 and 43, respectively. This agrees with Abrusán et al.'s (2007) observation that pure *Cryptomonas* diets cannot support *Daphnia* in the long run. The low content of n-6 PUFA (particularly of ARA) in *Cryptomonas* does not seem to be favourable for hatching daphnid broods.

In another feeding experiment where *D. pulex* was fed a mixture of three cyanobacteria species (Cyano-mix), a relatively high growth rate was observed for the Cyano-mix diet, a small not significant increase was seen with ALA additions to Cyano-mix, and a larger increase was produced with EPA (Ravet et al. 2003). The Cyano-mix contained some n-3 and n-6 PUFA and had an n-3/n-6 ratio of 0.4, which increased to 6 after the ALA and EPA additions. A feeding experiment with *D. pulex*, which also included chlorophytes, showed good agreement with the observations of Ravet et al. (2003) using Cyano-mix and Crypto-mix: three eggs for Cyanophytes, nine eggs for Cryptophytes and four eggs for Chlorophytes (Brett et al. 2006). When comparing these results with those of Von Elert and co-workers, where PUFA supplementation did not stimulate *Daphnia* growth, the most important hypothesis proposed to explain the differences was the way in which the PUFA supplements were delivered (Ravet et al. 2003). We suggest that their second hypothesis is also likely, i.e. that the effect was related to the differences in the Cyanophyte species used. Von Elert and Wolffrom (2001) used a *Synechococcus* strain that did not contain any PUFA and the additions of PUFA gave only low ratios of 1.5 (Table 7.5). In contrast, the Cyano-mix used by Ravet et al. (2003) and Brett et al. (2006) contained both C_{18} and C_{20} carbon n-3 and n-6 PUFA, and the additions of PUFA probably increased the ratios to more favourable levels. Therefore, variances in the n-3/n-6 ratios of the diets may explain the contrasting results.

In addition, the two *Daphnia* species used, *D. galeata* and *D. pulex*, probably differ in their food quality requirements. *D. pulex* was apparently unable to fully convert ALA and LIN to longer-chained PUFA at a rate required for high egg production in these experiments (Ravet et al. 2003; Brett et al. 2006). A 30% higher growth rate and the highest number of eggs were produced when *D. pulex* was fed cryptomonads for 6 days. Flagellates, including cryptomonads, have also been described as the best feed for zooplankters (e.g. Ahlgren et al. 1990; Von Elert and Stampfl 2000). We suggest that this is at least partly due to their favourable FA patterns, i.e. high contents of the biologically active EPA, DHA and ARA. However, some studies reveal that some cryptomonads have high levels of ALA and EPA, low levels of LIN, and only traces, or complete deficiencies of

ARA (e.g. Ravet and Brett 2006; Brett et al. 2006; Abrusán et al. 2007). Other studies show that some strains contain moderate levels of LIN and n-6DPA (Ahlgren et al. 1992) or high levels of LIN and small amounts of ARA (Müller-Navarra 2006). The n-6DPA is probably produced by an analogous route to that of EPA conversion to DHA using the Δ6-enzyme (Voss et al. 1991) and is easily retroconverted to ARA (cf. Koven et al. 2001). It remains to be shown if those cryptomonads which lack long-chained n-6 PUFA, can explain the low survival of daphnids in long-term feeding experiments.

Recently, Parrish and coworkers presented strong evidence that n-6DPA is essential to cod and scallop larvae (Milke et al. 2006; Parrish et al. 2007; see Chap. 13). However, the biochemical structures of the phospholipids are unique whereby it only fits into specific biochemical processes like a key in its keyhole (Crawford et al. 1999). Consequently, it is not likely that n-6DPA with five double bonds can replace the shorter molecule of ARA unless reconverted to ARA.

Thus, the laboratory experiments with crustacean zooplankton indicate that the optimal n-3/n-6 ratio would be ≥ 3 for *D. galeata*, ≥ 5 for *D. magna* and ≥ 10 for *D. pulex*. Although there is considerable variation in fatty acid composition, both within and among crustaceans, the variation in the n-3/n-6 ratio appears to be much smaller in zooplankton than in their food (see Chap. 6). This quasi-homeostasis in PUFA composition suggests that the n-3/n-6 ratio is also regulated within an optimal range.

7.3.3 Freshwater Zoobenthos

7.3.3.1 In Situ Benthos

Amongst insects, both filter-feeding and deposit-feeding chironomids contained nearly the same amounts of n-3 and n-6 PUFA (n-3/n-6 ratio = 1–2), whereas larvae of free-living, predatory chironomids of the genus *Procladius* had a higher n-3/n-6 ratio (2–3) (Bell et al. 1994; Goedkoop et al. 2000; Sushchik et al. 2003). Ephemerids had even higher n-3/n-6 ratios of about 6 (Bell et al. 1994; Sushchik et al. 2003; Torres-Ruiz et al. 2007). However, both filter-feeding and deposit-feeding chironomids and ephemerids contained ARA but had no, or only traces of, DHA. This suggests that these taxa need ARA but not DHA for growth and reproduction. Interestingly, the predators *Procladius* and *Chaoborus* contained both DHA and ARA with DHA/ARA ratios of ~1. Hydropsychidae, Corixidae, Notonectidae, Coleoptera, Plecoptera, and Trichopteridae had n-3/n-6 ratios that varied between 2 and 4 and DHA/ARA ratios ≤ 1 (Bell et al. 1994; Sushchik et al. 2003; Dahl 2006). The sediment-living *Oligochaeta* had n-3/n-6 ratios of 1 to 3.5 (Bell et al. 1994; Goedkoop et al. 2000; Torres-Ruiz et al. 2007) and only traces of DHA (Bell et al. 1994; Goedkoop et al. 2000) (Table 7.6).

Members of the crustacean families *Gammaridae* and *Asellidae* as well as two crayfish species (Bell et al. 1994; Ackefors et al. 1997; Makhutova et al. 2003; Sushchick et al. 2003; Dahl 2006; Maazouzi et al. 2007) had relatively low n-3/n-6 ratios ranging from 1 to 3 and a DHA/ARA ratio of ≤1, whereas the predator *Mysis relicta* had high ratios for both n-3/n-6 and DHA/ARA, on average 7 and 5, respectively (Schlechtriem et al. 2008). Eggs of the mollusc *Dreissena* had lower ratios for both n-3/n-6 and DHA/ARA in the epilimnion than in the hypolimnion (Wacker and Von Elert 2004).

In general, carnivorous zoobenthos have higher DHA than ARA, whereas herbivorous and/or detritivorous species frequently lack DHA, a pattern similar to that observed in mammals. The dietary source of DHA for predators like *Procladius* and *Chaoborus* is probably copepodite resting stages, harpacticoid copepods or microfauna in surficial profundal sediment, since deposit-feeding chironomids and oligochaetes generally lack DHA (Goedkoop et al. 2000). There seems to be a positive relationship between assumed food quality and n-3/n-6 ratios of zoobenthos taxa, with profundal, deposit-feeding, taxa that feed on partly decomposed settling algae having lower n-3/n-6 ratios than taxa in littoral or lotic habitats where fresh epibenthic algae dominate. For example, several genera of *Ephemeridae* feeding on algae in littoral habitats have markedly higher n-3/n-6 ratios than profundal species, but also lack DHA. Somewhat surprisingly, however, the amphipod *Diporeia* is relatively rich in n-3 PUFA (Schlechtriem, Arts and Johannsson, pers. comm.), as are several *Chironomus* species and oligochaetes feeding in profundal sediments (Goedkoop et al. 2000). *Diporeia* is known to be a strong and highly efficient trophic link between settling diatoms, which are rich in EPA, and fish (Gardner et al. 1990). These observations reveal that not only the feeding mode, but also feeding selectivity and phylogeny are important for contents and ratios of FA (Table 7.6).

Another ecologically relevant pattern is that freshwater insects such as some *Chironomus* species, with long larval stages in aquatic environments and a short terrestrial adult stage, have a PUFA composition similar to that of terrestrial insects, i.e. higher in n-6 than in n-3 FA, with an n-3/n-6 ratio <1 (e.g. Thompson 1973). This comparison may, however, only be valid for aquatic insects that have poor-quality diets, such as detritus, since different genera of *Ephemeridae*, and pooled *Plecoptera* and *Trichoperidae* feeding on fresh algae in shallow habitats all have n-3/n-6 ratios exceeding 1.

7.3.3.2 Feeding Experiments with Zoobenthos

Thus far, few studies have addressed the role of PUFA for the growth and recruitment success of benthic invertebrates. Surprisingly, experiments with larvae of the deposit-feeder *Chironomus riparius* fed oat flakes or the cyanobacteria *Spirulina*, both of which have very low n-3/n-6 ratios (0.03), gave only slightly lower growth rates than did the fish food Tetraphyll, with a much higher ratio of 0.36 (Goedkoop et al. 2007). In contrast, the green alga *Scenedesmus obliquus*, which is often used as a control in feeding experiments with zooplankton, supported poor growth in *Chironomus* that was only about half compared to the growth in the Tetraphyll treatment. One reason for the latter result is probably that *Chironomus* lacks carbohydrase activity (Bjarnov 1972) and thus has problems digesting *Scenedesmus* and other green algae. Adult emergence was significantly higher for the Tetraphyll diet than for the oats and *Spirulina* diets. Possibly, the reduced emergence in treatments with oats and *Spirulina*, particularly apparent at low food concentrations, can be attributed to the 10X lower n-3/n-6 ratio of these food types. No emergence and extremely low poor growth was also seen in a treatment with peat (essentially no food) (Table 7.7).

Zebra mussels, *Dreissena polymorpha*, fed with freshwater *Chlorella* produced 5X larger larvae than those fed a marine *Chlorella* species, but survival to the pediveliger stage was nearly the same (Vanderploeg et al. 1996). Larval survival on the freshwater *Chlorella* diet was also higher than when the mussels were fed the cryptophyte *Rhodomonas,* which together with *Cryptomonas*, is considered to be high-quality food for zooplankters. However, no survival was observed with *Cryptomonas* as a food, but this strain was probably too large to be ingested (Vanderploeg et al. 1996). Similar to the peat treatment (no food), treatments with several algal species, which all lacked EPA (e.g. *Synechococccus, Chlamydomonas, Nannochloris* no FA data available, and *Chromulina*), also resulted in eventual mortality. When fed the flagellate *Isochrysis* and the eustigmatophycee *Nannochloropsis, Dreissena* showed higher larval growth rate and greater survival compared to treatments with the green alga *Chlorella* and the cyanobacteria *Aphanothece* (Wacker et al. 2002). *Nannochloropsis* has an extremely high EPA content and additions of lipid extract from *Isochrysis*, DHA or EPA did not increase the specific growth rate of *Dreissena* substantially. Wacker and Von Elert (2004) conducted a 90-d long feeding experiment with *Dreissena*, long enough for egg production, and found that the *Nannochloropsis* diet gave the highest amounts of eggs per adult compared to the three other phytoplankters (Table 7.7). *Nannochloropsis* contains not only high amounts of EPA but also some ARA, giving a moderately high EPA/ARA ratio of 2.7. These data show that both EPA and ARA (or n-6DPA) must be present in the diet if ≥30% of *Dreissena* larvae are to survive to the pediveliger stage or stay alive at least 4 weeks. Without EPA and ARA in the diet only 50% of the larvae survived 2–3 weeks (Wacker and Von Elert 2004). This indicates that *Dreissena* larvae lack the capacity to convert C_{18} PUFA to longer-chain PUFA at a rate that is required to optimize survival. As in fish, n-3 PUFA (EPA and DHA) seem to stimulate growth in *Dreissena*, whereas n-6 PUFA (ARA) stimulate reproduction.

This comparison of studies with a deposit feeder and a filter feeder also suggests that there may be large functional-feeding guild-related differences in the dependence of specific PUFA or n-3/n-6 ratios. While *Dreissena* apparently is dependent on dietary supplies of EPA, deposit-feeding *Chironomus* larvae grow equally well if supplied only with shorter homologues. These interspecific differences may be related to differences in the feeding and life cycle strategies of primary consumers and may even vary among populations that have evolved under different food conditions. The scarcity of data and the diversity of feeding modes and life cycle strategies show that we are only just beginning to understand the role of specific FA or FA ratios for benthic invertebrate growth and reproduction.

7.4 Concluding Remarks

This review shows that there are specific differences in n-3/n-6 and DHA/ARA ratios amongst freshwater organisms. For fish species, the ratios were lowest and rather stable within species of herbivorous–omnivorous fish, intermediate in carnivorous–piscivorous fish, and highest and highly variable in carnivorous–benthivorous fish. All three groups were significantly different from each other. No changes could be observed during ontogeny for the three functional feeding groups. However, between single species, significant differences in the ratios were found in an oligotrophic lake, but not in a eutrophic lake. The largest effect on the PUFA ratios was observed when comparing wild and cultured fish. Together, these results suggest that diet has a strong influence on the PUFA composition of fish. This observation can also explain the relatively small variation among wild piscivorous fish.

No general differences in the n-3/n-6 and DHA/ARA ratios between herbivorous and carnivorous cladocerans could be seen, whereas significantly higher ratios were present in copepods. In zoobenthos, however, predators showed higher ratios than detrivores. All wild zooplankton contained EPA, DHA and ARA with DHA/ARA ratios <1 in cladocerans and >1 in copepods. PUFA ratios in wild zooplankton were quite stable, whereas the ratios in freshwater zoobenthos differed strongly among taxonomic groups, suggesting that there are species-specific PUFA ratios. This observation is further corroborated by the results from feeding experiments with invertebrates, in which many of the contrasting results could be explained by differences in the feed n-3/n-6 and DHA/ARA ratios. Ephemerids and chironomids contained ARA, but DHA was rare or completely lacking. All aquatic insect contained ARA but DHA was rare or lacking in non-predacious taxa with n-3/n-6 ratios <1, whereas both ratios exceeded 1 in benthic crustaceans. ARA seems to be important for successful reproduction in invertebrates as well as in fish.

This review also finds that there are several gaps in our knowledge regarding PUFA patterns and the role of PUFA in organisms in freshwater food webs, as well as a need for refined experimental designs that acknowledge the role of PUFA ratios:

- The physiological function of these PUFA is at present unclear in invertebrates common as fish feed. For example, the role of eicosanoid fatty acids as precursors of prostaglandins is biologically important, but can be difficult to study through dietary manipulations of individuals or small populations due to the low concentrations of these fatty acids per animal needed to synthesize sufficient levels of these hormones. Therefore, research on the physiological function of fatty acids in aquatic invertebrates and fish should primarily focus on their importance for growth and reproduction and their role as structural components (e.g. in cell membranes and eggs).
- There is a conspicuous lack of experimental studies of freshwater copepods and several functional feeding-groups of zoobenthos. Similarly, there is little information on PUFA in wild zooplankton from lake types other than oligotrophic clear water lakes. Also, among the zoobenthos, general patterns in PUFA composition and content can only be identified by increasing the database of

structurally and/or functionally different species across a range of habitats. Controlled experiments must be done with designs that specifically test for optimum PUFA ratios. Therefore, PUFA supplementation should be made to create gradients in PUFA ratios, including manipulations of both n-3 and n-6 PUFA, rather than using absence/presence designs as has been the common practice thus far. Although much of what we have inferred is based on such experiments, none of them has been specifically designed to address optimum PUFA ratios. It is also important to present data on both n-3 and n-6 PUFA both in the feed types and in the experimental animals.
- The response variable in feeding experiments has often been growth rate or the size increment of animals, but this may not be the best measure of fitness. Other criteria such as stress tolerance, occurrence of deformities, swimming activity, survival, egg production, and/or recruitment success may be more appropriate for assessing the effects of varying feed PUFA contents on fitness. In addition, food quality may affect juveniles and adults differently and may differ for populations that have evolved under different food regimes. As growth and reproduction are controlled by different eicosanoids, experiments should also be sufficiently long to include egg laying and preferably even recruitment success.
- Considering the fact that there are obvious gaps in the accumulated knowledge on this topic, we recommend that in order to better understand the role of PUFA in freshwater food webs, future study designs need to acknowledge and take into consideration that both the amount of PUFA in the diet and the specific proportions of the types of PUFA are important to the performance of organisms. By acknowledging this, we should be able to achieve a better understanding of key regulatory processes that affect consumer growth and eventually food web dynamics.

Acknowledgements We thank J. Johansson for chemical analyses, I. Ahlgren for help with collecting references and comments on the early manuscript, and several colleagues, such as J. Persson, M.T. Brett, M. Kainz and C, Schlechtriem, M.T. Arts and O.E. Johannsson, for the use of their data. We also bothered D.C. Müller-Navarra and E. von Elert several times with questions about original data. Important remarks of an anonymous referee improved an earlier version of this work. We are also deeply thankful to the editors, M.T. Arts, M.T. Brett and M. Kainz, whose support and encouragement made this chapter possible.

References

Abrusán, G., Fink, P., and Lampert, W. 2007. Biochemical limitation of resting egg production in *Daphnia*. Limnol. Oecanogr. 52:1724–1724
Acharya, K., Jack, J. D., and Bukaveckas, P. 2005. Dietary effects on life history traits of riverine *Bosmina*. Freshw. Biol. 50:965–975
Ackefors, A., Castell, J., and Örde-Öström, I.-L. 1997. Preliminary results on the fatty acid composition of freshwater crayfish, *Astacus astacus* and *Pacifastacus leniusculus*, held in captivity. J. World Aquac. Soc. 28:97–105
Ahlgren, G., Lundstedt, L., Brett, M., and Forsberg, C. 1990. Lipid composition and food quality of some freshwater phytoplankton for cladoceran zooplankters. J. Plankton Res. 12:809–818

Ahlgren, G., Gustafsson, I.-B., and Boberg, M. 1992. Fatty acid content and chemical composition of freshwater microalgae. J. Phycol. 28:37–50

Ahlgren, G., Blomqvist, P., Boberg, M., and Gustafsson, I. B. 1994. Fatty acid content of the dorsal muscle – an indicator of fat quality in freshwater fish. J. Fish Biol. 45:131–157

Ahlgren, G., Sonesten, L., Boberg, M., and Gustafsson, I. B. 1996. Fatty acid content of some freshwater fish in lakes of different trophic levels – a bottom-up effect? Ecol. Freshw. Fish 5:15–27

Ahlgren, G., Carlstein, M., and Gustafsson, I.-B. 1999. Effects of natural and commercial diets on the fatty acid content of European grayling. J. Fish Biol. 55:1142–1155

Ahlgren, G., Hyenstrand, P., Vrede, T., Karlsson, E., and Zetterberg, S. 2000. Nutritional quality of *Scenedesmus quadricauda* (Chlorophyceae) grown in different nitrogen regimes and tested on *Daphnia*. Verh. Internat. Verein. Limnol. 27:1234–1238

Ahlgren, G., Van Nieuwerburgh, L., Wänstrand, I., Pedersén, M., Boberg, M., and Snoeijs, P. 2005. Imbalance of fatty acids in the base of the Baltic Sea food web – a mesocosm study. Can. J. Fish. Aquat. Sci. 62:2240–2253

Arts, M. T., Ackman, R. G., and Holub, B. J. 2001. "Essential fatty acids" in aquatic ecosystems: a crucial link between diet and human health and evolution. Can. J. Fish. Aquat. Sci. 58:122–137

Becker, C., and Boersma, M. 2005. Differential effects of phosphorus and fatty acids on *Daphnia magna* growth and reproduction. Limnol. Oceanogr. 50:388–397

Bell, M. V., Henderson, R. J., and Sargent, J. R. 1986. Minireview. The role of polyunsaturated fatty acids in fish. Comp. Biochem. Physiol. 83B:711–719

Bell, J. G., Ghioni, C., and Sargent, J. R. 1994. Fatty acid compositions of 10 freshwater invertebrates which are natural food organisms of Atlantic salmon parr (*Salmo salar*); a comparison with commercial diets. Aquaculture 128:301–313

Bell, M. V., Dick, J. R., Thrush, M., and Navarro, J. C. 1996. Decreased 20:4n–6/20:5n–3 ratio in sperm from cultured sea bass, *Dicentrarchus labrax*, broodstock compared with wild fish. Aquaculture 144:189–199

Bjarnov, N. 1972. Carbohydrases in *Chirononmus*, *Gammarus* and some trichopteran larvae. Oikos 23:261–263

Boersma, M. 2000. The nutritional quality of P-limited algae for *Daphnia*. Limnol. Oceanogr. 45:1157–1161

Brett, M. T., and Müller-Navarra, D. C. 1997. The role of highly unsaturated fatty acids in aquatic foodweb processes. Freshw. Biol. 38:483–499

Brett, M. T., Müller-Navarra, D. C., Ballantyne, A. P., Ravet, J. L., and Goldman, C. R. 2006. *Daphnia* fatty acid composition reflects that of their diet. Limnol. Oceanogr. 51:2428–2437

Broadhurst, C. L., Cunnane, S. C., and Crawford, M. A. 1998. Rift valley lake fish and shellfish provided brain-specific nutrition for early *Homo* (review article). Br. J. Nutr. 79:3–21

Castell, J. D., Bell, J. G., Tocher, D. R., and Sargent, J. R. 1994. Effects of purified diets containing different combinations of arachidonic and docosahexaenoic acid on survival, growth and fatty acid composition of juvenile turbot (*Scrophthalmus maximus*). Aquaculture 128:315–333

Copeman, L. A., Parrish, C. C., Brown, J. A., and Harel, M. 2002. Effects of docosahexaenoic, ecosapentaenoic, and arachidonic acids on the early growth, survival, lipid composition and pigmentation of yellowtail flounder (*Limanda ferruginea*): a live food enrichment experiment. Aquaculture 210:285–304

Crawford, M., and Marsh, D. 1989. The Driving Force. Heinemann, London

Crawford, M. A., Casperd, N. M., and Sinclair, A. J. 1976. The long chain metabolites of linoleic and linolenic acids in liver and brain in herbivores and carnivores. Comp. Biochem. Physiol. 54B:395–401

Crawford, M. A., Doyle, W., Williams, G., and Drury, P. J. 1989. The role of fats and EFAs for energy and cell structures in the growth of fetus and neonates, pp. 81–115. In A. J. Vergroesen, and M. Crawford [eds.], The role of fats in human nutrition. Academic Press, London

Crawford, M. A., Bloom, M., Broadhurst, C. L., Schmidt, W. F., Cunnane, S. C., Galli, C., Gehbremeskel, K., Linseisen, F., Lloyd-Smith, L., and Parkinton, J. 1999. Evidence for the unique function of docosahexaenoic acid during the evolution of the modern hominid brain. Lipids 34:S39–S47

Dahl, J. 2006. Functional feeding groups of benthic macro-invertebrates in Swedish lakes and streams and the importance of spatial scale. MSc-thesis. Swedish University of Agricultural Sciences, Department of Environmental Assessment, Report 1999:4

Dalsgaard, J., StJohn, M., Kattner, G., Müller-Navarra, D., and Hagen, W. 2003. Fatty acid trophic markers in the pelagic marine environment. Adv. Mar. Biol. 46:225–246

De Lange, H. J., and Van Donk, E. 1997. Effects of UVB-irradiated algae on life history traits of *Daphnia pulex*. Freshw. Biol. 38:711–720

DeMott, W. R. 1986. The role of taste in food selection by freshwater zooplankton. Oecologia 69:334–340

DeMott, W. R., and Müller-Navarra, D. C. 1997. The importance of highly unsaturated fatty acids in zooplankton nutrition: evidence from experiments with *Daphnia*, a cyanobacterium and lipid emulsions. Freshw. Biol. 38:649–664

Elendt, B. -P. 1990. Nutritional quality of a microencapsulated diet for *Daphnia magna*. Effects on reproduction, fatty acid composition, a midgut ultrastructure. Arch. Hydrobiol. 118:461–475

Gardner, W. S., Quigley, M. A., Fahnenstiel, G. L., Scavia, D., and Frez, W. A. 1990. *Pontoporeia hoyi* – a direct trophic link between spring diatoms and fish in Lake Michigan, pp. 632–644. *In* M. M. Tiller, and C. Serruya [eds.], Large lakes – ecological structure and function. Springer, New York

Goedkoop, W., Sonesten, L., Markensten, H., and Ahlgren, G. 2000. Fatty acids in profundal benthic invertebrates and their major food resources in Lake Erken, Sweden: seasonal variation and trophic indications. Can. J. Fish. Aquat. Sci. 57:2267–2279

Goedkoop, W., Demandt, M., and Ahlgren, G. 2007. Interactions between food quantity and quality (long-chain PUFA concentrations) effects on growth and development of the midge *Chironomus riparius* Meigen. Can. J. Fish. Aquat. Sci. 64:425–436

Harrison, K. E. 1990. The role of nutrition in maturation, reproduction and embryonic development of decapod crustaceans: a review. J. Shellfish Res. 9:1–28

Henderson, R. J., Tillmanns, M. M., and Sargent, J. R. 1996. The lipid composition of two species of Serasalmid fish in relation to dietary polyunsaturated fatty acids. J. Fish Biol. 48:522–538

Hessen, D. O., and Leu, E. 2006. Trophic transfer and trophic modification of fatty acids in high Arctic lakes. Freshw. Biol. 51:1987–1998

Higgs, D. A., Macdonald, J. S., Levings, C. D., and Dosanjh, B. S. 1995. Nutrition and feeding habits in relation to life history stage, pp. 200–280. In C. Groot, L. Margolis, and W. C. Clarke [eds.], Physiological ecology of pacific salmon. UBC Press, Vancouver

Kainz, M., Arts, M. T., and Mazumder, A. 2004. Essential fatty acids in the planktonic food web and their ecological role for higher trophic levels. Limnol. Oceanogr. 49:1784–1793

Kainz, M., Telmer, K., and Mazumder, A. 2006. Bioaccumulation patterns of methyl mercury and essential fatty acids in lacustrine planktonic food webs and fish. Sci. Total Environ. 368:271–282

Koussoroplis, A. M., Lemarchand, C., Bec, A., Desvilettes, C., Amblard, C., Fournier, C., Berny, P., and Bourdier, G. 2008. From aquatic to terrestrial food webs: decrease of the docosahexaenoic acid/linoleic acid ratio. Lipids 43:461–466

Koven, W., Barr, Y., Lutzky, S., Ben-Atia, I., Weiss, R., Harel, M., Behrens, P., and Tandler, A. 2001. The effect of dietary arachidonic acid (20:4n–6) on growth, survival and resistance to handling stress in gilthead seabream (*Sparus aurata*) larvae. Aquaculture 193:107–122

Lürling, M., and Van Donk, E. 1997. Life history consequences for *Daphnia pulex* feeding on nutrient-limited phytoplankton. Freshw. Biol. 38:693–709

Maazouzi, C., Masson, G., Izquierdo, M. S., and Pihan, J. C. 2007. Fatty acid composition of the amphipod invader *Dikerogammarus villosus*: feeding strategies and feeding strategies and trophic links. Comp. Biochem. Physiol. A 147:868–875

Makhutova, O., Kalachova, G. S., and Gladyshev, M. I. 2003. A comparison of the fatty acid composition of *Gammarus lacustris* and its food sources from a freshwater reservoir, Bugach, and the saline Lake Shira in Siberia, Russia. Aquat. Ecol. 37:159–167

Milke, L. M., Bricelj, V. M., and Parrish, C. C. 2006. Comparison of early history stages of the bay scallop, *Argopecten irradians*: effects of microalgal diets on growth and biochemical composition. Aquaculture 260:272–289

Müller-Navarra, D. C. 2006. The nutritional importance of polyunsaturated fatty acids and their use as trophic markers for herbivorous zooplankton: does it contradict? Arch. Hydrobiol. 167:501–513

Navas, J. M., Thrush, M. A., Ramos, J., Zanuy, S., Carrillo, M., and Bromage, N. 1993. Calidad de puesta y niveles plasmaticos de vitelogenina en reproductores de lubina (*Dicentrarchus labrax*) mantenidos con diferentes dietas. Actas IV Congreso Nac. Acuicult. 19–24.

Olsen, Y. 1999. Lipids and essential fatty acids in aquatic food webs: what can freshwater ecologists learn from mariculture, pp. 161–202. In M. T. Arts, and B. C. Wainman [eds.], Lipids in freshwater ecosystems. Springer, New York

Otwell, W. S., and Richards, W. L. 1981/1982. Cultured and wild American eels, *Anguilla rostrata*: fat content and fatty acid composition. Aquaculture 26:67–76

Parrish, C. C., Whiticar, M., and Puvanendran, V. 2007. Is w6 docosapentaenoic acid an essential fatty acid during early ontogeny in marine fauna? Limnol. Oceanogr. 52:476–479

Persson, J., and Vrede, T. 2006. Polyunsaturated fatty acids in zooplankton: variation due to taxonomy and trophic position. Freshw. Biol. 51:887–900

Pickova, J., Kiessling, A., Pettersson, A., and Dutta, P. C. 1999. Fatty acid and carotenoid composition of eggs from two nonanadromous Atlantic salmon stocks of cultured and wild origin. Fish Physiol. Biochem. 21:147–156

Ravet, J. L., Brett, M. T., and Müller-Navarra, D. C. 2003. A test of the role of polyunsaturated fatty acids in phytoplankton food quality for *Daphnia* using liposome supplementation. Limnol. Oceanogr. 48:1938–1947

Ravet, J. L., and Brett, M. T. 2006. Phytoplankton essential fatty acid and phosphorus content constraints on *Daphnia* somatic growth and reproduction. Limnol. Oceanogr. 51:2438–2452

Repka, S. 1997. Effects of food type on the life history of *Daphnia* clones from lakes differing in trophic state. I. *Daphnia galeata* feeding on *Scenedesmus* and *Oscillatoria*. Freshw. Biol. 37:675–683

Sargent, J. R. 1995. Origins and functions of egg lipids: nutritional implications, pp. 353–372. In N. R. Bromage, and R. J. Robert [eds.], Brood stock managements and egg and larval quality. Blackwell Science, Cambridge

Sargent, J. R., Parkes, R. J., Mueller-Harvey, I., and Henderson, R. J. 1987. Lipid markers in marine ecology, pp. 119–138. In M. A. Sleigh [ed.], Microbes in the sea. Ellis Horwood Ltd, Chichester

Sargent, J. R., Bell, J. G., Bell, M. V., Henderson, R. J., and Tocher, D. R. 1995. Requirement criteria for essential fatty acids. J. Appl. Ichthyol. 11:183–198

Sargent, J. R., McEvoy, L. A., and Bell, J. G. 1997. Requirements, presentation and sources of polyunsaturated fatty acids in marine fish larval feeds. Aquaculture 155:117–127

Sargent, J. R., McEvoy, L., Estevez, A., Bell, G., Bell, M., Henderson, J., and Tocher, D. 1999. Lipid nutrition of marine fish during early development: current status and future directions. Aquaculture 179:217–229

Schlechtriem, C., Arts, M. T., and Johannsson, O. E. 2008. Effect of long-term fasting on the use of fatty acids as trophic markers in the opossum shrimp *Mysis relicta*. A laboratory study. J. Great Lakes Res. 34:143–152

Stanley-Samuelson, D. W. 1994. Prostaglandins and related eicosanoids in insects. Adv. Insect Physiol. 24:115–212

Sushchik, N. N., Gladyshev, M. I., Moskvichova, A. V., Makhutova, O. N., and Kalachova, G. S. 2003. Comparison of fatty acid composition in major lipid classes of the dominant benthic invertebrates of the Yenisei River. Comp. Biochem. Physiol. B 134:111–122

Thompson, S. N. 1973. A review and comparative characterization of the fatty acid compositions of seven insect orders. Comp. Biochem. Physiol. 45B:467–482

Thrush, M., Navas, J. M. Ramos, J., Bromage, N., Carrillo, M., and Zanuy, S. 1993. The effect of artificial diets on lipid class and total fatty acid composition on cultured sea bass (*Dicentrarchus labrax*) eggs. Actas IV Congreso Nac. Acuicult. 37–42.

Torres-Ruiz, M., Wehr, J. D., and Perrone, A. A. 2007. Trophic relations in a stream food web: importance of fatty acids for macroinvertebrate consumers. J. N. Am. Benthol. Soc. 26:509–522.

Vanderploeg, H. A., Liebig, J. R., and Gluck, A. A. 1996. Evaluation of different phytoplankton for supporting development of zebra mussel larvae (*Dreissena polymorpha*): the importance of size and polyunsaturated fatty acid content. J. Great Lakes Res. 22:36–45

Van Vliet, T., and Katan, M. B. 1990. Lower ratio of n-3 to n-6 fatty acids in cultured than in wild fish. Am. J. Clin. Nutr. 51:1–2

Von Elert, E. 2002. Determination of limiting polyunsaturated fatty acids in *Daphnia galeata* using a new method to enrich food algae with single fatty acids. Limnol. Oceanogr. 47:1764–1773

Von Elert, E., and Stampfl, P. 2000. Food quality for *Eudiaptomus gracilis*: the importance of particular highly unsaturated fatty acids. Freshw. Biol. 45:189–200

Von Elert, E., and Wolffrom, T. 2001. Supplementation of cyanobacterial food with polyunsaturated fatty acids does not improve growth of *Daphnia*. Limnol. Oceanogr. 46:1552–1558

Von Elert, E., Martin-Creuzburg, D., and Le Coz, J. R. 2003. Absence of sterols constrains carbon transfer between cyanobacteria and a freshwater herbivore (*Daphnia galeata*). Proc. Roy. Soc. B – Biol. Sci. 270:1209–1214

Voss, A., Reinhart, M., Sankarappa, S., and Sprecher, H. 1991. The metabolism of 7,10,13,16,19-docosapentaenoic acid to 4,7,10,13,19-docosahexaenoic acid in rat liver is independent of a 4-desaturase. J. Biol. Chem. 266:19995–20000

Wacker, A., and Von Elert, E. 2004. Food quality controls egg quality of the zebra mussel *Dreissena polymorpha*: the role of fatty acids. Limnol. Oceanogr. 49:1794–1801

Wacker, A., Becher, P., and Von Elert, E. 2002. Food quality effects of unsaturated fatty acids on larvae of the mussel *Deissena polymorpha*. Limnol. Oceanogr. 47:1242–1248

Weers, P. M. M., and Gulati, R. D. 1997a. Effect of addition of polyunsaturated fatty acids to the diet on the growth and fecundity of *Daphnia galeata*. Freshw. Biol. 38:721–729

Weers, P. M. M., and Gulati, R. D. 1997b. Growth and reproduction of *Daphnia galeata* in response to changes in fatty acids, phosphorus, and nitrogen in *Clamydomonas reinhardtii*. Limnol. Oceanogr. 42:1584–1589

Weissburg, M. J., Doall, M. H., and Yen, J. 1998. Following the invisible trail: kinematic analysis of mate-tracking in the copepod *Temora longicornis*. Proc. Roy. Soc. B – Biol. Sci. 353:701–712

Xu, X., Ji, W., Castell, J. D., and O'Dor, R. 1993. The nutritional value of dietary n-3 and n-6 fatty acids for the Chinese prawn (*Panaeus chinensis*). Aquaculture 118:277–285

Yang, X., and Dick, T. A. 1994. Arctic char (*Salvelinus alpinus*) and rainbow trout (*Oncorhynchus mykiss*) differ in their growth and lipid metabolism in response to dietary polyunsaturated fatty acids. Can. J. Fish. Aquat. Sci. 51:1391–1400. J. Plankton Res. 14:495–512

Yen, J., Lenz, P. H., Gassie, D. V., and Hartline, D. K. 1992. Mechanoreception in marine copepods: electrophysiological studies on the first antennae. J. Plankton Res. 14:495–512

Zenebe, T., Ahlgren, G., and Boberg, M. 1998a. Fatty acid content of some freshwater fish of commercial importance from tropical lakes in the Ethiopian Rift Valley. J. Fish Biol. 53:987–1005

Zenebe, T., Ahlgren, G., Gustafsson, I. B., and Boberg, M. 1998b. Fatty acid and lipid content of *Oreochromis niloticus* L. in Ethiopian lakes – dietary effects of phytoplankton. Ecol. Freshw. Fish 7:146–158

Zenebe, T., Boberg, M., Sonesten, L., and Ahlgren, G. 2003. Effects of algal diets and temperature on the growth and fatty acid content of the cichlid fish *Oreochromis niloticus* L. A laboratory study. Aquat. Ecol. 37:169–182

7 Fatty Acid Ratios in Freshwater Fish

Table 7.1 Polyunsaturated fatty acid (PUFA) ratios in feeding experiments with freshwater fish fed different kinds of feed

Species and feed	Feed		Fish muscles			Source
	n-3/n-6	DHA/ARA	n-3/n-6	DHA/ARA	Increase (g)	
Salvelinus alpinus (fed 12 weeks)						
PUFA deficient	0.25	0.0	5.2	21.1	1.7	Yang and Dick
Low PUFA	3.13	0.0	–	–	8.8	(1994)
High PUFA	3.12	0.0	3.4	22.1	11.2	
Commercial	2.30	13	2.9	27.4	14.1	
Oncorhynchus mykiss (fed 12 weeks)						
PUFA deficient	0.25	0.0	4.2	15.6	3.0	Yang and Dick
Low PUFA	3.13	0.0	–	–	5.8	(1994)
High PUFA	3.12	0.0	3.8	26.2	8.3	
Commercial	2.30	13	3.3	29.7	14.1	
Mylossoma aureum (herb.; fed 9 months)						
Oatflakes	0.03	0.0	0.04	0.0	15.6	Henderson et al. (1996)
Serrasalmus natterie (carn.; fed 9 months)						
Chironomus larvae	0.22	1.3	0.19	0.52	8.8	Henderson et al. (1996)
Oreochromis niloticus (fed 8 weeks)						
Cichlid pellets	0.61	0.0	1.18	7.29	9.91	
Microcystis (colonies)	0.33	0.0	2.13	9.50	0.68	
Microcystis (cells)	0.10	0.0	1.41	5.37	−0.27	
Arthrospira	0.01	0.0	1.49	5.38	−1.44	
Scen. quadricauda	2.1	0.0	1.42	5.46	−2.30	Zenebe et al.
Unfed	0.0	0.0	1.27	4.58	−2.49	(2003)

– indicates 'no values available'

Table 7.2 PUFA ratios in wild and reared salmonids from the rivers Vindelälven (RV), Gullspång (W1), Klarälven (W2), Norwegian rivers (NR), reared fish from the River Gullspång⁻ (F1), and from lakes in Canada (C)

Fish species	Lipids	n	n-3/n-6	DHA/ARA	Source
Wild					
Thymallus thymallus (RV)	Total	18	4.7 ± 0.6	6.3 ± 1.0	Ahlgren et al. (1999)
Salmo salar					
W1	PL	6	5.2	4.0	Pickova et al. (1999)
W2	PL	6	4.8	3.8	
W1	TAG	6	4.5	2.9	
W2	TAG	6	4.2	3.1	
Eggs	PL	12	5.0	3.9	
Average ± SD			4.7 ± 0.4	3.5 ± 0.5	
Salvelinus alpinus (C)	Total	3	1.8	2.2	Yang and Dick (1994)
Oncorhynchus mykiss (C)	Total	3	1.7	6.9	Yang and Dick (1994)
Atlantic *Salmo salar* (C)	Total	3	1.7	1.9	Yang and Dick (1994)
Average ± SD			1.7 ± 0.06	3.7 ± 2.8	
Salmo salar (NR)	Total	–	16	25	Olsen (1999)
Reared					
Thymallus thymallus	Total	9	9.8±2.6	50.9±5.9	Ahlgren et al. (1999)
Salmo salar					
F1	PL	6	13.9	9.7	Pickova et al. (1999)
F1	TAG	6	6.8	10.4	
Eggs	PL	6	13.9	9.6	
Average ± SD			11.5 ± 4.1	9.9 ± 0.4	
S. alpinus	Total	2	2.9	27	Yang and Dick (1994)
O. mykiss	Total	2	3.3	30	Yang and Dick (1994)
Average ± SD			3.1 ± 0.3	29 ± 2	
Salmo salar (NR)	Total	–	4	31	Olsen (1999)

Total total muscle fatty acids; *PL* phospholipids; *TAG* triacylglycerols; – indicates 'no values available'

7 Fatty Acid Ratios in Freshwater Fish

Table 7.3 PUFA ratios (average ± standard deviation) in freshwater zooplankters

Taxa	n-3/n-6	DHA/ARA	n	Lake type	Source
Cladocera					
(a) Filter-feeding Cladocerans					
Bosmina coregoni	3.3 ± 0.7	0.6 ± 0.2	4	Sub-alpine oligotrophic	Persson and Vrede (2006)
Daphnia cucullata	2.1 –	0.3 –	1		Higgs et al. (1995)
D. galeata and *D. cucullata*	3.4 ± 1.3	0.2 ± 0.1	5	Temperate eutrophic	Persson and Vrede (unpublished)
D. galeata, D. galeata–hyalina	2.8 ± 0.3	0.2 ± 0.1	7	Temperate mesotrophic	Müller-Navarra (2006)
Daphnia magna	2.3 –	0.0 –	1		Higgs et al. (1995)
Daphnia spp.	3.2 ± 0.7	0.2 ± 0.1	5	Sub-alpine oligotrophic	Persson and Vrede (2006)
Daphnia tenebrosa	4.7 ± 0.9	0.6 ± 0.7	6	Arctic ponds oligo-eutrophic	Hessen and Leu (2006)
Holopedium gibberum	3.4 ± 1.1	0.4 ± 0.2	15	Sub-alpine oligotrophic	Persson and Vrede (2006)
Cladocerans (8 species)	2.5 –	0.5 –	17		Higgs et al. (1995)
>70% Cladocera (average 87%)	2.6 ± 1.2	1.1 ± 1.7	10	Temperate oligotrophic	Kainz et al. (2006)
Filter-feeding Cladocerans, average	3.0 ± 0.8	0.5 ± 0.3			
(b) Predatory Cladocerans					
Bythotrephes longimanus	2.6 ± 0.3	0.2 ± 0.1	4	Sub-alpine oligotrophic	Persson and Vrede (2006)
Bythotrephes longimanus	2.2 –	0.2 –	1	Temperate eutrophic	Persson and Vrede (unpublished)
Predatory Cladocerans, average	2.4 ± 0.3	0.2 ± 0.02			
Copepoda					
(a) Calanoida					
Arctodiaptomus laticeps	4.0 ± 0.8	11.0 ± 7.6	8	Sub-alpine oligotrophic	Persson and Vrede (2006)
Eudiaptomus gracilis	2.1 –	2.1 –	1		Higgs et al. (1995)
Eudiaptomus gracioides	3.9 ± 1.1	5.6 ± 2.8	5	Temperate eutrophic	Persson and Vrede (unpublished)
E. gracilis and *E. gracioides*	4.0 ± 0.2	5.3 ± 2.9	7	Temperate mesotrophic	Müller-Navarra (2006)
Heterocope spp.	3.8 ± 0.9	5.7 ± 1.2	8	Sub-alpine oligotrophic	Persson and Vrede (2006)
(b) Cyclopoida					
Cyclops strenuus	4.5 –	5.8 –	1		Higgs et al. (1995)
C. strenuus and *C. vicinus*	5.2 –	6.1 –	2		Higgs et al. (1995)
(c) Mixed copepods					
>70% Copepoda (average 84%)	5.0 ± 2.7	6.3 ± 3.5	5	Temperate oligotrophic	Kainz et al. (2006)
Copepods, average	4.1 ± 1.0	6.0 ± 2.4			

– 'indicates no values available'

Table 7.4 PUFA ratios of chlorophytes in feeding experiments with daphnids

Treatment	Feed EPA (mg g^{-1} C)	Daphnia n-3/n-6	r-max (d^{-1})	Number of Individuals	Source
Daphnia galeata (fed 14 (4–6) days)					
Scenedesmus quadricauda					Ahlgren
NO_3-N 100% Z8	0.0	2.5	0.13 (0.30)	62 (35)	et al.
NO_3-N 10% Z8	0.0	1.7	neg (0.13)	2 (25)	(2000)
NH_4-N 100% Z8	0.0	3.0	0.13 (0.13)	61 (23)	
NH_4-N 10% Z8	0.0	1.4	neg (0.09)	1 (12)	
D. galeata (fed 8 days)					
Chlamydomonas reinhardtii					Weers and Gulati
Chlam. control	0.0	5.9	0.26	42	(1997b)
N-limited	0.0	1.7	0.27	44	
P-limited	0.0	1.2	0.06	3	
D. galeata (fed 4 days)					
Scenedesmus acutus					Weers and
Scen. (Control)	0.0	12	0.31	42	Gulati
Scen. + emulsion-B	0.0	10	0.29	37	(1997a)
Scen. + emulsion-A	260	15	0.35	52	
Cryptomonas	20	38	0.35	51	
Rhodomonas	40	18	0.35	53	
D. galeata (fed 4 days)					
Scenedesmus obliquus	0.0	2.9	0.47	–	
Scen. + 18:1n-7	0.0	3.0	0.50	–	Von Elert (2002)
Scen. + ALA	0.0	5.6	0.50	–	
Scen. + ARA	0.0	1.4	0.48	–	
Scen. + EPA	74	5.6	0.50	–	
Scen. + DHA	0.0	3.7	0.51	–	
Daphnia magna (fed 22 days)					
Scenedesmus subspicatus	0.0[a]	3.0	4.4[b]	165	Elendt
Scen. + Frippak 75:25	1.5[a]	1.9	4.5[b]	163	(1990),
Scen. + Frippak 50:50	3.0[a]	1.3	4.4[b]	160	(newborns)
Scen. + Frippak 25:75	4.5[a]	0.8	4.2[b]	85	
Frippak 100%	6.0[a]	0.5	3.5[b]	24	
D. magna (fed 6 days)					
Scenedesmus obliquus					Boersma
P+ (Contr.)	0.0	1.2	0.42	–	(2000)
+ HUFA-poor	0.0	0.2	0.42	–	
+ HUFA-rich	32	4.6	0.62	–	
P-lim	0.0	0.76	0.20	–	
+ HUFA-poor	0.0	0.42	0.23	–	
+ HUFA-rich	32	2.8	0.24	–	
P-pulsed	0.0	1.3	0.23	–	
+ HUFA-poor	0.0	0.46	0.26	–	
+ HUFA-rich	32	3.8	0.32	–	

(continued)

Table 7.4 (continued)

Treatment	Feed EPA (mg g^{-1} C)	n-3/n-6	Daphnia r-max (d^{-1})	Number of Individuals	Source
D. magna (fed 6 days)					
Scenedesmus obliquus					Becker and
Scen. P-lim	0.0	0.75	0.28	–	Boersma
Scen. + ARA	0.0	0.51	0.28–0.30	–	(2005)
Scen. + EPA	4.5	1.6	0.37	–	
Bosmina (sinobosmina) freyi (fed 22 (6) days)					
Scenedesmus acutus	0.25	0.7	0.11 (0.38)	25	Acharya et al. (2005), (neonates)
Ohio river seston	2.8	7.0	0.09 (0.41)	14	

r, specific growth rate; numerals within parentheses refer to the shorter feeding periods; Z8, algal medium; *neg*, negative growth (i.e. *Daphnia* population decreased after days 4–6); emulsion-A is rich in PUFA; emulsion-B is rich in MUFA and SAFA; Frippak is a microencapsulated diets for prawn larvae (# 1 CAR) in relative proportions to *Scenedesmus*; '–' indicates 'no values available'
[a]% of total FA
[b]Length (mm) at day 22

Table 7.5 PUFA ratios of cyanophytes in feeding experiments with daphnids

Treatment	Algal feed EPA (mg g^{-1} C)	n-3/n-6	Daphnia r (d^{-1})	Number of eggs	Source
D. galeata (fed 6–7 days)					
Synechococcus elongatus	0.11	3.7	0.04–0.14	0	DeMott and
Syn + oleic acid	0.42	3.7	0.0	0	Müller-Navarra
Syn + fish oils	7.0	6.5	0.22–0.26	6	(1997)
Scenedesmus acutus	0.84	**3.1**	0.38	6	
D. galeata (fed 4 days)					
Synechococcus elongatus	0.0	0.0	0.16	–	Von Elert and
Syn + beads	0.0	0.29	0.19	–	Wolffrom
Syn + ARA	0.0	<<0.29	0.13	–	(2001)
Syn + EPA	105	12.3	0.18	–	
Syn + PUFA-mix	15	1.5	0.17	–	
Syn + lipids from Scen	0.0	1.1	0.28	–	
Scenedesmus obliquus	0.0	**2.5**	0.39	–	
D. galeata (fed 4 days)					
Synechococcus elongatus	0.0	0.0	0.045	0	Von Elert et al.
Syn + ALA + EPA	15	1.5	0.054	–	(2003)
Syn + sterols	0.0	0.0	0.23	1.3	
Syn + sterols + ALA + EPA	15	1.5	0.27	–	
Scenedesmus obliquus	0.0	2.5	0.38	–	
Anabaena variabilis	0.0	8.1	0.14	0	
+ ALA + EPA	15	9.6	0.14	–	
+ Sterols	0.0	8.1	0.34	4.5	
+ Sterols + ALA + EPA	15	9.6	0.37	–	

(continued)

Table 7.5 (continued)

Treatment	Algal feed			Daphnia		Source
	EPA (mg g^{-1} C)	n-3/n-6	r (d^{-1})	Number of eggs		
D. galeata (fed 15 days)						
Oscillatoria limnetica	0.0	2.4	0.06–0.16	–		Repka (1997)
Scenedesmus obliquus	0.0	2.7	0.27–0.34	–		
D. magna (fed 6–7 days)						
Synechococcus elongatus	0.11	3.7	0.20–0.30	0.5–2		DeMott and
Syn + oleic acid	0.05	3.7	0.27	1		Müller-
Syn + fish oils	6.9	6.5	0.33–0.37	4		Navarra
Scenedesmus acutus	0.84	3.1	0.43	10		(1997)
D. pulicaria (fed 6–7 days)						
Synechococcus elongatus	0.11	3.7	0.04–0.08	0		DeMott and
+ Fish oils	13	0.0	0.23	0		Müller-
+ *Ankistrodesmus*	–	–	0.35	4.9		Navarra
Ankistrodesmus	0.18[a]	1.5[a]	0.36–0.40	5.9		(1997)
D. pulex (fed until third brood)						
Microcystis aeruginosa	0.0	1.6	0.16	8		Lürling and Van Donk (1997)
Scenedesmus acutus	0.1	2.6	0.42	47		De Lange and Van Donk (1997), (newborns)
Synedra tenuissima	10	0.0	0.44	54		
Cryptomonas pyrenoidifera	16	43	0.46	49		
Chlamydomonas reinhardtii	0.0	6.4	0.39–0.50	65–69		
Daphnia pulex (fed 6 days)						
Cyano mix (Micr 1 + Micr 2 + Syn)	0.5	0.3–0.4	0.37	2.5		Ravet et al. (2003), Exp. 2
Cyano mix + EPA	15	6	0.44	6		
Crypt 1 + Crypt 2 + Rhod	15	**10**	0.57	11		
Daphnia pulex (fed 6 days)						
Cyano mix	0.5	0.3–0.4	0.36	3		Ravet et al. (2003), Exp. 4 and 5
Cyano mix + ALA	0.5	6	0.38	3		
Cyano mix + DHA	0.5	6	0.38	4		
Cyano mix + EPA	15	6	0.42	7		
Cyano mix + ALA + DHA + EPA	5	6	0.42	7		
Cyano mix + 65% of Crypt FA	10	6	0.48	6		
Crypt 1 + Crypt 2 + Rhod	15	**10**	0.56	10–13		
Daphnia pulex (fed 6 days)						
Cyanophytes	1.5[b]	0.46	0.35	2		Brett et al. (2006)
Chlorophytes	0.1[b]	1.2	0.47	4		
Cryptophytes	9.9[b]	**12**	0.57	9		

r, specific growth rate; –, indicates 'no values available'; Micr 1 and 2 = *Microcystis aruginosa* strain 2063 and 2387, respectively; Syn = *Synechococcus elongatus*; Crypt 1 and 2 = *Cryptomonas ovata* strains 979/44 and 979/61, respectively; Rhod = *Rhodomonas minuta*; text in bold indicates a possible optimum

[a]Data from another strain (M. Brett, pers. comm.)
[b]% of total FA

Table 7.6 PUFA ratios in freshwater zoobenthos

Species	Feeding mode	n-3/n-6	DHA/ARA	Source
Aquatic insects				
Chironomidae (n = 3)		2.2	0.1	Bell et al. (1994)
Chironomidae (n = 6)		3.54	0.1	Sushchik et al. (2003)
Chironomus plumosus, (autumn cohort, n = 6)	Filter/deposit-feeder	1.03 ± 0.16	0.0	Goedkoop et al. (2000)
Chironomus plumosus, (spring cohort, n = 7)	Filter/deposit-feeder	1.09 ± 0.38	0.0	Goedkoop et al. (2000)
C. anthracinus (n = 5)	Deposit-feeder	0.74 ± 0.22	0.0	Goedkoop et al. (2000)
Average ± SD		1.7 ± 1.2	0	
Chironomus riparius males (n = 1)		0.19	0.8	Goedkoop (unpublished)
Chironomus riparius females (n = 2)		0.16 ± 0.003	0.67 ± 0.015	Goedkoop (unpublished)
Chironomus riparius larvae (n = 2)	Deposit-feeder	0.22 ± 0.003	0.35 ± 0.35	Goedkoop (unpublished)
Chironomus riparius eggs		0.12	0.38	Goedkoop (unpublished)
Average ± SD		0.17	0.55	
Procladius sp. (n = 6)	Predator	1.77 ± 0.23	0.38	Goedkoop et al. (2000)
Chaoborus flavicans (n = 6)	Predator	3.54 ± 0.36	1.26	
Average ± SD		2.7 ± 1.2	0.8 ± 0.6	
Ephemeridae (n = 5)	Filter-feeder	2.2	0.02	Sushchik et al. (2003)
Ecdyonurus (n = 3)	Grazer/detritivore	8.2	0.0	Bell et al. (1994)
Ephemerella (n = 3)	Grazer/detritivore	6.3	0.0	Bell et al. (1994)
Caenis (n = 2)	Detritivore	6.2	0.0	Bell et al. (1994)
Average ± SD		5.7 ± 2.5	0	
Hydropsychidae				
Hydropsyche spp. (n = 3)	Collector-gatherer	1.6	0.8	Dahl (2006)
Corixidae (n = 3)	Mostly detritivores	3.3	0.4	Bell et al. (1994)
Notonectidae				Bell et al. (1994)
Notonecta sp. (n = 3)	Predator	3.7	0.1	Bell et al. (1994)
Coleoptera (n = 3)	Mostly predators	3.6	0.0	Bell et al. (1994)
Plecoptera (n = 3)	Highly variable	4.1	0.2	Bell et al. (1994)
Trichoptera (n = 4)	Highly variable	3.22	1.1	Sushchik et al. (2003)
Average ± SD		3.3 ± 0.9	0.5 ± 0.4	
Oligochaeta (n = 2)	Subsurface deposit-feeder	2.2	0.2	Bell et al. (1994)
Oligochaeta (n = 6)	Subsurface deposit-feeder	1.03 ± 0.07	0.0	Goedkoop et al. (2000)

(continued)

Table 7.6 (continued)

Species	Feeding mode	n-3/n-6	DHA/ARA	Source
Crustacea				
Gammaridae ($n = 3$)	Shredder/detritivore	2.3	0.4	Bell et al. (1994)
Gammaridae ($n = 6$)	Shredder/detritivore	3.5	1.3	Sushchik et al. (2003)
Gammarus lacustris	Shredder/detritivore			Makhutova et al. (2003)
August 4		2.3	1.4	
August 5		1.9	1.7	
August 8		1.7	1.6	
Gammarus pulex ($n = 2$)	Shredder/detritivore	1.1	0.4	Dahl (2006)
Dikerogammarus villosus	Shredder/detritivore/predator			Maazouzi et al. (2007)
July		2.1	1.0	
August		1.3	0.6	
September		1.8	0.6	
Asellidae				
Asellus aquaticus ($n = 2$)	Shredder/detritivore	0.8	0.3	Dahl (2006)
Astacus astacus ($n = 8$), (tail-muscle)	Detritivore/predator	1.1	0.2	Ackefors et al. (1997)
Pacifastacus leniusculus, (tail-muscle) ($n = 7$)	Detritivore/predator	1.2	0.3	Ackefors et al. (1997)
Average ± SD		1.8 ± 0.8	0.8 ± 0.5	
Mysidacae				Schlechtriem et al. (2008)
Mysis relicta	Predator			
Spring: male		8.2	6.0	
Female		8.2	5.7	
Juvenile		10.1	8.7	
Fall: male		4.6	2.9	
Female		4.9	2.9	
Diporeia spp.	Deposit-feeder	6.7	5.5	Schlechtriem et al. (pers. comm.)
Average ± SD		7.1 ± 2.1	5.3 ± 2.2	
Bivalvia				
Dreissena polymorpha	Filter-feeder			Wacker and Von Elert (2004)
May (surface)		4.6	4.4	
June (surface)		4.1	4.2	
July (surface)		3.6	3.4	
August (surface)		2.8	2.5	
Average ± SD		3.8 ± 0.8	3.6 ± 0.8	
July (15 m)		5.8	5.8	
August (15 m)		5.8	5.7	
Average ± SD		5.8 ± 0.1	5.7 ± 0.1	

'–' indicates 'no values available'

7 Fatty Acid Ratios in Freshwater Fish

Table 7.7 PUFA ratios in feeding experiment with freshwater zoobenthos

Treatment	PUFA in feed					Larvae			Source
	EPA (mg g^{-1} C)	DHA (mg g^{-1} C)	ARA or DPA	n-3/n-6	DHA/ARA or DHA/DPA	Size (mm) (1)	(2)	Survival (%)	
Chironomus riparius fed 10d from 1st instar									Goedkoop et al. (2007)
peat (28)	0.0	0.0	0.0	0.0	0.0	<2		0	
Oats	0.0	0.0	0.0	0.03	0.0	4.88	8.45	40	
Scenedesmus obliquus	0.0	0.0	0.0	6.2	0.0	1.69	5.28	0	
Scenedesmus + EPA	5.4	0.0	0.0	2.9	0.0	1.95	5.98	0	
Spirulina	0.2	0.0	0.2	0.03	0.0	4.94	10.9	43	
Tetraphyll®	1.9	2.9	0.3	0.4	8.4	5.81	10.7	57	
Dreissena polymorpha (fed 15–22 days)									Vanderploeg et al. (1996), (survival to the pediveliger stage, %)
No food	0.0	0.0	0.0	0.0	0.0	0.12		0	
Synechococcus spp.	0.0	0.0	0.0	0.4	0.0	0.14		0	
Chlamydomonas sp.	0.0	0.0	0.0	2.5	0.0	0.12		0	
Chlorella minutissima (marine)	32	0.0	5.4	4.7	5.9[c]	0.23		27	
Chlorella minutissima (freshwater)	4.6	0.0	3.2	2.1	1.4[c]	1.12		34	
Nannochloris sp.	–	–	–	–	–	0.17		0	
Nannochloris sp.	–	–	–	–	–	0.11		0	
Chromulina chinophila[a]	0.0	4.5	5.5[b]	1.5	0.8[b]	0.17		0	
Cryptomonas sp.[a]	13	4.6	2.1[b]	6.1	2.2[b]	0.12		0	
Rhodomonas[a] (fed 18 days)	26	7.6	4.4[b]	18	1.7[b]	1.51		9	Wacker et al. (2002) (days when 50% of the larvae died)
Isochrysis galbana	3.71	22.1	0.7	5.2	32	0.18		27	
Nannochloropsis limnetica	87.7	0.0	14.6	3.5	6.0[c]	0.16		29	
Monoraphidium minutum	0.0	0.0	0.0	3.2	0.0	0.11		16	
Chlorella minutissima	0.0	0.0	0.0	6.3	0.0	0.12		17	

Aphanothece sp.	0.0	0.0	0.0	N/A	0.0	0.14	21	Wacker et al. (2002) (growth mm d⁻¹ × 11)
Fed 11 days								
Nannochloropsis limnetica	67	0.0	14.6	2.7	0.0	0.055		
N. limn. + extract	84	2.2	14.6	3.4	0.2	0.078		
N. limn. + DHA	75	5.8	14.6	3.0	0.4	0.073		
N. limn. + EPA	83	0.0	14.6	3.4	0.0	0.056		
Fed 90 days								
Aphanothece sp.	0.0	0.0	0.0	N/A	0.0		–	0.4	Wacker and Von Elert (2004) (eggs mg/mussel)
Cryptomonas erosa	37.5	3.95	0.44	17.3	9.0		–	0.9	
N. limnetica	103	0.0	15.9	4.3	6.5[c]		–	1.5	
Scenedesmus obliquus	0.0	0.0	0.0	2.2	0		–	0.5	

Extract = lipid extract from *Isochrysis galbana*; '–' indicates 'no values available'. Under the heading Larvae '(1)' and '(2)' indicate different amounts of carbon in the feed where (1) = 2–3 mg C and (2) = 23–27 mg C; N/A = the n-6 content was not detectable; hence, the n-3/n-6 could not be calculated

[a]Vanderploeg et al (1996) referred to data (% of total FA) that were originally presented in Ahlgren et al. (1990). However, to make these data consistent with the mg/g C units in this table the data presented here were derived from Ahlgren et al. (1992) (i.e. mg/g dry weight × 2)

[b]DPA (22:5n-6) or DHA/DPA

[c]EPA/ARA

Chapter 8
Preliminary Estimates of the Export of Omega-3 Highly Unsaturated Fatty Acids (EPA + DHA) from Aquatic to Terrestrial Ecosystems

M.I. Gladyshev, M.T. Arts, and N.N. Sushchik

8.1 Introduction

In recent decades polyunsaturated fatty acids (PUFA) have come to be recognized as compounds with considerable physiological importance for animals at all taxonomic levels, including humans. Animals do not have the enzymes necessary to insert double bonds in fatty acid molecules in positions closer than the 7th carbon (designated n-7 or ω7) from the methyl end of the molecule; therefore, 18-carbon-long PUFA such as linoleic acid (LIN; 18:2n-6) and α-linolenic acid (ALA; 18:3n-3) are essential dietary nutrients (Fig. 8.1). These two essential PUFA are primarily synthesized by plants (both vascular plants and algae) and by some fungi (Fig. 8.1). These PUFA are the biochemical precursors of the most physiologically active PUFA: arachidonic acid (ARA; 20:4n-6), eicosapentaenoic acid (EPA; 20:5n-3), and docosahexaenoic acid (DHA; 22:6n-3). Higher plants cannot desaturate and elongate ALA to EPA and DHA; however, many algae can perform these reactions (Fig. 8.1, and see Sect. 8.2 for details). Although animals, including humans, can desaturate and elongate the parent ALA to EPA and DHA (Fig. 8.1, Gerster 1998 concludes that this conversion is "unreliable and restricted" in humans; see also Plourde and Cunnane 2007) the efficiency of this process, and the tissue-specific requirements for particular PUFA, are quite variable among animal species and developmental stages within species. Nevertheless, the n-3 highly unsaturated fatty acids (HUFA[1]) are increasingly understood to play a key role

---

[1] The two long chain highly unsaturated fatty n-3 fatty acids (i.e. EPA and DHA) will be referred to in this chapter as n-3 HUFA or, for brevity, simply as HUFA.

M.I. Gladyshev (✉) and N.N. Sushchik
Institute of Biophysics, Siberian Branch of the Russian Academy of Sciences,
Krasnoyarsk, Akademgorodok, Russia, 660036
e-mail: glad@ibp.ru

M.T. Arts
National Water Research Institute, Environment Canada
867 Lakeshore Road, P.O. Box 5050 Burlington, Ontario Canada L7R 4A6

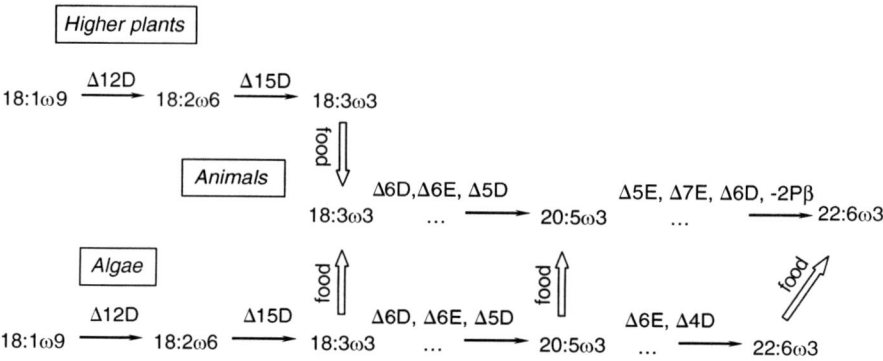

Fig. 8.1 Schematic of n-3 polyunsaturated fatty acids synthesis in different groups of organisms: black arrows – biosynthesis; open arrows – food intake; Δn – number of carbon from carbonyl end of the molecule; D desaturase; E elongase; $2P\beta$ peroxisomal system of enzymes for β oxidation. Note: for simplicity only one pathway of biosynthesis of 20:5ω3 and 22:6ω3 in algae is depicted, and a number of intermediate fatty acids are substituted by "..."

in the health of all organisms. They have a generally positive impact on animal (and human; see Chap. 14) health due to their effects in preventing/mitigating cardiovascular diseases, ontogenesis (particularly neural development), "dysfunctional" behaviors (e.g., aggression, homicide), atherosclerosis, neural disorders, and, potentially, some cancers, as well as autoimmune diseases (e.g., Arts et al. 2001; Lauritzen et al. 2001; Broadhurst et al. 2002; Copeman et al. 2002; Silvers and Scott 2002; Aktas and Halperin 2004; Hibbeln et al. 2004, 2006; Simopoulos 2004a). In addition, DHA is known to play a pivotal role in the health and function of the vertebrate retina and nervous tissues (SanGiovanni and Chew 2005).

In this chapter we consider evidence that algal-derived HUFA are an important, or, even the main source of dietary-derived HUFA for many terrestrial animals (Fig. 8.1). We also explore the alternative hypothesis that, in terrestrial food webs, HUFA in sufficient abundance may be commonly found, therefore lessening the importance of aquatic-derived HUFA. A second hypothesis, and one that we cannot fully resolve due mainly to a paucity of existing data, is that terrestrial animals can simply desaturate and elongate dietary ALA (Fig. 8.1) to meet whatever requirements they may have for HUFA.

8.2 HUFA in Aquatic and Terrestrial Ecosystems

Among organisms in the biosphere, algae, and, in particular, diatoms, cryptophytes, euglenoids, and dinoflagellates, can de novo synthesize high amounts of HUFA (Fig. 8.1). Once synthesized at the level of primary producers, HUFA are transferred and can accumulate, at progressively higher trophic levels, in the biomass of aquatic organisms. Therefore, aquatic ecosystems occupy the unique position on earth as the principal dietary source of n-3 HUFA for all animals, including inhabitants of terrestrial ecosystems. There are three critical assumptions that logically follow the previous statement, i.e., that terrestrial animals: (a) cannot obtain sufficient HUFA solely from terrestrially based foods, (b) cannot synthesize n-3 HUFA (from ALA) at levels sufficient to meet their needs, and (c) require adequate HUFA levels in the diet for optimal physiological performance, i.e., HUFA are beneficial. Several lines of evidence can be advanced to support to these three assumptions.

First, it is necessary to point out that the majority of unsaturated fatty acids in higher plants are the C_{18} compounds: oleic, linoleic, and α-linolenic acids (Shorland 1963; Harwood 1996). Higher plants generally cannot convert 18-carbon chain (C_{18}) PUFA to HUFA (Heinz 1993; Tocher et al. 1998, Fig. 8.1). In contrast, the lower plants, i.e., some groups of the eukaryotic microalgae, and a few fungi (e.g., *Saprolegnia sp.* and *Mortierella alpina*) possess the enzymatic systems for de novo biosynthesis of HUFA and can produce significant amounts of these FA (Cohen et al. 1995; Tocher et al. 1998; and, for fungi see Leonard et al. 2004, Fig. 8.1). Thus, even for some of the most oil-rich tissues of terrestrial plants (i.e., the seed crops and their wild relatives), genetic intervention is needed in order to achieve a substantial HUFA content (Robert 2006; Damude and Kinney 2007). Another example is that the seeds of common tree species (pine, spruce, larch), which are commonly consumed by terrestrial herbivores, also do not contain HUFA in any appreciable quantities (Wolff et al. 2001). Wild edible leafy plants can also be shown to contain only C_{18} PUFA (Simopoulos 2004b). Finally, studies on terrestrial vegetative matter entering streams and lakes have demonstrated that this material does not contain HUFA (Mills et al. 2001).

Second, many invertebrate species consume primary producers directly and may themselves be consumed by predators at higher trophic levels. Recognizing that there is some conservatism with respect to incorporation of dietary FA into invertebrate tissue (see Chap. 6), a comparison of FA profiles of aquatic versus terrestrial invertebrates should shed light on: (a) the relative differences in HUFA contents of aquatic versus terrestrially derived dietary plant materials, (b) their subsequent retention in primary consumers, and (c) the potential for their transfer to consumers at higher trophic levels. Although there are some exceptions (e.g., Nor Aliza et al. 2001), the majority of terrestrial insects examined so far have very low HUFA concentrations (Uscian and Stanley-Samuelson 1994; Howard and Stanley-Samuelson 1996) relative to aquatic insects, which have been shown to be generally rich in HUFA (Hanson et al. 1985). Thus, terrestrial insects, as one of the key food sources for terrestrial predators, may be, in general, unlikely candidates to supply the majority of the HUFA to terrestrial predators at higher trophic levels.

Third, terrestrial predators show reductions in the n-3 PUFA and HUFA concentrations as the quantity of aquatic food in their diets declines. Three examples include: (a) small carnivorous mammals show a significant decline in their DHA to linoleic acid (18:2n-6) ratios as their dependence on aquatic food webs decreases (Koussoroplis et al. 2008), (b) long-term decreases in the concentrations of C_{18} and C_{20} n-3 PUFA have been observed in herring gulls in the Great Lakes as their forage base shifted from primarily aquatic to terrestrial diet items (Hebert et al. 2008), and (c) carnivorous reptiles consuming seabirds on small islands have higher plasma EPA and DHA contents than conspecifics on nearby islands where rats (terrestrial prey) form a higher percentage of the diet (Blair et al. 2000). Taken together such observations suggest that n-3 HUFA concentrations in terrestrial predators will generally increase as their consumption of aquatic prey increases.

Finally, we must ask – do terrestrial animals actually need dietary HUFA? This is a question with far-reaching implications, and we provide several lines of reasoning to suggest that this may often be the case. It is clear that the ability to elongate and desaturate ALA to EPA and DHA, at levels that are sufficient to supply them with optimum concentrations of HUFA, is not equally prevalent and/or efficient among animal species and/or among tissues (Mitchell et al. 2007; Leonard et al. 2004) and also that HUFA requirements might differ seasonally (Pruitt and Lu 2008). African camilids and rodents, for example, have EPA and DHA concentrations that range from very low to nondetectable in their meat (Hoffman 2008), whereas domestic farm animals (cattle, sheep, and especially pigs) contain appreciable levels of both EPA and DHA in their muscle and adipose tissues (Wood et al. 2008). To add to the complexity it is also clear that diet affects the ultimate expression of HUFA concentrations in the meat of vertebrates (Wood et al. 2008) and that the specific requirement for HUFA may be more critical for some developmental stages and/or tissues than others. Further, energy is required to elongate and desaturate C_{18} PUFA to form HUFA and, therefore, conservatively acquiring these materials "pre-formed" in the diet is, at the very least, energetically advantageous. Also, the inherent biochemical properties of EPA and DHA, that give them their special properties, apply universally, i.e., their functional roles in the physiological competency of cell membranes and in maintaining a healthy immune system (e.g., Stanley-Samuelson et al. 1991) is not limited in their applicability to aquatic animals (e.g., see Chap. 10) but extend to terrestrial animals as well (e.g., Geiser et al. 1992; Stanley 2006).

It is already well established that clear physiological benefits accrue to both aquatic and terrestrial animals when they consume adequate levels of HUFA in their diets. This can be assessed in two ways: by examining the effects of offering animals diets that are either HUFA deficient or HUFA sufficient or by adding HUFA to the diet. Such techniques, combined with field studies, have clearly demonstrated the importance of food quality (i.e., of contents of essential nutrients, including HUFA) for aquatic animals, especially for daphnids (e.g., Müller-Navarra 1995; Gulati and DeMott 1997; Gladyshev et al. 2006a; Danielsdottir et al. 2007). Somatic growth of *Daphnia* correlates with EPA content in seston (i.e., with food quality), rather than with sestonic carbon content (i.e., with food quantity) (e.g., Müller-Navarra 1995; Gladyshev et al. 2006a) even for very low food levels (Boersma and Kreutzer 2002). Most of the evidence for the effects of HUFA content on terrestrial animals comes primarily from laboratory studies of mammals and birds. For example, n-3 HUFA-deficient rats exhibited significantly longer escape latency and poorer memory performance in maze experiments compared with n-3 HUFA sufficient rats (Lim et al. 2005a), and n-3 fatty-acid-deficient mice had impaired learning in a reference-memory version of circular maze, i.e., they spent more time and made more errors in search of an escape tunnel (Fedorova et al. 2007). In birds it has been shown that egg FA composition reflects diet FA composition (Farrell 1998) and birds such as herring gulls were in poorer condition and exhibited reduced reproductive success in areas where they consumed a higher proportion of terrestrial food (Hebert et al. 2002, 2008). The offspring of domestic cats (carnivores), whose mothers were fed corn-oil based diets, had insufficient levels of EPA and DHA in their brains and retinas (and consequently exhibited delayed photoreceptor responses) which suggested that they had a low biosynthetic capacity to produce these FA (Pawlosky et al. 1997). Rats (omnivores) have also been shown to have a requirement for dietary DHA in order to insure proper brain function (Lim et al. 2005b). Vegetarian, and especially vegan, humans must rely on the internal conversion of ALA to EPA and DHA and, these people have, in addition to lower total lipid levels in their plasma and erythrocytes, plasma EPA and DHA levels that are only 12–15% and 32–35%, respectively, as high as those of nonvegetarians (cited in Davis and Kris-Etherton 2003). Tissue and/or developmental stage-specific needs are also recognized such as the critical requirement for DHA for vision in vertebrates (Politia et al. 2001).

In conclusion, due to the paucity of data and limited number of species examined we can, at present, neither absolutely confirm nor deny a universal dependency for dietary HUFA among terrestrial animals. However, a growing body of evidence suggests that: (a) given the apparent scarcity of these compounds in terrestrial ecosystems, (b) their many recognized physiological benefits, and (c) the probability that they are required by least at some tissues, and/or some developmental stages, and/or during different seasons, this at least remains a real and likely possibility.

Thus, we postulate that the well-established function of waterbodies as a source of drinking water should be augmented by an explicit recognition of their role in supplying terrestrial ecosystems with biochemically and physiologically essential lipids. It is also important to note that aquatic ecosystems differ in their ability to produce HUFA. For instance, water bodies dominated by cyanobacteria have significantly lower relative HUFA production than those dominated by diatoms (Müller-Navarra et al. 2004). Moreover, algae, such as the HUFA-rich diatoms (e.g., Sushchik et al. 2004), are known to accumulate higher levels of HUFA as temperatures decrease – a process referred to as cold adaptation (see also Chaps. 1 and 10). Thus, large-scale processes such as eutrophication and global warming may act either independently, or in concert, to effect an overall decrease in HUFA production in aquatic ecosystems with possible negative implications for surrounding terrestrial ecosystems. At present it is not possible to assess the potential ecological risks associated with decreased HUFA production as a result of anthropogenic impacts and global warming because there are, as yet, no global, or even regional, estimates of the amount of HUFA that is exported from aquatic to terrestrial ecosystems. Thus, we suggest that a concerted effort to quantify HUFA export from aquatic to terrestrial ecosystems, in geographically and climatically diverse regions with different levels of anthropogenic impact, should be attempted. Although we view such an attempt as necessary we also clearly acknowledge that estimates of HUFA export are, by their very nature, and also largely because of the paucity of data, preliminary and incomplete. Thus, this chapter is also intended to stimulate further research on quantifying the role of aquatic ecosystems as *producers* and *providers* of HUFA to terrestrial organisms.

In aquatic ecosystems, HUFA produced by microalgae are transferred to primary consumers such as zooplankton and zoobenthos. Naturally, zooplankton and zoobenthos comprise several trophic levels, i.e., carnivorous animals consume HUFA along with their prey and thus HUFA are bioaccumulated within aquatic food webs (Kainz et al. 2004), as well as recycled within plankton and benthic communities. Nevertheless, although these functional food web links are crucial for aquatic ecosystems, we will not refer to internal aquatic food web dynamics here because such processes are not directly related to HUFA export. Aquatic animals including zooplankton, zoobenthos, and fish can be consumed by water birds (e.g., Hebert et al. 2008) and riparian animals (e.g., Koussoroplis et al. 2008), and their HUFA are thus exported to terrestrial ecosystems. Water birds and riparian animals, in turn, may be consumed by other terrestrial predators or by soil organisms after death. Nevertheless, here we will only focus on the potential export of HUFA from aquatic to terrestrial ecosystems. Besides direct consumption of animals in the aquatic environment by terrestrial animals (e.g., Hilderbrand et al. 1999a), HUFA are also exported from aquatic to terrestrial ecosystems through aquatic insect and amphibian emergence (Fig. 8.2).

These considerations may be attributed primarily to inland waters. Export of HUFA from marine waters to terrestrial ecosystems is believed to be primarily due to shore drift of carrion and seaweeds, and anadromous salmon migrating in rivers (Fig. 8.2). Here we omit the input through seabirds, since colonies of these birds represent closed systems, which are usually somewhat separated from other terrestrial communities. Seabirds usually nest on coastal rocks, or live on ice

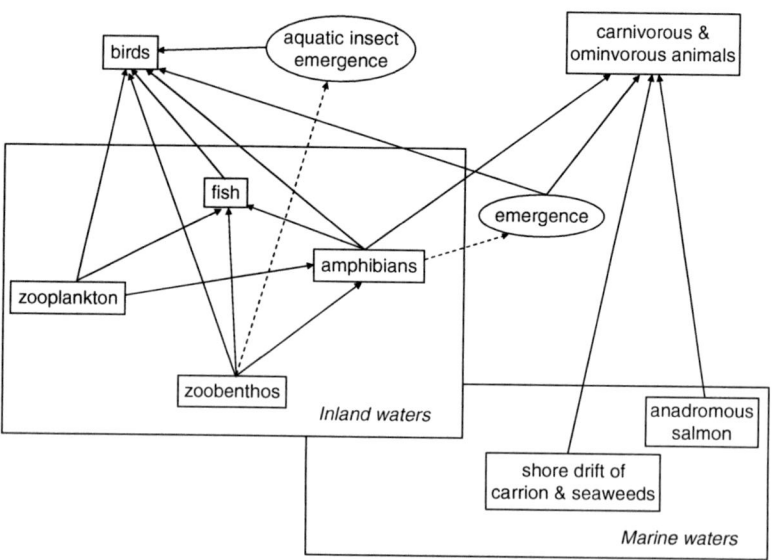

Fig. 8.2 Schematic flow diagram depicting directionality of HUFA fluxes from aquatic to terrestrial ecosystems

8   Preliminary Estimates of the Export

shelves (penguins), and therefore exhibit a lower level of connectivity (although, clearly, predation on seabirds by terrestrial animals occurs) with terrestrial systems. Because of this we here consider seabirds as a primarily oceanic rather than a terrestrial fauna.

8.3   Required Measurements

Based on general considerations (Fig. 8.2), we can begin to identify the type of data needed in order to obtain preliminary estimates of the amount of HUFA exported from a particular aquatic ecosystem to adjacent terrestrial ecosystems. Such calculations require knowledge of:

(a) HUFA composition of aquatic animals (e.g., zooplankton, zoobenthos, and fish)
(b) Rations of water birds and riparian animals, fed with aquatic animals
(c) Areal biomass of water birds and riparian animals that feed on aquatic animals
(d) HUFA content of emerging aquatic animals (insects and amphibians)
(e) Emergence intensity
(f) Surface area of the aquatic ecosystems of interest
(g) Surface area of the adjacent terrestrial ecosystems
(h) Metabolic HUFA requirements of carnivorous and omnivorous terrestrial animals and their biomass.

The final point is required in order to gauge whether or not terrestrial organisms might ever be limited by their access to HUFA derived from aquatic ecosystems.

Unfortunately, at present, there are no systematic, quantitative, or even semiquantitative estimates reported for the export of the two essential n-3 HUFA (EPA and DHA; the focus of this chapter) from aquatic to terrestrial ecosystems despite growing evidence that these compounds contribute significantly to maintaining animals in a state of optimal physiological competency (see Chaps. 10, 13, 14). We are aware that our literature survey and resulting estimates will not be comprehensive. There are two main reasons for this. First, there is a general paucity of data in the primary literature in relation to the eight steps outlined earlier, especially for some groups of animals (e.g., amphibians). Second, global estimates, by their very nature, will probably always lack the specificity, completeness, and refinement that more regional estimates may, one day, be able to provide.

Does this mean that the prospect of generating the first ever global estimate of HUFA export from aquatic to terrestrial systems is not worth attempting? We argue against this for two reasons. First, it is important to formally acknowledge, the heretofore unrecognized function that aquatic ecosystems provide as the producers of essential nutrients for the whole biosphere. In order to provide meaning and context to this concept estimates of HUFA flux rates (however preliminary) are required. Second, we seek the indulgence of our readers to allow us this opportunity to develop a conceptual framework from which future estimates may be developed and refined as more data becomes available.

8.4 HUFA Content of Aquatic Organisms

A subset of the available literature data on the HUFA content in aquatic animals and seaweed is summarized in Tables 8.1–8.4. We restrict the freshwater zooplankton data to two locations (Table 8.1; but see Chap. 6 for greater detail). For Cladocera

Table 8.1 Average contents of EPA and DHA, mg g^{-1} of dry weight, in zooplankton organisms of inland waters: n = number of samples

Species (taxon)	EPA	DHA	Sum	Ecosystem, Region	n	References
Cladocera[a]						
Daphnia spp.	6.3	4.8	11.1	Lakes in north-western Sweden	5	Persson and Vrede (2006)
Bosmina coregoni	7.2	1.3	8.5	Lakes in north-western Sweden	4	Persson and Vrede (2006)
Holopedium gibberum	7.4	0.9	8.3	Lakes in north-western Sweden	15	Persson and Vrede (2006)
Bythotrephes longimanus	8.7	0.8	9.5	Lakes in north-western Sweden	4	Persson and Vrede (2006)
Calanoida[a]						
Arctodiaptomus laticeps	5.6	11.1	16.7	Lakes in north-western Sweden	8	Persson and Vrede (2006)
Heterocope spp.	4.3	8.3	12.6	Lakes in north-western Sweden	8	Persson and Vrede (2006)
Zooplankton >500 μm (*Daphnia, Holopedium,* calanoid copepods)	10.8	2.2	13.0	Lakes and reservoirs in British Columbia, Canada	nr	Kainz et al. (2004)
Average for zooplankton			11.4			

[a]Recalculated from the reference data using ratio for zooplankton (Cladocera and Copepoda) 1 g C = 2.75 g dry mass (Alimov 1989)
nr = not reported

Table 8.2 Average contents of EPA and DHA, mg g^{-1} of dry weight, in zoobenthos organisms of inland waters: n = number of samples

Species (taxon)	EPA	DHA	Sum	Ecosystem, Region	n	References
Insecta						
Chaoborus flavicans	23.6	5.5	29.1	Lake Erken, Sweden	6	Goedkoop et al. (2000)
Trichoptera[a]	10.8	0.8	11.6	Yenisei river, Siberia, Russia	4	Sushchik et al. (2003)
Trichoptera	2.37	0.15	2.5	3 streams in Quebec, Canada	46	A. Mazumder, pers. comm.
Trichoptera, Glossosomatidae	16.1	0.2	16.3	6 streams in Washington State, USA	12	C. Volk, pers. comm.
Ephemeroptera[a]	12.8	tr	12.8	Yenisei river, Siberia, Russia	5	Sushchik et al. (2003)
Ephemeroptera	2.12	0.02	2.1	3 streams in Quebec, Canada	46	A. Mazumder, pers. comm.
Ephemeroptera, Baetidae	13.9	0.3	14.2	6 streams in Washington State, USA	11	C. Volk, pers. comm.

(continued)

8 Preliminary Estimates of the Export

Table 8.2 (continued)

Species (taxon)	EPA	DHA	Sum	Ecosystem, Region	n	References
Ephemeroptera, Heptagenidae	9.36	0.15	9.5	6 streams in Washington State, USA	15	C. Volk, pers. comm.
Diptera, Simulidae	3.00	0.08	3.1	3 streams in Quebec, Canada	15	A. Mazumder, pers. comm.
Plecoptera, Pteronarcidae	0.72	0.00	0.7	3 streams in Quebec, Canada	15	A. Mazumder, pers. comm.
Chironomidae						
Chironomus spp.	4.0	tr	4.0	Lake Erken, Sweden	18	Goedkoop et al. (2000)
Procladius sp.	6.4	0.6	7.0	Lake Erken, Sweden	6	Goedkoop et al. (2000)
Diamesa baikalensis and unidentified spp.[a]	7.7	tr	7.7	Yenisei river, Siberia, Russia	6	Sushchik et al. (2003)
Oligochaeta[a]	6.3	0.5	6.8	Yenisei river, Siberia, Russia	10	Sushchik et al. (2006)
Amphipoda						
Gammarus lacustris[a]	2.6	1.8	4.4	Wetland ponds, Saskatchewan, Canada	nr	Arts et al. (2001)
Hyalella azteca	3.3	0.4	3.7	Wetland ponds, Saskatchewan, Canada	nr	Arts et al. (2001)
Gammarus fossarum[b]	10.2	2.3	12.5	Struga Dobieszkowska River, Poland	nr	Kolanowski et al. (2007)
Gammarus pulex[b]	3.9	1.2	5.1	Stradanka River, Poland	nr	Kolanowski et al. (2007)
Gammarus roeseli[b]	9.6	3.8	13.4	Notec River, Poland	nr	Kolanowski et al. (2007)
Pontogammarus robustoides[b]	20.9	4.3	25.2	Wloclawski Reservoir, Poland	nr	Kolanowski et al. (2007)
Dikerogammarus haemobaphes[b]	12.6	2.7	15.3	Wloclawski Reservoir, Poland	nr	Kolanowski et al. (2007)
Gammaridae[a]	8.1	1.2	9.3	Yenisei river, Siberia, Russia	6	Sushchik et al. (2003)
Average for zoobenthos			9.8			

tr tracers (<0.1 mg g^{-1} DM); nr not reported
[a]Recalculated from the reference using moisture contents Gammaridae 75.3%, Thichoptera 83.8%, Chironomidae 78.0%, Ephemeroptera – 80% and Oilgochaeta – 78%
[b]Recalculated from Figs. 3 and 4 of the reference

averaging the mean values of EPA + DHA of the different species give the following general mean value and standard error (SE) 9.4 ± 0.6 mg g^{-1} of dry weight (DW). For the zooplankton species and groups listed in Table 8.1 the general EPA + DHA average is 11.4 ± 1.1 mg g^{-1} DW. It is clear that increasing the number of studied zooplankton species and water bodies will expand increase our understanding variation of HUFA contents. Nevertheless, at present there is no reason to consider

Table 8.3 Average contents of EPA and DHA, mg g^{-1} of dry weight, in fish of inland waters: n = number of samples

Species	EPA	DHA	Sum	Ecosystem, region	n	References
Abramis brama L. (bream)	2.02	3.25	5.27	Lakes in Sweden	4	Ahlgren et al. (1994)
Alosa pseudoharengus (alewife)	13.7	11.9	25.6	Hamilton Harbour, Ontario, Canada	5	M.T. Arts and M. Koops (pers. comm.)
Ameiurus nebulosus (brown bullhead)	2.39	3.91	6.3	Hamilton Harbour, Ontario, Canada	21	M.T. Arts and M. Koops (pers. comm.)
Anguilla anguilla L. (eel)	4.98	6.86	11.84	Lakes and brackish Baltic, Sweden	8	Ahlgren et al. (1994)
Astyanax fasciatus Cuvier and *Melaniris sardina* Meek	1.2	5.0	6.2	Lakes Xolotlan and Cocibolca, Nicaragua	11	Ahlgren et al. (2002)
Barbus sp. (barbs)	2.17	4.48	6.65	Lakes in Ethiopia	7	Zenebe et al. (1998)
Blicca bjoerkna L. (white bream)	1.88	4.10	5.98	Lakes in Sweden	2	Ahlgren et al. (1994)
Carassius carassius L. (crucian carp)	1.05	3.03	4.08	Lakes in Sweden	2	Ahlgren et al. (1994)
Clarias gariepinus Burchell (catfish)	1.43	4.45	5.88	Lakes in Ethiopia	13	Zenebe et al. (1998)
Cyprinus carpio (common carp)	2.29	4.49	6.78	Hamilton Harbour, Ontario, Canada	9	M.T. Arts and M. Koops (pers. comm.)
Esox lucius L. (pike)	1.31	5.21	6.52	Lakes and brackish Baltic, Sweden	7	Ahlgren et al. (1994)
Gymnocephalus cernuus L. (ruffe)	2.11	4.04	6.15	Lakes in Sweden	5	Ahlgren et al. (1994)
Lates niloticus L. (Nile perch)	0.64	3.62	4.26	Lakes in Ethiopia	3	Zenebe et al. (1998)
Lepomis gibbosus (pumpkinseed)	1.90	6.04	7.94	Hamilton Harbour, Ontario, Canada	4	M.T. Arts and M. Koops (pers. comm.)
Leuciscus idus L. (ide)	2.28	4.94	7.22	Lakes in Sweden	1	Ahlgren et al. (1994)
Lota lota L. (burbot)	2.58	4.57	7.15	Lakes in Sweden	4	Ahlgren et al. (1994)
Lucioperca lucioperca L. (pike-perch)	1.07	4.18	5.25	Lakes in Sweden	2	Ahlgren et al. (1994)
Micropterus salmoides (large mouth bass)	1.62	4.61	6.23	Hamilton Harbour, Ontario, Canada	9	M.T. Arts and M. Koops (pers. comm.)
Notropis atherinoides (emerald shiner)	4.59	5.70	10.29	Hamilton Harbour, Ontario, Canada		M.T. Arts and M. Koops (pers. comm.)
Oncorhynchus mykiss W. (rainbow trout)	3.38	10.18	13.56	Lakes and reservoirs in British Columbia, Canada	12	Kainz et al. (2004)
Oreochromis niloticus L. (tilapia)	0.86	4.80	5.66	Lakes in Ethiopia	18	Zenebe et al. (1998)
Perca fluviatilis L. (perch)	1.27	4.49	5.76	Lakes in Sweden	9	Ahlgren et al. (1994)

8 Preliminary Estimates of the Export

Rutilus rutilus L. (roach)	2.54	5.00	7.54	Lakes in Sweden	9	Ahlgren et al. (1994)
Tinca tinca L. (tench)	1.62	2.67	4.29	Lakes in Sweden	1	Ahlgren et al. (1994)
Thymallus arcticus Pallas (Siberian grayling)	3.61	8.68	12.29	Yenisei river, Siberia, Russia	44	Sushchik et al. (2007)
Thymallus thymallus L. (European grayling)	4.65	9.37	14.02	Rivers in Sweden	8	Ahlgren et al. (1994)
Thymallus thymallus L. (European grayling)	3.56	6.97	10.53	River Vindelalven, Sweden	18	Ahlgren et al. (1999)
Average for fish			8.1			

Table 8.4 Average contents of EPA and DHA, mg · g^{-1} of dry weight, in marine organisms; n = number of samples.

Species (taxon)	EPA	DHA	Sum	Ecosystem, Region	n	References
Macroalgae						
11 species of seaweeds	1.3	<0.01	1.3	French Brittany coast	nr	Fleurence et al. (1994)
Invertebrates						
Copepods[a]	13.8	10.9	24.7	Norwegian Sea	14	Ahlgren et al. (2005)
Copepods[a]	6.2	9.1	15.3	Southern Baltic Sea	16	Ahlgren et al. (2005)
Euphausia superba (Euphausiacea)[b]	1.6	0.6	2.6	South Georgia, South Atlantic	20	Cripps et al. (1999)
Mysis mixta (Mysidacea)	13.7	9.5	23.2	Conception Bay, Newfoundland	174	Richoux et al. (2005)
Acanthostepheia malmgreni (Amphipoda)	8.7	5.7	14.4	Conception Bay, Newfoundland	119	Richoux et al. (2005)
Shellfish (12 species of Gastropoda and Bivalvia)	12.9	17.6	30.5	South Korea	81	Surh et al. (2003)
Brittle star *Amphiura elandiformis*	0.23	0.02	0.25	Southern Tasmania	2	Mansour et al. (2005)
Fish						
Cod (*Gadus morhua maris-albi*)	3.1	7.6	10.7	White Sea, north west of Russia	3	Gladyshev et al. (2007)
Gilthead sea bream (*Sparus aurata*)[c]	3.6	8.3	11.9	Mediterranean coast of Turkey	12	Ozyurt et al. (2005)
Herring (*Clupea harengus pallasi*)	5.6	12.3	17.9	Pacific Ocean, far east of Russia	3	Gladyshev et al. (2007)
Rock sole (*Lepidopsetta bilineata*)	7.2	4.6	11.8	Pacific Ocean, far east of Russia	3	Gladyshev et al. (2007)
Sea bass (*Dicentrarchus labrax*)[d]	7.5	14.2	21.7	Iskenderun bay, Mediterranean	48	Ozyurt and Polat (2006)
White sea bream (*Diplodus sargus*)[c]	4.4	10.2	14.6	Mediterranean coast of Turkey	9	Ozyurt et al. (2005)
Average for animals			15.4			

nr not reported

[a]Recalculated from Fig. 5 of the reference, using for copepods 1 g C = 2.75 g dry mass (from Alimov 1989) – see above table for zooplankton
[b]Recalculated from Tables 3 and 4 of the reference, using moisture content 10% (Alimov 1989)
[c]Recalculated from Tables 1 and 4 of the reference
[d]Recalculated from Tables 1 and 4 of the reference

that the average values will vary by more than a factor of 3 from the estimate presented here.

For freshwater zoobenthos we report data from four locations (Table 8.2) with an EPA + DHA average of 9.8 ± 1.6 mg g^{-1} DW. The number of HUFA measurements of zoobenthos species from diverse locations, like those of zooplankton, should be increased significantly in the future.

Data on HUFA content of freshwater fish from six locations are provided in Table 8.3. The average HUFA concentration is 8.1 ± 0.9 mg g^{-1} DW. Because of commercial interests the number of studies reporting HUFA concentrations in freshwater fish is higher than for zooplankton and zoobenthos (Tables 8.1 and 8.2) and, correspondingly, SE values are comparatively smaller. It is clear that data on HUFA content in as yet unmeasured freshwater fish species will still have to be obtained for more reliable generalizations.

In Table 8.4 very sparse data on HUFA content in only a few marine organisms are provided. The HUFA contents span two orders of magnitude, from 0.25 mg g^{-1} DW for the brittle star to 30.5 mg g^{-1} DW for shellfish (Table 8.4). Evidently, to calculate a more meaningful contribution of marine organisms to the global HUFA export, significantly more quantitative data are required. Furthermore, if one excludes humans, HUFA export from marine systems to terrestrial ecosystems would not reasonably be expected to include largely inaccessible deep water or open-water pelagic animals but would be restricted to marine organisms that inhabit the near-shore regions of marine and brackish systems.

In summary, one can readily see that comprehensive datasets on HUFA contents in groups of aquatic organisms are not yet available for the purposes of providing a more precise calculation of the global HUFA exports. Instead we will highlight some preliminary case studies for particular water-land HUFA fluxes.

8.5 Case Study I: Estimating the Export of Aquatic HUFA to Terrestrial Predators (Bears) in the Pacific Rim

The most conspicuous and well-documented aspect of terrestrial animals feeding on the aquatic organisms is the consumption of spawning Pacific salmon by terrestrial predators. This feeding represents an export of marine-derived energy and matter (and HUFA) to terrestrial ecosystems (Willson et al. 1998; Naiman et al. 2002; Helfield and Naiman 2006). Bears are believed to be the keystone consumer of these fish. It is necessary to emphasize that fish comprise a negligible part of bear rations outside of the area of Pacific Rim, where there are no anadromous salmon (e.g., Felicetti et al. 2004; Dobey et al. 2005). For example, in the vicinity of Yellowstone Lake consumption of cutthroat trout (*Oncorhynchus clarki*) by grizzly bears is ~18 kg trout bear^{-1} year^{-1} (Felicetti et al. 2004).

The average abundance of bears in the Pacific Rim is 0.19 ± 0.11 ind km^{-2}, and the average consumption of salmon by these bears is 416 ± 42 kg DW bear^{-1} year^{-1}

(Table 8.5). Thus, bears convey to terrestrial ecosystems via consumption 79 kg DW km^{-2} year^{-1} of aquatic productivity. Bears, on average, consume only ~50% of biomass of each salmon they kill (Willson et al. 1998; Gende et al. 2001; Reimchen 2000; Hilderbrand et al. 2004); the remaining parts of carcasses are included in terrestrial food webs via consumption by insects, birds, and small mammals. Thus, the total export of salmon biomass to terrestrial ecosystems can be estimated as $79 \times 2 = 158$ kg DW km^{-2} year^{-1}. It should be noted that the first line in Table 8.5 refers to measurements made at a location where the density of bears is known to be the highest in the world (southeastern Alaska), therefore adopting an average bear density 0.19 ind km^{-2} for the global calculations may lead to an overestimation of the export of salmon biomass by bears by a factor of 2.

Data on absolute EPA + DHA contents in Pacific salmon are very sparse. Gladyshev et al. (2006b) reported EPA and DHA concentrations of 0.536 ± 0.121 g 100 g^{-1} DW and 1.056 ± 0.152 g 100 g^{-1} DW, respectively, for humpback (pink) salmon (*Oncorhynchus gorbuscha*). Thus, the EPA + DHA concentration of pink salmon was, on average, 16.2 mg g^{-1} DW. This figure is close to that for land-locked rainbow trout (*O. mykiss*), given in Table 8.3 (Kainz et al. 2004). Using a salmon EPA + DHA concentration of 16 mg g^{-1} DW, we obtain a HUFA export to bears in terrestrial ecosystems of 2.5 kg km^{-2} year^{-1}. Although bears are the most visible consumers of salmon, they are not the only species to make use of this HUFA-rich food source. More than two dozen mammal and bird species are known to consume salmon carcasses especially in places where bears are relatively scarce (Szepanski et al. 1999; Darimont et al. 2003; Helfield and Naiman 2006). Thus, the total export of HUFA from Pacific salmon consumption by terrestrial predators and omnivores (bears and other animals consuming the remaining parts of carcasses) can be estimated to be ~5.0 kg DW km^{-2} year^{-1}.

Table 8.5 Abundance (N, ind. km^{-2}) of bears and their consumption of spawning salmon (C, kg DW bear^{-1} year^{-1})

Species	N	C	Region	References
Brown bear (*Ursus arctos*)	0.4	nr	Alexander Archipelago, Alaska	Ben-David et al. (2004)
Brown bear	0.15	nr	Wood River Lakes, Alaska	Helfield and Naiman (2006)
Grizzly bear (*Ursus arctos horribilis*)	0.03	391[a]	British Columbia	Hilderbrand et al. (2004)
Brown bear	nr	358[b]	Kenai Peninsula, Alaska	Hilderbrand et al. (1999b)
Black bear (*Ursus americanus*)	nr	498[a]	Haida Gwaii Islands, British Columbia	Reimchen (2000)
Average	0.19	416		

nr not reported
[a]Recalculated from wet weight using moisture content of salmon 71% (Winder et al. 2005)
[b]Recalculated from nitrogen consumption, using nitrogen content in fish 10% of DW (Vanni et al. 1997; Naiman et al. 2002)

Although we did not find quantitative estimates of the extent of spawning area in the available literature, we tried to estimate the global contribution of Pacific salmon to HUFA export to terrestrial ecosystems using the yearly oceanic production of Pacific salmon which migrate into rivers. The catch of Pacific salmon is used for the production estimation because exploitation rates are high, ranging from ~40 to 80% (Beamish et al. 1999). An average of the total catches (Canada, United States, Japan, and Russia) of pink (*O. gorbuscha*), chum (*O. keta*), and sockeye (*O. nerka*) salmon in 1926–1995 can be obtained from Beamish et al. (1999; Fig. 6) as ~500 000 ton year^{-1}. Since the catches of these species account for about 90% of the total catch of all salmon species (Beamish et al. 1999), the total catch of Pacific salmon can be estimated at ~555,000 ton year^{-1}. Using the average exploitation rate (40 + 80)/2 = 60%, the annual biomass of salmon population migrating into rivers is estimated to yield 925,000 tons. On the basis of data of (Reimchen 2000; Quinn et al. 2003; Winder et al. 2005; Helfield and Naiman 2006) we estimated that bears, on average, kill ~47% of salmon spawning in rivers. The salmon die after spawning, but many of their carcasses stay within aquatic ecosystems. We take into account only those directly exported to terrestrial ecosystems. Thus, the total annual export of salmon to terrestrial ecosystems via bear predation is 434,000 ton wet weight (126,000 ton dry weight). Considering the HUFA content in pink salmon (~16 mg g^{-1} DW; Gladyshev et al. 2006b) as a proxy for other Pacific salmon, the annual HUFA export to terrestrial ecosystems via bear predation on Pacific salmon is estimated to be 2×10^6 kg year^{-1}. In the following we will try to compare the export of HUFA from aquatic ecosystems via bears with the other fluxes, such as insect emergence, bird feeding (Fig. 8.1), and human fisheries.

8.6 Case Study II: Humans and Fisheries

Humans, although not explicitly depicted in Fig. 8.1, are terrestrial predators that actively consume HUFA-rich resources from aquatic ecosystems. The total world catch (1998–2003; marine and freshwater) of fish and shellfish was ~92.2×10^6 ton year^{-1} (FAO 2004). Using the data in Tables 8.3 and 8.4, and assuming a moisture content of 80%, the average HUFA content in the captured wet biomass can be estimated as 2 mg g^{-1}. Thus, man withdraws from aquatic ecosystems ~1.8 10^8 kg year^{-1} of EPA + DHA. This number is ~90× higher than the HUFA export through bears hunting salmon in Pacific Rim. Human consumption of wild and farmed fish and aquatic invertebrates is ~16 kg per person per year (FAO 2004). This means that the global average personal daily consumption of EPA + DHA is about 0.1 g, i.e., about ten times lower than is currently recommended by the World Health Organization. Worldwide exploitation rates of fisheries have been shown to be too high, even at present (Pauly et al. 2002). Thus, the further development of aquaculture and/or genetic engineering, for example, the insertion of genes directing HUFA synthesis into terrestrial oil-seed producing plants, may be required to supply mankind with the appropriate quantity of essential HUFA in order to avoid overexploiting natural aquatic ecosystems in the future (see also Chap. 9).

8.7 Case Study III: Estimation of the Export of HUFA Through Aquatic Insect Emergence

The emergence of aquatic insects is one of nature's most conspicuous displays of the export of aquatic-derived biomass to the terrestrial landscape. Insects from the orders Ephemeroptera (mayflies), Diptera (midges, black flies), Trichoptera (caddisflies), Plecoptera (stoneflies) lay their eggs in rivers, ponds, and lakes where the larvae then develop and accumulate HUFA throughout the various larval stages until they emerge as adults and enter the riparian zone and adjacent regions where mating occurs. In productive environments this emergence can be so prodigious that it can readily be detected by land-based radar, a tool with the potential to reveal both spatial and temporal variations in the strength of insect emergence (Fig. 8.3).

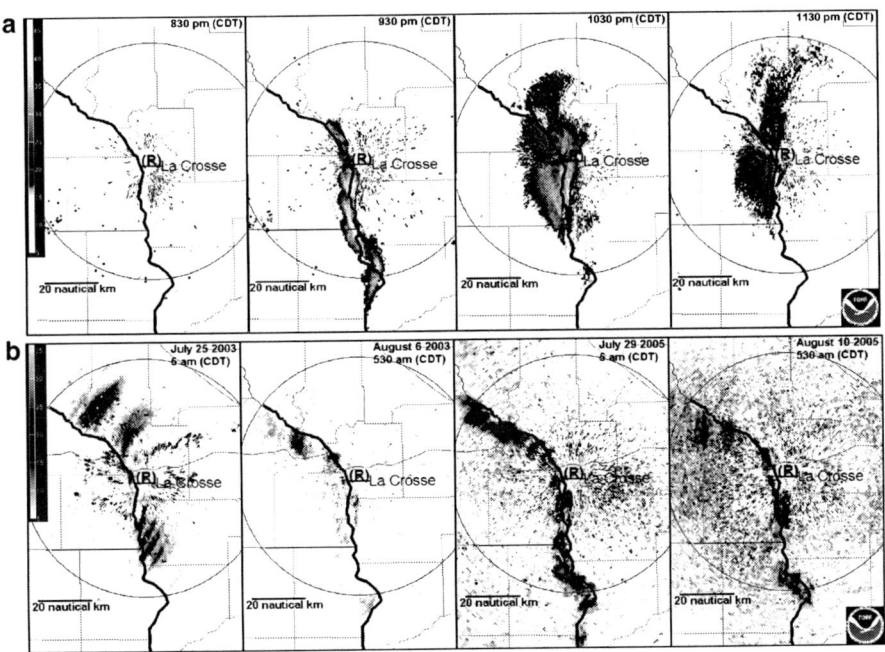

Fig. 8.3 Mayfly (Ephemeroptera) hatches from the Mississippi River centered on the city of La Crosse, Wisconsin, USA, as captured by radar images taken by the National Weather Service Office, National Oceanographic and Atmospheric Administration, La Crosse, Wisconsin. Evolution of the mayfly hatch: (**a**) during the evening of June 30, 2006; wind from the south, and (**b**) on, from left to right, July 25, 2003; August 6, 2003; July 29, 2005; and August 10, 2005 with accompanying wind directions of; SW, NE, NE, and NE, respectively. The decibels that are returned are energy returned which is a function of density and target dimensions. Since the same target dimensions are roughly the same (although the mayfly bodies are randomly oriented to the beam), the scale on the left of each figure provides a measure of relative insect density. The circle (ring) indicates where the radar beam, originating at (*R* radar source) and emanating away at an angle of 0.5° above the surface, reaches 1 km above the river valley floor. With radar, a greater distance away from the radar indicates a greater elevation of the echo

Emergence of aquatic insects, calculated from diverse literature data (Table 8.6), is, on average, 4.1 ± 1.9 g DW m^{-2} year^{-1}. Our calculations give the same value as an independent global estimate made by Baxter et al. (2005). Variations of the emergence value in diverse ecosystems were from 0.2 to 23.1 g DW m^{-2} year^{-1}, i.e., spanning two orders of magnitude. The return of aquatic adults to streams was estimated at ~1.5% of the emergence (Ballinger and Lake 2006). Therefore, the net export of biomass of emergent aquatic insect can be estimated as ~4.3 g DW m^{-2} year^{-1}. To calculate the export of HUFA, we need to know the EPA and DHA content of the imagoes of aquatic insects. Such data are not common in the available literature. Nevertheless, we had, available to us, one sample of the imago of chironomids collected near the site where we studied zoobenthos (Sushchik et al. 2003). Concentrations of EPA and DHA in these imagoes were 17.8 mg g^{-1} DW and 0.3 mg g^{-1} DW, respectively, using a moisture content of 72% (see footnote to Table 8.2). The sum of these two values was of the same order of magnitude as the average HUFA (EPA + DHA) content in the larvae of aquatic insects, 9.3 ± 2.2 mg g^{-1} DW (Table 8.2).

Table 8.6 Aquatic insect emergence, g DW m^{-2} year^{-1}, from diverse ecosystems

Dominant taxa	Emergence	Ecosystem, Region	References
Diptera, Ephemeroptera, Plecoptera, Odonata	4.5	Global generalization	Baxter et al. (2005)
Ephemeroptera, Plecoptera, Trichoptera	0.382	Pick Creek, southwestern Alaska	Francis et al. (2006)
Chironomidae (>50%)	2.1[a]	Rivers of Philippines island	Freitag (2004)
Many taxa	10[b]	Global generalization	Huryn and Wallace (2000)
Chironomidae	1.1[a]	Temporary wetland pond, South Carolina	Leeper and Taylor (1998)
Chironomidae	0.2[a]	Salt marsh, southern Maine	MacKenzie (2005)
Odonata, Ephemeroptera	0.5	Lake Michigan coastal wetland	MacKenzie and Kaster (2004)
Many taxa	1.2[c]	Stream, Hokkaido, Japan	Nakano and Murakami (2001)
Plecoptera, Ephemeroptera, Chironomidae	6.3	Tagliamento River (NE-Italy)	Paetzold et al. (2005)
Diptera, Trichoptera, Ephemeroptera	1.7	Outlet of Lake Belau, Germany	Poepperl (2000)
Chironomidae, Trichoptera	23.1	Sycamore Creek, Arizona	Sanzone et al. (2003)
Chironomidae	1.5	Wetland in Alabama	Stagliano et al. (1998)
Many taxa	1.3[d]	Lake Esrom, Denmark	Woollhead (1994)
Average	4.1		

[a]Recalculated using an average specimen dry mass of 150 μg (Stagliano et al. 1998)
[b]Recalculated from aquatic insect production
[c]Recalculated form Fig. 1C of the reference
[d]Recalculated using energy equivalent 1 g DM = 23 kJ (Alimov 1989)

Therefore, we make the assumption that mean EPA and DHA concentration of emerging insects is roughly equal to that in their aquatic larvae. Using the average HUFA concentration in larvae of aquatic insects (i.e., 9.3 mg g^{-1} DW), the export via emerging aquatic insects can be estimated as 4.3 × 9 = 40 mg m^{-2} (kg km^{-2}) year^{-1}.

The two orders of magnitude difference in the reported emergence values among different ecosystems and the paucity of data on HUFA content in imagoes mean that our first-ever estimates of HUFA export from aquatic to terrestrial systems should be regarded as very coarse. Using the lowest and highest values (worst- and best-case scenarios) of insect larvae HUFA contents (Table 8.2) and those of emergence (Table 8.6) we may suppose that, for the particular ecosystems cited here, the HUFA export values can vary from as little as 0.1 mg m^{-2} year^{-1} to as high as 672.2 mg m^{-2} year^{-1}. Clearly, more studies reporting on the abundance and HUFA content of larval insects from riverine and wetland environments as well as the emergence densities of adult forms are necessary in order to calculate a more reliable estimate of the contribution that emerging insects make to the export of HUFA from aquatic to terrestrial ecosystems.

8.8 Case Study IV: Estimation of Aquatic HUFA Import to Terrestrial Ecosystems Through Birds

Birds import a significant amount of aquatic HUFA to terrestrial ecosystems. While ducks, loons, mergansers, etc. clearly fall into the category of "water birds" there are other birds that are usually thought of as "terrestrial" but which, depending on the season, do consume a large number of aquatic organisms. For example, swallows (Hirundinidae) consume great quantities of mosquitoes, black flies, mayflies, and caddisflies during peak insect emergence periods. Many other terrestrial birds, e.g., songbirds, rely to varying extents on the emergence of insects from aquatic systems. Finally, HUFA-rich amphipods which are often abundant in littoral areas of lakes and in ponds may become infected with acanthocephalan parasites causing a well-described positive phototaxis (Benesh et al. 2005). This infection, in turn, makes them more widely susceptible to predation by shorebirds such as terns (Sternidae) and sandpipers (Scolopacidae).

Water bird abundance (biomass) depends on lake trophic status and morphometry, i.e., surface area, shoreline length (e.g., Hoyer and Canfield 1994; Suter 1994). Nevertheless, by generalizing data from diverse water bodies and locations (Table 8.7), one can see that annual abundances are very close to each other and on average can be estimated as ~0.4 ± 0.1 ind 10^{-3} m^{-2}.

The energy content of aquatic organisms, consumed by an "average" bird (mean individual weight = ~0.7 kg) from a community with 19 different species, can be estimated from the data of Gardarsson and Einarsson (2002) as 227,000 kJ ind.$^{-1}$ year^{-1}. The 1 g DW = 23 kJ conversion factor provided by Alimov (1989) yields an annual intake of aquatic organisms by the "average bird" of 9.9 kg DW ind.$^{-1}$ year^{-1}.

Table 8.7 Abundance (N, ind. $\times 10^{-3}$ m^{-2}) of water birds in diverse ecosystems and their consumption of aquatic invertebrates and fish (C, g m^{-2} year^{-1} dry mass)

Dominant taxa (number of species)	N	C	Ecosystem, Region	References
Pelicans, cormorants, herons, royal tern (23)	0.4	10.4[a]	Lagoon on the Pacific Coast of Mexico	Acuna et al. (1994)
Ducks *Anas* spp. (4)	0.7[b]	18.2[a]	Boundary Bay, British Columbia, Canada	Baldwin and Lovvorn (1994)
Tufted ducks, greater scaup, merganser (19)	0.5	4.9	Lake Myvatn, Iceland	Gardarsson and Einarsson (2002)
Mallard, coot, red-winged blackbird (50)	0.2	5.2[a]	46 Florida lakes	Hoyer and Canfield (1994)
Mallard, cormorant (not reported)	0.1[c]	2.6[a]	Lake Grand-Lieu, France	Marion et al. (1994)
Anas platyrhynchos, Aythya fuligula (63)	0.25	6.5[a]	158 fishponds, Bohemia, Czech Republic	Musil and Fuchs (1994)
Anser, Aythya (7)	0.7[d]	18.2[a]	Lake Balaton, Hungary	Ponyi (1994)
Tufted duck, coot, mallard, pochard (29)	0.1[e]	2.6[a]	20 major Swiss lakes, north of the Alps	Suter (1994)
Anas, Aythya, Fulica, Cygnus, Podiceps (26)	nr	2.1	Lake Esrom and Lake Sjelso, Denmark	Woollhead (1994)
Average	0.4	7.8		

nr not reported
[a]Calculated using N and the intake of aquatic organisms by the "average" bird, 26 kg DM ind.$^{-1}$ year^{-1} (see text for details)
[b]Estimated from Fig. 2 of the reference.
[c]Calculated from Table 1 of the reference (only birds, feeding in the lake)
[d]Calculated from Table 4 of the reference
[e]Estimated from Fig. 2 of the reference

On the basis of data from Gere and Andrikovics (1994) for ducks from Lake Balaton, Hungary, intake can be estimated as 43 kg DW ind.$^{-1}$ year^{-1}. Thus, for the following calculation, we use the mean value obtained from these two estimates, i.e., 26 kg DW ind.$^{-1}$ year^{-1}. It is important to note that average individual bird biomass, calculated on the basis of data from 46 Florida lakes where the bird community comprised 50 species, was 0.6 kg (Hoyer and Canfield 1994), i.e., similar to the value given earlier. Thus, our calculations of intake, based on the "average" bird at ~0.7 kg, seem reasonable for many ecosystems. On average then, the annual consumption of aquatic animals by water birds is 7.8 g ± 2.1 DW m^{-2} year^{-1} (Table 8.7).

Using an average EPA + DHA content for aquatic animals of 9.2 mg g^{-1} DW (calculated from Tables 8.1–8.3) the HUFA export from a "typical" aquatic ecosystem by water birds can be estimated as ~7.8 × 9.2 ≈ 72 kg km^{-2} year^{-1}. Based on the observed variation in this preliminary dataset, global HUFA export rates can be expected to range anywhere from 19.3 to 167.4 kg km^{-2} year^{-1}. These values, although imprecise, provide some idea of the scale of HUFA export from aquatic systems to aquatic or riparian birds.

8.9 Case Study V: Can HUFA Export from Aquatic Ecosystems Meet the HUFA Requirements of Terrestrial Animals?

Using available data, we carry out the first calculations of HUFA export from selected aquatic to terrestrial ecosystems. We suggest that, except for particular locations in the Pacific Rim where bears intensively consume salmon, the main pathways of HUFA export are emerging insects and water birds. It is also likely that, in some ecosystems, emerging amphibians can make a substantial contribution to the HUFA export. However, we failed to find in the available literature the necessary quantitative data on amphibian abundance, biomass, emergence, and HUFA content. We measured the EPA + DHA content in two specimens of frog (*Rana ridibunda*) (Palla) caught in the vicinity of Krasnoyarsk city (Siberia, Russia) and obtained a value of 2.0 mg g^{-1} DW. The content of essential HUFA in these amphibians is lower than in many fish and aquatic invertebrates (Tables 8.1–8.3), but nonetheless significant.

Hence, at present we will for the following considerations only use the HUFA export rates due to emerging insects and water birds. As mentioned earlier, to correlate HUFA export with the requirements of terrestrial animals one must know, besides the fluxes (i.e., g of HUFA m^{-2} year^{-1}), the area of the adjacent terrestrial ecosystem, metabolic HUFA requirements of carnivorous and omnivorous terrestrial animals and their biomass. At present no detailed comprehensive studies have been conducted with these specific parameter estimates. Nevertheless, since we here have estimated the first coarse average figures (Tables 8.1–8.7) it is very tempting to carry out some calculations for an "average" ecosystem. Since there are no available relevant data for any particular ecosystem, we decided to approach the task from the other side and model an "average" ecosystem using available global-scale data.

8.9.1 Area of Terrestrial Ecosystems and Inland Aquatic Ecosystems and Average HUFA Export

First, we need to develop a relationship between terrestrial and aquatic ecosystem areas. The total landmass area of terrestrial ecosystems has been estimated at 99.5 × 10^6 km^2 (Alimov 1989). In comparison, the total area of large lakes and rivers has been estimated at 2.0 × 10^6 km^2 (Alimov 1989; Raven and Maberly 2004). The number of small waterbodies (0.1–1.0 ha), characteristic of wetlands in subarctic regions for example, is large and has been estimated at between 5 × 10^8 and 8 × 10^7 (Wetzel 1992) providing a vast surface area of 20 × 10^6 km^2 (upper limit). Therefore, by including waterbodies with a surface area between 0.001 and 1 km^2 the total landlocked waterbody area can be estimated as ~22 × 10^6 km^2. Subtracting the surface area of small lakes (1 km^2 and less, i.e., 20 × 10^6 km^2) yields a net landmass area of terrestrial ecosystems of 79.5×10^6 km². Because of their physical proximity and connectivity with terrestrial systems we also include the surface area of estuaries, i.e., 1.4×10^6 km² (Alimov 1989; Raven and Maberly 2004) in our calculations and adding it to the total area of inland waters. Hence, the total surface area of aquatic ecosystems, i.e., freshwater and estuaries, which can potentially supply terrestrial ecosystems with HUFA, is 23.4×10^6 km². Thus, on a surface area basis, production of HUFA from 1 m² of such model inland aquatic ecosystems (lakes, rivers and estuaries) can potentially supply 3.4 m² (i.e., 79.5/23.4) of terrestrial ecosystems.

One recent calculation (Downing et al. 2006) provides a contrasting estimate of total inland water area of $\sim 4.6 \times 10^6$ km². Although there might be some underestimation, e.g., too low of a predicted density of small waterbodies in tundra regions, we will use the Downing et al. figure as our "lower limit" estimate. Therefore, by subtracting the additional surface area of lakes, 2.6×10^6 km², calculated in the study, cited earlier, from the total landmass area we get a net landmass area of terrestrial ecosystems of $\sim 96.9 \times 10^6$ km². The lower limit of total surface area of aquatic ecosystems, which can potentially supply terrestrial ecosystems with HUFA, is 4.6×10^6 km² $+ 1.4 \times 10^6$ km² (estuaries) $= 6.0 \times 10^6$ km². Thus, according to the lower limit estimate, production of HUFA from 1 m² of possible inland aquatic ecosystems supplies $96.9 \times 10^6/6.0 \times 10^6 = 16.2$ m² of terrestrial ecosystems. Therefore, for all the following calculations, which refer to the area of aquatic ecosystems, the lower limit estimate will be 3.9× lower (i.e., 23.4×10^6 km²/6.0×10^6 km²) than the upper limit estimate. For the ratio between areas of terrestrial and aquatic ecosystems, the lower limit estimate yields a 4.8× lower value (i.e., 16.2 m²/3.4 m² of terrestrial ecosystems supplied with HUFA from 1 m² of aquatic ecosystems).

Thus, 1 m² of surface area of the model "average" aquatic ecosystem can potentially provide HUFA to between 3.4 and 16.2 m² of the model "average" adjacent terrestrial ecosystem. The sum of the average values of the two HUFA fluxes through emerging insects and water birds, calculated earlier, is $40 + 72 = 112$ mg m^{-2} (kg km^{-2}) year^{-1}. Therefore, HUFA supply to the model "average" terrestrial ecosystem can be from 6.9 to 32.9 kg km^{-2} year^{-1}. The HUFA supply due to emerging insects, which will be used for the following case study, ranges from a low of 2.5 to a high of 11.8 kg km^{-2} year^{-1}.

8.9.2 Relation of HUFA Export to Biomass of Terrestrial Animals

In Sect. 8.4 we attempted to relate HUFA supply to the requirements of the human population and found that a potentially significant deficiency exists with respect to the continued supply of essential fatty acids to humans. Next we try to answer a similar question: is the HUFA export from the model "average" aquatic ecosystem high enough to support terrestrial animals, which we suggest must also, to varying degrees, obtain HUFA from their food? To answer this question, we need to know the average metabolic requirements and biomass of omnivorous and carnivorous terrestrial animals in the "average" model terrestrial ecosystem.

We take into consideration secondary consumers only. Herbivorous animals, e.g., ruminants, hardly consume the production of aquatic ecosystems, and more likely obtain HUFA from their precursor ALA which is quantitatively abundant in higher plants, or through synthesis by symbiotic intestinal micro-organisms. Although gut microbes are an additional potential source of HUFA, very little is known of the overall importance of this source (Hulbert et al. 2002).

Next we need to obtain an estimate of the average HUFA requirements for omnivorous and carnivorous terrestrial animals. The requirements for humans are in the range of 180–1,000 mg person^{-1} d^{-1} (Garg et al. 2006). If we use 70 kg as an average weight of a human, the HUFA requirements may be expressed as 2.6–14.3 mg kg^{-1} WW d^{-1}, or, on average, ~8 mg kg^{-1} WW d^{-1}. Data on HUFA requirements for animals are very sparse in the available literature. A daily supplementation of EPA + DHA of 22 mg kg^{-1} WW d^{-1} for dogs with early stages atopy gave a good clinical effect (Abba et al. 2005). For rats, a diet with a HUFA intake of 0.6 mg g^{-1} WW d^{-1} (600 mg kg^{-1} WW d^{-1}) was found to be most advantageous to long bone density (Green et al. 2004).

For the case study we choose omnivorous rodents, assuming that they have HUFA requirements, close to that of rat, mentioned earlier. One widespread and comparatively well-studied species is the deer mouse (*Peromyscus maniculatus*). It has an average individual biomass of ~17 g WW (Merritt et al. 2001; Stapp and Polis 2003), and insects are a staple in their diet (Merritt et al. 2001). Their densities ranged from 1.0 to 13 ind. ha^{-1} in prairie of Central Plains, Colorado, USA (Stapp and Van Horne 1997), to 9.0–16.9 ind. ha^{-1} in forests of west-central British Columbia (Sullivan et al. 1999) and to 1.0–17.6 ind. ha^{-1} in the Kananaskis Valley, southwestern Alberta, Canada (Millar and McAdam 2001). Thus, the average biomass of these mice in diverse ecosystems ranged from 17 to 299.2 g ha^{-1} or from 1.7 to 29.9 kg km^{-2}. HUFA (EPA + DHA) requirements of these rodents, using the earlier assumption, ranged from 1.0 g km^{-2} d^{-1} to 17.9 g km^{-2} d^{-1}, or 0.37–6.53 kg km^{-2} year^{-1}. The supply of HUFA due to emerging insects, which contribute to the food items of these mice, from the "average" aquatic to the "average" terrestrial ecosystem, calculated earlier, is from 2.5 to 11.8 kg km^{-2} year^{-1}. Thus, according to these extremely simplified considerations, there exists at least the potential for situations in which there may be a shortage of HUFA in deer mice populations. Clearly, more focused research is needed to ascertain the contributions and effects of aquatic versus terrestrially derived HUFA in deer mice population dynamics. It is an equally valid question to ask if other omnivorous and carnivorous terrestrial animals, besides deer mice, also depend, to various extents, on aquatic-source-derived HUFA.

8.10 Ocean Contribution

Biomass from marine ecosystems enters the coastal-terrestrial ecosystems through shore drift of algal wrack and carrion (Polis and Hurd 1996). There are many terrestrial consumers of such marine-derived inputs: spiders, scorpions, ants, lizards, landbirds, coyotes, foxes, jackals, etc. (Polis and Hurd 1996). In marine islands there are also deer mice, described in the case study in Sect. 8.7.2., which consume littoral detritus (Stapp and Polis 2003).

Polis and Hurd (1996) estimated that the average dry mass of algal drift arriving per linear meter of supralittoral was 27.6 kg m^{-1} $year^{-1}$. A large quantity of algae washes ashore as a consequence of either storms or seasonal mortality and breakup of algal beds. Using the data of Fleurence et al. (1994; Table 8.4) on the HUFA content of seaweeds, 1.3 mg g^{-1} DW, import of EPA + DHA through the shore drift to a coastal ecosystem could be as high as 36 g m^{-1} $year^{-1}$.

Carrion drift, estimated from the data of Polis and Hurd (1996), is 0.3 kg DW m^{-1} $year^{-1}$. Using the average content of EPA + DHA in marine animals (Table 8.4), the HUFA export to a coastal terrestrial ecosystem could be ~5 g m^{-1} $year^{-1}$, and thus total export through the shore drift (plants + animals) is 36 + 5 = 41 g m^{-1} $year^{-1}$. This input of HUFA to terrestrial ecosystems is evidently very important for local coastal areas, especially for small islands. Unfortunately, there are no quantitative data for any particular ecosystem on the biomass of terrestrial animals, which consume the drift products and therefore, at present, we can say very little about the HUFA deficiency potential for coastal ecosystems.

In order to compare export of HUFA from the oceanic drift contribution with those from inland waters we need to use a coarse, global-scale analysis. The total linear distance of global coastlines has been estimated as 594,000 km (Polis and Hurd 1996). Thus, global HUFA export from the drift, estimated in the case study of Polis and Hurd (1996) in conjunction with the average HUFA content of marine organisms (Table 8.4), may be ~24 × 10^6 kg $year^{-1}$. If we use the surface area estimate of inland waters and estuaries as 23.4 × 10^6 km^2 and the average HUFA export through emerging insects and birds as 112 kg km^{-2} $year^{-1}$ (see Sect. 8.7.1), then the global HUFA export from the inland waters appears to be ~3 × 10^9 kg $year^{-1}$. Thus, the ocean contribution to the global HUFA export could be ~100 times lower than that of inland waters. The cause of this phenomenon is the comparatively higher amount of interface (ecotone zone) between inland waters and terrestrial ecosystems. Indeed, if we take into consideration lakes with a surface area of 1 ha (0.01 km^2) the perimeter of each lake would be at least 0.35 km (L = $2\pi r$ = 0.35 km), if one equates their shape to a circle. The number of such lakes, as mentioned in Sect. 8.7.1, is ~1 × 10^8 (Wetzel 1992); thus, their total perimeter is 35 × 10^6 km. If one were to add the shoreline of large lakes the total perimeter of inland waters will be at least twice as high (~70 × 10^6 km). This figure is much larger than the length of ocean shoreline (~0.6 × 10^6 km). Thus, it is not surprising that the interaction between inland waters and terrestrial ecosystems is much higher than that between the ocean and terrestrial ecosystems. It is interesting to note that if one relates the estimated HUFA export from the perimeters (shores) of inland waters (i.e., 3 × 10^9 kg $year^{-1}$/70 × 10^6 km = 43 g m^{-1} $year^{-1}$) the resulting estimate is very similar to the HUFA export estimated to be due to the contribution from ocean shoreline perimeters (41 g m^{-1} $year^{-1}$).

8.11 Assumptions and Underestimates

Our attempt to calculate HUFA export from aquatic to terrestrial ecosystems is the first one, and obviously there are many underestimates caused, in part, by an absence of relevant data in the available literature (see Sect. 8.1). First, we could not calculate the export of HUFA through amphibian emergence. In some ecosystems amphibians are very abundant, and their contribution to food webs is quantitatively important (Burton and Likens 1975; Ballinger and Lake 2006).

Second, we took into consideration only consumption of salmon by bears around the Pacific Rim. However, there are many other terrestrial "interface specialists" – primary and secondary consumers, such as beetles, spiders, lizards, etc., which scavenge aquatic organisms, washed up on the banks, as well as large mobile mammals, such as the hippopotamus which are important conduits for energy exchange across the aquatic – terrestrial interface in some regions (Ballinger and Lake 2006). Third, for inland waters we do not take into account flooding and drying events which also directly subsidize terrestrial ecosystems with fresh biomass of aquatic organisms (Ballinger and Lake 2006). Fourth, we omitted colonies of seabirds on rocks and shelf ice. Thus, our conclusion about possible limitation of the "average" terrestrial ecosystem with HUFA export from the adjacent "average" aquatic ecosystem may be underestimating the true potential for limitation for many ecosystems. However, it is also clear that there are very little available data (with the exception of some commonly studied species such as rats, dogs, and cats) on the innate ability of terrestrial animals to desaturate and elongate ALA to EPA and DHA. This ability, if present at a reasonably ubiquitous level in terrestrial organisms and/or their symbiotic microbial gut communities, would have the opposite effect on our estimates of the potential for HUFA limitation in terrestrial systems. Clearly much more work needs to be done.

8.12 Conclusions and Perspectives

According to our coarse average estimations at least some components of terrestrial ecosystems have the potential to be limited by the supply of essential HUFA from adjacent aquatic ecosystems. On a global scale it seems reasonable to conclude that the main flux of aquatic HUFA to terrestrial ecosystems (excluding humans) originates from inland waters and estuaries, rather than from the ocean. Nevertheless, our estimate of HUFA fluxes are inevitably coarse because detailed data on HUFA synthesis, transfer, and retention in organisms at the water-land interface are scarce and limited to incomplete studies conducted in only a few ecosystems. Thus, we suggest that future studies should be aimed at obtaining more comprehensive quantitative estimates of (a) specific fluxes of HUFA from particular aquatic ecosystems to surrounding terrestrial ecosystems, (b) abilities of terrestrial organisms to synthesize n-3 HUFA (from ALA), and (c) level of HUFA required from the diet in order to maintain optimal physiological performance.

It is likely that in different types of biomes (e.g., tundra, taiga, rain forests, savanna, steppe, deserts, etc.) there are different ecological roles in the context of HUFA export and that some terrestrial ecosystems may be more limited than others by the quantity of essential HUFA exported from adjacent aquatic ecosystems. Moreover, the role of anthropogenic pollution and/or climate change in affecting aquatic HUFA production and export and hence the functioning of terrestrial ecosystems (from the perspective of creating potential HUFA deficiencies) should be investigated.

Aquatic ecologists today are faced with global challenges and must integrate knowledge from a variety of disciplines. One such challenge involves quantifying the production, storage, and movements of HUFA in aquatic ecosystems of different types leading to better estimates of their contribution to the health and ecological integrity of aquatic and terrestrial ecosystems. A particularly important task is to estimate the potential role of different organisms and aquatic ecosystems as sources and sinks of healthy, biochemically valuable food for human nutrition. This is critical in that the consumption of the n-3 HUFA by humans is suspected as being insufficient even in Western developed countries (see Chap. 14). Current evidence from nutritional, epidemiological, and clinical studies shows that a regular shortage of n-3 HUFA in the diet aggravates cardiovascular diseases in humans and may limit normal neonatal and infant brain growth and perhaps intellectual development (see Chap. 14). However, such studies should be well grounded/founded on a detailed knowledge of fatty acid contents in natural fish and aquatic invertebrate populations in relation to the taxonomic affiliation, trophic position, age, and diet of the main contributing species. The HUFA accumulation and transfer rates within food webs should be traced and measured leading to better estimates of the potential total harvest of HUFA in different aquatic ecosystems. HUFA are also very important to the aquaculture industry which face the constant challenge of maintaining adequate concentrations of essential fatty acids in the diets of cultured organisms (e.g., Atlantic salmon, shrimp). This is because overfishing for forage fish is rapidly increasing the cost, and simultaneously threatening to decrease the quality, of fish-meal-based feeds (see Chap. 9).

Second, potential hazards which may reduce HUFA production must be studied, forecasted, and, where possible, mitigated. For instance, we postulate that anthropogenically induced processes such as eutrophication and global climate change (see Chap. 11) may, either qualitatively or quantitatively (or both), lead to decreased HUFA production and storage in aquatic ecosystems. This is because: (a) eutrophication in aquatic ecosystems favors cyanobacteria (blue-green algae) which often contain very little, if any, HUFA (see Chap. 7) and, (b) warmer temperatures have the general effect of reducing the concentrations of long-chain fatty acids such as EPA and DHA in biomembranes in a wide range of aquatic organisms (see Chap. 10; Schlechtriem et al. 2006). Such threats coupled with the critical need for these substances in animal and human diets underline the urgency of this new direction of research on HUFA fluxes. Given the importance of essential fatty acids for animal and human health and nutrition and for existing and emerging aquaculture facilities, follow-up studies are necessary in order to more precisely identify the sources' sinks and flows of these compounds as well as to characterize the risks to their continued production.

In summary, to further improve the component estimates required for a more reliable world-wide estimate we suggest that the following quantitative studies are required: (1) insect emergence in conjunction with HUFA measurements in their biomass, (2) water bird abundance and ration, including HUFA content of their food, (3) amphibian emergence and contents of HUFA in their biomass, (4) abundance and ration of riparian animals, including HUFA content of the portion of their food obtained from water, (5) abundance and ration of terrestrial predators, consuming emergent insects, amphibians and water birds, and (6) HUFA requirements of diverse terrestrial animals. The studies, proposed earlier, should be carried out in diverse aquatic ecosystems, pools, swamps, lakes, rivers, seas, situated in different landscapes: tundra, taiga, steppe, mountains, desert, etc. Additional and more precise estimates of the surface area of aquatic ecosystems and adjacent terrestrial ecosystems are also essential.

Finally, it is now well recognized that, beyond the obvious provisioning of drinking water, navigation, and flood control, aquatic ecosystems provide a variety of additional and highly valuable "ecosystem services." These include esthetic and recreational services as well as their recently recognized role in mitigating the effects of a wide variety of organic contaminants. To this list of services we must now add one heretofore unrecognized service, namely, the provision of essential HUFA to adjacent terrestrial systems. This newly recognized service of aquatic ecosystems provides conservationists and resource managers with a new outlook and justification for preserving lakes, rivers and wetlands.

Acknowledgments The authors are grateful to Dan Baumgardt (Science and Operations Officer, National Weather Service, La Crosse, Wisconsin) for providing the Doppler radar images of mayfly emergence from the upper headwaters of the Mississippi River. The work was supported by Russian Foundation for Basic Research (RFBR) and Krasnoyarsk Science Foundation grant # 07-04-96803-r_yenisei and by RFBR grant # 07-05-00076 (M.I.G. and N.N.S) and by the National Water Research Institute, Environment Canada (M.T.A.). We are grateful to Drs. M. Kainz and M.T. Brett for useful comments on previous drafts of this chapter and to our anonymous reviewer whose insightful comments greatly improved our chapter.

References

Abba, C., Mussa, P.P., Vercelli, A. and Raviri, G. 2005. Essential fatty acids supplementation in different-stage atopic dogs fed on a controlled diet. J. Anim. Physiol. Anim. Nutr. 89:203–207

Acuna, R., Contreras, F. and Kerekes, J. 1994. Aquatic bird densities in two coastal lagoon systems in Chiapas State, Mexico, a preliminary assessment. Hydrobiologia. 279/280:101–106

Ahlgren, G., Blomqvist, P., Boberg, M. and Gustafsson, I.-B. 1994. Fatty acid content of the dorsal muscle – an indicator of fat quality in freshwater fish. J. Fish Biol. 45:131–157

Ahlgren, G., Carlstein, M. and Gustafsson, I.-B. 1999. Effects of natural and commercial diets on the fatty acid content of European grayling. J. Fish. Biol. 55:1142–1155

Ahlgren, G., Ahlgren, I., Hernandez, S. and Mejia, M. 2002. Fatty acid quality of seston in the Lakes Xolotlan and Cocibolca, Nicaragua. Verh. Internat. Verein. Limnol. 28:786–791

Ahlgren, G., Van Nieuwerburgh, L., Wanstrand, I., Pedersen, M., Boberg, M. and Snoeijs, P. 2005. Imbalance of fatty acids in the base of the Baltic Sea food web – a mesocosm study. Can. J. Fish. Aquat. Sci. 62:2240–2253

Aktas, H. and Halperin, J.A. 2004. Translational regulation of gene expression by ω-3 fatty acids. J. Nutr. 134:2487S–2491S

Alimov, A.F. 1989. An Introduction to Production Hydrobiology. Leningrad: Gidrometeoizdat (in Russian). 152 pgs

Arts, M.T., Ackman, R.G. and Holub, B.J. 2001. "Essential fatty acids" in aquatic ecosystems: a crucial link between diet and human health and evolution. Can. J. Fish. Aquat. Sci. 58:122–137

Baldwin, J.R. and Lovvorn, J.R. 1994. Habitats and tidal accessibility of the marine foods of dabbling ducks and brant in Boundary Bay, British Columbia. Mar. Biol. 120:627–638

Ballinger, A. and Lake, P.S. 2006. Energy and nutrient fluxes from rivers and streams into terrestrial food webs. Mar. Freshw. Res. 57:15–28

Baxter, C.V., Fausch, K.D. and Saunders, W.C. 2005. Tangled webs: reciprocal flows of invertebrate prey link streams and riparian zones. Freshw. Biol. 50:201–220

Beamish, R.J., Noakes, D.J., McFarlane, G.A., Klyashtorin, L., Ivanov, V.V. and Kurashov, V. 1999. The regime concept and natural trends in the production of Pacific salmon. Can. J. Fish. Aquat. Sci. 56:516–526

Ben-David, M., Titus, K. and Beier, L.R. 2004. Consumption of salmon by Alaskan brown bears: a trade-off between nutritional requirements and the risk of infanticide? Oecologia 138:465–474

Benesh, D.P., Duclos, L.M. and Nichol, B.B. 2005. The behavioral response of amphipods harboring *Corynosoma constrictum* (acanthocephala) to various components of light. J. Parasitol. 91:731–736

Blair, T.A., Cree, A. and Skeaff, C.M. 2000. Plasma fatty acids, triacylglycerol and cholesterol of the tuatara (*Sphenodon punctatus punctatus*) from islands differing in the presence of rats and the abundance of seabirds. J. Zool. 252:463–472

Boersma, M. and Kreutzer, C. 2002. Life at the edge: is food quality really of minor importance at low quantities? Ecology 83:2552–2561

Broadhurst, C.L., Wang, Y., Crawford, M.A., Cunnane, S.C., Parkington, J.E. and Schmidt, W.F. 2002. Brain-specific lipids from marine, lacustrine, or terrestrial food resources: potential impact on early African *Homo sapiens*. Comp. Biochem. Physiol. B. 131:653–673

Burton, T.M. and Likens, G.E. 1975. Energy flow and nutrient cycling in salamander population in the Hubbard Brook experimental forest, New Hampshire. Ecology. 56:1068–1080

Cohen, Z., Norman, H.A. and Heimer, Y.M. 1995. Microalgae as a source of ω3 fatty acids. In: Plants in human nutrition. World review of nutrition and dietetics, Vol. 77. Edited by Simopoulos, A.P. Basel: Karger, pp. 1–31

Copeman, L.A., Parrish, C.C., Brown, J.A. and Harel, M. 2002. Effects of docosahexaenoic, eicosapentaenoic, and arachidonic acids on the early growth, survival, lipid composition and pigmentation of yellowtail founder (*Limanda ferruginea*): a live food enrichment experiment. Aquaculture 210:285–304

Cripps, G.C., Watkins, J.L., Hill, H.J. and Atkinson, A. 1999. Fatty acid content of Antarctic krill *Euphausia superba* at South Georgia related to regional populations and variations in diet. Mar. Ecol. Prog. Ser. 181:177–188

Damude, H.G. and Kinney, A.J. 2007. Engineering oilseed plants for a sustainable, land-based source of long chain polyunsaturated fatty acids. Lipids 42:179–185

Danielsdottir, M.G., Brett, M.T. and Arhonditsis, G.B. 2007. Phytoplankton food quality control of planktonic food web processes. Hydrobiologia 589:29–41

Darimont, C.T., Reimchen, T.E. and Paquet, P.C. 2003. Foraging behaviour by gray wolves on salmon streams in coastal British Columbia. Can. J. Zool. 81:349–353

Davis, B.C. and Kris-Etherton, P.M. 2003. Achieving optimal essential fatty acid status in vegetarians: current knowledge and practical implications. Am. J. Clin. Nutr. 78(Suppl):640S–646S

Dobey, S., Masters, D.V., Scheick, B.K., Clark, J.D., Pelton, M.R. and Sunquist, M.E. 2005. Ecology of Florida black bears in the Okefenokee-Osceola ecosystem. Wildlife Monogr. 158:1–41

Downing, J.A., Prairie, Y.T., Cole, J.J., Duarte, C.M., Tranvik, L.J., Striegl, R.G., McDowell, W.H., Kortelainen, P., Caraco, N.F., Melack, J.M. and Middelburg, J.J. 2006. The global abundance and size distribution of lakes, ponds, and impoundments. Limnol. Oceanogr. 51:2388–2397

FAO. 2004. The State of World Fisheries and Aquaculture. Rome: FAO Fisheries Department

Farrell, D.J. 1998. Enrichment of hen eggs with n-3 long-chain fatty acids and evaluation of enriched eggs in humans. Am. J. Clin. Nut. 68:538–544

Fedorova, I., Hussein, N., Di Martino, C., Moriguchi, T., Hoshiba, J., Majchrzak, S. and Salem, N. Jr. 2007. An n-3 fatty acid deficient diet affects mouse spatial learning in the Barnes circular maze. Prostaglandins Leukot. Essent. Fatty Acids 77:269–277

Felicetti, L.A., Schwartz, C.C., Rye, R.O., Gunther, K.A., Crock, J.G., Haroldson, M.A., Waits, L. and Robbins, C.T. 2004. Use of naturally occurring mercury to determine the importance of cutthroat trout by Yellowstone grizzly bears. Can. J. Zool. 82:493–501

Fleurence, J., Gutbier, G., Mabeau, S. and Leray, C. 1994. Fatty acids from 11 marine macroalgae of the French Brittany coast. J. Appl. Phycol. 6:527–532

Francis, T.B., Schindler, D.E. and Moore, J.W. 2006. Aquatic insects play a minor role in dispersing salmon-derived nutrients in southwestern Alaska. Can. J. Fish. Aquat. Sci. 63:2543–2552

Freitag, H. 2004. Composition and longitudinal patterns of aquatic insect emergence in small rivers of Palawan Island, the Philippines. Int. Rev. Hydrobiol. 89:375–391

Gardarsson, A. and Einarsson, A. 2002. The food relations of the waterbirds of Lake Myvatn, Iceland. Verh. Internat. Verein. Limnol. 28:754–763

Garg, M.L., Wood, L.G., Singh, H. and Moughan, P.J. 2006. Means of delivering recommended levels of long chain n-3 polyunsaturated fatty acids in human diets. J. Food Sci. 71:R66–R71

Geiser, F., Firth, B.T. and Seymour, R.S. 1992. Polyunsaturated dietary lipids lower the selected body temperature of a lizard. J. Comp. Physiol. B 162:1–4

Gende, S.M., Quinn, T.P. and Willson, M.F. 2001. Consumption choice by bears feeding on salmon. Oecologia. 127:372–382

Gere, G. and Andrikovics, S. 1994. Feeding of ducks and their effects on water quality. Hydrobiologia. 279/280:157–161

Gerster, H. 1998. Can adults adequately convert α-linoleic acid (18:3n–3) to eicosapentaenoic acid (20:5n–3) and docosahexaenoic acid (22:6n–3)? Int. J. Vitam. Nutr. Res. 68:159–173

Gladyshev, M.I., Sushchik, N.N., Kalachova, G.S., Dubovskaya, O.P. and Makhutova, O.N. 2006a. Influence of sestonic elemental and essential fatty acid contents in a eutrophic reservoir in Siberia on population growth of *Daphnia* (*longispina* group). J. Plankton Res. 28:907–917

Gladyshev, M.I., Sushchik, N.N., Gubanenko, G.A., Demirchievam, S.M. and Kalachova, G.S. 2006b. Effect of way of cooking on content of essential polyunsaturated fatty acids in muscle tissue of humpback salmon (*Oncorhynchus gorbuscha*). Food Chem. 96:446–451

Gladyshev, M.I., Sushchik, N.N., Gubanenko, G.A., Demirchievam, S.M. and Kalachova, G.S. 2007. Effect of boiling and frying on the content of essential polyunsaturated fatty acids in muscle tissue of four fish species. Food Chem. 101:1694–1700

Goedkoop, W., Sonesten, L., Ahlgren, G. and Boberg, M. 2000. Fatty acids in profundal benthic invertebrates and their major food resources in Lake Erken, Sweden: seasonal variation and trophic indications. Can. J. Fish. Aquat. Sci. 57:2267–2279

Green, K.H., Wong, S.C.F. and Weiler, H.A. 2004. The effect of dietary n-3 long-chain polyunsaturated fatty acids on femur mineral density and biomarkers of bone metabolism in healthy, diabetic and dietary-restricted growing rats. Prostaglandins Leukot. Essent. Fatty Acids 71:121–130

Gulati, R.D. and DeMott, W.R. 1997. The role of food quality for zooplankton: remarks on the state-of-the-art, perspectives and priorities. Freshw. Biol. 38:753–768

Hanson, B.J., Cummins, K.W., Cargill, A.S. and Lowry, R.R. 1985. Lipid content, fatty acid composition, and the effect of diet of fats of aquatic insects. Comp. Biochem. Physiol. B 80:257–276

Harwood, J.L. 1996. Recent advances in the biosynthesis of plant fatty acids. Biochim. Biophys. Acta. 1301:7–56

Hebert, C.E., Shutt, J.L. and Ball, R.O. 2002. Plasma amino acid concentrations as an indicator of protein availability to breeding herring gulls (*Larus argentatus*). Auk 119:185–200

Hebert, C.E., Weseloh, D.V.C., Idrissi, A., Arts, M.T., O'Gorman, R., Gorman, O.T., Locke, B., Madenjian, C.P. and Roseman, E.F. 2008. Restoring piscivorous fish populations in the Laurentian Great Lakes causes seabird dietary change. Ecology 89:891–897

Heinz, E. 1993. Biosynthesis of polyunsaturated fatty acids. In: Lipid metabolism in plants. Edited by Moore, T.S. Boca Raton, USA: CRC Press, pp. 34–89

Helfield, J.M. and Naiman, R.J. 2006. Keystone interactions: salmon and bear in riparian forests of Alaska. Ecosystems 9:167–180

Hibbeln, J.R., Nieminen, L.R.G. and Lands, W.E.M. 2004. Increasing homicide rates and linoleic acid consumption among five western countries, 1961–2000. Lipids 39:1207–1213

Hibbeln, J.R., Ferguson, T.A. and Blasbalg, T.L. 2006. Omega-3 fatty acid deficiencies in neurodevelopment, aggression and autonomic dysregulation: opportunities for intervention. Int. Rev. Psychiatry 18:107–118

Hilderbrand, G.V., Jenkins, S.G., Schwartz, C.C., Hanley, T.A. and Robbins, C.T. 1999a. Effect of seasonal differences in dietary meat intake on changes in body mass and composition in wild and captive brown bears. Can. J. Zool. 77:1623–1630

Hilderbrand, G.V., Hanley, T.A., Robbins, C.T. and Schwartz, C.C. 1999b. Role of brown bears (*Ursus arctos*) in the flow of marine nitrogen into a terrestrial ecosystem. Oecologia 121:546–550

Hilderbrand, G.V., Farley, S.D., Schwartz, C.C. and Robbins, C.T. 2004. Importance of salmon to wildlife: implications for integrated management. Ursus 15:1–9

Hoffman, L.C. 2008. The yield and nutritional value of meat from African ungulates, camelidae, rodents, ratites and reptiles. Meat Sci. 80:94–100

Howard, R.W. and Stanley-Samuelson, D.W. 1996. Fatty acid composition of fat body and Malpighian tubules of the tenebrionid beetle, *Zophobas atratus*: significance in eicosanoid-mediated physiology. Comp. Biochem. Physiol. B 115:429–437

Hoyer, M.V. and Canfield, D.E. Jr. 1994. Bird abundance and species richness on Florida lakes: influence of trophic status, lake morphology, and aquatic macrophytes. Hydrobiologia 297/280:107–119

Hulbert, A.J., Rana, T. and Couture, P. 2002. The acyl composition of mammalian phospholipids: an allometric analysis. Comp. Biochem. Physiol. B. 132:515–527

Huryn, A.D. and Wallace, J.B. 2000. Life history and production of stream insects. Annu. Rev. Entomol. 45:83–110

Kainz, M., Arts, M.T. and Mazumder, A. 2004. Essential fatty acids in the planktonic food web and their ecological role for higher trophic levels. Limnol. Oceanogr. 49:1784–1793

Kolanowski, W., Stolyhwo, A. and Grabowski, M. 2007. Fatty acid composition of selected fresh water gammarids (Amphipoda, Crustacea): a potentially innovative source of omega-3 LC PUFA. J. Am. Oil. Chem. Soc. 84:827–833

Koussoroplis, A.M., Lemarchand, C., Bec, A., Desvilettes, C., Amblard, C., Fournier, C., Berny, P. and Bourdier, G. 2008. From aquatic to terrestrial food webs: decrease of the docosahexaenoic acid/linoleic acid ratio. Lipids 43:461–466

Lauritzen, L., Hansen, H.S., Jorgensen, M.H. and Michaelsen, K.F. 2001. The essentiality of long chain n-3 fatty acids in relation to development and function of the brain and retina. Prog. Lipid Res. 40:1–94

Lim, S.-Y., Hoshiba, J., Moriguchi, T. and Salem, N. Jr. 2005a. N-3 fatty acid deficiency induced by a modified artificial rearing method leads to poorer performance in spatial learning tasks. Pediatr. Res. 58:741–748

Lim, S.-Y., Hoshiba, J. and Salem, N. Jr. 2005b. An extraordinary degree of structural specificity is required in neural phospholipids for optimal brain function: n-6 docosapentaenoic acid substitution for docosahexaenoic acid leads to a loss in spatial task performance. J. Neurochem. 95:848–857

Leeper, D.A. and Taylor, B.E. 1998. Insect emergence from a South Carolina (USA) temporary wetland pond, with emphasis on the Chironomidae (Diptera). J. North Am. Benthol. Soc. 17:54–72

Leonard, A.E., Pereira, S.L., Sprecher, H. and Huang, Y.S. 2004. Elongation of long-chain fatty acids. Prog. Lipid Res. 43:36–54

MacKenzie, R.A. 2005. Spatial and temporal patterns in insect emergence from a southern Marine salt marsh. Am. Midl. Nat. 153:257–269

MacKenzie, R.A. and Kaster, J.L. 2004. Temporal and spatial patterns of insect emergence from a Lake Michigan coastal wetland. Wetlands 24:688–700

Mansour, M.P., Holdsworth, D.G., Forbes, S.E., Macleod, C.K. and Volkman, J.K. 2005. High contents of 24:6(n-3) and 20:1(n-13) fatty acids in the brittle star *Amphiura elandiformis* from Tasmanian coastal sediments. Biochem. Syst. Ecol. 33:659–674

Marion, L., Clergeau, P., Brient, L. and Bertru, G. 1994. The importance of avian-contributed nitrogen (N) and phosphorus (P) to Lake Grand-Lieu, France. Hydrobiologia 279/280:133–147

Merritt, J.F., Lima, M. and Bozinovic, F. 2001. Seasonal regulation in fluctuating small mammal populations: feedback structure and climate. Oikos 94:505–514

Millar, J.S. and McAdam, A.G. 2001. Life on the edge: the demography of short-season populations of deer mice. Oikos 93:69–76

Mills, G.L., McArthur, J.V., Wolfe, C., Aho, J.M. and Rader, R.B. 2001. Changes in fatty acid and hydrocarbon composition of leaves during decomposition in a southeastern blackwater stream. Arch. Hydrobiol. 152:315–328

Mitchell, T.W., Ekroos, K., Blanksby, S.J., Hulbert, A.J. and Else, P.L. 2007. Differences in membrane acyl phospholipid composition between an endothermic mammal and an ectothermic reptile are not limited to any phospholipid class. J. Exp. Biol. 210:3440–3450

Musil, P. and Fuchs, R. 1994. Changes in abundance of water birds species in southern Bohemia (Czech Republic) in the last 10 years. Hydrobiologia. 279/280:511–519

Müller-Navarra, D.C. 1995. Evidence that a highly unsaturated fatty acid limits *Daphnia* growth in nature. Arch. Hydrobiol. 132:297–307

Müller-Navarra, D.C., Brett, M.T., Park, S., Chandra, S., Ballantyne, A.P., Zorita, E. and Goldman, C.R. 2004. Unsaturated fatty acid content in seston and tropho-dynamic coupling in lakes. Nature 427:69–72

Naiman, R.J., Bilby, R.E., Schindler, D.E. and Helfield, J.M. 2002. Pacific salmon, nutrients, and the dynamics of freshwater and riparian ecosystems. Ecosystems 5:399–417

Nakano, S. and Murakami, M. 2001. Reciprocal subsidies: dynamic interdependence between terrestrial and aquatic food webs. Proc. Natl. Acad. Sci. U.S.A. 98:166–170

Nor Aliza, A.R., Bedick, J.C., Rana, R.L., Tunaz, H., Wyatt Hoback, W. and Stanley, D.W. 2001. Arachidonic and eicosapentaenoic acids in tissues of the firefly, *Photinus pyralis* (Insecta: Coleoptera). Comp. Biochem. Physiol. A 128:251–257

Ozyurt, G. and Polat, A. 2006. Amino acid and fatty acid composition of wild sea bass (*Dicentrarchus labrax*): a seasonal differentiation. Eur. Food Res. Technol. 222:316–320

Ozyurt, G., Polat, A. and Ozkutuk, S. 2005. Seasonal changes in the fatty acids of gilthead sea bream (*Sparus aurata*) and white sea bream (*Diplodus sargus*) captured in Iskenderun Bay, eastern Mediterranean coast of Turkey. Eur. Food Res. Technol. 220:120–124

Paetzold, A., Schubert, C.J. and Tockner, K. 2005. Aquatic terrestrial linkages along a braided-river: riparian arthropods feeding on aquatic insects. Ecosystems 8:748–759

Pauly, D., Christensen, V., Guénette, S., Pitcher, T.J., Sumaila, U.R., Walters, C.J., Watson, R. and Zeller, D. 2002. Towards sustainability in world fisheries. Nature 418:689–695

Pawlosky, R.J., Denkins, Y., Ward, G. and Salem, N. Jr. 1997. Retinal and brain accretion of long-chain polyunsaturated fatty acids in developing felines: the effects of corn oil-based maternal diets. Am. J. Clin. Nutr. 65:465–472

Persson, J. and Vrede, T. 2006. Polyunsaturated fatty acids in zooplankton: variation due to taxonomy and trophic position. Freshw. Biol. 51:887–900

Plourde, M. and Cunnane, S.C. 2007. Extremely limited synthesis of long chain polyunsaturates in adults: implications for their dietary essentiality and use as supplements. Appl. Physiol. Nutr. Metabol. 32:619–634

Poepperl, R. 2000. Benthic secondary production and biomass of insects emerging from a northern German temperate stream. Freshw. Biol. 44:199–211

Polis, G.A. and Hurd, S.D. 1996. Linking marine and terrestrial food webs: allochthonous input from the ocean supports high secondary productivity on small islands and coastal land communities. Am. Nat. 147:396–423

Politia, L., Rotsteina, N. and Carrib, N. 2001. Effects of docosahexaenoic acid on retinal development: cellular and molecular aspects. Lipids 36:927–935

Ponyi, J.E. 1994. Abundance and feeding of wintering and migrating aquatic birds in two sampling areas of Lake Balaton in 1983–1985. Hydrobiologia 279/280:63–69

Pruitt, N.L. and Lu, C. 2008. Seasonal changes in phospholipid class and class-specific fatty acid composition associated with the onset of freeze tolerance in third-instar larvae of *Eurosta solidaginis*. Physiol. Biochem. Zool. 81:226–234

Quinn, T.P., Gende, S.M., Ruggerone, G.T. and Rogers, D.E. 2003. Density-dependent predation by brown bears (*Ursus arctos*) on sockeye salmon (*Oncorhynchus nerka*). Can. J. Fish. Aquat. Sci. 60:553–562

Raven, J.A. and Maberly, S.C. 2004. Plant productivity of inland waters. In: Chlorophyll *a* fluorescence. Advances in photosynthesis and respiration, Vol. 19. Edited by Papageorgiou, G.C. and Govindjee. Dordrecht: Springer, pp. 779–793

Reimchen, T.E. 2000. Some ecological and evolutionary aspects of bear–salmon interactions in coastal British Columbia. Can. J. Zool. 78:448–457

Richoux, N.E., Deibel, D., Thompson, R.J. and Parrish, C.C. 2005. Seasonal and developmental variation in the fatty acid composition of *Mysis mixta* (Mysidacea) and *Acanthostepheia malmgreni* (Amphipoda) from the hyperbenthos of a cold-ocean environment (Conception Bay, Newfoundland). J. Plankton Res. 27:719–733

Robert, S.S. 2006. Production of eicosapentaenoic and docosahexaenoic acid-containing oils in transgenic land plants for human and aquaculture nutrition. Mar. Biotech. 8:103–109

SanGiovanni, J.P. and Chew, E.Y. 2005. The role of omega-3 long-chain polyunsaturated fatty acids in health and disease of the retina. Prog. Retinal Eye Res. 24:87–138

Sanzone, D.M., Meyer, J.L., Marti, E., Gardiner, E.P., Tank, J.L. and Grimm, N.B. 2003. Carbon and nitrogen transfer from a desert stream to riparian predators. Oecologia. 134:238–250

Schlechtriem, C., Arts, M.T. and Zellmer, I.D. 2006. Effect of temperature on the fatty acid composition and temporal trajectories of fatty acids in fasting *Daphnia pulex* (Crustacea, Cladocera). Lipids. 41:397–400

Shorland, F.B. 1963. The distribution of fatty acids in plant lipids. In: Chemical plant taxonomy. Edited by Swain, T. London, New York: Academic Press, pp. 253–311

Silvers, K.M. and Scott, K.M. 2002. Fish consumption and selfreported physical and mental health status. Public Health Nutr. 5:427–431

Simopoulos, A.P. 2004a. The traditional diet of Greece and cancer. Eur. J. Cancer Prevent. 13:219–230

Simopoulos, A.P. 2004b. Omega-3 fatty acids and antioxidants in edible wild plants. Biol. Res. 37:263–277

Stagliano, D.M., Benke, A.C. and Anderson, D.H. 1998. Emergence of aquatic insects from 2 habitats in a small wetland of the southeastern USA: temporal patterns of numbers and biomass. J. North Am. Benthol. Soc. 17:37–53

Stanley, D. 2006. Prostaglandins and other eicosanoids in insects: biological significance. Ann. Rev. Entomol. 51:25–44

Stanley-Samuelson, D.W., Jensen, E., Nickerson, K.W., Tiebel, K., Ogg, C.L. and Howard, R.W. 1991. Insect immune response to bacterial infection is mediated by eicosanoids. Proc. Natl. Acad. Sci. U.S.A. 88:1064–1068

Stapp, P. and Polis, G.A. 2003. Marine resources subsidize insular rodent populations in the Gulf of California, Mexico. Oecologia 134:496–504

Stapp, P. and Van Horne, B. 1997. Response of Deer Mice (*Peromyscis maniculatus*) to shrubs in shortgrass prairie: linking small-scale movements and spatial distribution of individuals. Funct. Ecol. 11:644–651

Sullivan, T.P., Lautenschlager, R.A. and Wagner, R.G. 1999. Clearcutting and burning of northern spruce-fir forests: implications for small mammal communities. J. Appl. Ecol. 36:327–344

Surh, J., Ryu, J.S. and Kwon, H. 2003. Seasonal variations of fatty acid compositions in various Korean shellfish. J. Agric. Food Chem. 51:1617–1622

Sushchik, N.N., Gladyshev, M.I., Moskvichova, A.V., Makhutova, O.N. and Kalachova, G.S. 2003. Comparison of fatty acid composition in major lipid classes of the dominant benthic invertebrates of the Yenisei River. Comp. Biochem. Physiol. B 134:111–122

Sushchik, N.N., Gladyshev, M.I., Makhutova, O.N., Kalachova, G.S., Kravchuk, E.S. and Ivanova, E.A. 2004. Associating particulate essential fatty acids of the ω3 family with phytoplankton species composition in a Siberian reservoir. Freshw. Biol. 49:1206–1219

Sushchik, N.N., Gladyshev, M.I., Kalachova, G.S., Makhutova, O.N. and Ageev, A.V. 2006. Comparison of seasonal dynamics of the essential PUFA contents in benthic invertebrates and grayling *Thymallus arcticus* in the Yenisei River. Comp. Biochem. Physiol. B. 145:278–287

Sushchik, N.N., Gladyshev, M.I. and Kalachova, G.S. 2007. Seasonal dynamics of fatty acid content of a common food fish from the Yenisei River, Siberian grayling, *Thymallus arcticus*. Food Chem. 104:1353–1358

Suter, W. 1994. Overwintering waterfowl on Swiss lakes: how are abundance and species richness influenced by trophic status and lake morphology? Hydrobiologia 279/280:1–14

Szepanski, M.M., Ben-David, M. and Van Ballenberghe, V. 1999. Assessment of anadromous salmon resources in the diet of the Alexander Archipelago wolf using stable isotope analysis. Oecologia 120:327–335

Tocher, D.R., Leaver, M.J. and Hodson, P.A. 1998. Recent advances in the biochemistry and molecular biology of fatty acyl desaturases. Prog. Lipid Res. 37:73–117

Uscian, J.M. and Stanley-Samuelson, D.W. 1994. Fatty acid compositions of phospholipids and triacylglycerols from selected terrestrial arthropods. Comp. Biochem. Physiol. B 107:371–379

Vanni, M.J., Layne, C.D. and Arnott, S.E. 1997. "Top-down" trophic interactions in lakes: effects of fish on nutrient dynamics. Ecology 78:1–20

Wetzel, R.G. 1992. Gradient-dominated ecosystems: sources and regulatory function of dissolved organic matter in freshwater ecosystems. Hydrobiologia 229:181–198

Willson, M.F., Gende, S.M. and Marston, B.H. 1998. Fishes and the forest. BioScience. 48:455–462

Winder, M., Schindler, D.E., Moore, J.W., Johnson, S.P. and Palen, W.J. 2005. Do bears facilitate transfer of salmon resources to aquatic macroinvertebrates? Can. J. Fish. Aquat. Sci. 62:2285–2293

Wolff, R.L., Lavialle, O., Pëdrono, F., Pasquier, E., Deluc, L.G., Marpeau, A.M. and Aitzetmüller, K. 2001. Fatty acid composition of Pinaceae as taxonomic markers. Lipids 36:439–451

Wood, J.D., Enser, M., Fisher, A.V., Nute, G.R., Sheard, P.R., Richardson, R.I., Hughes, S.I. and Whittington, F.M. 2008. Fat deposition, fatty acid composition and meat quality: a review. Meat Sci. 78:343–358

Woollhead, J. 1994. Birds in the trophic web of Lake Esrom, Denmark. Hydrobiologia. 279/280:29–38

Zenebe, T., Ahlgren, G. and Boberg, M. 1998. Fatty acid content of some freshwater fish of commercial importance from tropical lakes in the Ethiopian Rift Valley. J. Fish Biol. 53:987–1005

Chapter 9
Biosynthesis of Polyunsaturated Fatty Acids in Aquatic Ecosystems: General Pathways and New Directions

Michael V. Bell and Douglas R. Tocher

9.1 Introduction

It is now well established that the long-chain, omega-3 (ω3 or n-3) polyunsaturated fatty acids (PUFA) are vitally important in human nutrition, reflecting their particular roles in critical physiological processes (see Chap. 14). In comparison to terrestrial ecosystems, marine or freshwater ecosystems are characterised by relatively high levels of long-chain n-3PUFA and, indeed, fish are the most important source of these vital nutrients in the human food basket. Virtually all PUFA originate from primary producers but can be modified as they pass up the food chain. This is generally termed trophic upgrading, and various aspects of these phenomena have been described in Chaps. 2, 6 and 7 (this volume). However, while qualitative aspects of essential fatty acid production and requirements in aquatic ecosystems are relatively well understood, in order to fully understand and model ecosystems, quantitative information is needed on synthesis and turnover rates of n-3PUFA at different trophic levels in the food web. The present chapter describes the biochemistry and molecular biology involved in the various pathways of PUFA biosynthesis and interconversions in aquatic ecosystems.

To appreciate the biochemical mechanisms involved in their biosynthesis, some understanding of fatty acid chemistry is required. Fatty acids are designated on the basis of their chain lengths, degree of unsaturation (number of ethylenic or 'double' bonds) and the position of their ethylenic bonds. Thus, 18:0 designates a fatty chain with 18 carbon atoms and no ethylenic bonds. In the n- (or omega, ω) nomenclature, the position of double bonds is designated by counting from the methyl terminus, and so 16:1n-7 designates a fatty acid with 16 carbon atoms whose single ethylenic bond is 7 carbon atoms from the methyl end. In the alternative delta (Δ) nomenclature, the position of the ethylenic bond is counted from the carboxyl end, and so 16:1n-7 is written as $16:1\Delta^9$. The n- or ω-nomenclatures are

M.V. Bell and D.R. Tocher (✉)
Institute of Aquaculture, University of Stirling, Stirling, FK9 4LA, UK
e-mail: d.r.tocher@stir.ac.uk

those now in general use, but the Δ nomenclature remains relevant as it has traditionally been used for characterising fatty acyl desaturase activities. PUFA are defined as fatty acids containing two or more ethylenic bonds and a common example is eicosapentaenoic acid (EPA), 20:5n-3 and 20:5ω3 or, 20:5$^{\Delta5,8,11,14,17}$ in the Δ nomenclature. Highly unsaturated fatty acid (HUFA) is a term being used increasingly, most often without proper definition, mainly to distinguish the major bioactive long-chain PUFA such as EPA, docosahexaenoic acid (DHA; 22:6n-3) and arachidonic acid (ARA; 20:4n-6) from the shorter-chain C_{18} PUFA such as linoleic (LIN; 18:2n-6) and ALA (α-linolenic; 18:3n-3) acids. In our laboratory we define HUFA as fatty acids with ≥20 carbons and ≥3 double bonds. A further term requiring careful definition is essential fatty acid (EFA; and see Chap. 13). PUFA are essential dietary components for most animals, and all vertebrates, as they cannot synthesise PUFA de novo from monounsaturated fatty acids. However, which specific PUFA can satisfy the EFA requirements (and prevent EFA deficiency symptoms) in a particular species is entirely dependent upon its endogenous capacity to convert C_{18} PUFA to the biologically active HUFA: ARA, EPA and DHA. Therefore, EFA requirements will vary qualitatively as well as quantitatively among different animal species.

9.2 Primary Production of Polyunsaturated Fatty Acids

9.2.1 Bacteria

The presence of n-3PUFA in heterotrophic prokaryotes was first reported by Johns and Perry (1977), who found EPA in the marine bacterium *Flexibacter polymorphus*. Subsequently EPA and DHA were found in a number of bacteria isolated from cold marine habitats, but particularly from intestines of deep-sea fish and invertebrates (Yano et al. 1994). Bacteria producing EPA and DHA were found in the culturable intestinal flora of all species of a selection of ten Arctic and subArctic invertebrates and one of four fish species (Jøstensen and Landfald 1997). In total, 103 out of 330 strains of bacteria tested contained n-3PUFA. The highest prevalences, in >50% of bacterial isolates, were from two species of bivalve *Chlamys islandica* and *Astarte* sp. and in the amphipod *Gammarus wilkitzkii*. PUFA producers clustered into eight groups depending on PUFA profile, six groups had 3.8–18.7% EPA and two groups had 7.1 and 13.5% DHA, both by weight total fatty acid (Jøstensen and Landfald 1997). Interestingly the bacteria contained either EPA or DHA, but not both. Some bacteria associated with coastal Antarctic sea-ice diatom assemblages were found to synthesise DHA (Bowman et al. 1998). Eight strains were identified by 16S rRNA sequence analysis as belonging to the genus *Colwellia*. All exhibited psychrophilic and facultative anaerobic growth and produced 0.7–8.0% of total fatty acids as DHA (Bowman et al. 1998).

Vibrio sp. and *Shewanella* sp. comprise the majority of the PUFA-producing bacterial species isolated from the guts of fish and invertebrates. However, there have been a number of misidentifications, which are discussed in a detailed review of PUFA in marine bacteria by Russell and Nichols (1999). This review also discussed pathways of fatty acid synthesis by bacteria and concluded that conventional aerobic pathways must be used to synthesise PUFA. The discovery of the polyketide pathway allows an anaerobic alternative, which may be more appropriate with respect to the general metabolism of these microorganisms (see later).

9.2.2 Photosynthetic Organisms (Microalgae)

De novo synthesis of n-3PUFA in microalgae involves sequential addition of double bonds to saturated fatty acids via $\Delta 9$, $\Delta 12$ and $\Delta 15$ (or $\omega 3$) desaturases to give ALA. A sequence of front-end desaturases (inserting double bonds between the $\Delta 9$ bond and the carboxyl terminus) acting with elongases then produces EPA and DHA. Conventionally this sequence is $\Delta 6$ desaturase – elongase – $\Delta 5$ desaturase – elongase – $\Delta 4$ desaturase, but in some species the initial step is elongation to 20:3n-3 followed by $\Delta 8$ desaturation. Unlike the situation in vertebrates the last step appears to involve a direct $\Delta 4$ desaturation (Meyer et al. 2003; Tonon et al. 2005) rather than the 'Sprecher' shunt operating via 24:5n-3 and 24:6n-3 intermediates.

However, some PUFA that are abundant in some classes of marine microalgae do not fit on the conventional pathway for synthesizing n-3PUFA (e.g. 16:4n-3 and most notably 18:5n-3). Octadecapentaenoic acid ($18:5^{\Delta-3,6,9,12,15}$, 18:5n-3) was first reported by Joseph (1975) who found 3.8–22.2% in total lipid from 11 species of marine dinoflagellate. Early studies probably overlooked the presence of 18:5n-3 since on polar GC columns it elutes very close to, or coelutes with, 20:1n-9. It has subsequently been shown to be particularly abundant in prymnesiophytes and dinoflagellates where it may comprise up to 43% of total fatty acid (Volkman et al. 1981; Nichols et al. 1984; Okuyama et al. 1993; Bell et al. 1997) and together with DHA can be the main PUFA. In *Emiliania huxleyi*, eight strains contained 13.7–22.0% 18:5n-3 in the stationary phase, and all cultures showed accumulation of 18:5n-3 during the growth phase (Pond and Harris 1996). This was mainly in digalactosyldiacylglycerol and monogalactosylacylglycerol varying from 33.1–62.1% to 40.9–50.5% of fatty acids, respectively, over the growth cycle (Bell and Pond 1996). Two other algal groups are now known to contain 18:5n-3. Five species of green algae from the *Prasinophyceae* contained 2.2–10.6% of 18:5n-3 in total lipid (Dunstan et al. 1992), while two strains of the raphidophyte *Heterosigma akashiwo* contained 4.6 and 5.2% 18:5n-3 (Nichols et al. 1987).

To the best of our knowledge HUFA synthesis pathways have not been studied in macroalgae/seaweeds. Based on fatty acid compositions, seaweeds can produce ARA and EPA at quite high levels, but generally lack DHA (Sanchez-Machado et al. 2004; Dawczynski et al. 2007). The lack of data precludes any firm conclusions, but the pathways are possibly essentially the same as those found in microalgae of the same group.

9.2.3 Protozoans

A protozoan parasite of oysters, *Perkinsus marinus*, has been shown to synthesise LIN and ARA from acetate indicating de novo synthesis of PUFA (Chu et al. 2002). Synthesis of ARA appears to be via a Δ8 pathway, but also involves a C_{18} Δ9-elongating activity catalysed by a FAE1 (*fatty acid elongation 1*)-like 3-ketoacyl-CoA synthase previously only reported in higher plants and algae (Venegas-Caleron et al. 2007). *P. marinus* represents a key organism in the taxonomic separation of the single-celled eukaryotes, the alveolates, and these data imply ancestral endosymbiotic acquisition of plant-like genes. Little is known about PUFA synthesis in other protozoans.

9.2.4 Heterotrophic Organisms

The Thraustochytrids are eukaryotic protists in the phylum labyrinthulomycota, the slime nets. They are of interest to the fatty acid biochemist since they produce HUFA-rich oils containing 22:5n-6 and/or DHA. A *Thraustochytrium* sp. was also one of the first organisms in which the presence of a Δ4 desaturase was demonstrated (Qiu et al. 2001). A related *Schizochytrium* sp. has recently been shown to use a polyketide-like pathway for HUFA synthesis (Hauvermale et al. 2006).

9.2.5 The Polyketide Pathway and PUFA Synthase

One of the most exciting developments in the field of fatty acid metabolism in recent years was the discovery that PUFA could be synthesized in both prokaryotes and eukaryotes via a completely novel anaerobic pathway using polyketide synthases (Metz et al. 2001). Yazawa (1996) identified five open reading frames[1] (ORFs) from bacterial *Shenawella* sp. strain SCRC2738 found in the intestine of Pacific mackerel that were necessary and sufficient for EPA production in *Escherichia coli*. Metz and colleagues noted that eight of the PUFA-synthesizing domains within the *Shewanella* ORFs were more closely related to polyketide synthase (PKS) than to fatty acid synthase, aerobic desaturases, or elongases (Metz et al. 2001). These included domains of PKS such as acyl carrier protein (ACP), 3-ketoacyl synthase, malonyl-CoA:ACP acyltransferase, 3-ketoacyl-ACP-reductase, chain length factor and acyl transferases. A definitive demonstration of the significance

---

[1] An open reading frame is the portion of mRNA located between the translation start-code sequence (initiation codon) and the stop-code sequence (termination codon) containing the protein-coding sequence.

of these *Shewanella* ORFs was obtained by heterologous expression in *E. coli* cultured under aerobic and anaerobic conditions. EPA synthesis was found under both conditions ruling out a role for aerobic desaturases and indicating a PKS-like system (Metz et al. 2001).

Schizochytrium is a thraustrochytrid-like marine protist that accumulates large amounts of C_{22} HUFA in triacylglycerol (Barclay et al. 1994). Additional support for a PKS-based pathway was provided by sequencing selected clones from a *Schizochytrium* cDNA library. Sequences showing homology to 8 of the 11 domains of the *Shewanella* PKS genes were identified. Further sequencing of cDNA and genomic clones allowed the identification of three ORFs containing domains with homology to those in *Shewanella*. These proteins may constitute a PKS that catalyses DHA and 22:5n-6 synthesis in *Schizochytrium*. The homology between the prokaryotic *Shewanella* and the eukaryotic *Schizochytrium* genes suggests that the PUFA PKS has undergone a lateral gene transfer (Metz et al. 2001). Subsequent work confirmed that the enzyme complex in *Schizochytrium* (renamed PUFA synthase) comprised three genes which account for the production of DHA and 22:5n-6 (Hauvermale et al. 2006). Additionally it was found that the other two abundant fatty acids in *Schizochytrium*, 14:0 and 16:0, were the products of a separate fatty acid synthase which, in terms of sequence homology and domain organisation, resembled those found in fungi (Hauvermale et al. 2006). The presence of this alternative PUFA pathway in *Schizochytrium* was somewhat surprising since a Δ4 desaturase has been cloned from a thraustrochytrid species closely related to *Schizochytrium* showing that the enzymes of the aerobic pathway are also present in these organisms (Qiu et al. 2001).

The polyketide pathway requires six enzyme proteins: 3-ketoacyl synthase (KS), 3-ketoacyl-ACP-reductase (KR), dehydrase (DH), enoyl reductase (ER), dehydratase/2-trans, 3-cis isomerase (DH/2,3I), dehydratase/2-trans and 2-cis isomerase (DH/2,2I). A defined sequence of steps then adds C2 units and double bonds. An important point about the PKS pathway is that it adds double bonds to nascent acyl chains, whereas the desaturase pathway inserts double bonds into intact acyl chains. This makes the PKS pathway more efficient energetically since the ATP used by the desaturase steps in the conventional aerobic pathway is not required. The scheme starts with acetyl-CoA and malonyl-CoA. KS and KR add C2 units while DH and ER or DH/2,3I and DH/2,2I control the positioning of double bonds. Since KS adds two carbon units and double bonds are inserted at three-carbon intervals, two different dehydrase isomerases are required to produce the methylene-interrupted pattern of double bonds characteristic of n-3 and n-6 PUFA (Fig. 9.1).

An extension of the scheme presented by Metz et al. (2001) is shown in Fig. 9.2 illustrating how n-3PUFA could be synthesized by this pathway. The main n-3 pathway goes 16:4n-3, 18:5n-3, 20:6n-3, DHA (Fig. 9.2). The PUFA, ALA, 18:4n-3, 20:4n-3 and EPA do not lie on this pathway (Fig. 9.2). EPA can be produced from 18:5n-3 with a KS, KR, DH, ER step. Omega-6 PUFA are produced by altering the second step in the scheme from KS, KR, DH/2,3I which introduces the Δ3 double bond to give 6:1n-3 to a KS, KR, DH, ER step to give 6:0. The next step then introduces a Δ2 double bond to give 8:1n-6.

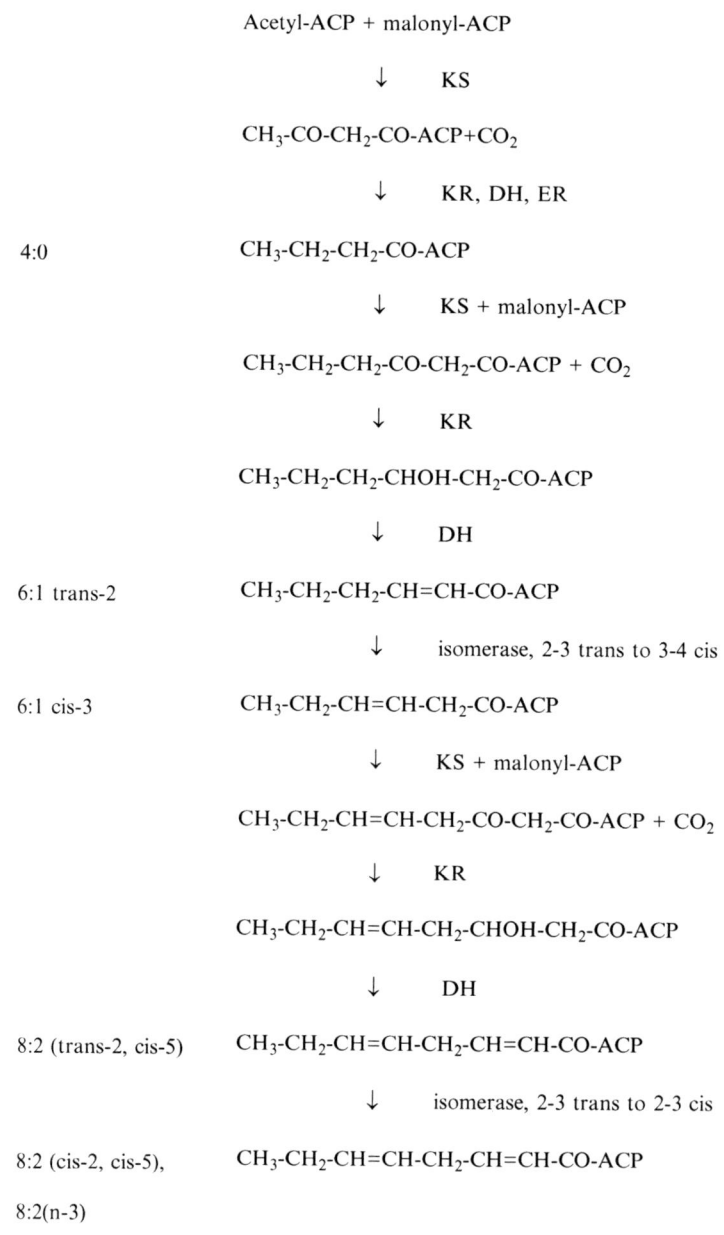

Fig. 9.1 Fatty acid synthesis using the polyketide pathway (from Metz et al. 2001). *DH* dehydrase; *ER* enoyl reductase; *KR* 3-ketoacyl-ACP-reductase; *KS* 3-ketoacyl synthase

A major question following on from this work is to what extent other marine microorganisms might be synthesizing PUFA via a PKS pathway rather than a conventional desaturase-elongase pathway. Since a $\Delta 3$ desaturase was not known, initial suggestions were that 18:5n-3 could be produced by chain-shortening EPA

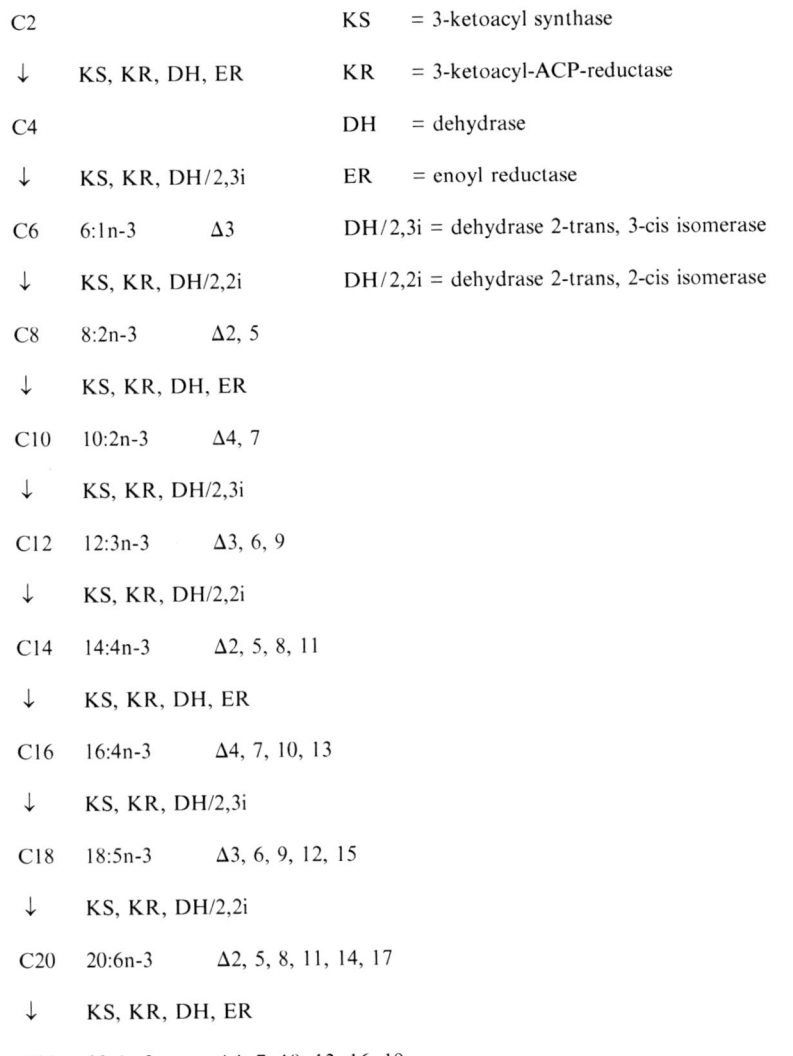

Fig. 9.2 A scheme showing the synthesis of 22:6n-3 using PKS (adapted from Metz et al. 2001). *DH* dehydrase; *DH/2,2i* dehydratase/2-trans, 2-cis isomerase; *DH/2,3i* dehydratase/2-trans, 3-cis isomerase; *ER* enoyl reductase; *KR* 3-ketoacyl-ACP-reductase; *KS* 3-ketoacyl synthase

(Joseph 1975; Volkman et al. 1981). This hypothesis became more plausible with the discovery that the last step in the synthesis of DHA in mammals is a chain shortening of 24:6n-3 to give DHA (Sprecher 2000). However, 16:4n-3 and 18:5n-3 are mainstream fatty acids on the PKS pathway.

This scheme also answers another problem with algal fatty acid composition, that of PUFA with >22 carbon atoms. These PUFA are under-reported in the literature because most workers stop their GC runs after the elution of DHA on polar-type columns.

However, workers who have used longer time programs have found C_{26} and C_{28} PUFA (e.g. 28:7n-6 and 28:8n-3, Mansour et al. 1999) in some species of marine dinoflagellates. These two fatty acids are on the PKS pathway (Fig. 9.3). Such fatty acids can be derived via conventional elongase, desaturase and chain-shortening steps but this is a cumbersome pathway.

The PKS pathway could thus offer an alternative solution to PUFA synthesis in some species of microalgae to the conventional pathway, but at present there is no direct evidence to support the presence of this pathway in microalgae. However, there have also been difficulties in demonstrating the conventional pathway. The heterotrophic dinoflagellate *Crypthecodinium cohnii* did not convert ALA to DHA, but de novo synthesis from acetate was rapid (Henderson and Mackinlay 1991). The dinoflagellate *Amphidinium carterae* did not convert the deuterated tracer D_5-17,17,18,18,18-linolenic acid (D_5-ALA) to longer-chain n-3PUFA (M.V. Bell, unpublished). Polyketide-derived metabolites are widespread in bacteria, fungi, microalgae and plants (O'Hagan 1995), and the dinoflagellates produce a wide variety of very complex toxins (e.g. brevetoxins) that are believed to be synthesised by a polyketide pathway (reviewed by Shimizu 1996; see Chap. 4). Thus, some of the PKS pathway enzymes are present in phytoplankton and involved in the synthesis of other metabolites.

Some parts of this pathway are speculative and experiments are needed to confirm the pathway in different organisms, especially those marine flagellates containing 18:5n-3. These could take the form of identifying the full array of necessary genes or identifying fatty acid intermediates. The latter may be difficult since intermediates may not be released from the enzyme protein into the fatty acid pool and would therefore be present in very small amounts.

```
C22    22:6n-3        Δ 4, 7, 10, 13, 16, 19

 ↓     KS, KR, DH/2,3i

C24    24:7n-3        Δ 3, 6, 9, 12, 15, 18, 21

 ↓     KS, KR, DH/2,2i

C26    26:8n-3        Δ 2, 5, 8, 11, 14, 17, 20, 23

 ↓     KS, KR, DH, ER

C28    28:8n-3        Δ 4, 7, 10, 13, 16, 19, 22, 25
```

Fig. 9.3 A scheme showing the synthesis of 28:8n-3 from 22:6n-3 using PKS. *DH* dehydrase; *DH/2,2i* dehydratase/2-trans, 2-cis isomerase; *DH/2,3i* dehydratase/2-trans, 3-cis isomerase; *ER* enoyl reductase; *KR* 3-ketoacyl-ACP-reductase; *KS* 3-ketoacyl synthase

9.3 Polyunsaturated Fatty Acid Metabolism in Invertebrates

9.3.1 Zooplankton

Marine copepods are thought to be unable to synthesise HUFA, and growth rates and reproductive success have been linked to the availability of these fatty acids in the phytoplankton (e.g. Pond and Harris 1996). The ability of four species of marine zooplankton to synthesise HUFA from ALA was tested directly using liposomes containing D_5-ALA. Female *Calanus finmarchicus, Calanoides acutus, Dropanopus forcipatus* and calyptopus larvae of *Euphausia superba* readily ingested the liposomes and incorporated D_5-ALA into their somatic lipid pool, but only negligible amounts of desaturation products were detected after 96-h incubation in *C. finmarchicus, D. Forcipatus and E. superba* with none in *C. acutus* (Bell et al. 2007). It was concluded that these four species were indeed unable to synthesise PUFA at ecologically significant rates under the conditions of this experiment. However, feeding studies have suggested that some other species of copepod may be able to synthesise HUFA from ALA. The harpacticoid copepods *Tisbe holothuriae* and *Tisbe* sp. accumulated substantial amounts of EPA and DHA when fed algae containing small amounts of these fatty acids implying synthesis (Nanton and Castell 1998). The freshwater copepod *Eucyclops serrulatus* (Desvilettes et al. 1997) accumulated DHA and *Daphnia pulex* (Schlechtriem et al. 2006) accumulated EPA when fed green algae lacking this fatty acid.

9.3.2 Other Invertebrates

There is little information available on PUFA synthesis across the wide range of other aquatic invertebrate genera. The freshwater rotifer *Brachionus plicatilis* was able to synthesise n-3PUFA de novo when fed diets lacking these fatty acids (Lubzens et al. 1985). The sea urchin *Psammechinus miliaris* was able to convert dietary D_5-ALA to D_5-EPA, but the rate of conversion was very slow, 0.09 µg g tissue^{-1} mg^{-1} ALA eaten over 14 days (Bell et al. 2001a).

There is rather more information available, especially in the older literature, concerning PUFA biosynthesis in terrestrial invertebrates. Twelve species of insect have been identified which can synthesise LIN de novo (e.g. Cripps et al. 1986). *Caenorhabditis elegans* can synthesise ARA and EPA using the pathway as described later for vertebrates, and thus has been used as a model organism in studies of HUFA synthesis in animals (Watts and Browse 1999). It seems likely, therefore, that the ability to synthesise PUFA is widespread in the invertebrate kingdom. Studies are required to determine which pathway is used and the contribution of invertebrates to the aquatic PUFA pool.

9.4 Production of Highly Unsaturated Fatty Acids in Fish

9.4.1 Pathways for Biosynthesis of Highly Unsaturated Fatty Acids

Vertebrates, including fish, lack the $\Delta 12$ and $\omega 3$ ($\Delta 15$) desaturases and so cannot form LIN and ALA, respectively, from 18:1n-9 and, therefore, PUFA are essential dietary components. However, dietary LIN and ALA can, with varying efficiencies, be further desaturated and elongated in vertebrates to form HUFA, including ARA, EPA and DHA (Fig. 9.4). Fatty acyl desaturation is an aerobic reaction catalysed by a terminal oxygenase (the 'desaturase') requiring reducing equivalents, derived from NADPH, delivered via an electron transport chain including cytochrome b_5 and a reductase. Elongation is effected in four steps each catalysed by a specific enzyme. The first step is a condensation reaction of the precursor fatty acyl chain with malonyl-CoA to produce a β-ketoacyl chain that is subsequently hydrogenated in three successive steps. The condensation step determines the substrate specificity and is the rate-limiting step of the process and is therefore regarded as being the 'elongase' enzyme. The main features of the HUFA synthesis pathway, most fully studied in rats (Sprecher 2000), are summarised later.

With one exception, the reactions occur in the smooth endoplasmic reticulum with the same enzymes acting on both n-3 and n-6 fatty acids, although the affinity of the enzymes is generally higher for the n-3 series. DHA rather than EPA is the main end product of desaturation and elongation of ALA, whereas ARA rather than 22:5n-6 is the primary end product of desaturation and elongation of LIN.

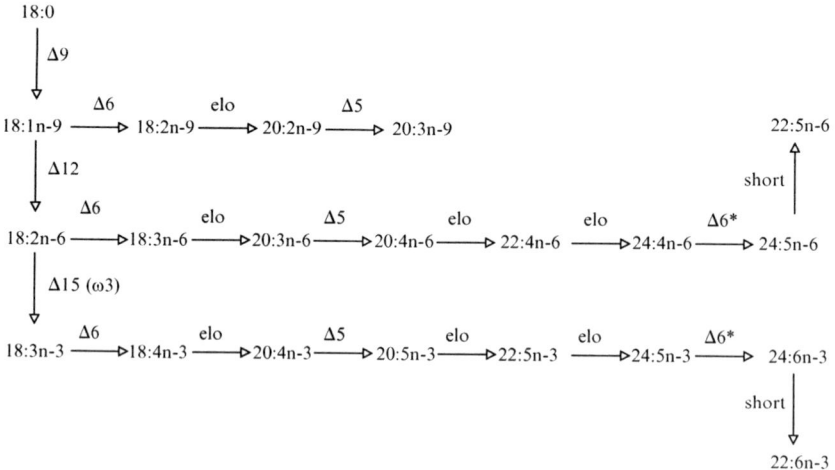

Fig. 9.4 Pathways of biosynthesis of C_{20} and C_{22} highly unsaturated fatty acids from n-3, n-6 and n-9 C_{18} precursors. $\Delta 5$, $\Delta 6$, $\Delta 6^*$, $\Delta 9$, $\Delta 12$, $\Delta 15$ ($\omega 3$), fatty acyl desaturases; elo, fatty acyl elongases; short, chain shortening

The insertion of the last, Δ4, ethylenic bond in DHA and 22:5n-6 does not occur through direct Δ4 desaturation of their immediate precursors 22:5n-3 and 22:4n-6. Rather, these intermediates are chain elongated to C_{24} fatty acids, 24:5n-3 and 24:4n-6, which are then converted by Δ6 desaturation to 24:6n-3 and 24:5n-6, respectively. The Δ6-desaturated C_{24} fatty acids are then chain shortened in the peroxisomes to DHA and 22:5n-6. Thus, Δ5 fatty acid desaturation occurs at one step in the pathway, involving 20:3n-6 or 20:4n-3, whereas Δ6 fatty acid desaturation occurs at two steps, first involving LIN or ALA and second involving 24:4n-6 or 24:5n-3. Heterologous expression studies of human and rat Δ6 desaturases showed that the same enzymes are active on C_{18} and C_{24} fatty acids (De Antueno et al. 2001; D'Andrea et al. 2002).

In common with all vertebrates, PUFA are essential dietary components for fish. However, the biologically active and physiologically important PUFA in fish are the HUFA, ARA, EPA and DHA, as LIN and ALA have no specific or unique metabolic role in themselves (Sargent et al. 2002; Tocher 2003). Indeed, the great majority of dietary ALA is catabolised in fish (Bell et al. 2001; Bell and Dick 2004) as it is in mammals. Therefore, the major role of dietary C_{18} PUFA in fish is to function as precursors for C_{20} and C_{22} HUFA, highlighting the importance of the HUFA synthesis pathway. Early nutritional studies suggested that ALA and/or LIN could satisfy the EFA requirements of freshwater fish, whereas the n-3HUFA, EPA and DHA were required to satisfy the EFA requirements of marine fish (see Sargent et al. 2002). Dietary conversion studies performed in turbot using radioactive substrates in vivo strongly suggested that this marine species was unable to produce EPA and ARA from ALA and LIN, respectively, although these experiments were unable to determine precisely the deficiency in the HUFA synthesis pathway (see Tocher 2003).

The HUFA pathway has been most extensively studied in hepatocytes from the salmonids, rainbow trout (*Oncorhynchus mykiss*) and Atlantic salmon (*Salmo salar*). Biochemical studies established the presence of the entire HUFA synthesis pathway from ALA to DHA, including the role of C_{24} intermediates, in rainbow trout hepatocytes (Buzzi et al. 1996, 1997). There is also extensive evidence, based on the conversion of radioisotopes, that EPA and DHA are produced from ALA in hepatocytes from Atlantic salmon (Tocher et al. 1997), and several other species of freshwater fish including Arctic charr (*Salvelinus alpinus*), brown trout (*Salmo trutta*) (Tocher et al. 2001a), tilapia (*Oreochromis niloticus*) and zebrafish (*Danio rerio*) (Tocher et al. 2001b). In contrast, studies in hepatocytes from marine fish have consistently shown very little desaturation of ALA occurs with no production of EPA or DHA (Sargent et al. 2002; Tocher 2003). Early studies with cell lines established that the inability of marine fish cells to produce EPA and DHA was due either to limited activities of C_{18} to C_{20} elongase, or fatty acyl Δ5 desaturase (Tocher et al. 1989; Tocher and Sargent 1990). Subsequent studies utilising isotopically labelled substrates for the C_{18-20} elongase ([U-^{14}C]18:4n-3 and (D5)18:4n-3), and Δ5 desaturase ([U-^{14}C]20:4n-3 and (D5)20:4n-3) showed that a turbot cell line had low C_{18-20} elongase activity (Ghioni et al. 1999), whereas a gilthead sea bream cell line had very low Δ5 desaturase activity (Tocher and Ghioni 1999).

Although results from the cell line studies are entirely consistent with data from feeding studies, they require confirmation by in vivo studies as it is not certain that desaturase and elongase enzymes continue to be expressed in cultured cells exactly as they are in vivo. It is also difficult to extrapolate whole body nutritional and physiological requirements from such in vitro studies. The degree of HUFA synthesis from C_{18} PUFA is dependent upon the activities of fatty acyl desaturases and elongases, and in turn these may be dependent on the extent to which HUFA are readily available in natural diets. For instance, carnivores such as cats, which can obtain abundant preformed HUFA from their natural prey, have a very poor ability to form HUFA and appear to show very low $\Delta 6$ and $\Delta 5$ desaturase activities (see Sargent et al. 2002). Similarly, marine fish have large amounts of EPA and DHA in their natural diets, whereas the natural prey of many freshwater fish, particularly their invertebrate prey, may be much less rich in n-3HUFA, especially DHA (see Chapter 6 for detailed discussion). Thus, although freshwater fish originally evolved in the ocean, they moved to the terrestrial freshwater ecosystem where conversion of ALA to DHA may be more necessary, whereas marine fish remained in an environment where such conversion is less advantageous.

9.4.2 Molecular Biology of Fatty Acyl Desaturases and Elongases

The biochemistry of HUFA synthesis, including pathways and reaction mechanisms, has been well described for a number of years but, until recently, little was known of the genes and gene products involved and of the factors affecting their expression and function(s), respectively. Significant progress has now been made in characterising the desaturases and elongases involved in HUFA synthesis in animals (Tocher et al. 1998). Full-length cDNAs for $\Delta 6$ desaturases have been isolated from the nematode worm *Caenorhabditis elegans* (Napier et al. 1998), rat (Aki et al. 1999), mouse and human (Cho et al. 1999a). Fatty acyl $\Delta 5$ desaturase genes have been isolated from *C. elegans* (Michaelson et al. 1998; Watts and Browse 1999) and human (Cho et al. 1999b; Leonard et al. 2000). Genes involved in the elongation of PUFA have been cloned and characterised from *C. elegans* (Beaudoin et al. 2000), human (Leonard et al. 2000), mouse (Leonard et al. 2002) and rat (Inagaki et al. 2002).

Desaturase cDNAs from fish were first cloned from zebrafish and rainbow trout (Hastings et al. 2001; Seiliez et al. 2001). Functional characterisation of the zebrafish desaturase cDNA by heterologous expression in the yeast *Saccharomyces cerevisiae*, which lacks the ability to synthesise HUFA, showed the enzyme to have both $\Delta 6$ and $\Delta 5$ desaturase activity. Desaturases cloned from other freshwater fish, including rainbow trout and common carp (*Cyprinus carpio*), were shown to be unifunctional $\Delta 6$ desaturases (Zheng et al. 2004a). In contrast, separate cDNAs for $\Delta 6$ and $\Delta 5$ desaturases have been cloned from Atlantic salmon (Hastings et al. 2004; Zheng et al. 2005a), whereas only $\Delta 6$ desaturase cDNAs have been cloned from marine fish, including gilthead sea bream (*Sparus aurata*), turbot (*Psetta maximus*) and cod (*Gadus morhua*) (Seiliez et al. 2003; Zheng et al. 2004a; Tocher et al. 2006). Thus, the zebrafish desaturase is unique, being the only bifunctional desaturase involved in HUFA synthesis so far isolated, not only in fish, but, in general, in vertebrates studied to date. Similar to human and rat Δ6 desaturases, zebrafish desaturase also showed a low level of activity towards C_{24} fatty acids indicating that it could be responsible for all three desaturation steps required for the production of DHA. All of the fish desaturases were more active towards the n-3 fatty acid than the equivalent n-6 substrate (Table 9.1).

The fish desaturase cDNAs encode proteins of between 444 and 454 amino acids with the zebrafish and carp desaturase cDNAs encoding proteins of 444 amino acids similar to mammalian desaturases, whereas marine fish desaturase proteins contain 1–3 additional amino acids and salmonids, an additional 8–10 amino acids. The protein sequences of fish desaturases possess all the characteristic features of microsomal fatty acid desaturases, including three histidine boxes, two transmembrane regions and an N-terminal cytochrome b_5 domain containing the haem-binding motif, HPGG. Thus, the fish desaturases are fusion proteins presumably containing both desaturase and cytochrome b_5 functions. The phylogenetic sequence analyses grouped the fish desaturases largely as expected based on classical phylogeny with the carp and zebrafish (Ostariophysi; cyprinids), trout and salmon (Salmoniformes; salmonidae), and tilapia, sea bream and turbot (Acanthopterygia; cichlids, perciformes and pleuronectiformes) appearing in three distinct clusters with the cod (Paracanthopterygii; Gadiformes) branching from the Acanthopterygia line (Fig. 9.5) (Nelson 1994). The function of Δ6 desaturase in species that do not readily convert ALA to EPA is not known. In evolutionary terms the fact that a functional Δ6 desaturase gene has been retained implies a biologically meaningful function. As the same Δ6 desaturase likely also operates in the pathway from EPA to DHA, it is possible that a functional Δ6 has been retained to enable manipulation and 'fine tuning' of membrane EPA/DHA ratios in fish, such as marine species, that can obtain both fatty acids from the diet, but not necessarily always at optimal ratios.

Although the Δ6 desaturases cloned from marine fish show similar activities in the yeast expression system to the Δ6 desaturases cloned from salmonids and freshwater fish (Zheng et al. 2004a), the Δ6 activities measured in hepatocytes and enterocytes from cod and sea bass are very low (Mourente et al. 2005; Tocher et al. 2006). Gene expression studies investigating tissue distributions may offer an explanation. In Atlantic salmon, the enzymes of HUFA synthesis, Δ6 and Δ5 desaturases and PUFA elongase, are all expressed most highly in liver > intestine > brain, with much lower levels in other tissues (Zheng et al. 2005a). In contrast, in cod, the highest expression of Δ6 desaturase and elongase was in the brain, where expression exceeded that in liver by over 50-fold (Tocher et al. 2006). Thus, the major role for Δ6 desaturase in marine fish may be the maintenance of optimal brain and neural tissue DHA levels by conversion of dietary EPA.

Elongases of HUFA biosynthesis have been cloned from a number of fish species including the freshwater fish zebrafish, carp and tilapia, the salmonids, rainbow trout and anadromous Atlantic salmon, and marine fish, cod, turbot and sea bream (Hastings et al. 2004; Agaba et al. 2004, 2005). The elongase cDNAs encode

Table 9.1 Functional characterisation of fish fatty acyl desaturase cDNAs

	Zebrafish Δ6/Δ5		Carp Δ6		Rainbow trout Δ6		Atlantic salmon Δ5		Atlantic salmon Δ6		Sea bream Δ6		Turbot Δ6		Atlantic cod Δ6	
	n-3	n-6	n-3	n-6	n-3	n-6	n-3	n-6	n-3	n-6	n-3	n-6	n-3	n-6	n-3	n-6
Δ6	29.4	11.7	7.1	1.5	31.5	3.6	0.6	0.4	60.1	14.4	23.1	12.2	59.5	31.2	59.5	31.2
Δ5	20.4	8.3	0.5	0.5	0.6	0.0	10.2	0.9	2.3	0.0	1.0	0.5	0.0	0.0	0.0	0.0
Δ4	0.0	0.0	0.0	0.0	nd	nd	0.0	0.0	0.0	0.0	nd	nd	nd	nd	nd	nd
Δ6*	~5	~2	nd	nd	nd	nd	0.0	0.0	tr	0.0	nd	nd	nd	nd	nd	nd

Results are presented as percentage of substrate fatty acid converted to desaturated product. Fatty acid substrates were Δ6 activity, 18:3n-3 and 18:2n-6; Δ5 activity, 20:4n-3 and 20:3n-6; Δ4 activity, 22:5n-3 and 22:4n-6; Δ6*activity, 24:5n-3 and 24:4n-6
nd not determined; *tr* trace

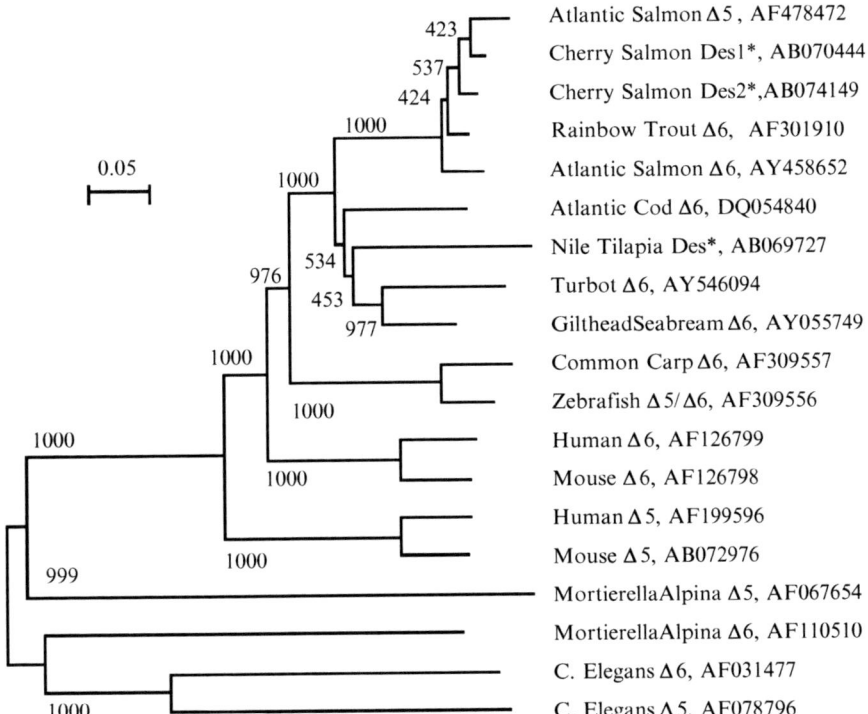

Fig. 9.5 Phylogenetic tree comparing amino acid sequences of fatty acyl desaturases from fish, mammals, fungus (*Mortierella alpina*) and nematode (*Caenorhabditis elegans*). The tree was constructed using the Neighbour Joining method using *CLUSTALX* and *NJPLOT* (Perrière and Gouy 1996). The horizontal branch length is proportional to amino acid substitution rate per site. The numbers represent the frequencies with which the tree topology presented here was replicated after 1,000 bootstrap iterations. Sequences marked with an asterisk are not functionally characterized

proteins of 288–294 amino acids that are highly conserved among the fish species. The predicted polypeptides included characteristic features of microsomal elongases, including a single histidine box redox centre motif, a canonical ER retention signal (carboxyl-terminal dilysine targeting signal), multiple transmembrane regions, and KXXEXXDT. QXXFLHXYHH (which contains the histidine box), NXXXHXXNYXYY and TXXQXXQ motifs, which are highly conserved in all PUFA elongases cloned to date (Meyer et al. 2004). Phylogenetic analysis comparing all the elongase sequences cloned from fish, along with a range of elongases from mammals, bird, insect, fungus and nematode, grouped the fish elongases into four clusters largely as expected based on the main groups of modern teleosts identified in classical phylogeny and similar to the grouping observed with desaturases, with the clusters being catfish and zebrafish, cod, trout and salmon, and tilapia, sea bream and turbot (Fig. 9.6). The fish elongases clustered most closely with the mammalian ELOVL5/elovl5 and ELOVL2/elovl2 elongases that have been functionally

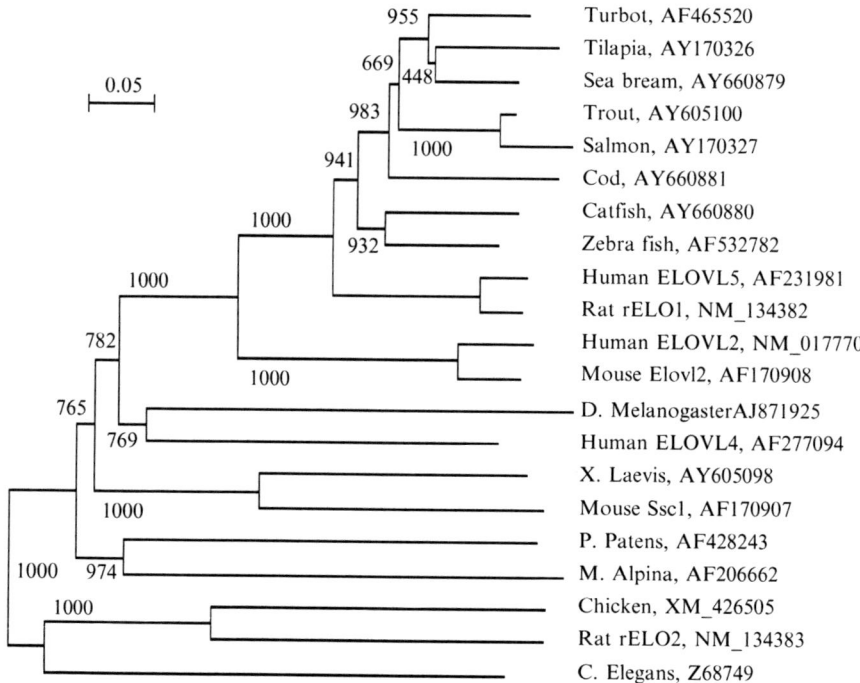

Fig. 9.6 Phylogenetic tree comparing amino acid sequences of fatty acyl elongases from fish with those of elongases from mammals, bird (chicken, *Gallus gallus*), amphibian (*Xenopus laevis*), insect (*Drosophila melanogaster*), fungus (*Mortierella alpina*), nematode (*Caenorhabditis elegans*) and plant (moss, *Physcomitrella patens*). Tree construction and description as in Fig. 9.5

characterised and shown to be PUFA elongases (Leonard et al. 2000, 2002). They were more distant from other elongases including rat rELO1 (also known as rat Elovl6), that has been shown to be predominantly a 16:0 to 18:0 elongase (Inagaki et al. 2002), and uncharacterised chicken (an elovl6 homologue), human ELOVL4 and mouse Sscl elongases that are presumed to be involved in sphingolipid metabolism and the synthesis of very long-chain fatty acids such as 26:0. Heterologous expression in yeast, *S. cerevisiae*, demonstrated that the zebrafish elongase had the ability to lengthen PUFA with chain lengths from C_{18} to C_{22} and also monounsaturated, but not saturated fatty acids. Most of the fish elongases showed a pattern of activity towards PUFA in the rank order $C_{18} > C_{20} > C_{22}$, although the tilapia and turbot elongases had similar activity towards 18:4n-3 and 20:5n-3 (Table 9.2). The fish elongases also generally showed greater or similar activities with n-3 than with n-6 homologues, with the exception of the cod enzyme, which was more active towards n-6 fatty acids. Thus, in fish, a single PUFA elongase gene is all that is required to perform all the elongations required for the full functioning of the HUFA synthesis pathway from ALA to DHA. Whether this will actually be the case can only be speculated when dealing with species with unsequenced genomes. However,

Table 9.2 Functional characterisation of fish fatty acyl elongase cDNAs

	Zebrafish		Catfish		Tilapia		Atlantic salmon		Sea bream		Turbot		Atlantic cod	
	n-3	n-6	n-3	n-6	n-3	n-6	n-3	n-6	n-3	n-6	n-3	n-6	n-3	n-6
C_{18-20}	85.4	70.7	87.5	78.3	82.6	71.2	71.8	42.1	85.9	84.3	49.0	22.8	7.4	15.7
C_{20-22}	49.4	25.6	48.1	40.8	86.6	61.8	38.9	22.6	66.8	43.9	39.3	21.3	0.8	1.5
C_{22-24}	4.9	0.5	2.8	0.9	1.0	tr	0.7	0.3	4.7	0.8	0.6	0.6	nd	nd

Results are presented as percentage of substrate fatty acid converted to elongated product. Fatty acid substrates were C_{18-20} elongase activity, 18:4n-3 and 18:3n-6; C_{20-22} elongase activity, 20:5n-3 and 20:4n-6; C_{22-24} elongase activity, 22:5n-3 and 22:4n-6
nd not detected; *tr* trace

for zebrafish, whose genome is largely sequenced, it does appear that only two genes, the bifunctional Δ6/Δ5 desaturase and the single PUFA elongase, are involved in the production of DHA from dietary ALA as no other PUFA desaturases or elongases appear to be present.

9.4.3 Regulation of Highly Unsaturated Fatty Acid Biosynthesis in Fish

Both Δ6 and Δ5 desaturase activities are reported to be under nutritional and/or endocrine control in mammals, although Δ6 desaturase is regarded as the rate-limiting step in the biosynthesis of ARA. Regulation of desaturase activity in mammals may involve transcriptional control of gene expression. Thus, the expression of Δ6 desaturase in liver was increased in mice fed triolein (18:1n-9), an EFA-deficient diet, compared to mice fed corn oil, a diet rich in LIN (Cho et al. 1999a). Similarly, the expression of both Δ6 and Δ5 desaturases was fourfold higher in rats fed a fat-free diet or a diet containing triolein compared to that in rats fed either safflower oil (LIN) or menhaden oil (n-3HUFA) (Cho et al. 1999b).

The expression of HUFA biosynthesis genes in fish has been investigated using a range of different techniques including Northern blotting, quantitative real-time polymerase chain reaction (Q-PCR) and cDNA microarrays. The expression of Δ5 desaturase and elongase genes in liver of salmon, as determined by Q-PCR, could be increased in a graded manner by increasing dietary linseed oil (Zheng et al. 2004b). Expression of both Δ5 desaturase and elongase genes was positively and negatively correlated with dietary ALA and n-3 HUFA, respectively. In salmon fed rapeseed oil (RO), the expression of the Δ5 desaturase and PUFA elongase in liver was determined using a cDNA mini microarray containing around 70 genes involved in lipid metabolism (Jordal et al. 2005). Of the genes in the array, the Δ5 desaturase gene showed the greatest degree of regulation and was one of only three genes that were up-regulated in fish fed RO compared to fish fed fish oil (FO). The up-regulation of the expression of the Δ5 desaturase gene in liver of fish fed RO was confirmed by Q-PCR. Although around 25 other genes were down-regulated in fish fed RO compared to fish fed FO, the expression of the PUFA elongase was unaffected by dietary oil. Recently, the expression of all the genes of HUFA synthesis in salmon fed FO and vegetable oil (VO) was investigated using an Atlantic salmon 17K cDNA microarray. Of the fifteen gene features that showed the strongest regulation, seven were due to fatty acyl desaturases including both Δ6 and Δ5 desaturase sequences included on the array as candidate genes along with three other ESTs that were annotated as Δ6/Δ5 desaturase. The expression of the PUFA elongase was not affected significantly by dietary VO diet in the Jordal et al. (2005) study.

The expression of both Δ6 and Δ5 desaturase genes and the PUFA elongase gene has also been determined by Q-PCR at various points during the entire 2-year production cycle in salmon fed diets containing either FO or a VO blend (Zheng et al.

2005b). Gene expression of Δ6 desaturase was highest around the point of seawater transfer and lowest during the seawater phase. In addition, the expression of both Δ6 and Δ5 desaturase genes was generally higher in fish fed VO compared to fish fed FO, particularly in the seawater phase. The results were consistent with HUFA biosynthesis activity, which peaked around seawater transfer, was lower in seawater, and higher in fish fed VO compared to fish fed FO. The expression of PUFA elongase was also investigated, but no up-regulation was observed. In all the studies of gene expression, the expression of the desaturases was increased by up to twofold in fish fed VO compared with fish fed FO (Zheng et al. 2004b, 2005a, b; Jordal et al. 2005). This compares well with biochemical studies of enzyme activities and HUFA synthesis, which show generally a twofold to threefold increase in activity in fish fed VO at the highest inclusion levels (see Tocher 2003). Taken together, these studies show that both nutritional and environmental modulation of HUFA biosynthesis in Atlantic salmon involves regulation of fatty acyl desaturase gene expression.

9.4.4 Use of Stable Isotopes for In Vivo Studies of HUFA Biosynthesis

Although DHA is a functional EFA in all vertebrates, including fish, there is little quantitative information available on rates of formation of DHA in any species or on the bioequivalence of ALA and DHA. The use of stable isotope tracers and GC–MS has allowed these problems to be addressed at the whole animal level, and has given vital information on nutritional and physiological requirements for PUFA at the organismal level which can be difficult to obtain from in vitro studies on tissues or cells where large extrapolations are necessary. The principle of using stable isotope tracers to investigate n-3PUFA metabolism in fish is straightforward though a number of practical difficulties have to be overcome in relation to diet preparation and presentation to the fish. Administering tracer via the diet is the most natural way of delivery and minimises stress due to handling which may alter lipid metabolism if fish are given tracer by injection.

In summary, the method is to add a known amount of labelled fatty acid to the diet, feed a known amount of diet to the fish, and determine quantitatively the fatty acid profile and tissue distribution of labelled product fatty acids at various times post-dose. The commercially available deuterated fatty acid, D_5-ALA, is an ideal tracer for this type of work. The mass increment of five is readily detected and quantitated by GC-negative chemical ionisation MS of the pentafluorobenzyl ester. This methodology was deployed in experiments with trout and gave information on rates of synthesis, which tissues are important in HUFA biosynthesis, changes during development, and tissue distribution of intermediate and product fatty acids. A time course of DHA synthesis from 3 to 35 days showed that synthesis was slow. Whole body accumulation of D_5-DHA was linear over the first 7 days corresponding to a rate of 0.54 ± 0.12 µg D_5-DHA g wet weight of fish^{-1} mg D_5-ALA consumed^{-1} day^{-1} (Bell et al. 2001b). Maximum accretion of D_5-DHA was 4.3 ± 1.2 µg g fish^{-1} mg

D_5-ALA consumed^{-1} after 14 days. The great majority of the D_5-tracer was catabolised with the combined recovery of D_5-ALA plus D_5-DHA being 2.6%. One of the most significant findings was that the concentration of DHA in the fish decreased during the 13-week period on the experimental diet and the amount of DHA synthesised from ALA was only 5% of that obtained directly from the fish meal in the diet (Bell et al. 2001b), even though DHA synthesis was fully induced under these conditions.

Experiments over a shorter time course showed that pyloric ceca were more active than liver in DHA synthesis, and that the great majority of D_5-ALA was catabolised very rapidly (Bell et al. 2003a). Subsequent studies with intestinal enterocytes confirmed directly that they are active in DHA synthesis (Tocher et al. 2002). Deuterated intermediate pathway fatty acids, including 24:5n-3 and 24:6n-3, were identified in liver, ceca, brain and eyes, but while D_5-DHA was the main product in liver and ceca, in neural tissue over a longer time course there was a build up of pentaene intermediates D_5-EPA and D_5-22:5n-3 (Bell et al. 2003b). The results showed that the kinetics of accumulation and depletion of the various n-3 PUFA differ among tissues. The presence of pathway intermediate fatty acids provided further evidence that liver and ceca possess the full metabolic pathway for synthesis of DHA, whereas brain and eyes are less active with an accumulation of pentaene intermediate fatty acids.

During these studies it was noted that smaller fish within each cohort always gave the highest rate of DHA synthesis. A study was undertaken to measure the rate of DHA synthesis in trout from first feeding up to 10 g weight in fish fed either FO or a VO diet consisting of soybean lecithin, linseed oil and high oleic acid sunflower oil providing sufficient LIN and ALA to satisfy EFA requirements (Bell and Dick 2004). It was found that trout fed a VO diet from first feeding (~0.2 g weight) gave an initial rate of 5.4 µg D_5-DHA g fish^{-1} mg D_5-ALA consumed^{-1} 7 days^{-1} which increased rapidly to a peak activity of ~50 µg D_5-DHA g fish^{-1} mg D_5-ALA consumed^{-1} 7 days^{-1} at around 1 g weight then declined rapidly to ~12 µg D_5-DHA g fish^{-1} mg D_5-ALA consumed^{-1} 7 days^{-1} at 2 g weight and continued to fall thereafter. Fish fed a FO diet showed the same pattern, but DHA synthesis was repressed and the rate was approximately ten times lower. Fish fed the VO diet were unable to maintain their DHA concentration, the tissue content falling from 6.2 mg DHA g fish^{-1} to 2.4 by ~5 g weight (Bell and Dick 2004). These results have important implications for the replacement of FO and fish meal in diets by VO.

9.5 Concluding Remarks

Aquatic ecosystems produce the majority of long-chain n-3HUFA, and therefore studies of the molecular mechanisms involved in their biosynthesis have not been conducted for purely scientific interest alone. Such studies are of considerable importance to human health, as there is simply not enough n-3HUFA available to meet the human population's dietary requirements (see Chap. 8). Man evolved with a diet having an n-6:n-3 PUFA ratio of about 1-2:1, but this ratio is now as high as 15–25:1 in the west (Simopoulos 2000). This imbalance is implicated in many pathological conditions, and it is widely accepted that we must reduce the n-6:n-3 ratio in our diet (see Chap. 14), with n-3HUFA being more effective than ALA in balancing excess n-6PUFA (Simopoulos 2000). Fish is the major source of n-3HUFA in our diet but, with a burgeoning human population and declining capture fisheries, we will be increasingly dependent on aquaculture for our supplies of fish and n-3HUFA. The paradox is that, traditionally, aquaculture feeds have used marine raw ingredients, fish meal and oil, themselves rendered from reduction fisheries (Arts et al. 2001). VOs may be sustainable alternatives, but they do not contain n-3HUFA, leaving ALA as the only source of n-3PUFA in plant-based diets (Sargent et al. 2002). Hence the considerable interest in determining pathways for the synthesis of n-3HUFA from ALA in fish with the aim being to develop diets to optimally switch on the genes necessary for EPA and DHA production (Tocher 2003). Transgenics are also an option, and a study has already shown that overexpression of Δ6 activity in zebrafish, transformed with a salmon desaturase, led to modestly increased tissue levels of EPA and DHA (Alimuddin et al. 2005). The transgenic approach may be the only option in marine fish with an apparently incomplete HUFA synthesis pathway. Even if successful, these approaches alone cannot solve our n-3HUFA supply problems as VOs rich in ALA, such as perilla, linseed, camelina and hemp, are currently limiting, with LIN supply in total global fat and oil production exceeding that of ALA by about 24-fold.

In the wild, fish naturally obtain n-3HUFA from the food chain, with microalgae and other unicellular organisms being the major primary producers. Therefore, this is also where we must look for provision of the n-3HUFA in the future. The cloning of genes from marine microalgae and the discovery of the role of the PKS pathway-like PUFA synthase are perhaps the truly exciting advances in this area. However, the marine heterotrophs, *Crypthecodinium cohnii*, a dinoflagellate, and *Schizochytrium sp.*, thraustochytrids, are currently being used to produce DHA for supplementing infant formulae and short shelf-life food products, respectively, through traditional fermentor technology. However, whether fermentation and direct algal culture can be a solution is debatable, and many see the production of EPA and DHA in transgenic oil seed crops as more promising. The molecular tools, the biosynthetic genes, are becoming available and initial transgenic trials have demonstrated the viability of the approach (Abbadi et al. 2004; Robert et al. 2005). However, fluidity and oxidation issues remain significant scientific problems in developing oil seeds capable of storing high levels of HUFA, with consumer acceptance of transgenics a further hurdle. Providing these problems can be overcome, our view is that fish will remain a major source of n-3HUFA in the human food basket, but that fish will be farmed using diets containing transgenic seed oils. In conclusion, studies of the molecular mechanisms of HUFA synthesis in aquatic environments have produced not only a great deal of scientifically highly interesting information, but also a suite of potentially very valuable molecular tools in the form of genes involved in HUFA biosynthesis.

References

Abbadi, A., Domergue, F., Bauer, J., Napier, J.A., Welti, R., Zahringer, U., Cirpus, P., and Heinz, E. 2004. Biosynthesis of very-long-chain polyunsaturated fatty acids in transgenic oilseeds: constraints on their accumulation. Plant Cell. 16:1–15

Agaba, M., Tocher, D.R., Dickson, C., Dick, J.R., and Teale, A.J. 2004. A zebrafish cDNA encoding a multifunctional enzyme involved in the elongation of polyunsaturated, monounsaturated and saturated fatty acids. Mar. Biotechnol. 6:251–261

Agaba, M.K., Tocher, D.R., Dickson, C.A., Zheng, X., Dick, J.R., and Teale, A.J. 2005. Cloning and functional characterisation of polyunsaturated fatty acid elongases from marine and freshwater teleost fish. Comp. Biochem. Physiol. 142B:342–352

Aki, T., Shimada, Y., Inagaki, K., Higashimoto, H., Kawamoto, S., Shiget, S., Ono, K., and Suzuki, O. 1999. Molecular cloning and functional characterisation of rat $\Delta 6$ fatty acid desaturase. Biochem. Biophys. Res. Commun. 255:575–579

Alimuddin, Y.G., Kiron, V., Satoh, S., and Takeuchi, T. 2005. Enhancement of EPA and DHA biosynthesis by over-expression of masu salmon delta-6-desaturase-like gene in zebrafish. Transgenic Res. 14:159–165

Arts, M.T., Ackman, R.G., and Holub, B.J. 2001. "Essential fatty acids" in aquatic ecosystems: a crucial link between diet and human health and evolution. Can. J. Fish. Aquat. Sci. 58:122–137

Barclay, W.R., Meager, K.M., and Abril, J.R. 1994. Heterotrophic production of long chain omega-3 fatty acids utilizing algae and algae-like microorganisms. J. Appl. Phycol. 6:123–129

Beaudoin, F., Michaelson, L.V., Lewis, M.J., Shewry, P.R., Sayanova, O., and Napier, J.A. 2000. Production of C20 polyunsaturated fatty acids (PUFAs) by pathway engineering: identification of a PUFA elongase component from *Caenorhabditis elegans*. Biochem. Soc. Trans. 28:661–663

Bell, M.V., and Dick, J.R. 2004. Changes in capacity to synthesise 22:6n-3 during early development in rainbow trout (*Oncorhynchus mykiss*). Aquaculture 235:393–409

Bell, M.V., and Pond, D.W. 1996. Lipid composition during growth of motile and coccolith forms of *Emiliania huxleyi*. Phytochemistry 41:465–471

Bell, M.V., Dick, J.R., and Pond, D.W. 1997. Octadecapentaenoic acid in a raphidophyte alga, *Heterosigma akashiwo*. Phytochemistry 45:303–306

Bell, M.V., Dick, J.R., and Kelly, M.S. 2001a. Biosynthesis of eicosapentaenoic acid in the sea urchin *Psammechinus miliaris*. Lipids 36:79–82

Bell, M.V., Dick, J.R., and Porter, A.E.A. 2001b. Biosynthesis and tissue deposition of docosahexaenoic acid (22:6n-3) in rainbow trout (*Oncorhynchus mykiss*). Lipids 36:1153–1159

Bell, M.V., Dick, J.R., and Porter, A.E.A. 2003a. Pyloric ceca are a major site of 22:6n-3 synthesis in rainbow trout (*Oncorhynchus mykiss*). Lipids 38:39–44

Bell, M.V., Dick, J.R., and Porter, A.E.A. 2003b. Tissue deposition of n-3 FA pathway intermediates in the synthesis of DHA in rainbow trout (*Oncorhynchus mykiss*). Lipids 38:925–931

Bell, M.V., Dick, J.R., Anderson, T.R., and Pond, D.W. 2007. Application of liposome and stable isotope tracer techniques to study polyunsaturated fatty acid biosynthesis in marine zooplankton. J. Plankton Res. 29:417–422

Bowman, J.P., Gosink, J.J., McCammon, S.A., Lewis, T.E., Nichols, D.S., Nichols, P.D., Skerratt, J.H., Staley, J.T., and McMeekin, T.A. 1998. *Colwellia demingiae* sp. nov., *Colwellia hornerae* sp. nov., *Colwellia rossensis* sp. nov. and *Colwellia psychrotropica* sp. nov.: psychrophilic Antarctic species with the ability to synthesize docosahexaenoic acid (22:6ω3). Int. J. Syst. Bacteriol. 48:1171–1180

Buzzi, M., Henderson, R.J., and Sargent, J.R. 1996. The desaturation and elongation of linolenic acid and eicosapentaenoic acid by hepatocytes and liver microsomes from rainbow trout (*Oncorhyncus mykiss*) fed diets containing fish oil or olive oil. Biochim. Biophys. Acta 1299:235–244

Buzzi, M., Henderson, R.J., and Sargent, J.R. 1997. Biosynthesis of docosahexaenoic acid in trout hepatocytes proceeds via 24-carbon intermediates. Comp. Biochem. Physiol. 116:263–267

Cho, H.P., Nakamura, M.T., and Clarke, S.D. 1999a. Cloning, expression and nutritional regulation of the human Δ6 desaturase. J. Biol. Chem. 274:471–477

Cho, H.P., Nakamura, M.T., and Clarke, S.D. 1999b. Cloning, expression and nutritional regulation of the human Δ5 desaturase. J. Biol. Chem. 274:37335–37339

Chu, F.L.E., Lund, E., Soudant, P., and Harvey, E. 2002. De novo arachidonic acid synthesis in *Perkinsus marinus*, a protozoan parasite of the eastern oyster *Crassostrea virginica*. Mol. Biochem. Parasitol. 119:179–190

Cripps, C., Blomquist, G.J., and de Renobales, M. 1986. De novo synthesis of linoleic acid in insects. Biochim. Biophys. Acta 876:572–580

D'Andrea, S., Guillou, H., Jan, S., Catheline, D., Thibault, J.-N., Bouriel, M., Rioux, V., and Legrand, P. 2002. The same rat Δ6-desaturase not only acts on 18- but also on 24-carbon fatty acids in very-long-chain polyunsaturated fatty acid biosynthesis. Biochem. J. 364:49–55

Dawczynski, C., Schibert, R., and Jahreis, G. 2007. Amino acids, fatty acids, and dietary fibre in edible seaweed products. Food Chem. 103:891–899

De Antueno, R.J., Knickle, L.C., Smith, H., Elliot, M.L., Allen, S.J., Nwaka, S., and Winther, M.D. 2001. Activity of human Δ5 and Δ6 desaturases on multiple n-3 and n-6 polyunsaturated fatty acids. FEBS Lett. 509:77–80

Desvilettes, C., Bourdier, G., and Breton, J.C. 1997. On the occurrence of a possible bioconversion of linolenic acid into docosahexaenoic acid by the copepod *Eucyclops serrulatus* fed on microalgae. J. Plankton Res. 19:273–278

Dunstan, G.A., Volkman, J.K., Jeffrey, S.W., and Barrett, S.M. 1992. Biochemical composition of microalgae from the green algal classes Chlorophyceae and Prasinophyceae. 2. Lipid classes and fatty acids. J. Exp. Mar. Biol. Ecol. 161:115–134

Ghioni, C., Tocher, D.R., Bell, M.V., Dick, J.R., and Sargent, J.R. 1999. Low C_{18} to C_{20} fatty acid elongase activity and limited conversion of stearidonic acid, 18:4n-3, to eicosapentaenoic acid, 20:5n-3, in a cell line from the turbot, *Scophthalmus maximus*. Biochim. Biophys. Acta 1437:170–181

Hastings, N., Agaba, M., Tocher, D.R., Leaver, M.J., Dick, J.R., Sargent, J.R., and Teale, A.J. 2001. A vertebrate fatty acid desaturase with Δ5 and Δ6 activities. Proc. Natl. Acad. Sci. U.S.A. 98:14304–14309

Hastings, N., Agaba, M.K., Tocher, D.R., Zheng, X., Dickson, C.A., Dick, J.R., and Teale, A.J. 2004. Molecular cloning and functional characterization of fatty acyl desaturase and elongase cDNAs involved in the production of eicosapentaenoic and docosahexaenoic acids from α-linolenic acid in Atlantic salmon (*Salmo salar*). Mar. Biotechnol. 6:463–474

Hauvermale, A., Kuner, J., Rosenzweig, B., Guerra, D., Diltz, S., and Metz, J.G. 2006. Fatty acid production in *Schizochytrium* sp.: involvement of a polyunsaturated fatty acid synthase and a type 1 fatty acid synthase. Lipids 41:739–747

Henderson, R.J., and Mackinlay, E.E. 1991. Polyunsaturated fatty acid metabolism in the marine dinoflagellate *Crypthecodinium cohnii*. Phytochemistry 30:1781–1787

Inagaki, K., Aki, T., Fukuda, Y., Kawamoto, S., Shigeta, S., Ono, K., and Suzuki, O. 2002. Identification and expression of a rat fatty acid elongase involved the biosynthesis of C18 fatty acids. Biosci. Biotechnol. Biochem. 66:613–621

Johns, R.B., and Perry, G.J. 1977. Lipids of the bacterium *Flexibacter polymorphus*. Arch. Microbiol. 114: 267–271

Jordal, A.-E.O., Torstensen, B.E., Tsoi, S., Tocher, D.R., Lall, S.P., and Douglas, S. 2005. Profiling of genes involved in hepatic lipid metabolism in Atlantic salmon (*Salmo salar* L.) – Effect of dietary rapeseed oil replacement. J. Nutr. 135:2355–2361

Joseph, J.D. 1975. Identification of 3, 6, 9, 12, 15-octadecapentaenoic acid in laboratory-cultured photosynthetic dinoflagellates. Lipids 10: 395–403

Jøstensen, J.P., and Landfald, B. 1997. High prevalence of polyunsaturated-fatty-acid producing bacteria in Arctic invertebrates. FEMS Microbiol. Lett. 151: 95–101

Leonard, A.E., Bobik, E.G., Dorado, J., Kroeger, P.E., Chuang, L.-T., Thurmond, J.M., Parker-Barnes, J.M., Das, T., Huang, Y.-S., and Murkerji, P. 2000. Cloning of a human cDNA encoding a novel enzyme involved in the elongation of long chain polyunsaturated fatty acids. Biochem. J. 350:765–770

Leonard, A.E., Kelder, B., Bobik, E.G., Chuang, L.-T., Lewis, C.J., Kopchick, J.J., Murkerji, P., and Huang, Y.-S. 2002. Identification and expression of mammalian long-chain PUFA elongation enzymes. Lipids 37:733–740

Lubzens, E., Marko, A., and Tietz, A. 1985. De novo synthesis of fatty acids in the rotifer, *Brachionus plicatilis*. Aquaculture 47: 27–37

Mansour, M.P., Volkman, J.K., Holdsworth, D.G., Jackson, A.E., and Blackburn, S.I. 1999. Very-long chain (C_{28}) highly unsaturated fatty acids in marine dinoflagellates. Phytochemistry 50:541–548

Metz, J.G., Roessler, P., Facciotti, D., Levering, C., Dittrich, F., Lassner, M., Valentine, R., Lardizabel, K., Domergue, F., Yamada, A., Yazawa, K., Knauf, V., and Browse, J. 2001. Production of polyunsaturated fatty acids by polyketide synthases in both prokaryotes and eukaryotes. Science 293:290–293

Meyer, A., Cirpus, P., Ott, C., Schlecker, R., Zahringer, U., and Heinz, E. 2003. Biosynthesis of docosahexaenoic acid in *Euglena gracilis*: biochemical and molecular evidence for the involvement of a Δ4-fatty acyl group desaturase. Biochemistry 42:9779–9788

Meyer, A., Kirsch, H., Domergue, F., Abbadi, A., Sperling, P., Bauer, J., Cirpus, P., Zank, T.K., Moreau, H., Roscoe, T.J., Zähringer, U., and Heinz, E. 2004. Novel fatty acid elongases and their use for the reconstitution of docosahexaenoic acid biosynthesis. J. Lipid Res. 45:1899–1909

Michaelson, L.V., Napier, J.A., Lewis, M., Griffiths, G., Lazarus, C.M., and Stobart, A.K. 1998. Functional identification of a fatty acid Δ5 desaturase gene from *Caenorhabditis elegans*. FEBS Lett. 439:215–218

Mourente, G., Dick, J.R., Bell, J.G., and Tocher, D.R. 2005. Effect of partial substitution of dietary fish oil by vegetable oils on desaturation and oxidation of [1-14C] 18:3n-3 and [1-^{14}C]20:5n-3 in hepatocytes and enterocytes of European sea bass (*Dicentrarchus labrax* L.). Aquaculture 248:173–186

Nanton, D.A., and Castell, J.D. 1998. The effects of dietary fatty acids on the fatty acid composition of the harpacticoid copepod, *Tisbe* sp., for use in live food for marine fish larvae. Aquaculture 163:251–261

Napier, J.A., Hey, S.J., Lacey, D.J., and Shewry, P. 1998. Identification of a *Caenorhabditis elegans* Δ6 fatty acid – desaturase by heterologous expression in *Saccharomyces cerevisiae*. Biochem. J. 330:611–614

Nelson, J.S. 1994. Fishes of the World. 3rd edition. New York: John Wiley

Nichols, P.D., Jones, G.J., de Leeuw, J.W., and Johns, R.B. 1984. The fatty acid and sterol composition of two marine dinoflagellates. Phytochemistry 23:1043–1047

Nichols, P.D., Volkman, J.K., Hallegraeff, G.M., and Blackburn, S.I. 1987. Sterols and fatty acids of the red tide flagellates *Heterosigma akashiwo* and *Chattonella antiqua* (Raphidophyceae). Phytochemistry. 26:2537–2541

O'Hagan, D. 1995. Biosynthesis of fatty acid and polyketide metabolites. Nat Prod Rep. 12:1–32

Okuyama, H., Kogame, K., and Takeda, S. 1993. Phylogenetic significance of the limited distribution of octadecapentaenoic acid in prymnesiophytes and photosynthetic dinoflagellates. Proc. NIPR Symp. Polar Biol. 6:21–26

Perrière, G., and Gouy, M. 1996. WWW-Query: an on-line retrieval system for biological sequence banks. Biochimie. 78:364–369

Pond, D.W., and Harris, R.P. 1996. The lipid composition of the coccolithophore *Emiliania huxleyi* and its possible ecophysiological significance. J. Mar. Biol. Assoc. U.K. 76:579–594

Qiu, X., Hong, H.P., and Mackenzie, S.L. 2001. Identification of a Δ4 fatty acid desaturase from *Thraustochytrium* sp. involved in biosynthesis of docosahexaenoic acid by heterologous expression in *Saccharomyces cerevisiae* and *Brassica juncea*. J. Biol. Chem. 276:31561–31566

Robert, S.S., Singh, S.P., Zhou, X.R., Petrie, J.R., Blackburn, S.I., Mansour, P.M., Nichols, P.D., Liu, Q., and Green, A.G. 2005. Metabolic engineering of *Arabidopsis* to produce nutritionally important DHA in seed oil. Funct. Plant Biol. 32:473–479

Russell, N.J., and Nichols, D.S. 1999. Polyunsaturated fatty acids in marine bacteria – a dogma rewritten. Microbiology 145:765–779

Sanchez-Machado, D.I., Lopez-Cervantes, J., Lopez-Hernandez, J., and Paseiro-Losada, P. 2004. Fatty acids, total lipid, protein and ash contents of processed edible seaweeds. Food Chem. 85:439–444

Sargent, J.R., Tocher, D.R., and Bell, J.G. 2002. The Lipids, pp. 181–257, In J.E. Halver and R.W. Hardy (eds.), Fish Nutrition, 3rd edition. Academic Press, San Diego

Schlechtriem, C., Arts, M.T., and Zellmer, I.D. 2006. Effect of temperature on the fatty acid composition and temporal trajectories of fatty acids in fasting *Daphnia pulex* (Crustacea, Cladocera). Lipids 41:397–400

Seiliez, I., Panserat, S., Corraze, G., Kaushik, S., and Bergot, P. 2003. Cloning and nutritional regulation of a Δ6-desaturase-like enzyme in the marine teleost gilthead seabream (*Sparus aurata*). Comp. Biochem. Physiol. 135B:449–460

Seiliez, I., Panserat, S., Kaushik, S., and Bergot, P. 2001. Cloning, tissue distribution and nutritional regulation of a D6-desaturase-like enzyme in rainbow trout. Comp. Biochem. Physiol. 130B:83–93

Shimizu, Y. 1996. Microalgal metabolites:a new perspective. Annu. Rev. Microbiol. 50:431–465

Simopoulos, A.P. 2000. Human requirement for n-3 polyunsaturated fatty acids. Poult. Sci. 79:961–970

Sprecher, H. 2000. Metabolism of highly unsaturated n-3 and n-6 fatty acids. Biochim. Biophys. Acta 1486:219–231

Tocher, D.R. 2003. Metabolism and functions of lipids and fatty acids in teleost fish. Rev. Fisheries Sci. 11:107–184

Tocher, D.R., and Ghioni, C. 1999. Fatty acid metabolism in marine fish: low activity of Δ5 desaturation in gilthead sea bream (*Sparus aurata*) cells. Lipids 34:433–440

Tocher, D.R., and Sargent, J.R. 1990. Effect of temperature on the incorporation into phospholipid classes and the metabolism via desaturation and elongation of (n-3) and (n-6) polyunsaturated fatty acids in fish cells in culture. Lipids 25:435–442

Tocher, D.R., Carr, J., and Sargent, J.R. 1989. Polyunsaturated fatty acid metabolism in cultured cell lines: differential metabolism of (n-3) and (n-6) series acids by cultured cells originating from a freshwater teleost fish and from a marine teleost fish. Comp. Biochem. Physiol. 94B:367–374

Tocher, D.R., Bell, J.G., Dick, J.R., and Sargent, J.R. 1997. Fatty acid desaturation in isolated hepatocytes from Atlantic salmon (*Salmo salar*): stimulation by dietary borage oil containing γ-linolenic acid. Lipids 32:1237–1247

Tocher, D.R., Leaver, M.J., and Hodgson, P.A. 1998. Recent advances in the biochemistry and molecular biology of fatty acyl desaturases. Prog. Lipid Res. 37:73–117

Tocher, D.R., Bell, J.G., MacGlaughlin, P., McGhee, F., and Dick, J.R. 2001a. Hepatocyte fatty acid desaturation and polyunsaturated fatty acid composition of liver in salmonids: effects of dietary vegetable oil. Comp. Biochem. Physiol. 130B:257–270

Tocher, D.R., Agaba, M., Hastings, N., Bell, J.G., Dick, J.R., and Teale, A.J. 2001b. Nutritional regulation of hepatocyte fatty acid desaturation and polyunsaturated fatty acid composition in zebrafish (*Danio rerio*) and tilapia (*Oreochromis nilotica*). Fish Physiol. Biochem. 24:309–320

Tocher, D.R., Fonseca-Madrigal, J., Bell, J.G., Dick, J.R., Henderson, R.J., and Sargent, J.R. 2002. Effects of diets containing linseed oil on fatty acid desaturation and oxidation in hepatocytes and intestinal enterocytes in Atlantic salmon (*Salmo salar*). Fish Physiol. Biochem. 26:157–170

Tocher, D.R., Zheng, X., Schlechtriem, C., Hastings, N., Dick, J.R., and Teale, A.J. (2006). Highly unsaturated fatty acid synthesis in marine fish; cloning, functional characterisation and nutritional regulation of fatty acid Δ6 desaturase Atlantic cod (*Gadus morhua* L.). Lipids 42:1003–1016

Tonon, T., Sayanova, O., Michaelson, L.V., Qing, R., Harvey, D., Larson, T.R., Li, Y., Napier, J.A., and Graham, I.A. 2005. Fatty acid desaturases from the microalga *Thalassiosira pseudonana*. FEBS J. 272:3401–3412

Venegas-Calerón, M., Beaudoin, F., Sayanova, O., and Napier, J.A. 2007. Co-transcribed genes for long chain polyunsaturated fatty acid biosynthesis in the protozoon *Perkinsus marinus* include a plant-like FAE1 3-ketoacyl coenzyme A synthase. J. Biol. Chem. 282:2996–3003

Volkman, J.K., Smith, D.J., Eglington, G., Forsberg, T.E.V., and Corner, E.D.S. 1981. Sterol and fatty acid composition of four marine haptophycean algae. J. Mar. Biol. Assoc. U.K. 61:509–527

Watts, J.L., and Browse, J. 1999. Isolation and characterisation of a Δ5 fatty acid desaturase from *Caenorhabditis elegans*. Arch. Biochem. Biophys. 362:175–182

Yano, Y., Nakayama, A., Saito, H., and Ishihara, K. 1994. Production of docosahexaenoic acid by marine bacteria isolated from deep sea fish. Lipids 29:527–528

Yazawa, K. 1996. Production of eicosapentaenoic acid from marine bacteria. Lipids 31:S297–S300

Zheng, X., Seiliez, I., Hastings, N., Tocher, D.R., Panserat, S., Dickson, C.A., Bergot, P., Teale, A.J. 2004a. Characterisation and comparison of fatty acyl Δ6 desaturase cDNAs from freshwater and marine teleost fish species. Comp. Biochem. Physiol. 139B:269–279

Zheng, X., Tocher, D.R., Dickson, C.A., Bell, J.G., and Teale, A.J. 2004b. Effects of diets containing vegetable oil on expression of genes involved in polyunsaturated fatty acid biosynthesis in liver of Atlantic salmon (*Salmo salar*). Aquaculture 236:467–483

Zheng, X., Tocher, D.R., Dickson, C.A., Dick, J.R., Bell, J.G., and Teale, A.J. 2005a. Highly unsaturated fatty acid synthesis in vertebrates: new insights with the cloning and characterisation of a Δ6 desaturase of Atlantic salmon. Lipids 40:13–24

Zheng, X., Torstensen, B.E., Tocher, D.R., Dick, J.R., Henderson, R.J., and Bell, J.G. 2005b. Environmental and dietary influences on highly unsaturated fatty acid biosynthesis and expression of fatty acyl desaturase and elongase genes in liver of Atlantic salmon (*Salmo salar*). Biochim. Biophys. Acta 1734:13–24

Chapter 10
Health and Condition in Fish: The Influence of Lipids on Membrane Competency and Immune Response

Michael T. Arts and Christopher C. Kohler

10.1 The Influence of Lipids on Health and Condition

Traditionally fisheries biologists have used various metrics to indicate the condition and, by implication, health of fish. These indices are usually based on relationships between length and weight (Anderson and Neumann 1996). Although such metrics can, under some circumstances, provide a quick estimate of a fish's condition, their ability to shed light on the underlying cause-and-effect relationship(s) governing a fish's health and nutritional status are limited. Biochemical measures (e.g. lipids including fatty acids (FA) and sterols, proteins and their constituent amino acids, and trace elements) offer complimentary measures to assess, in a more specific way, the condition and underlying health of fish. Fatty acids and other lipids affect the health of fish in many ways; including, but not limited to, their effects on growth, reproduction, behavior, vision, osmoregularity, membrane fluidity (thermal adaptation), and immune response. In this review, we focus on the latter two roles that lipids play in mediating the health and condition of fish.

10.2 The Influence of Lipids on Membrane Fluidity and Other Membrane Properties

10.2.1 Homeoviscous Adaptation

Aquatic organisms are exposed to varying and sometimes extreme environmental conditions (e.g., marked changes in temperature) that can induce strong and often debilitating effects on their physiology. Fish in temperate regions and at high altitudes must adapt to changing temperatures throughout the year. Behavioral and physiological

---

M.T. Arts (✉) and C.C. Kohler
Aquatic Ecosystems Management Research Division, National Water Research Institute - Environment Canada, P.O. Box 5050, 867 Lakeshore Road, Burlington, OntarioCanada, L7R 4A6
e-mail: Michael.Arts@ec.gc.ca

adaptations provide an effective response to such stressors. Although in some circumstances, fish may be able to adapt behaviorally (e.g. by moving to warmer or colder water), biochemical and physiological adaptations, especially at the level of the cell membrane, provide the most specific, enduring, response to sustained changes in temperature. In effect, temperature can be regarded as a stressor to which cells must respond to establish a new equilibrium between their environment and the physiochemical properties of their membranous structures; a response termed "homeoviscous adaptation" by Sinensky (1974).

Fatty acids play important structural and functional roles in cell membranes. Although the composition of the cell membrane is specific to each cell type, all membranes must maintain an adequate level of "fluidity," particularly at colder temperatures (Fig. 10.1). The first targets of temperature changes are the biomembranes, which exist in a liquid-crystalline state at body temperatures (Singer and Nicholson 1972). In their free form, highly unsaturated fatty acids (HUFA; FA with ≥20 carbons and ≥3 double bonds; a subset of PUFA; FA with ≥2 double bonds) have a very low melting point (approaching −50°C) and thus have a much greater tendency to remain fluid in situ than do saturated FA (SAFA), which have much higher melting points (ranging from +58°C to +77°C for myristic (12:0) and arachidic (18:0) acid, respectively). Thus, the relative proportions of fatty acids matters a great deal because the lipid bilayer of cell membranes requires a certain degree of rigidity but must also be sufficiently fluid to permit lateral movement of the constituent lipids and embedded proteins. The large numbers of double bonds in HUFA enhance the ability of such FA to "bend" thereby increasing their flexibility, reducing order, and, at least in principle, leading to an increase in the fluidity of cell membranes (Fig. 10.2; Eldho et al. 2003).

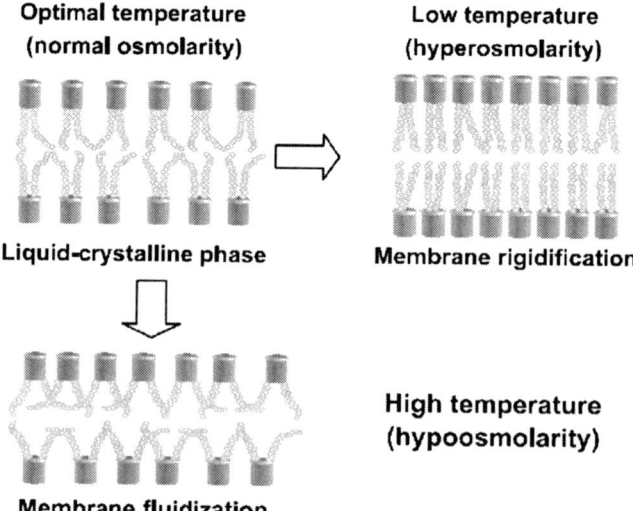

Fig. 10.1 Changes in membrane structure and the behavior of lipid bilayers under low and high-temperature stress. Low temperatures cause "rigidification" of membranes, while high temperatures cause "fluidization" of membranes. From Los and Murata (2004)

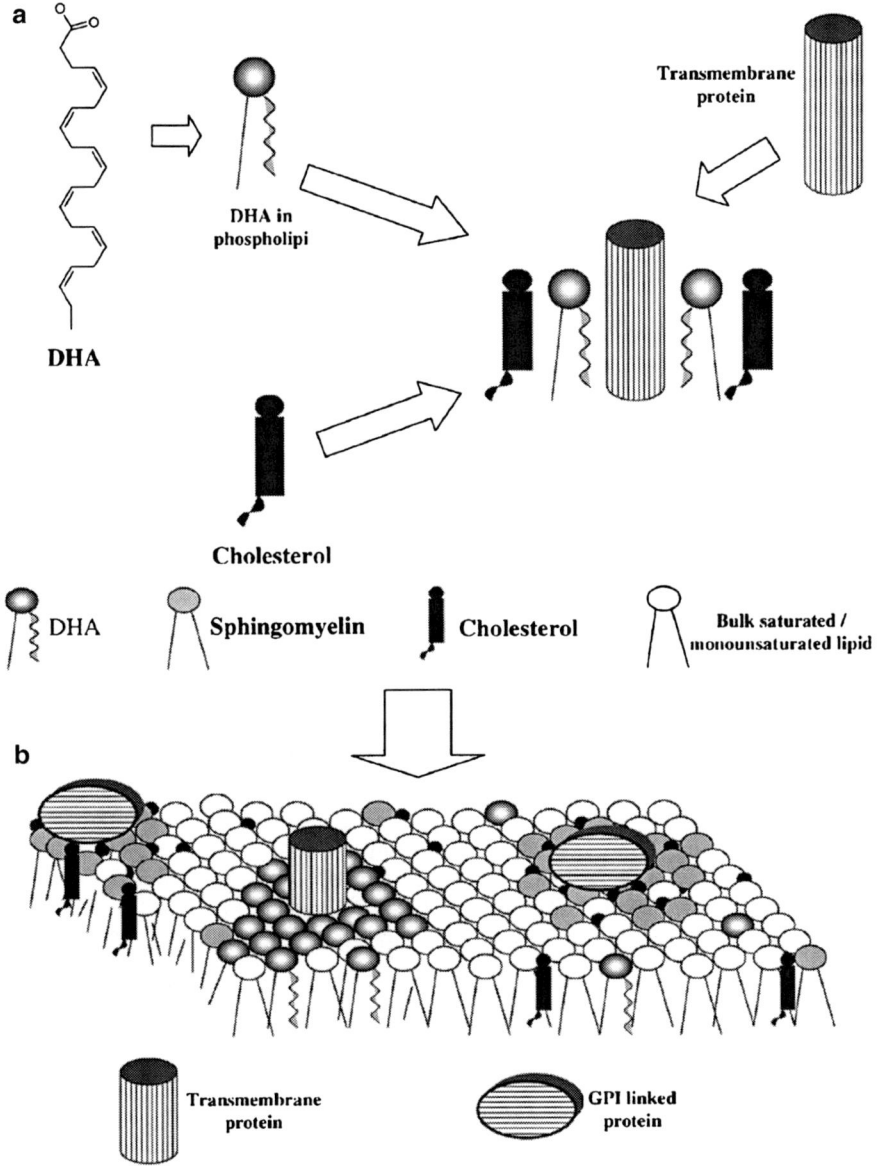

Fig. 10.2 (**a**) Schematic demonstrating the incorporation of docosahexaenoic acid (DHA) into the *sn-2* chain of a phospholipid and the association of this phospholipid with cholesterol. Also included is a transmembrane protein (e.g., rhodopsin) that has a high affinity for DHA. (**b**) Membrane phase separation into DHA-rich/cholesterol-poor (liquid disordered) and DHA-poor/cholesterol-rich (liquid ordered) domains. Different proteins partition into each domain. The liquid ordered domains are often referred to as "lipid rafts." *GPI* = Glycosylphosphatidyl-inositol. From Stillwell and Wassall (2003)

Docosahexaenoic acid (DHA; 22:6n-3) is especially important in this regard because it is the longest and most unsaturated FA commonly found in biological systems. The extremely high degree of acyl chain flexibility (Feller et al. 2002; Huber et al. 2002) conferred by the six double bonds in DHA are believed to be, at least in part, responsible for its effect on several important membrane properties including, but not limited to; membrane permeability, membrane fusion, and elasticity and vesicle formation (Stillwell and Wassall 2003; Wassall et al. 2004; and references therein). In addition, ion-transport processes of membranes as well as the functions of membrane-embedded, temperature-sensitive, enzymes (e.g. succinate hydrogenase, cytochrome oxidase and Na+/K+-ATPase) must be sustained (Hazel and Williams 1990; Hazel 1993; Adams 1999). Membrane competency is thus highly dependant on fluidity, ion transport and enzyme function, with temperature having a profound effect, both direct and indirect, on all of these properties.

Much of what is known concerning effects of specific FA and molecular species such as phospholipids on membrane fluidity come from studies on invertebrates, fish, and other vertebrates, especially mammals. Although ectothermic animals appear to increase the membrane content of unsaturated FA in response to colder temperatures, a clear and direct relationship between specific unsaturated FA and quantitative measurements of membrane fluidity has not commonly been demonstrated (but see Hall et al. 2002). However, such correlations do exist for some vertebrates. For example, experiments on rat platelet cells (Hashimoto et al. 2006), where the concentration of membrane-hardening cholesterol was manipulated, demonstrated that DHA has a more potent impact on membrane fluidity than eicosapentaenoic acid (EPA; 20:5n-3). Also, as vertebrates age, platelet membrane fluidity decreases as the DHA to arachidonic acid (ARA; 20:4n-6) ratio increases further supporting the critical role DHA plays in maintaining membrane fluidity (Hossain et al. 1999). Finally, there are cases where indirect evidence strongly points to a connection between n-3 HUFA in the diet and enhanced membrane fluidity and/or cold tolerance in fish (Kelly and Kohler 1999, Snyder and Hennessey 2003) and other vertebrates (Hagve et al. 1998) including humans (Lund et al. 1999).

10.2.2 Measuring Fluidity

In the ecological literature, the concept of membrane "fluidity" (or, more correctly, average membrane lipid order) is often invoked to explain mortality of fish due to cold exposure, however; actual physical measurements of membrane lipid order are often not done (Hall et al. 2002). Three main quantitative methods to assess membrane lipid order at the cellular level include steady-state fluorescence anisotropy measurements, electron spin resonance (ESR) spectroscopy, and attenuated total reflection-Fourier transform infrared spectroscopy (ATR-FT-IR). The first method relies on the degree of polarization of various synthetic fluorescent probes where the degree of polarization of the probe is inversely related to rotational fluidity (Kitajka et al. 1996, Hossain et al. 1999). The second method involves the incorporation of an electron spin label and subsequent measurements of the rate of motion of the spin label using an electron spin resonance spectrometer (e.g. Buda et al. 1994). The third method is used in conjunction with fluorescence recovery after photo bleaching (FRAP) to determine the fluidity of the sample (Hull et al. 2005) and provides a more quantitative measure of the physical state of lipids embedded within the membrane. The interaction between lipids and proteins in the membrane is analyzed by monitoring the disorder of fatty acid acyl chains in terms of the stretching of bonds, in particular CH2 stretching in phospholipids. The fluid and rigidified states of lipids are represented by high and low frequencies of CH2 stretching, respectively; in other words, the higher the frequency, the more fluid the membrane lipids and the higher the degree of disorder.

10.2.3 Controlling Fluidity via Lipid Compositional Changes

It might appear that the most straightforward way to increase membrane fluidity would be to increase the overall concentration of HUFA, and especially DHA, in cell membranes. Although there is some evidence of modest increases in the proportion of DHA in fish in response to decreasing temperatures (Farkas et al. 2000) such a straightforward explanation does not take into account the structural details of the biologically relevant molecules (i.e., the phospholipids).

The situation is clearly more complex and must be viewed in the context of structural and chemical properties of the membrane phospholipids and their constituent parts. For example, it has been shown that the inclusion and position of the first unsaturated bond (especially in the *cis* configuration) in one of the two FA of the phospholipid molecule produces the greatest change in the physical properties of the molecule (e.g. melting point), with additional double bonds in the FA comprising the phospholipid molecule having progressively less effect (Coolbear et al. 1983; Stubbs and Smith 1984). Thus, the overall level of FA unsaturation in biomembranes is now known to play only a subordinate role in determining their overall fluidity. This is because it is now recognized that the physical properties of each molecular species of phospholipid, and their subsequent effects on membrane fluidity, largely depend on several interacting factors including; (a) the chemical structure of the headgroup, (b) FA chain length, (c) degree of unsaturation, and (d) positional distribution (*sn-1* vs *sn-2*) of the two FA within the molecules (Brooks et al. 2002). Perhaps most importantly (and, as yet, only partially understood) are the interactions between specific molecular phospholipids species and the configuration (packing) of membrane-bound proteins that contribute the greatest degree of membrane ordering state (i.e., a higher degree of disorder thereby resisting contraction) of cell membranes as temperatures fall (Buda et al. 1994).

Recent studies have revealed that organisms can adjust the fluidity of their membranes by modifying the proportions of specific combinations of monounsaturated (in the *sn-1*) and polyunsaturated (in the *sn-2* position) FA in phospholipid molecules such as phosphatidylcholine (PC), phosphatidylinositol (PI), and phosphatidylethanolamine (PE). For example, Farkas et al. (2000) concluded that the relative amounts of these molecular species and their relative proportions are the major factors contributing to the maintenance of proper fluidity relationships in the brains of fish as well as helping maintain important brain functions such as signal transduction and membrane permeability. Similarly, Buda et al. (1994) found that, although the overall FA composition, including ARA and DHA, of carp (*Cyprinus carpio*) brain did not vary as an adaptive response to declining temperature, separation of PC and PE into their specific molecular species revealed a two to threefold accumulation of 18:1/22:6, 18:1/20:4, and 18:1/18:1 species in the latter, but no significant change in the molecular species composition of PC. Dey et al. (1993), in a study of the composition and physical state of fish liver phospholipids, found a two to threefold and a tenfold increase in the levels of 18:1/22:6 and 18:1/20:5 species of PE, respectively. Subsequent work with carp liver and with synthetic vesicles revealed that the placement of a *cis* Δ9 monounsaturated FA (MUFA) in the *sn-1* position, especially in PE, is amongst the most important factor determining the efficacy of the response of biomembranes to cold temperatures (Fodor et al. 1995). Similarly, exposure to cold induced precise changes to a limited number of phospholipid species in liver microsomes of carp including increases in the proportions of 16:1/22:6 and 18:1/22:6 for PC and PE, respectively (Brooks et al. 2002). The response for PI is a general increase in molecular species that had a MUFA in the *sn-1* position and, unlike the situation with PC and PE, a concomitant increase in either ARA or EPA in the *sn-2* position. Taken together, these studies point to a special and precise relationship between the molecular species composition of phospholipids and temperature, and especially to the critical role of the specific placement of particular MUFA in the *sn-1* position of PC and PE.

Biological membranes are no longer considered to be homogeneous mixtures of lipid and protein, but rather are now known to be composed of lateral patches of PUFA-rich domains and cholesterol-rich membrane rafts (Wassall et al. 2004; Fig. 10.2). Sterols and, in particular, cholesterol in animal biomembranes interact closely with phospholipids in a way that moderates fluidity, i.e., they are thought to have an ordering influence (inducing a tighter lateral packing of phospholipids) on membrane fluidity when the membrane lipids are in their liquid-crystalline state (Cooper et al. 1978). This is known as the condensing effect i.e., the total area occupied by cholesterol plus phospholipids is less than the sum of the area occupied by each molecular type separately (Haines 2001). The kinks in unsaturated acyl chains limit the formation of bonds between those fatty acids and cholesterol resulting in a lipid driven mechanism to explain why there is lateral phase separation into PUFA-poor/sterol-rich and PUFA-rich (especially DHA-rich)/sterol-poor microdomains (Ohvo-Rekila et al. 2002; Wassall et al. 2004; Fig. 10.2). Thus, the extent to which cholesterol is present in membranes can have a pronounced effect on fluidity. For example, rainbow trout fed menhaden oil (containing cholesterol and HUFA) demonstrated reduced fluidity in their macrophages compared with macrophages of trout containing sunflower oil containing no cholesterol or HUFA (Bowden et al. 1994). A second important role of cholesterol in animal cells is to diminish energy loss by reducing cation (Na^+) leakage through lipid bilayers of biomembranes (Haines 2001). In summary, sterols are now known to be essential nutrients (Martin-Creuzberg and von Elert – Chap. 3) since the deprivation of sterols is lethal to all animal cells. Although some of their functions are now known, currently the precise roles of cholesterol and phytosterols are only partially understood (Haines 2001).

10.2.4 Environmental Effects on HUFA Supply

In fish, depot triacylglycerols and flesh phospholipids typically contain ~95 and ~75% FA by weight, respectively (Arts et al. 2001). However, a review of fish oil triacylglycerols demonstrates that only a small proportion of these molecules contain EPA and DHA (Moffat 1995). This occurs because there is a fair degree of specificity in FA assembly in the triacylglycerol depot fats of fish (Brockerhoff and Hoyle 1963; Brockerhoff et al. 1964). Specifically, triacylglycerol synthesis starts with a phospholipid that usually has any one of five common FA (e.g., myristic, 14:0; palmitic, 16:0; palmitoleic, 16:1n-7; stearic, 18:0; or oleic, 18:1n-9) in the *sn–1* position. The placement of FA in the center (*sn–2*) position of the phospholipid molecule is directed by enzymes that preferentially put a HUFA (e.g., EPA or DHA) in this position but, depending on whether the fish is in either a biosynthesis or the catabolism energy status, the assembly process may have to again use a common FA. Fatty acids in the last (*sn–3*) position of the triacylglycerol molecule derive from whatever surplus FA are in the diet and are circulating in the blood when the triacylglycerol molecules are assembled (Arts et al. 2001). Thus, the FAs available in the diet when triacylglycerols and phospholipids are synthesized play an important role in their final molecular configuration.

Since diet affects the availability of specific FA and since this availability affects the final molecular configuration of membrane phospholipids, it follows that environmental perturbations that affect the availability of HUFA for fish may, in turn, influence a fish's ability to maintain adequate membrane fluidity and perhaps other aspects of membrane competency. Changes in the FA composition of phospholipids in response to a cold challenge take place quite quickly; on the order of days to weeks (Trueman et al. 2000).

Large-scale ecosystem perturbations in the Laurentian Great Lakes provide examples of how such changes have the potential to affect membrane fluidity in fish and their predators (Hebert et al. 2008). For instance, populations of the lipid-rich amphipod *Diporeia* sp., which can attain densities of >1,000 animals per m^2, have declined sharply in many areas of the Great Lakes, especially in Lakes Michigan and Huron (Nalepa et al. 2006, and references therein) potentially as a consequence of a not fully understood interactions with invasive mussels. *Diporeia* sp. are rich in EPA and DHA (Arts, unpublished data) and this fact, combined with their historically high densities, suggest that this species provides an important role in contributing essential n-3 HUFA, obtained from settling algae, to bottom foraging fish such as the commercially important lake whitefish (*Coregonus clupeaformis*). Lake whitefish spends a significant amount of time feeding in deep cold hypolimnetic waters where the maintenance of membrane fluidity is likely important. Thus, the widespread loss of *Diporeia* sp. from their historic ranges is a key issue of concern in the Great Lakes in part because their loss is expected to have a strong effect on the availability of n-3 HUFA for bottom-feeding fish predators such as lake whitefish.

A second example is the case of Chinook salmon (*Oncorhynchus kisutch*) and alewife (*Alosa pseudoharengus*) in Lake Michigan. Some fish species (e.g., salmonids) cannot elongate and further desaturate shorter chain (18 carbons) FA (ALA; 18:3n-3 and LIN; 18:2n-6) to their longer chain homologues (ARA, EPA, and DHA) in amounts sufficient to meet their needs (Sargent et al. 1995; and see Chap. 9). Historically, alewife comprised the main food source for Chinook salmon (Madenjian et al. 2005). This forage fish is known to be very rich in n-3 HUFA (Snyder and Hennessey 2003) and therefore likely supplied the abundant populations of Chinook salmon with much of their required dietary n-3 HUFA. Although the dramatic decline in the Chinook population in the late 1980s was associated with an epizootic disease outbreak, the underlying cause was thought to be nutritionally related (Benjamin and Bence 2003; Holey et al. 1998). It is therefore tempting to speculate that the underlying basis of this decline may have been, in part, due to a dietary n-3 HUFA deficiency, which contributed to the poor condition of the Chinook salmon rendering them more vulnerable to disease. Clearly, we require more information on the HUFA biosynthetic capabilities of key fish species and also on the distribution and relative abundance of these essential FA in their prey. Such information could be used to develop and refine specific biochemical indices of nutritional status and physiological competency, which, at least in theory, could provide a predictive tool to fisheries managers, forewarning them of impending fishery collapses.

Cultural eutrophication and climate change provide two other examples of large-scale perturbations with the potential to affect n-3 HUFA availability in aquatic ecosystem. Eutrophication generally results in a replacement of HUFA-rich taxa such as diatoms, cryptophytes, and dinoflagellates (Brett and Müller-Navarra 1997) with HUFA-poor taxa such as bloom forming chlorophytes and especially cyanobacteria (Müller-Navarra et al. 2004). Variations in the HUFA content of algae (i.e., availability of different taxa for consumption) can induce up to an eightfold difference in the PUFA concentration of the herbivore *Daphnia pulex* (Brett et al. 2006). Thus, cultural eutrophication can cause a marked impoverishment in the relative availability of n-3 HUFA rich food for uptake into the biomembrane phospholipids of herbivorous zooplankton, and, subsequently, for fish. This has important implications for fish condition and health since the availability of n-3 HUFA in fish diets is reflected in n-3 HUFA concentrations in fish tissues. For example, Estévez et al. (1999) found that the incorporation of ARA and EPA into fish eyes, brains, livers, and white muscle of turbot (*Scophthalmus maximus*) reflected the percentage of these FA in their diet. Masuda et al. (1999) demonstrated that ^{14}C-labeled DHA in the diet was incorporated directly to the central nervous system of yellowtail (*Seriola quinqueradiata*).

Climate warming also has the potential to affect the production of n-3 HUFA at the base of the food chain in cold water ecosystems. Algae react to lower temperatures by increasing the proportion of n-3 HUFA in their biomembranes (Chap. 1).

Compounding this, zooplankton, which are consumed by planktivorous fish and many larval fish, also demonstrate reduced concentrations of n-3 HUFA at higher temperatures in freshwater systems (and see Chap. 11 for a marine perspective). For example, members of the genus *Daphnia*, a keystone prey species for fish, demonstrate reduced EPA concentrations at higher temperatures (Schlechtriem et al. 2006). Thus, increased water temperatures are predicted, based on laboratory experiments, to exert a negative effect on the underlying availability of growth enhancing n-3 HUFA in the ecosystem. This also has implications for terrestrial systems because of their dependence on obtaining growth enhancing n-3 HUFA from aquatic systems (Chap. 8).

In conclusion, a fish's innate biochemical strategies and capabilities interact with the availability of key PUFA in its diet to maintain membrane competencies in response to varying environmental temperatures. As discussed earlier, the availability of some PUFA, in particular the n-3 HUFA, can be compromised under certain situations potentially leading to detrimental effects on membrane structure and, by consequence, fish health and condition. Clearly, this ability of fish to maintain membrane lipid order at physiologically advantageous levels affects their ability to withstand stress and disease. It is also clear that a fish's ability to withstand stress and disease is predicated on having a healthy, robust, immune system.

10.3 Modulatory Effects of Dietary Fatty Acids on Teleost Immune Response

10.3.1 Introduction

Lipids, FA, and their derivatives play a role in virtually every physiological process that occurs in vivo (Higgs and Dong 2000; Tocher 2003; Trushenski et al. 2006). For example, it has long been recognized in humans and other mammals that high dietary intake of saturated fats and cholesterol increase the likelihood of arteriosclerosis and heart attacks (Ulbricht and Southgate 1991). Saturated fats may have similar negatives effects on teleosts, as Atlantic salmon *Salmo salar* fed diets containing excessive quantities of saturated fats developed severe cardiomyopathy (Bell et al. 1991; 1993). However, it is with respect to immunity where the role of lipids and their constitutive FA in teleost health has been most clearly demonstrated (Rowley et al. 1995; Balfry and Higgs 2001; Sargent et al. 2002). Most of what we know about the immunomodulatory effects of dietary FA in teleosts has come from research using prepared diets with aquaculturally important fishes, most of which are carnivores. Because differences likely exist among species and trophic levels, major gaps exist in this body of research (see, generally, Blazer 1992; Rowley et al. 1995; Balfry and Higgs 2001). Nevertheless, the information gleaned in this manner provides useful insights on teleosts in the wild. Moreover, it would be nearly impossible to have obtained the current level of knowledge of teleost immunity and

FA if it were not for manipulative studies with aquaculture fish. Before reviewing the literature on the subject, a brief review of fish immunity is provided, with particular reference to those aspects dealing with FA derivatives.

10.3.2 Teleost Immune System

The immune system can simply be defined as a collection of mechanisms providing protection for an organism against pathogenic vectors (viruses, bacteria, fungi, parasites, etc.). To successfully meet the challenge of a vast array of potential pathogens, mechanisms must exist to distinguish pathogens from the organism's own proteins, and then to attack the pathogens before they manifest into debilitating disease. Immune strategies can generally be classified as innate (also called nonspecific) or adaptive (specific). Innate immune mechanisms are the more ancestral of the two, and function via self/nonself recognition of pathogens. Adaptive immune responses are based on production of pathogen-specific "populations" of antibodies, which can change in response to an individual's history of pathogen exposure. However, teleosts, though capable of adaptive immune responses, are more reliant on innate mechanisms due to the challenges of the aquatic environment (Tort et al. 2004).

10.3.3 Physical and Chemical Barriers

Mucous, scales (when present) and skin serve to seal the teleost and form the first line of defense against pathogens. These physical barriers may be breached by pathogens following injury or, in the case of stress, from impaired osmoregulatory ability. In the latter case, osmoregulatory failure leading to unregulated movement of water across the gill membranes provides pathogens, such as bacteria and viruses, with unimpeded access to the bloodstream. The skin may also secrete chemicals, antimicrobial peptides, and enzymes, to protect against infections. Antimicrobial peptides are short proteins, usually less than 50 amino acids long, which kill bacteria by disrupting membranes, interfering with metabolism, and/or targeting cytoplasmic components. Lysozyme is an enzyme that is sometimes classified as an antimicrobial peptide, though it is much longer in length. It acts as a self-induced antibiotic, destroying cell walls by hydrolyzing the polysaccharide component.

10.3.4 Innate and Adaptive Immune Systems

Those pathogens successfully invading a teleost will subsequently have to battle the innate immune system, including humoral (chemical secretions) and cellular agents. Upon penetration of the epithelium, inflammation occurs as a result of

10 Health and Condition in Fish

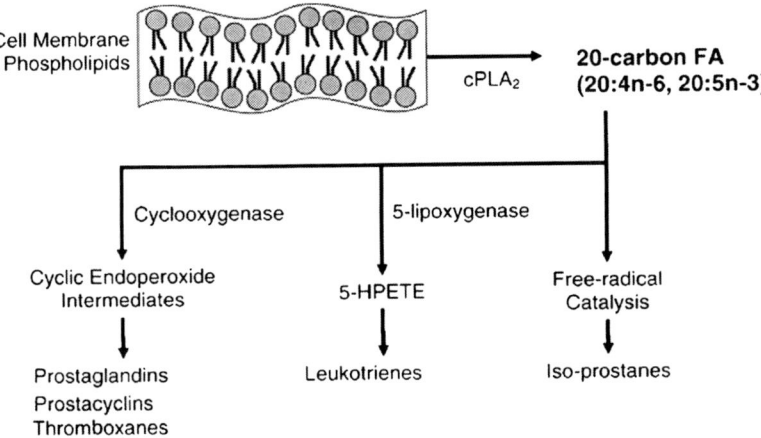

Fig. 10.3 Abbreviated diagram of eicosanoid synthesis pathways (*cPLA2* cytosolic phospholipase, *FA* fatty acid, *5-HPETE* hydroperoxyeicosatetraenoic acid). From Trushenski and Kohler (2008)

cytokines and eicosanoids being released by injured cells. Cytokines (e.g., interleukins, chemokines, and interferons) consist of proteins and peptides that bind to specific cell-surface receptors, each causing a cascade of intracellular signaling to alter cell functions. Eicosanoids, including prostaglandins and leukotrienes, are autocrines (hormone-like compounds), which serve, among other things, as modulators of inflammation and specific immunity responses. Eicosanoids are derived from 20-carbon n-3 and n-6 FA (Fig. 10.3), and are given specific attention in a subsequent section of this chapter. Teleosts possess leucocytes, (i.e. granulocytes, monocytes/macrophages, thrombocytes, nonspecific cytotoxic cells, and lymphocytes), all of which play vital roles in the innate immune system by patrolling the body in search of pathogens to be terminated. The adaptive immune response involves another class of leucocytes, called lymphocytes, which comprise a strong, complex defense mechanism, while also employing immunological memory. However, no immune response relationship of dietary FA to leucocytes, including lymphocytes, is apparent in teleosts.

10.3.5 Eicosanoids

Eicosanoids are a class of biologically active, oxygenated molecules that have been incorporated as esters in phospholipids and diacylglycerols of the cell and nuclear membranes. Along with serving as messengers in the central nervous system, eicosanoids act like local hormones or signaling molecules to control inflammation and immunity. Eicosanoids largely consist of prostaglandins, prostacyclins, thromboxanes, and leukotrienes, with each having different biosynthetic pathways and physiological effects (Fig. 10.3).

Eicosapentaenoic acid, ARA, and dihomo-gamma-linolenic acid (DG-LIN; 20:3n-6) are the primary FA from which eicosanoids are derived via two enzymes, cyclooxygenase and lipoxygenase, both of which exist in several forms. Although a normal physiological product, excess eicosanoid biosynthesis is initiated by extreme stress/trauma, cytokines, or other stimuli, which trigger release of phospholipase at the cell wall. The phospholipase migrates to the nuclear membrane catalyzing ester hydrolysis of phospholipid or diacylglycerol to free EPA, ARA, and/or DG-LIN. These FA are subsequently oxygenated via a number of pathways into prostanoids and leukotrienes. Prostanoids, in general, mediate inflammation by acting upon platelet aggregation and vasoconstriction/relaxation, and can also directly influence gene transcription. Leukotrienes have a more neuroendocrine role in inflammation, particularly with respect to leucocyte activity.

Arachidonic acid-derived eicosanoids promote inflammation (vasodilators) while those from EPA and DG-LIN are described as less inflammatory, inactive, or anti-inflammatory (vasoconstrictors). Accordingly, dietary n-3 FA can counteract ARA-derived eicosanoids (see Bell et al. 1994) by (1) displacement (higher n-3/n-6 ratio in phospholipids), (2) competition (EPA and DG-LIN compete for cyclooxygenases and lipoxygenase), and (3) opposition (eicosanoids derived from EPA and DG-LIN counter/impede actions of ARA-derived eicosanoids).

10.3.6 Effects of Diet on Tissue Fatty Acid Composition

Marine oils derived from feed-grade, reduction fisheries often serve as the primary lipid constituents of aquaculture diets because of their high palatability to cultured fishes, attractant properties, and historically widespread availability and competitive pricing (Lane et al. 2006). On the one hand, tissue FA composition of fishes largely reflects the diet (Shearer 1994; Jobling 2003), and thus fishes fed marine-derived oils contain substantial amounts of bioactive HUFA, namely DHA, EPA, and ARA. On the other hand, tissues of fish fed diets containing plant or livestock-derived lipids will be higher in saturated fats. Several studies have shown reductions in HUFA when these latter diets have been fed to channel catfish (*Ictalurus punctatus*) (Manning et al. 2007; O'Neal and Kohler 2008), sunshine bass (*Morone saxatilis* x *M. chrysops*) (Kelly and Kohler 1999; Wonnacott et al. 2004; Lewis and Kohler 2008), rainbow trout (*Oncorhynchus mykiss*) (Caballero et al. 2002), Atlantic salmon (Bransden et al. 2003), gilthead seabream (*Sparus aurata*) (Izquierdo et al. 2003), and red seabream (*Pargrus auratus*) (Glencross et al. 2003). Marine oil-based "finishing diets" have been used to restore the tissue FA profiles of turbot (*Psetta maxima*) (Regost et al. 2003) and Atlantic salmon (Bell et al. 2003; 2004; Torstensen et al. 2004) previously fed nonmarine derived oils for the majority of the production cycle.

A simple dilution model describes FA turnover for medium-fat species following dietary fat modification (Robin et al. 2003; Jobling 2003, 2004a, b). Conversely, the dilution model did not adequately describe FA turnover in a "lean-flesh" fish (<2%

lipid), e.g., the sunshine bass (Lane et al. 2006). The deviations from the model in this lean fish suggest selective retention of n-3 HUFA along with preferential catabolism of saturates. The aforementioned studies have direct bearing on fishes in the wild in that they demonstrate tissue FA profiles which will essentially mirror seasonal profiles of the food chain, but that some selective retention of HUFA may occur on a species-specific basis (Dalsgaard et al. 2003).

10.3.7 Tissue Fatty Acid Composition and Teleost Immunity

In addition to altering tissue FA profiles, a number of manipulative dietary lipid studies have noted attendant influences in immune system functioning, particularly with respect to eicosanoid production. For example, Bell et al. (1992) found diet related increases in ARA in post-smolt Atlantic salmon phospholipids induced production of ARA-derived eicosanoids. In a subsequent study, Bell et al. (1993) found, though the metabolic pathways involved are not fully clear, that a diet high in linolenic acid (ALA; 18:3n-3) (the biosynthetic precursor to EPA) resulted in higher levels of EPA in Atlantic salmon leucocyte phospholipids with a concomitant reduction in ARA-derived eicosanoids, ultimately resulting in increased antiinflammatory response and a reduction in cardiac lesions. Similar findings were obtained in related studies with juvenile turbot (*Scophthalmus maximus*) (Bell et al. 1994). Moreover, when diets high in ARA were used, ARA-derived prostaglandins were found to be significantly higher in tissue homogenates (Bell et al. 1995). These studies suggest dietary fats may have profound effects on nonspecific immune factors. In particular, diets high in n-3 FA will generally result in relatively higher levels of the antiinflammatory 3-series prostaglandins and 5-series leukotrienes and lipoxins derived from EPA, whereas high dietary n-6 promotes relatively higher levels of proinflammatory two-series prostaglandins and four-series leukotrienes and lipoxins derived from ARA (Balfry and Higgs 2001). Moreover, EPA competitively interferes with eicosanoid production from ARA and, ultimately, eicosanoid actions are largely determined by the ratio of ARA to EPA in cellular membranes wherein an imbalanced ratio can be damaging to fish (Sargent et al. 2002). Lin and Shiau (2007) found three nonspecific immune response indicators (respiratory burst, lysozyme activity, alternative complement activity) were significantly higher in grouper (*Epinephelus malabaricus*) fed a diet with a blend of fish oil and corn oil as the lipid source than fish fed diets with either fish oil or corn oil as the sole lipid source. The blend of corn and fish oils likely proved positive in this regard because of the complexity of the immune response, i.e., contradictions exist in immunostimulation and suppression resulting from dietary source of lipid. Thus, as previously suggested by Fracalossi and Lovell (1994), a species-specific dietary n-6 and n-3 HUFA ratio should optimize immune function. A dietary n-6 to n-3 PUFA ratio no greater than 4:1 and, preferably, closer to 1:1, is recommended for human health (Lands 1992), but no specific ratios for teleosts have been determined.

10.3.8 Tissue Fatty Acid Composition and Disease Susceptibility

Information on the relationship of dietary lipids to specific disease resistance is limited and seemingly contradictory. For example, Li et al. (1994) reported negative effects of high dietary n-3 HUFA levels in channel catfish, while Sheldon and Blazer (1991) found in the same species that bactericidal activity was positively correlated with dietary HUFA. Lodemel et al. (2001) observed higher survival in Artic charr (*Salvelinus alpinus*) fed dietary soybean oil (high in n-6) than fish oil, whereas Montero et al. (2003) found in gilthead seabream the inclusion of soybean oil reduced both serum alternative complement pathway activity and head kidney phagocytic activity. Montero et al. (2003) also determined in gilthead bream that rapeseed oil affected head kidney macrophages activity, while linseed oil altered stress response. In actuality, such conflicting results may be due to the different modes in which different dietary lipids and FA ratios may affect disease susceptibility in teleosts. For example, Sheldon and Blazer (1991) examined the effects of lipid types primarily on the specific immune response while Lodemel et al. (2001) evaluated immunological effects indirectly via influence of dietary lipid on gut microflora. A greater incidence of goblet cells in the midgut region of infected fish fed soybean oil was observed, prompting Lodemel et al. (2001) to theorize mucous sloughing is a protective response against bacterial infection, similar to what has been shown in the gills (Ferguson et al. 1992). These results further emphasize the importance of a proper dietary n-3 to n-6 ratio for health maintenance in teleosts.

10.4 Future Directions

A solid body of evidence exists for teleosts demonstrating that: (1) tissue lipids largely reflect their dietary intake; (2) several FA play pivotal roles in the health of fish and other aquatic organisms including, as reviewed here, effects of lipids on membrane competency and immunity; and (3) some species-specific balance of n-3 and n-6 FA is necessary to optimize immune functioning and general health (and see Chap. 7 for specific examples pertaining to fish and other aquatic organisms). For fishes in captivity, future research should be directed at determining the interaction between cholesterol (and other sterols) and fatty acid profiles on membrane fluidity and thermal tolerance through careful dietary manipulations. Specific attention should be paid to determining dietary thresholds of specific FA or combinations of FA and sterols on thermal tolerance and other aspects of membrane competency. In an aquaculture context, strides can be made in further boosting immunity via dietary lipid manipulation, particularly with respect to optimizing species-specific balances for n-6:n-3 FA. For fishes in the wild, trophic studies should focus on identifying how seasonal changes in the food chain create a cascade of FA alterations at each trophic level, ultimately altering tissue compositions of top carnivores. Research on how anthropogenic alterations of ecosystems, including eutrophication as well as global warming and exotic invaders, impact the bottom-up transfer of FA and associated health factors should also prove quite illuminating. Clearly, great strides could be made if fish nutritionists, physiologists, and ecologists collaborated on such endeavors.

References

Adams, S. M. 1999. Ecological role of lipids in the health and success of fish populations. *In* M.T. Arts and B.C. Wainman [eds.] , Lipids in Freshwater Ecosystems. Springer, New York, pp. 132–160

Anderson, R. O. and Neumann, R. M. 1996. Length, weight, and associated structures. *In* B.R. Murphy and D.W. Willis [eds.], Fisheries Techniques, 2nd edition. American Fisheries Society, Bethesda, pp. 447–481

Arts, M. T., Ackman, R. G., and Holub, B. J. 2001. "Essential fatty acids" in aquatic ecosystems: a crucial link between diet and human health and evolution. Can. J. Fish. Aquat. Sci. 58:122–137

Balfry, S. and Higgs, D. A. 2001. Influence of dietary lipid composition on the immune system and disease resistance of finfish. *In* C. Lim and C.D. Webster [eds.], Nutrition and fish health. Food Products Press, Binghampton, pp. 213–243

Bell, J. G., McVicar, A. H., Park, M. T., and Sargent, J. R. 1991. High dietary linoleic acid affects the fatty acid compositions of individual phospholipids from tissues from Atlantic salmon (*Salmo salar*): association with stress susceptibility and cardiac lesion. J. Nutrition 121:1163–1172

Bell, J. G., Sargent, J. R., and Raynard, R. S. 1992. Effects of increasing dietary linoleic acid on phospholipid fatty acid composition and eicosanoid production in leucocytes and gill cells in Atlantic salmon (*Salmo salar*). Prostaglandins, Leukotrienes, and Essential Fatty Acids 45:197–206

Bell, J. G., Dick, J. R., McVicar, A. H., Sargent, J. R., and Thompson, K. D. 1993. Dietary sunflower, linseed, and fish oils affect phospholipid fatty acid composition, development of cardiac lesions, phospholipase activity, and eicosanoid production in Atlantic salmon (*Salmo salar*). Prostaglandins, Leukotrienes, and Essential Fatty Acids 49:665–673

Bell, J. G., Tocher, D. R., MacDonald, F. M., and Sargent, J. R. 1994. Effects of diets rich in linoleic (18:2n-6) and -linolenic (18:3n-3) acids on growth, lipid class and fatty acid compositions and eicosanoid production in juvenile turbot (*Scophthalmus maximus* L.) Fish Physiol. Biochem. 13:105–118

Bell, J. G., Castell, J. D., Tocher, D. R., MacDonald, F. M., and Sargent, J. R. 1995. Effects of different dietary arachidonic acid: docosahexaenoic acid ratios on phospholipid fatty acid compositions and prostaglandin production in juvenile turbot (*Scophthalmus maximus*). Fish Physiol. Biochem. 14:139–151

Bell, J. G., Tocher, D. R., Henderson, R. J., Dick, J. R., and Crampton, V. O. 2003. Altered fatty acid composition in Atlantic salmon (*Salmo salar*) fed diets containing linseed and rapeseed oils can be partially restored by a subsequent fish oil finishing diet. J. Nutrition 133:2793–2801

Bell, J. G., Henderson, R. J., Tocher, D. R., McGhee, F., and Sargent, J. R. 2004. Replacement of dietary fish oil with increasing levels of linseed oil: modification of flesh fatty acid compositions in Atlantic salmon (*Salmo salar*) using a fish oil finishing diet. Lipids 39:223–232

Blazer, V. S. 1992. Nutrition and disease resistance in fish. Ann. Rev. Fish Dis. 2:309–323

Bowden, L. A., Weitzel, B., Ashton, I. P., Secombs, C. J., Restall, C. J., Walton, T. J., and Rowley, A. F. 1994. Effect of dietary cholesterol on membrane properties and immune functions in rainbow trout. Biochem. Soc. Trans. 22:339S

Bransden, M. P., Carter, C. G., and Nichols, P. D. 2003. Replacement of fish oil with sunflower oil in feeds for Atlantic salmon (*Salmo salar* L.): effect on growth performance, tissue fatty acid composition, and disease resistance. Comp. Biochem. Physiol. 135B:611–625

Brett, M. T. and Müller-Navarra, D. C. 1997. The role of highly unsaturated fatty acids in aquatic food web processes. Freshw. Biol. 38:483–499

Brett, M. T., Muller-Navarra, D. C., Ballantyne, A. P., Ravet, J. L., and Goldman, C. R. 2006. *Daphnia* fatty acid composition reflects that of their diet. Limnol. Oceanogr. 51:2428–2437

Brockerhoff, H. and Hoyle, R. J. 1963. On the structure of the depot fats of marine fish and mammals. Arch. Biochem. Biophys. 102:452–455

Brockerhoff, H., Hoyle, R. J., and Ronald, K. 1964. Retention of the fatty acid distribution pattern of a dietary triglyceride in animals. J. Biol. Chem. 239:735–739

Brooks, S., Clark, G. T., Wright, S. M., Trueman, R. J., Postle, A. D., Cossins, A. R., and Maclean, N. M. 2002. Electrospray ionisation mass spectrometric analysis of lipid restructuring in the carp (*Cyprinus carpio* L.) during cold acclimation. J. Exp. Biol. 205:3989–3997

Buda, C. I., Dey, I., Balogh, N., Horvath, L. I., Maderspach, K., Juhasz, M., Yeo, Y. K., and Farkas, T. 1994. Structural order of membranes and composition of phospholipids in fish brain cells during thermal acclimatization. Proc. Natl. Acad. Sci. USA. 91:8234–8238

Caballero, M. J., Orbach, A., Rosenlund, G., Montero, D., Grisvold, M., and Izquierdo, M. S. 2002. Impact of different dietary lipid sources on growth, lipid digestibility, tissue fatty acid composition and histology of rainbow trout, *Oncorhychus mykiss*. Aquaculture 214:253–271

Coolbear, K. P., Berde, C. B., and Keough, K. M. W. 1983. Gel to liquid-cystalline phase transitions of aqueous dispersions of polyunsaturated mixed-acid phosphatidylcholines. Biochem. 22:1466–1473

Cooper, R. A., Leslie, M. H., Fischkoff, S., Shinitzky, M., and Shattil, S. J. 1978. Factors influencing the lipid composition and fluidity of red cell membranes in vitro: Production of red cells possessing more than two cholesterols per phospholipid. Biochem. 17:327–331

Dalsgaard, J., St. John, M., Kattner, G., Muller-Navarra, D., and Hagen, W. 2003. Fatty acid trophic markers in the pelagic marine environment. Adv. Mar. Biol. 46:225–340

Dey, I., Buda, C., Wiik, T., Halver, J. E., and Farkas, T. 1993. Molecular and structural composition of phospholipid membranes in lives of marine and freshwater fish in relation to temperature, Proc. Nat. Acad. Sci. 90:7498–7502.

Eldho, N. V., Feller, S. E., Tristram-Nagle, S., Polozov, I. V., Gawrisch, K. 2003. Polyunsaturated docosahexaenoic vs docosapentaenoic acid-differences in lipid matrix properties from the loss of one double bond. J Am. Chem. Soc. 125:6409–6421

Estévez, A., McEvoy, L. A., Bell, J. G., and Sargent, J. R. 1999. Growth, survival, lipid composition and pigmentation of turbot (*Scophthalmus maximus*) larvae fed live-prey enriched in arachidonic and eicosapentaenoic acids. Aquaculture 180:321–343

Farkas, T., Kitajka, K., Fodor, E., Csengeri, I., Landes, E., Yeo, Y. K., Krasznai, Z., and Halver, J. E. 2000. Docosahexaenoic acid-containing phospholipid molecular species in brains of vertebrates. Proc. Natl. Acad. Sci. USA. 97:6362–6366

Feller, S. E., Gawrisch, K., MacKerrell, Jr., A. D. 2002. Polyunsaturated fatty acids in lipid bilayers: intrinsic and environmental contributions to their unique physical properties. J. Am. Chem. Soc. 124:318–326

Ferguson, H. W., Morrison, D., Ostland, V. E., Lumsden, J., and Bryne, P. 1992. Responses of mucus-producing cells in gill disease of rainbow trout (*Oncorhynchus mykiss*). J. Comp. Pathol. 106:255–265

Fodor, E., Jones, R. H., Buda, C., Kitajka, K., Dey, I., and Farkas, T. 1995. Molecular architecture and biophysical properties of phospholipids during thermal adaptation in fish: an experimental and model study. Lipids 30:1119–1125

Fracalossi, D. M. and Lovell, R. T. 1994. Dietary lipid sources influence responses of channel catfish (*Ictalurus punctatus*) to challenge with the pathogen *Edwardsiella ictaluri*. Aquaculture 119:287–298

Glencross, B. D., Hawkins, W. E., and Curnow, J. G. 2003. Restoration of the fatty acid composition of red seabream (*Pagrus auratus*) using a fish oil finishing diet after growout on plant oil based diets. Aquacult. Nutr. 9:409–418

Hagve, T. A., Woldseth, B., Brox, J., Narce, M., and Poisson, J. P. 1998. Membrane fluidity and fatty acid metabolism in kidney cells from rats fed purified eicosapentaenoic acid or purified docosahexaenoic acid. Scand. J. Clin. Lab. Invest. 58:187–194

Haines, T. H. 2001. Do sterols reduce proton and sodium leaks through lipid bilayers. Prog. Lipid Res. 40:299–324

Hall, J. M., Parish, C. C., and Thompson, R. J. 2002. Eicosapentaenoic acid regulates scallop (*Placopecten magellanicus*) membrane fluidity in response to cold. Biol. Bull. 202:201–203

Hashimoto, M., Hossain, S., and Shido, O. 2006. Docosahexaenoic acid but not eicosapentaenoic acid withstands dietary cholesterol-induced decreases in platelet membrane fluidity. Mol. Cell. Biochem. 293:1–8

Hazel, J. R. 1993. Thermal Biology. *In* D.H. Evans [ed.], The Physiology of Fishes, CRC Press, Boca Raton, pp. 427–467

Hazel, J. R. and Williams, E. E. 1990. The role of alterations in membrane lipid composition in enabling physiological adaptation of organisms to their physical environment. Prog. Lipid Res. 26:281–347

Hebert, C. E., Weseloh, D. V. C., Idrissi, A., Arts, M. T., O'Gorman, R., Gorman, O. T., Locke, B., Madenjian, C. P., and Roseman, E. F. 2008. Restoring piscivorous fish populations in the Laurentian Great Lakes causes seabird dietary change. Ecology 89:891–897

Higgs, D. A. and Dong, F. M. 2000. Lipids and fatty acids. *In* R.R. Stickney [ed.], Aquaculture Encyclopedia. Wiley, New York, pp. 476–496

Holey, M. E., Elliot, R. F., Marcquenski, S. V., Hnath, J. G., and Smith, K. D. 1998. Chinook salmon epizootics in Lake Michigan: possible contributing factors and management implications. J. Aquat. Animal Health 10:201–210

Hossain, M. S., Hashimoto, M., Gamoh, S., and Masumura, S. 1999. Association of age-related decrease in platelet membrane fluidity with platelet lipid peroxide. Life Sci. 64:135–143

Huber, T., Rajamoorthi, K., Kurze, V. F., Beyer, K., Brown, M. F. 2002. Structure of docosahexaenoic acid-containing phospholipid bilayers as studied by 2H NMR and molecular dynamics simulations. J. Am. Chem. Soc. 124:298–309

Hull, M. C., Cambrea, L. R., Hovis, J. S. 2005. Infrared spectroscopy of fluid lipid bilayers. Anal. Chem. 77:6096–6099

Izquierdo, M. S., Obach, A., Arantzamendi, L., Montero, D., Robaina, L., and Rosenlund, G. 2003. Dietary lipid sources for seabream and seabass: Growth performance, tissue composition and flesh quality. Aquaculture Nutr. 9:397–407.

Jobling, M. 2003. Do changes in Atlantic salmon, *Salmo salar* L., fillet fatty acids following a dietary switch represent wash-out or dilution? Test of a dilution model and its application. Aquaculture Res. 34:1215–1221

Jobling, M. 2004a. Are modifications in tissue fatty acid profiles following a change in diet the result of dilution? Test of a simple dilution model. Aquaculture 232:551–562

Jobling, M. 2004b. "Finishing" feeds for carnivorous fish and the fatty acid dilution model. Aquaculture Res. 35:706–709

Kitajka, K., Buda, C. S., Fodor, E., Halver, J. E., and Farkas, T. 1996. Involvement of phospholipid molecular species in controlling structural order of vertebrate brain synaptic membranes during thermal evolution. Lipids 31:1045–1050

Kelly, A. M. and Kohler, C. C. 1999. Cold tolerance and fatty acid composition of striped bass, white bass, and their hybrids. N. Am. J. Aquaculture 61:278–285

Lands, W. E. M. 1992. Biochemistry and physiology of n-3 fatty acids. Fed. Am. Soc. Exp. Biol. 6:2530–2536

Lane, R. L., Trushenski, J. T., and Kohler, C. C. 2006. Modification of fillet composition and evidence of differential fatty acid turnover in sunshine bass *Morone chrysops* x *M. saxatilis* following change in dietary lipid source. Lipids 41:1029–1038

Lewis, H. A. and Kohler, C. C. 2008. Corn gluten meal partially replaces dietary fish meal without compromising growth or the fatty acid composition of sunshine bass. N. Am. J. Aquaculture 70:50–60

Li, M. H., Wise, D. J., Johnson, M. R., and Robinson, E. H. 1994. Dietary menhaden oil reduced resistance of channel catfish (*Ictalurus punctatus*) to *Edwardsiella ictaluri*. Aquaculture 128:335–344

Lin, Y. -H. and Shiau, S. -Y. 2007. Effect of dietary blend of fish oil with corn oil on growth and non-specific immune responses of grouper, *Epinephelus malabaricus*. Aquaculture Nutr. 13:137–144

Lodemel, J. B., Mayhew, T. M., Myklebust, R., Olsen, R. E., Espelid, S., and Ringo, E. 2001. Effect of three dietary oils on disease susceptibility in arctic charr (*Salvelinus* alpinis L.) during cohabitant challenge with *Aeromonus salmonicida* ssp. *salmonicida*. Aquaculture Res. 32:935–945

Los, D. A. and Murata, N. 2004. Membrane fluidity and its roles in the perception of environmental signals. Biochimica et Biophysica Acta 1666:142–157

Lund, E. K., Harvey, L. J., Ladha, S., Clark, D. C., and Johnson, I. T. 1999. Effects of dietary fish oil supplementation on the phospholipid composition and fluidity of cell membranes from human volunteers. Ann. Nutr. Metab. 43:290–300

Madenjian, C. P., Höök, T. O., Rutherford, E. S., Mason, D. M., Croley II, T. E., Szalai, E. B., and Bence, J. R. 2005. Recruitment variability of alewives in Lake Michigan. Trans. Am. Fish Soc. 134:218–230

Manning, B. B., Li, M. H., and Robinson, E. H. 2007. Feeding channel catfish, *Ictalurus punctatus*, diets amended with refined marine fish oil elevates omega-3 highly unsaturated fatty acids in fillets. J. World Aquaculture Soc. 38:49–58

Masuda, R., Takeuchi, T., Tsukamoto, K., Sato, H., Shimizu, K., and Imaizumi, K. 1999. Incorporation of dietary docosahexaenoic acid into the central nervous system of the yellowtail *Seriola quinqueradiata*. Brain Behav. Evol. 53:173–179

Moffat, C. F. 1995. Fish oil triglycerides: a wealth of variation. Lipid Technol. 7:125–129

Montero, D., Kalinowski, T., Obach, A., Robaina, L., Tort, L., Caballero, M. J., and Izquierdo, M. S. 2003. Vegetable lipid sources for gilthead seabream (*Sparus aurata*): effects on fish health. Aquaculture 225:353–370

Müller-Navarra, D. C., Brett, M. T., Park, S., Chandra, S., Ballantyne, A. P., Zorita, E., and Goldman, C. R. 2004. Unsaturated fatty acid content in seston and tropho-dynamic coupling in lakes. Nature 427:69–72

Nalepa, T. F., Fanslow, D. L., Foley, A. J. III, Lang, G. A., Eadie, B. J., and Quigley, M. A. 2006. Continued disappearance of the benthic amphipod *Diporeia* spp. in Lake Michigan: is there evidence for food limitation? Can. J. Fish Aquat. Sci. 63:872–890

Ohvo-Rekila, H., Ramstedt, B., Leppimaki, P., and Slotte, J. P. 2002. Cholesterol interactions with phospholipids in membranes. Prog Lipid Res. 41:66–97

O'Neal, C. C. and Kohler, C. C. 2008. Effects of replacing menhaden oil with catfish oil on the fatty acid composition of juvenile channel catfish, *Ictalurus punctatus*. J. World Aquaculture Soc. 39:62–71

Regost, C., Arzel, J., Robin, J., Rosenlund, G., and Kaushik, S. J. 2003. Total replacement of fish oil by soybean or linseed oil with a return to fish oil in turbot (*Psetta maxima*), 1. Growth performance, flesh fatty acid profile, and lipid metabolism. Aquaculture 217:465–482

Robin, J. H., Regost, C., Arzel, J., and Kaushik, S. J. 2003. Fatty acid profile of fish following a change in dietary fatty acid source: model of fatty acid composition with a dilution hypothesis. Aquaculture 225:283–293

Rowley, A. F., Knight, J., Lloyd-Evans, P., Holland, J. W., and Vickers, P. J. 1995. Eicosanoids and their role in immune modulation in fish—a brief overview. Fish and Shellfish Immunol. 5:549–567

Sargent, J. R., Bell, J. G., Bell, M. V., Henderson, R. J., and Tocher, D. R. 1995. Requirement criteria for essential fatty acids. J. Appl. Ichthyol. 11:183–198

Sargent, J. R., Tocher, D. R., and Bell, J. G. 2002. The lipids. In J.E. Halver R.W. Hardy [eds.] Fish nutrition, 3rd edition. Academic Press, San Diego, pp. 181–257

Schlechtriem, C., Arts, M. T., and Zellmer, I. D. 2006. Effect of temperature on the fatty acid composition and temporal trajectories of fatty acids in fasting *Daphnia pulex* (Crustacea, Cladocera). Lipids 41:397–400

Shearer, K. D. 1994. Factors affecting the proximate composition of cultured fishes with emphasis on salmonids. Aquaculture 119:63–88

Sheldon, Jr., W. M. and Blazer, V. S. 1991. Influence of dietary lipid and temperature on bactericidal activity of channel catfish macrophages. J. Aquat. Animal Health 3:87–93

Sinensky, M. 1974. Homoviscous adaptation – a homeostatic process that regulates the viscosity of membrane lipids in *Escherichia coli*. Proc. Natl. Acad. Sci. USA 71:522–525

Singer, S. J. and Nicholson, G. L. 1972. The fluid mosaic model of the structure of cell membranes. Science 175:720–731

Stillwell, W. and Wassall, S. R. 2003. Docosahexaenoic acid: membrane properties of a unique fatty acid. Chem. Phys. Lipids. 126:1–27

Stubbs, C. D. and Smith, A. D. 1984. The modification of mammalian membrane polyunsaturated fatty acid composition in relation to fluidity and function. Biochim. Biophys. Acta. 779:89–137

Snyder, R. J. and Hennessey, T. M. 2003. Cold tolerance and homeoviscous adaptation in freshwater alewives (*Alosa pseudoharengus*). Fish Physiol. Biochem. 29:117–126

Tocher, D. R. 2003. Metabolism and functions of lipids and fatty acids in teleost fish. Rev. Fish Sci. 11:107–184

Torstensen, B. E., Froyland, L., Ornsrud, R., and Lie, O. 2004. Tailoring of a cardioprotective muscle fatty acid composition of Atlantic salmon (*Salmo salar*) fed vegetable oils. Food Chem. 87:567–580

Tort, L., Balasch, J. C., and MacKenzie, S. 2004. Fish health challenge after stress. Indicators of immunocompetence. Contrib. Sci. 2:443–454

Trueman, R. J., Tiku, P. E., Caddick, M. X., and Cossins, A. R. 2000. Thermal thresholds of lipid restructuring and 9-desaturase expression in the liver of carp (*Cyprinus carpio*). J. Exp. Biol. 203:641–650

Trushenski, J. T., Kasper, C. S., and Kohler, C. C. 2006. Challenges and opportunities in finfish nutrition. N. Am. J. Aquaculture 68:122–140

Trushenski, J. T. and Kohler, C. C. 2008. Influence of stress, exertion, and dietary natural source vitamen E on prostaglandin synthesis, hematology, and tissue fatty acid composition of sunshine bass. N. Am. J. Aquaculture 70:251–265

Ulbricht, T. L. V. and Southgate, D. A. T. 1991. Coronary heart disease: seven dietary factors. Lancet 338:985–992

Wassall, S. R., Brzustowicz, M. R., Shaikh, S. R., Cherezov, V., Caffrey, M., and Stillwell, W. 2004. Order from disorder, corralling cholesterol with chaotic lipids. The role of polyunsaturated lipids in membrane raft formation. Chem. Phys. Lipids 132:79–88

Wonnacott, E. J., Lane, R. L., and Kohler, C. C. 2004. Influence of dietary replacement of menhaden oil with canola oil on fatty acid composition of sunshine bass. N. Am. J. Aquaculture 66:243–250

Chapter 11
Lipids in Marine Copepods: Latitudinal Characteristics and Perspective to Global Warming

Gerhard Kattner and Wilhelm Hagen

11.1 Introduction

Marine zooplankton represent a very diverse group in the world's oceans, with numerous taxa of high abundance and biomass. Many of these zooplankton species, especially the dominating copepods, are able to accumulate large reserves of energy-rich lipids, exhibiting some of the highest lipid levels in organisms on earth. Their unusual way to store these lipids, namely as wax esters, is another particularity of many zooplankton species. It is generally accepted that wax esters serve as long-term metabolic reserves, whereas triacylglycerols are utilized for short-term demands, although the physiological advantage of wax esters as long-term deposits over triacylglycerols is still unclear. The geographical distribution of wax esters in marine zooplankton was first studied in detail by Lee and co-authors in the 1970s (Lee et al. 1971; Lee and Hirota 1973). They showed that especially herbivorous calanoid copepods from habitats with a marked seasonality intensely synthesize wax esters, which in many herbivorous species consist, to a large degree, of specific long-chain monounsaturated fatty acids (MUFA) and alcohols (reviewed by Sargent and Henderson 1986; Dalsgaard et al. 2003; Lee et al. 2006).

In the transfer of energy through the food web, zooplankton species, especially copepods, play an important role as converters of usually rather lipid-poor phytoplankton (10–20% of dry mass DM) to lipid-rich herbivorous species (>50% DM). Copepods develop from eggs via six nauplii stages (NI-NVI) and five copepodite stages to adult males or females (CI-CVI). Although the earlier developmental stages invest energy into somatic growth, the older copepodids (CIV-CVI) exhibit a massive lipid build-up. This lipid accumulation is usually accomplished via wax ester biosynthesis, at least in most of the larger calanoid species.

G. Kattner (✉)
Alfred Wegener Institute for Polar and Marine Research, Ecological Chemistry,
Am Handelshafen 12, 27570, Bremerhaven, Germany
e-mail: gerhard.kattner@awi.de

W. Hagen
Marine Zoology, Faculty of Biology/Chemistry, University of Bremen, P.O. Box 330440,
28334 Bremen, Germany

These enormous amounts of lipids fuel major pathways of marine food webs, at the same time essential fatty acids are transferred from primary producers via zooplankton to higher trophic levels.

Fatty acids of dietary origin can be incorporated unchanged into copepod lipids. These so-called fatty acid trophic markers (FATM; Dalsgaard et al. 2003) are useful in the elucidation of dietary relationships. However, differentiation is only possible between larger groups of phytoplankton, e.g., diatoms or dinoflagellates, but not at the species level. Typical fatty acids of diatoms are 16:1n-7 and 20:5n-3 (e.g., Ackman et al. 1968; Kattner et al. 1983). Flagellates and also the haptophyte *Phaeocystis* spp. contain large amounts of 18:4n-3 and 22:6n-3 fatty acids (Harrington et al. 1970; Sargent et al. 1985). The 16:1n-7 and 18:4n-3 fatty acids are mainly incorporated into storage lipids and indicate dietary preferences and thus feeding behavior of the copepods (Graeve et al. 1994a; Dalsgaard et al. 2003). The essential fatty acids, 20:5n-3 (eicosapentaenoic acid, EPA) and 22:6n-3 (docosahexaenoic acid, DHA), are major components of membranes and are also necessary as precursors to bioactive compounds, e.g., eicosanoids. It is still under debate whether copepods are able to produce polyunsaturated fatty acids by de novo synthesis or by chain elongation and desaturation of dietary fatty acids. The ability for these biosynthetic steps seems to exist in copepods, but the de novo biosynthesis route appears not to be utilized due to sufficient amounts of polyunsaturated fatty acids in the diet.

Zooplankton species have developed specific adaptations to cope with the different oceanic regimes that exist between the tropics and higher latitudes. In the tropics, the high continuous turnover of carbon at high temperatures and thus high metabolic rates but generally low primary productivity results in usually lipid-poor zooplankton living in the epipelagic waters (Lee and Hirota 1973). In contrast, polar ocean zooplankton, mainly calanoid copepods and krill, seasonally accumulate enormous lipid deposits, which signifies a major specialization in polar bio-production (Kattner and Hagen 1995). The very effective lipid storage via wax esters is primarily related to a herbivorous life strategy and is not an adaptation to high latitude environments per se. Seasonality in food supply is an important factor determining the life-history traits of these copepods. In temperate and high latitudes, coastal, and in upwelling regions, where pulsed phytoplankton production (blooms) is most common, typical wax-ester storing copepods of the genus *Calanus* and *Calanoides* prevail and play a key role in the short and efficient energy flux to higher trophic levels associated with a high productivity (Hagen and Auel 2001; Lee et al. 2006).

Lipids may also play an important role in buoyancy regulation because of its lower density than seawater. The motionless over-wintering of copepods in diapause at depth may be regulated by lipids (Visser and Jónasdóttir 1999), although small changes in the proportion of lipid classes may have a strong impact on buoyancy (Campbell and Dower 2003).

The large (2–10 mm), lipid-rich *Calanus* species have been studied in great detail, but few lipid studies exist on smaller-sized copepods. Lipid class and fatty acid compositions have been determined for small (ca. 1 mm) calanoid species from Kongsfjorden, Svalbard (Lischka and Hagen 2007), the northern Norwegian Balsfjorden (Norrbin et al. 1990), as well as from temperate regions (Kattner et al. 1981; Peters et al. 2006).

Even less is known about the so-called micro-copepods (<1 mm), the cyclopoids *Oithona* and *Oncaea* (Kattner et al. 2003; Lischka and Hagen 2005).

This review will focus on copepods from the Atlantic Ocean ranging from the tropics to the polar regions. Our knowledge of zooplankton lipids is clearly dominated by species from high latitudes, since lipid accumulation and storage is most pronounced in these extreme environments. The latitudinal comparison will give a general field-oriented overview about the significance and characteristics of copepod lipids in different climatic zones. The discussion about lipid transfer within the food web and some ideas concerning the potential impact of global warming on zooplankton complete this review.

11.2 Lipid Patterns of Copepods from Different Latitudinal Regions

Copepod lipids of the world's ocean exhibit various special characteristics. In spite of substantial diversity, there are clear trends in distributional patterns of lipid composition, which are highlighted for the various biomes along a latitudinal gradient.

11.2.1 High-Latitude Copepods

Polar ocean ecosystems are typically characterized by a strong seasonality in light regime and ice cover. Light as a limiting factor allows primary production only during the short summer season, when the pack-ice recedes, and intense phytoplankton blooms develop at the marginal ice zones (Sakshaug 1997; Falk-Petersen et al. 2000; Smetacek and Nicol 2005). Although polar oceans have a rather low overall annual primary production of <20 g C m^{-2} year^{-1}, these blooms nourish food webs containing lipid-rich species from zooplankton to fish, seals, and whales. Typically, polar oceans are known for their short food chains from diatoms via zooplankton to higher trophic levels. The short pulses of intense primary production are sufficient to convert algae biomass into enormous amounts of lipids (50–65% of dry mass) by herbivorous copepods (Fig. 11.1). Lipids represent the most efficient energy stores, because they contain about double the calorific density of proteins or carbohydrates, thus providing a compact source of energy-rich food for higher trophic level organisms.

11.2.1.1 Herbivorous Copepods

The life cycle of the herbivorous Arctic copepods is characterized by massive lipid accumulation during the productive season, followed by a descent of older developmental stages to deeper waters, where they survive the food-limited winter period in diapause and reascend in spring (Smith and Schnack-Schiel 1990; Conover and Huntley 1991; Hirche 1996). In the Arctic, the large calanoid copepod *Calanus*

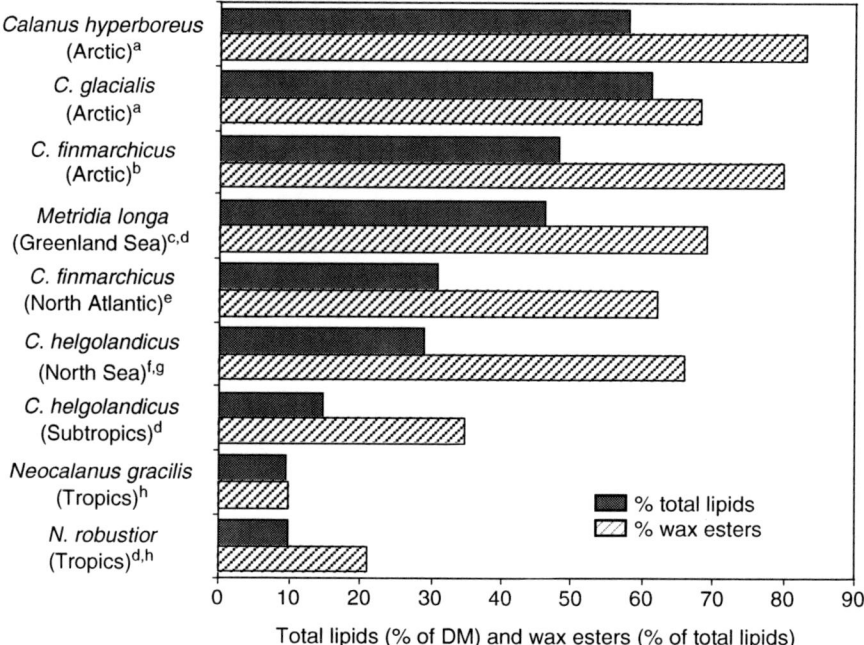

Fig. 11.1 General trends of total lipid content (% of dry mass) and wax esters (% of total lipids) of large copepod species (CV to females) from the Arctic to the tropics. Data compiled from [a]Lee et al. (2006), [b]Hagen (unpublished data), [c]Lee (1975), [d]Lee et al. (1971), [e]Jónasdóttir (1999), [f]Gatten et al. (1979), [g]Kattner and Krause (1989), [h]Kattner (unpublished data)

hyperboreus is a dominant zooplankton species, which represents a prime example of pronounced ontogenetic and seasonal lipid accumulation (Lee 1974; Kattner et al. 1989; Scott et al. 2002). The other herbivorous Arctic species, *C. glacialis*, also accumulates large amounts of lipids. The boreal North Atlantic species, *C. finmarchicus*, which is transported far to the north by Atlantic currents, occurs also in the pack-ice regions of the Greenland Sea (Hirche 1989), and it is present as an expatriate even in the central Arctic Ocean (Mumm 1993). Lipid storage in *C. finmarchicus* is only slightly less pronounced than in the more polar congeners (Fig. 11.1).

These northern *Calanus* species exhibit quite different degrees of independence from primary production to start reproduction. In contrast to *C. glacialis* and *C. finmarchicus*, *C. hyperboreus* relies on its lipids to fuel maturation and egg production independent of the phytoplankton bloom (Conover and Sifferd 1993). The calanoid *C. hyperboreus* reproduces in winter and early spring before the onset of the spring phytoplankton bloom and utilizes lipids accumulated during the previous spring/summer feeding period. Lipid levels of *C. hyperboreus* may decrease from 50 to 15% during reproduction (Hagen unpublished data). In contrast, species such as *C. finmarchicus* usually rely on phytoplankton for successful reproduction (Lee et al. 2006 and references therein), although early in the season initial reproductive processes may also be possible prior to the phytoplankton bloom (Richardson et al. 1999; Mayor et al. 2006).

Total lipids of the calanoid copepods are composed of up to 90% wax esters, which have a very high energy content (Fig. 11.1) (reviewed by Kattner and Hagen 1995; Lee et al. 2006) and consist mainly of long-chain MUFA and fatty alcohols (20:1n-9, 22:1n-11), as well as of fatty acids of dietary origin (Tables 11.1 and 11.2). Up to 50% of the wax esters, biosynthesized de novo by *C. hyperboreus*, may exhibit chain lengths from 40 to 44 carbon atoms, which derive from the combination of 20:1 and 22:1 fatty acid or alcohol units (Kattner and Graeve 1991). C. hyperboreus wax esters reach a higher calorific content when compared with the depot lipids of the other dominant herbivorous species, *C. glacialis* and *C. finmarchicus* from the Arctic and *Calanoides acutus* from the Antarctic (Albers et al. 1996), which have wax esters with shorter chain lengths (34–42 carbon atoms) (Kattner and Graeve 1991). These long-chain monounsaturated fatty acids and alcohols are not present in phytoplankton. They are synthesized de novo only by these herbivorous copepods, which, therefore, exhibit a very characteristic lipid signature (Tables 11.1 and 11.2).

11.2.1.2 Omnivorous and Carnivorous Copepods

Comparable to the herbivorous *Calanus* species, two other highly abundant omnivorous copepod species, *Metridia longa* from the Arctic and *M. gerlachei* from the Antarctic, store lipids mainly as wax esters. However, instead of long-chain monounsaturated fatty acids and alcohols, their major wax ester alcohols are dominated by 14:0 and 16:0 (Table 11.2). This deviation from the typical pattern of herbivorous calanoids probably reflects the more opportunistic feeding behavior and deviating over-wintering strategy (with year-round activity instead of diapause) (Hopkins et al. 1984; Båmstedt and Tande 1988). *M. longa* specimens may reach high lipid and wax ester levels, although not as high as the *Calanus* and *Calanoides* species (Lee 1975; Falk-Petersen et al. 1987; Albers et al. 1996). Fatty acid and alcohol composition are similar to other omnivorous and carnivorous species dominated by shorter-chain fatty acids and alcohols with 14–18 carbon atoms, but usually low percentages of long-chain wax ester moieties (Table 11.1). However, the lipid compositions of *M. longa* are quite variable, and specimens collected during summer contained substantial amounts of, probably dietary, monounsaturated alcohols with 20 and 22 carbon atoms (Table 11.2; Albers et al. 1996). *M. gerlachei* does not strongly rely on internal energy reserves, and the females accumulate only moderate lipid and wax ester levels with maximum lipid values later in the season (Graeve et al. 1994b; Schnack-Schiel and Hagen 1995).

The carnivorous species *Paraeuchaeta antarctica* accumulates large amounts of wax esters, almost exclusively composed of 14:0 and 16:0 alcohol moieties. Another characteristic often found in carnivorous species is the high proportion of the 18:1n-9 fatty acids, which can account for >50% of total fatty acids (Lee et al. 1974; Hagen et al. 1995).

Although wax ester storage is characteristic of most of the copepod species from higher latitudes, there are some exceptions, where triacylglycerols are the principal storage lipid. Copepod species such as *Calanus propinquus*, *C. simillimus*, and

Table 11.1 Large Atlantic copepod species from high to low latitudes. Compositions of major fatty acids and alcohols (mass% of total fatty acids or alcohols) of total lipids

	Calanus hyperboreus[a]	Calanus glacialis[a]	Calanoides acutus[b]	Calanus finmarchicus[a]	Metridia longa[c]	Calanus finmarchicus[d]	Calanus helgolandicus[e]	Neocalanus gracilis[c]	Neocalanus robustior[c]
Location	Arctic	Arctic	Antarctic	Arctic	Arctic	N. Atlantic	North Sea	Tropics	Tropics
Fatty acids									
14:0	3.7	9.8	3.6	16.9	4.1	8.2	11.1	3.3	4.0
16:0	4.3	6.9	3.2	12.7	6.8	15.5	12.5	20.1	22.8
16:1n-7	10.6	25.2	8.5	6.2	29.5	2.7	5.4	2.3	3.1
18:0	0.4	0.4	–	1.5	0.1	1.6	1.5	5.0	6.0
18:1n-9	3.2	3.7	5.0	5.3	29.3	4.6	4.9	7.7	7.9
18:1n-7	0.9	1.0	1.0	0.4	1.1	1.4	–	2.0	1.6
18:2n-6	1.7	0.9	1.7	1.8	2.0	4.3	1.0	1.7	2.5
18:3n-3	0.7	0.5	0.8	1.1	0.5	5.2	2.0	0.5	0.9
18:4n-3	10.3	3.2	9.1	9.5	1.6	12.8	3.4	1.1	1.4
20:1n-9	19.8	12.3	17.9	7.7	3.3	4.1	4.1	1.5	2.0
20:1n-7	1.9	1.0	0.9	1.0	–	–	–	0.5	0.3
20:5n-3	14.1	16.0	20.1	13.2	9.8	14.9	17.4	9.1	9.9
22:1n-11	15.0	7.1	9.1	8.0	1.4	5.0	7.7	6.9	4.6
22:1n-9	3.5	1.1	4.0	0.3	0.1	0.5	–	3.2	2.0
22:5n-3	1.0	0.6	1.4	0.3	–	0.4	–	2.1	1.7
22:6n-3	7.8	5.2	10.1	11.6	8.8	10.7	18.2	20.1	17.6
Alcohols									
14:0	2.8	3.2	7.5	1.7	54.2	2.0	2.5	n.d.	n.d.
16:0	6.1	11.2	8.4	9.6	20.6	16.7	17.0	n.d.	n.d.
16:1n-7	3.6	7.1	4.3	3.2	6.6	1.9	1.4	n.d.	n.d.
18:1n-9	0.4	–	1.6	1.7	–	3.0	1.7	n.d.	n.d.
18:1n-7	0.5	2.1	–	2.6	–	–	1.7	n.d.	n.d.
20:1n-9	32.6	43.4	49.5	36.6	10.6	30.9	27.1	n.d.	n.d.
22:1n-11	55.0	30.4	28.7	44.6	8.0	38.1	46.6	n.d.	n.d.

n.d. not determined; – not present or below 0.05%
Compiled from [a]Dalsgaard et al. (2003), [b]Hagen et al. (1993), [c]Kattner et al. (unpublished data), [d]Kattner (1989), [e]Kattner and Krause (1989)

11 Lipids in Marine Copepods

Table 11.2 Fatty acid and alcohol compositions of wax esters and phospholipids, respectively, of Arctic copepod species (data from Albers et al. 1996)

	Wax esters				Phospholipids			
	Calanus hyperboreus	Calanus glacialis	Calanus finmarchicus	Metridia longa	Calanus hyperboreus	Calanus glacialis	Calanus finmarchicus	Metridia longa
Fatty acids								
14:0	6.4	13.1	26.3	0.7	4.0	5.2	3.3	2.3
16:0	5.8	6.1	9.8	1.5	25.5	25.8	25.8	20.7
16:1n-7	11.7	32.9	6.7	21.4	2.5	4.6	1.1	1.9
18:0	0.6	–	0.9	0.7	2.3	2.3	3.6	2.6
18:1n-9	5.8	5.5	5.3	30.3	5.1	6.4	2.5	5.3
18:1n-7	1.6	1.1	0.3	1.1	1.4	3.7	1.0	2.2
18:2n-6	3.6	1.0	1.2	2.5	1.8	2.0	1.5	1.4
18:3n-3	1.6	0.3	1.5	0.5	0.5	0.2	0.6	0.2
18:4n-3	6.2	0.5	13.7	1.4	0.8	–	2.5	0.3
20:1n-9	19.0	23.0	7.8	19.5	0.6	0.9	0.2	1.5
20:1n-7	1.5	1.0	0.9	–	–	–	–	–
20:5n-3	7.0	2.7	11.4	7.1	18.6	16.5	19.2	15.9
22:1n-11	17.3	8.3	7.0	9.4	1.2	–	0.2	–
22:1n-9	3.2	2.0	0.2	–	–	–	–	–
22:5n-3	0.5	–	0.2	–	–	–	0.2	–
22:6n-3	2.4	0.8	2.2	0.5	33.8	30.5	37.4	43.9
Alcohols								
14:0	4.4	2.1	3.9	46.1	–	–	–	–
16:0	11.2	9.3	14.6	17.7	–	–	–	–
16:1n-7	1.6	5.3	3.4	0.9	–	–	–	–
20:1n-9	27.8	58.4	39.3	15.5	–	–	–	–
22:1n-11	55.0	25.0	38.8	19.8	–	–	–	–

Euchirella rostromagna store large amounts of triacylglycerols (Hagen et al. 1993, 1995; Kattner et al. 1994; Ward et al. 1996). The primarily herbivorous Antarctic copepod, *C. propinquus*, exclusively stores triacylglycerols, which seems to be related to a more opportunistic feeding behavior and active overwintering without diapause (Hopkins and Torres 1989; Schnack-Schiel and Hagen 1995). This species has, however, developed a very effective fatty acid biosynthesis, which at least partially compensates for the lack of wax esters. This is achieved by producing long-chain MUFA comparable to wax ester-storing copepods, made even more effective by elongating the 20:1n-9 fatty acid to 22:1n-9 (Hagen et al. 1993). Usually, this 22:1 fatty acid isomer is either absent or only found in very small amounts in marine zooplankton organisms. The addition of two carbon atoms results in a higher calorific content and allows for more efficient lipid utilization, since every fatty acid has to be activated by one ATP for catabolism (β-oxidation), independent of its chain length.

Few lipid analyses exist of small and micro-copepods (Fig. 11.2, Table 11.3). Only lipids of the abundant Arctic cyclopoids *Oithona similis* and *Oncaea borealis* have been studied in detail (Kattner et al. 2003; Lischka and Hagen 2007). The latter species appears to be a genuine cold-water species and is suggested to be the only true Arctic *Oncaea* species (Sewell 1947; Richter 1994; Auel and Hagen 2002), while in the Antarctic, *Oithona similis*, *Oncaea curvata*, and *O. antarctica* dominate the copepod assemblages (Schnack et al. 1985; Hopkins and Torres 1989; Metz 1995).

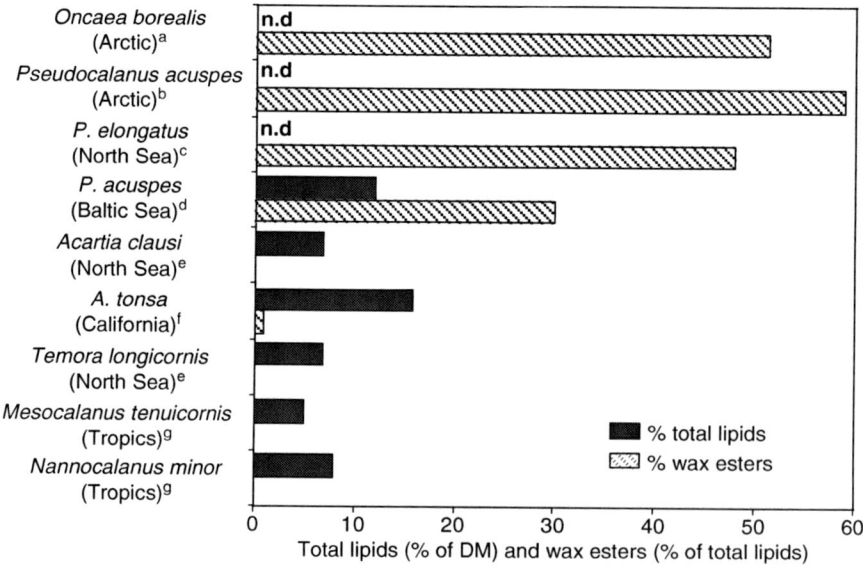

Fig. 11.2 General trends of total lipid content (% of dry mass) and wax esters (% of total lipid) of small copepod species (CV to females) from the Arctic to the tropics; no bar means no wax esters; *n.d.* not determined. Data compiled from [a]Kattner et al. (2003), [b]Norrbin et al. (1990), [c]Kattner and Krause (1989), [d]Peters et al. (2006), [e]Laakmann (2004), [f]Lee and Hirota (1973), [g]Kattner (unpublished data)

Table 11.3 Small Atlantic copepod species from high to low latitudes. Compositions of major fatty acids and alcohols (mass% of total fatty acids or alcohols) of total lipids

	Oncaea borealis[a]	Oithona similis[b]	Pseudocalanus minutus[b]	Pseudocalanus acuspes[c]	Acartia clausi[d]	Temora longicornis[d]	Mesocalanus tenuicornis[e]	Nannocalanus minor[e]
Location	Arctic	Arctic	Arctic	Baltic Sea	North Sea	North Sea	Tropics	Tropics
Fatty acids								
14:0	3.9	4.0	1.8	1.0	2.2	2.5	2.5	2.3
16:0	13.4	20.6	11.9	15.0	14.7	15.5	17.1	17.6
16:1n-7	5.0	8.1	14.4	2.4	1.5	3.3	1.3	1.3
18:0	3.0	5.4	0.7	2.5	4.7	3.1	8.5	8.5
18:1n-9	27.9	21.2	26.2	18.9	5.0	3.1	5.0	4.6
18:1n-7	3.4	2.9	2.6	2.0	1.4	2.9	1.2	1.2
18:2n-6	2.0	2.7	0.2	5.6	0.9	0.8	2.0	1.8
18:3n-3	0.8	–	–	3.3	1.1	0.9	0.5	0.4
18:4n-3	2.6	–	–	3.6	4.0	4.1	0.8	0.6
20:1n-9	10.0	–	–	–	0.9	1.2	0.4	0.4
20:1n-7	0.4	–	–	–	0.8	0.7	–	–
20:5n-3	13.9	16.7	24.1	15.2	17.9	22.4	11.8	12.1
22:1n-11	6.7	–	–	–	0.5	0.6	–	0.2
22:1n-9	1.7	–	–	–	0.5	0.8	–	0.1
22:5n-3	1.4	–	–	–	0.6	0.6	1.0	0.9
22:6n-3	13.3	16.9	12.4	22.0	38.4	30.9	37.7	39.3
Alcohols								
14:0	23.3	22.3	50.0	26.7	–	–	–	–
16:0	46.2	40.0	47.7	62.7	–	–	–	–
16:1n-7	0.9	–	–	–	–	–	–	–
18:0	–	8.2	0.4	5.1	–	–	–	–
18:1n-9	–	–	–	5.2	–	–	–	–
20:1n-9	15.7	25.8	–	0.5	–	–	–	–
22:1n-11	14.1	–	–	–	–	–	–	–

Compiled from [a]Kattner et al. (2003), [b]Lischka and Hagen (2007), [c]Peters et al. (2006), [d]Laakmann (2004), [e]Kattner et al. (unpublished data)

In addition, the lipid class and fatty acid compositions of *Pseudocalanus acuspes* and *Acartia longiremis* have been studied in Balsfjorden, northern Norway (Norrbin et al. 1990), and the seasonal lipid dynamics of *Pseudocalanus minutus* in the Arctic Kongsfjorden, Svalbard (Lischka and Hagen 2007).

The accumulation of considerable wax ester deposits (in % total lipids) in these smaller copepods (Fig. 11.2) is comparable to that of the larger polar species. Small copepods may use their lipid reserves during periods of food shortage in winter and for somatic growth, whereas reproductive processes are usually fuelled by the spring phytoplankton bloom, although in *Oithona similis* overwintering and gonad maturation seems to rely on internal lipid reserves (Lischka and Hagen 2007). Their lipids are not composed of long-chain fatty acids and alcohols typical of the larger species. The wax ester moieties are usually dominated by the shorter-chain alcohols 14:0 and 16:0, which is also indicative of opportunistic feeding (Table 11.3) (Graeve et al. 1994b). High percentages of the 18:1n-9 fatty acid during all seasons (Table 11.3) also point to a generally non-selective diet (Dalsgaard et al. 2003; Kattner et al. 2003; Lischka and Hagen 2007). This suggests that the small copepods are generally opportunistic feeders, as proposed by Paffenhöfer (1993), although their life cycles are related to the pronounced seasonality of food availability in polar regions (Metz 1995). It should be noted that higher proportions of long-chain monounsaturated fatty acids/alcohols in some species probably also reflect carnivory on other zooplankton organisms, including early stages of calanoid copepods.

11.2.2 Mid-Latitude Copepods

From polar to temperate regions, there is a gradual transition in the lipid characteristics among copepod species, which inhabit both areas (Fig. 11.1). This holds true for several smaller species as well as for *Calanus finmarchicus*, a species, in which these differences also depend on water mass distribution (Heath et al. 2004). In temperate regions, phytoplankton production is not as limited as in the Arctic due to the less seasonal light regime, which, together with the elevated temperatures, supports life-cycle strategies of herbivorous copepods characterized by shorter generation times. Primary production cycles in the temperate North Atlantic, for example, are governed mostly by nutrient supply and classically exhibit a spring bloom dominated by diatoms. Thereafter, a phytoplankton community exists at lower levels throughout the summer and in autumn a second smaller bloom may occur due to vertical mixing of the nutrient-depleted surface waters with nutrient-rich deeper waters.

11.2.2.1 Large Copepods

Similar to the Arctic Ocean, phytoplankton blooms are utilized by the larger copepods from temperate regions, such as *Calanus finmarchicus*, *C. helgolandicus*, *C. pacificus*, and *Neocalanus* spp., to accumulate lipids, although total lipid levels are lower than in the polar species (Fig. 11.1). These boreal copepods also exhibit a life cycle with seasonal vertical migration, diapause, and deposition of mainly wax esters in an oil sac. Although not quite reaching the levels of polar species, wax ester values of temperate copepods are also high (mean of 65% of total lipids). During their ontogenetic development from copepodite stage CI to adult stage (*C. finmarchicus* s.l.), wax ester levels increased from a minimum of 20 to 87% of total lipids (Kattner and Krause 1987). Off California, CIII to adult specimens of *C. pacificus* showed wax ester accumulation from 1 to 50%, whereas younger stages had no wax esters (Lee et al. 1972). The later copepodite stages of *C. helgolandicus* may range seasonally between 23 and 90% wax esters of total lipids with means ranging from ~66% (Kattner and Krause 1989) to 80% (Gatten et al. 1979). Wax esters of these temperate species, although more variable, closely resemble those of polar copepods with high percentages of long-chain monounsaturated fatty acids and alcohols (Table 11.1). Since dietary fatty acids change seasonally with the availability of phytoplankton species, the fatty acid and alcohol patterns seem to be more dependent on phytoplankton peak events than on temperature regimes and latitude.

Comparing specimens of *C. finmarchicus* from high latitude and temperate regions, Kattner (1989) found that *C. finmarchicus* from temperate seas exhibited a significantly smaller size and dry mass. Specimens from warmer regions are often smaller and have more generation cycles than their congeners in cold regions (Mauchline 1998). In addition, the percentage of long-chain monounsaturated alcohols was lower in temperate specimens (Table 11.1). The de novo synthesis of these energy-rich lipids seems to be less pronounced in the temperate species, probably due to a more reliable year-round food availability. The portion of wax esters synthesized also tended to be slightly smaller than in the Arctic congeners (Fig. 11.1).

When compared with the North Atlantic, production cycles are different in the subarctic Pacific, and phytoplankton blooms do not develop in spring and autumn. Micrograzers, e.g., heterotrophic microflagellates and ciliates, seem to control the developing algal biomass, whereas the assumed top-down control by "major grazers," the dominant copepods of the genus *Neocalanus*, *Eucalanus*, and *Metridia*, could not be verified due to insufficient grazing rates (Miller 1993).

Copepods of the genus *Neocalanus* exhibit an interesting life-cycle strategy: they go into diapause in summer/fall and produce lipid-rich eggs fuelled by internal energy reserves, which are spawned at depth during winter. There is another interesting but not yet understood difference between the Atlantic/Arctic *Calanus* species and the Pacific calanoid copepods *Neocalanus cristatus* and *N. flemingeri*, which synthesize mainly 20:1n-11 wax ester moieties (Saito and Kotani 2000) rather than 20:1n-9.

The Pacific copepod *Metridia okhotensis* stores wax esters comparable in amount and composition with its Atlantic counterpart *M. longa*. Other Pacific copepod species, e.g., *Eucalanus bungi* and *Euchirella* spp., store primarily triacylglycerols (Lee et al. 1971; Saito and Kotani 2000), as do the high-latitude species *Calanus propinquus*, *C. simillimus*, and *Euchirella rostromagna* (Hagen et al. 1993, 1995; Ward et al. 1996).

11.2.2.2 Small Copepods

The herbivorous Atlantic copepod *Pseudocalanus elongatus* occupies an intermediate position between the large herbivorous northern and the small omnivorous copepods of the temperate regions. The lipid composition of *P. elongatus* is similar to the Arctic and northern North Atlantic species due to its wax ester storage (~50% of total lipids; Fig. 11.2). These wax esters are also primarily composed of the short-chain alcohol moieties 14:0 and 16:0 (Table 11.3) (Fraser et al. 1989; Kattner and Krause 1989). During a seasonal study in the Baltic Sea, Peters et al. (2006) found similar results for *P. acuspes* with wax ester levels between 17 and 44% and triacylglycerol levels between 15 and 35% of total lipids. Fatty alcohols were also clearly dominated by 14:0 and 16:0 and fatty acids by 18:1n-9. The feeding behavior for *P. acuspes* is suggested to be opportunistic, and important food items are ciliates and cyanobacteria, in addition to diatoms and flagellates. In contrast to the primarily herbivorous *Calanus* species, the smaller calanoid species of the genus *Acartia (longiremis, tonsa), Temora (longicornis),* and *Centropages (typicus)* favor life strategies which include opportunistic feeding behavior, high metabolic rates, a limited lipid accumulation and in some cases (e.g., *Acartia*) resting eggs as survival mechanisms (e.g., Peters et al. 2007). Therefore, reproductive effort is directly dependent on food availability and is not buffered by lipid reserves. The lipid compositions of these species are usually characterized by high phospholipid and low neutral lipid levels (little triacylglycerols, no wax esters; Fig. 11.2). Accordingly, fatty alcohols are missing and principal fatty acids include 20:5n-3, 22:6n-3, and 16:0 typical of biomembranes (Table 11.3). Very little is known about the fatty acid composition of the neutral lipid fraction due to the minute amounts available. Few fatty acid compositions have been determined of separate lipid classes of small species: *Temora longicornis* from the North Sea (Fraser et al. 1989), *P. acuspes, T. longicornis,* and *A. longiremis* from the Baltic Sea (Peters et al. 2006, 2007). Dietary fatty acids such as 16:1n-7 and 18:4n-3 are found to be quite abundant in the neutral lipid fraction. We assume that the lipid biochemistry of these smaller copepods is less important for survival and reproduction and is less dependent on bloom events when compared with the large *Calanus* species.

11.2.3 Low-Latitude Copepods

In one of the first latitudinal comparisons Lee and co-authors described the minor importance of lipids and wax esters in some copepod species, e.g., *Neocalanus gracilis, N. robustior, Euchaeta marina,* from the upper 250 m of tropical waters. Many species lack wax esters, some have minor to moderate amounts of wax esters (range: trace to 40% of total lipids, median: trace) and most species have moderate amounts of triacylglycerol (range: 3–18.5%, median 6%). In deeper layers (below 500 m), however, lipids and wax esters were increasingly accumulated (wax ester range: 11–72%, median 63%) (Lee and Hirota 1973).

In the oligotrophic, epipelagic tropical oceans lipid deposits are not a key component in the life strategies of copepods (Figs. 11.1 and 11.2) and correspondingly, the lipid data-base is rather limited. In these latitudes, evolutionary pressure has selected against a sophisticated lipid biochemistry and favored opportunistic feeding. Typical life strategies include continuous feeding and rapid utilization of available food items for growth and reproduction, which results in a high turnover with short generation times.

Our lipid data of typical epipelagic tropical and subtropical copepod species from the East Atlantic with different ecological niches and feeding behavior support the general patterns of low lipid contents and little storage lipids (Figs. 11.1 and 11.2). These species include herbivorous and omnivorous species of the genus *Neocalanus*, *Mesocalanus*, *Nannocalanus*, and *Clausocalanus*, which are all widespread in tropical and subtropical waters and well adapted to oligotrophic conditions (Bradford-Grieve et al. 1999). Their lipids are essentially composed of phospholipids, except for the larger *Neocalanus* species, which have small amounts of wax esters (~10% of total lipids). The lipid compositions are clearly dominated by the fatty acids 16:0, 22:6n-3, and 20:5n-3 typical of membrane phospholipids. Additional fatty acids include 18:0 and 18:1n-9 (Tables 11.1 and 11.3).

Similar results were reported for *Clausocalanus farrani*, *C. furcatus*, and *Ctenocalanus vanus* from the subtropical Gulf of Aqaba (Red Sea), which showed low wax ester levels. Wax ester moieties consisted of shorter-chain alcohols (14:0, 16:0, 18:0), and fatty acid compositions were similar to those described above (Cornils et al. 2007).

11.2.4 Upwelling-System Copepods

The limited lipid accumulation of low-latitude epipelagic copepods is in stark contrast to copepod species living in major coastal upwelling areas, as well as deeper-living species (>500-m depth) found in tropical and subtropical waters (Lee et al. 1971, 2006). The high-productivity regions periodically or seasonally generate strong pulses of primary production, usually diatom blooms, which typically support short food chains with calanoid copepods and clupeiform fishes (e.g., anchovies, sardines) as major components. This periodicity is comparable to the seasonal food supply at temperate and high latitudes and thus creates a similar strategy of lipid accumulation. The same might be true for deeper-living species outside of the upwelling zones, which feed infrequently during occasionally occurring sedimentation events, but generally experience low prey densities.

Major coastal upwelling regions include the California and Humboldt Current systems in the East Pacific, the Canary and Benguela Current systems in the East Atlantic and the monsoon-driven Somali Current system in the Indian Ocean. These areas are dominated by calanoid copepods, e.g., *Calanus marshallae*, *C. pacificus*, *C. chilensis*, and *Calanoides carinatus* (Mauchline 1998; Petersen 1998; Lee et al. 2006). The latter species, the herbivorous *C. carinatus*, is a key component of the Atlantic and Indian Ocean upwelling systems including the Benguela region, where it exhibits typical life-history traits of *Calanus* and *Calanoides* species from high latitudes. Major features include intense feeding on phytoplankton in productive surface waters, use of surplus energy for rapid biosynthesis of wax esters with long-chain monounsaturated fatty acids and alcohols and accumulation of these lipid deposits in oil sacs, as well as vertical migration to deeper layers >500 m and diapause at depth to survive periods of food paucity (Petersen 1998; Verheye et al. 2005).

11.3 Essential Fatty Acids and Phospholipid Structure

Essential fatty acids, transferred and exchanged within the food web, are crucial components not only in marine life but also for human nutrition (e.g., Arts et al. 2001, and see Chap. 14). The term essential means that animals are unable to synthesize these specific fatty acids, or that these fatty acids are produced in insufficient amounts by animals. There is still an ongoing discussion about the biosynthesis and amount of essential fatty acids produced by zooplankton and animals in general (see Chap. 13) and those from precursor fatty acids ingested with the diet.

In the marine environment, the omega-3 fatty acids, EPA and DHA, are generally thought to be essential fatty acids. Phytoplankton is the primary source of these fatty acids in the oceans. However, phytoplankton groups have clearly different fatty acid compositions. Diatoms, for example, are rich in EPA and dinoflagellates are rich in DHA. Hence, spring blooms will offer high amounts of EPA, whereas later in the year during the summer season with a lower standing stock of phytoplankton, higher proportions of DHA are available for secondary production. To satisfy their physiological demands for essential fatty acids, herbivorous copepods, which play a pivotal role in the initial transfer of essential fatty acids along the food web, need to feed on a variety of phytoplankton species throughout the year.

In copepods, but also in marine animals in general, EPA and DHA represent the dominant fatty acids in phospholipids, which are the principal membrane constituents and thus a major source of essential fatty acids for higher trophic levels. Together with the saturated 16:0 fatty acid, they can comprise up to 80% of total phospholipid fatty acids. Membrane fatty acids reflect structural requirements and vary only somewhat with dietary changes. Copepods have a quite uniform composition of phospholipid fatty acids (Table 11.2, Fig. 11.3), and variations mostly occur within the proportions of the three major fatty acids.

In general, phospholipid molecules are composed of a saturated or monounsaturated fatty acid at the first carbon atom of the glycerol backbone (position *sn-1*) and a polyunsaturated fatty acid at the second carbon atom (position *sn-2*). This asymmetric fatty acid distribution is thought to be of major importance in the functional and structural roles of biomembranes (see Chap. 10). In many copepod species, there is, however, a surplus of polyunsaturated fatty acids in the phospholipids, which means that a polyunsaturated fatty acid is also located in *sn-1* position. The major phospholipid class in *Calanus hyperboreus* and *C. finmarchicus* is phosphatidylcholine (PC) followed by phosphatidylethanolamine (PE). The fatty acid composition differs in that PC is composed of equal amounts of EPA and DHA (both ~35%), whereas in PE half of the fatty acids are composed of DHA with smaller contributions of EPA

11 Lipids in Marine Copepods

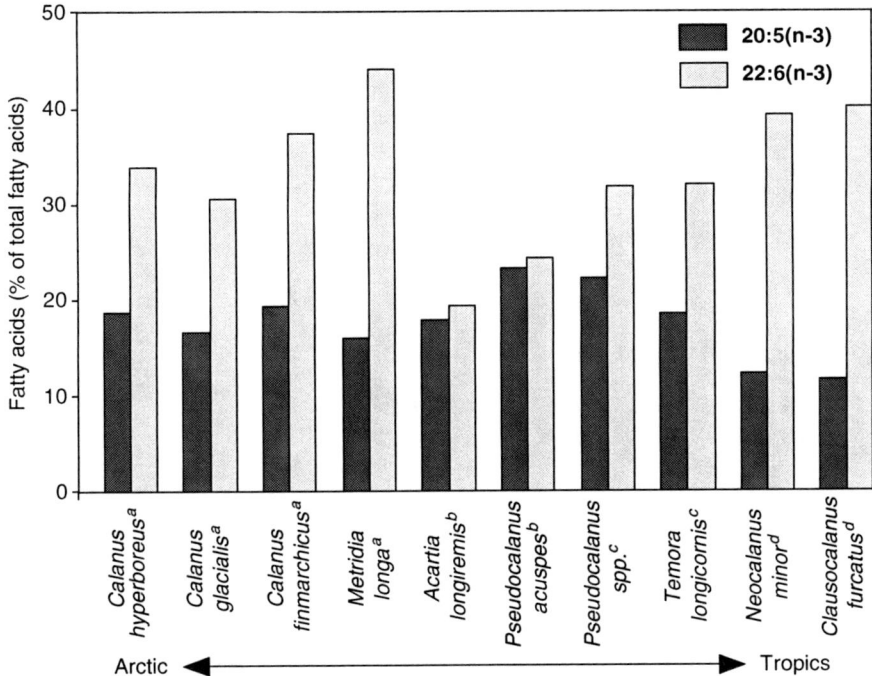

Fig. 11.3 Distribution of the polyunsaturated fatty acids 20:5n-3 and 22:6n-3 in phospholipids of copepods from high to low latitudes. Data compiled from [a]Albers et al. (1996), [b]Norrbin et al. (1990), [c]Fraser et al. (1989), [d]Kattner data of total lipids, mainly composed of phospholipids (unpublished)

(~10%; Table 11.4). Thus, the distribution of fatty acids in PE follows the general pattern of phospholipids, however, with mainly DHA in the *sn-2* position. In PC, a high proportion of EPA is also located in *sn-1* position (Fig. 11.4). Little is known about the advantages and reasons for this remarkable phospholipid pattern. The ratio of PE to PC reflects compensatory mechanisms that allow maintenance of physical–chemical membrane properties with changing temperatures (Shinitzky 1984).

Increasing amounts of polyunsaturated fatty acids in phospholipids are reported to be important in regulating membrane fluidity at cold temperatures (e.g., Farkas 1979; Hall et al. 2002). This regulatory mechanism is still under discussion, and it should be taken into account that the phase transition temperature of highly unsaturated phospholipids is not markedly lower than that of phospholipids composed of fatty acids with fewer double bonds (reviewed by Stillwell and Wassall 2003). The high proportions of EPA and DHA in tropical zooplankton phospholipids (Fig. 11.3) further challenge the reasoning with regard to membrane fluidity. There are still many open functional questions, which need to be studied in the context of biomembrane structure and function.

In addition, the dominant polar euphausiids accumulate PC simultaneously with increasing lipid levels (Hagen et al. 1996). Most of these phospholipids were found

Table 11.4 Fatty acid composition of phosphatidylcholine (PC) and phosphatidylethanolamine (PE) of *Calanus hyperboreus* and *C. finmarchicus* (Kattner and Farkas, unpublished data)

	Calanus hyperboreus		Calanus finmarchicus	
Phospholipid	PC	PE	PC	PE
Fatty acids				
14:0	3.6	1.1	2.7	0.6
16:0	12.4	23.2	10.8	24.7
16:1n-7	3.6	0.6	0.7	0.2
18:0	0.5	0.8	0.8	2.4
18:1n-9	2.5	2.7	1.5	0.4
18:1n-7	0.5	0.7	1.7	0.7
18:2n-6	1.0	0.3	1.1	0.1
18:3n-3	0.8	0.4	1.1	0.3
18:4n-3	2.4	0.8	3.0	0.6
20:4n-6	0.7	–	1.4	0.2
20:5n-3	36.5	12.1	35.3	8.0
22:5n-3	0.4	0.2	0.6	0.6
22:6n-3	32.6	54.7	37.6	57.3

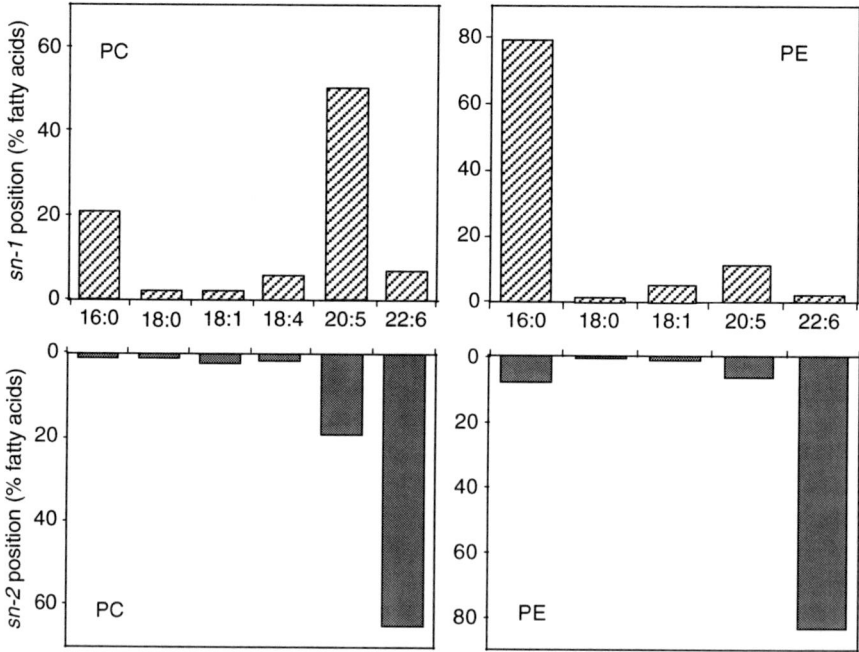

Fig. 11.4 Position-specific distribution of fatty acids of phosphatidylcholine (PC) and phosphatidylethanolamine (PE) of *Calanus finmarchicus*. The fatty acid distribution of *C. hyperboreus* phospholipids is similar (Kattner and Farkas, unpublished data)

in the free form in tissue surrounding the hepatopancreas, i.e., not incorporated into membranes (Stübing 2004). These lipids are an additional source of essential fatty acids because their high percentages of EPA and DHA are comparable to those of membrane lipids (Hagen et al. 2001).

The neutral lipids, wax esters and triacylglycerols, contain clearly lower percentages of EPA and DHA. EPA is usually more abundant, but comprises mostly <10%, whereas DHA occurs in even lower percentages (Table 11.2; Albers et al. 1996; Scott et al. 2002). Owing to the huge amounts of wax esters in high-latitude copepods, neutral lipids do improve the overall food quality (total essential fatty acid content) of copepods for higher trophic level organisms.

11.4 Impact of Global Warming on Lipid Dynamics

Polar oceans are characterized by an enormous lipid accumulation in major components of the pelagic food web with an intensive and fine-tuned energy transfer from copepods via fish to seabirds and marine mammals. This lipid-based energy flux in high-latitude ecosystems may be easily disturbed, since Polar Regions are especially sensitive to global changes (Flato et al. 2000). Recent investigations and modeling have shown that the Arctic ice cover is already strongly reduced in size and thickness (e.g., Serreze et al. 2007). The Fourth Assessment Report "Climate Change 2007" of the IPCC has unequivocally declared that global warming will likely continue in the near future, unless dramatic political measures are taken.

If we expect a poleward shift of ecosystem boundaries, i.e. tropical species expanding toward temperate regions and temperate species in turn moving to higher latitudes, then the dominance of lipid-rich species at high latitudes will probably decrease. This could have a tremendous impact on the food quantity and energy flux to higher trophic levels in polar oceans. In contrast, food quality will probably be less affected concerning essential fatty acids, since temperate and tropical species are also rich in these components. In the north, there might be competition between the three dominant *Calanus* species, as *C. finmarchicus* will probably shift toward the Arctic Ocean shelf areas competing with *C. hyperboreus* and *C. glacialis* during spring phytoplankton blooms (Gradinger 1995), which might then be more comparable to blooms in temperate regions. In some Arctic regions, these spring blooms might become less pronounced, since nutrient concentrations are currently low in Siberian shelf waters, and the nutrient input from the large Russian rivers is relatively small (Dittmar and Kattner 2003). However, the discharge of nutrients might also change due to melting of the huge permafrost areas (Wu et al. 2005). If nutrient supply by riverine discharge increases, phytoplankton blooms associated with river plumes may become more important in the Arctic Ocean than marginal ice zone blooms, and the whole ecosystem will change to a more temperate system. These shelf blooms may increasingly replace marginal ice zone blooms and, consequently, ice algae production will decline, a resource crucial to many herbivorous copepods.

The reduction in ice cover will have a general effect on the food supply of dominant zooplankton species in polar regions. At the underside of the ice, epontic phytoplankton communities adapted to very low light intensities grow early in the season, and various animals feed on these algae such as Antarctic krill and Arctic amphipods. Some zooplankton species, e.g., *Calanus glacialis* and *Pseudocalanus* spp., are known to migrate into the ice for feeding (Runge et al. 1991).

The light regime will still be a limiting factor at high latitudes, but an earlier break-up and later freezing or even disappearance of the sea-ice may prolong the productive season. It has been postulated that herbivorous copepods are affected primarily by changes in phytoplankton species composition and abundance, not so much by temperature shifts per se (Hirche 1987; Richardson and Schoeman 2004). These changes may have a strong impact on the timing of bloom events, which are again intertwined with the reproduction cycles of, for example, the *Calanus* species in the Arctic, which may no longer match with primary production cycles (Hansen et al. 2003). These modifications may in turn favor smaller and more flexible zooplankton taxa known from temperate waters such as copepods of the genus *Metridia, Temora,* and *Acartia.* It is difficult to predict the influence on the overall lipid flux within the food web, because this is dependent on the abundance of these species. Such a change in zooplankton species composition will obviously change the energetic value of the lipids because of the reduction of species with large amounts of lipids rich in long-chain wax ester moieties. The occurrence of essential fatty acids, such as EPA and DHA, is probably less impacted by global changes, since they are structural components of all marine bio-membranes, largely independent of latitudinal distribution. The amount of these essential lipids, however, is again dependent on zooplankton biomass, but the balance of food quality and food quantity is difficult to predict.

If global warming leads to the disappearance of key biomass species especially adapted to polar ice-covered regions, thus diminishing the present lipid-based energy flux (Falk-Petersen et al. 2007), this may induce a regime shift from a higher to lower-level energy system with cascading effects up the food web that will eventually threaten the entire intricately balanced system, in which lipids play a central role. Hence, biodiversity and energy flux of polar systems could be severely influenced by these changes.

The impact of global warming on the Southern Ocean ecosystem will be quite different, because the system is less pulsed when compared with the Arctic Ocean, due to differences in the light regime, iron-limitation of phytoplankton growth, and a more pronounced seasonal sea-ice cover, which is reduced to ~20% during summer. In contrast to the Arctic, the Southern Ocean is a krill-dominated region, and the life strategies of this key group are intensely linked to the seasonality of the ice regime. Hundred million tons of krill biomass strongly rely on the seasonal development of ice algae and exhibit a pronounced seasonal lipid accumulation, mainly triacylglycerols, and PUFA-rich phospholipids (Hagen et al. 2001). In the past, colder winters have been associated with stronger year classes of krill in the following year, indicating that the survival of the Antarctic krill, *Euphausia superba*, is strongly dependent on ice cover, which allows successful over-wintering of the larvae (Loeb et al. 1997; Siegel 2005). In contrast, warmer winters in the Antarctic with a less pronounced ice coverage are advantageous for salp stocks (*Salpa thompsoni*), which extend their distribution centers southward under favorable conditions. Of course, such a regime shift from lipid-rich krill to low-calorie gelatinous salps will lead to severe perturbations at higher trophic levels and a dramatic change in food availability for e.g., seals, whales, penguins, and other seabirds. The effects of such changes can already be observed today (Atkinson et al. 2004).

11.5 Conclusions and Perspectives

Lipids in marine organisms are highly diverse and complex and exhibit various special characteristics. Typical marine lipids include wax esters as energy depots, long-chain polyunsaturated omega-3 fatty acids synthesized by phytoplankton as well as long-chain monounsaturated fatty acids and alcohols synthesized by calanoid copepods. Even if we restrict our view to lipids in copepods, the numerous life cycle strategies of copepods make it difficult to establish general features. It is clear that high-latitude copepods are much more lipid-dependent than tropical species. However, copepods in upwelling regions and deeper-living species also accumulate huge amounts of lipids for survival. It is now generally accepted that the seasonality or periodicity in food supply is the driving force for the evolution of specific life-cycle strategies in these cold-water systems, including lipid accumulation and wax ester storage. The more pronounced the seasonality the higher the amounts of lipids produced and accumulated. Lipid levels of zooplankton in polar regions can comprise more than half of the total carbon, whereas zooplankton in warm epipelagic waters of the tropics store almost no lipids.

In spite of many detailed lipid studies of various zooplankton species, there are still important discoveries to be made. Examples are the finding that the important 20:1n-9 component in North Atlantic *Calanus* species is substituted by 20:1n-11 in North Pacific *Neocalanus* species (Saito and Kotani 2000). Reasons for this difference are unknown. Another example is the predominance of the 18:1n-9 and 18:1n-7 alcohols in the wax esters of the Antarctic *Thysanoessa macrura* (Kattner et al. 1996). These alcohols are usually only minor lipid components in zooplankton species. The lipid class compositions can also be very unusual: for instance, the pteropod *Clione limacina* produces significant amounts of diacylglycerol ethers in combination with odd-chain fatty acids (Kattner et al. 1998). These unusual biosynthetic pathways are still unexplained.

We suppose that many of the deeper-living zooplankton species have also developed special lipid adaptations, which still need to be elucidated. The lipid compositions of frequently occurring copepods, especially of small species, have not yet been determined, although we suspect that such species will probably exhibit a "normal" lipid composition. A large gap exists concerning the early developmental stages. Almost no lipid data are available for most of the nauplii and early copepodids, which all are, of course, difficult to collect in sufficient numbers and to differentiate into their respective stages. Data sets on the seasonality of lipid levels and composition need to be improved. Experimental studies following lipid assimilation via labeled diets (Graeve et al. 2005) will considerably improve our knowledge on pathways of lipid biosynthesis and the incorporation of essential fatty acids. We know that lipids undergo various changes during development and are influenced by environmental conditions, but the results are still fragmentary and sometimes contradictory. Lipid research is also needed on biosynthetic pathways, metabolism, and membrane structure and functioning.

The lipid data base for prognostic modeling is thus still insufficient, which makes it difficult or even impossible to make sound predictions. However, all these data are important to predict changes in zooplankton distribution and migration to higher-latitude regions, which become more favorable to temperate species due to future changes in climate. Because zooplankton, especially herbivorous copepods, represent the crucial link between primary producers and consumers at higher trophic levels, changes in species distribution and biochemical (lipid) composition of copepods will have a decisive effect on future life in the oceans.

Many questions have been summarized during a recent workshop on lipids in marine zooplankton (Kattner et al. 2007). A close cooperation between biologists, chemists, physicists, and modelers is necessary to improve and complete our knowledge on lipids and zooplankton in general. Interdisciplinary research including life cycles, migration, biochemical compositions, ecophysiological experiments, and climate scenarios should be better coordinated for optimal utilization of the available expertise.

Acknowledgements We are grateful to Sigrid Schnack-Schiel for her constructive support and her expertise in copepod biology. We thank Martin Graeve for helpful comments and sharing unpublished lipid data.

References

Ackman, R.G., Tocher, C.S., and McLachlan, J. 1968. Marine phytoplankter fatty acids. J. Fish. Res. Board Can. 25:1603–1620.
Albers, C.S., Kattner, G., and Hagen, W. 1996. The compositions of wax esters, triacylglycerols and phospholipids in Arctic and Antarctic copepods: evidence of energetic adaptations. Mar. Chem. 55:347–358.
Arts, M.T., Ackman, R.G., and Holub, B.J. 2001. "Essential fatty acids" in aquatic ecosystems: a crucial link between diet and human health and evolution. Can. J. Fish. Aquat. Sci. 58:122–137.
Atkinson, A., Siegel, V., Pakhomov, E., and Rothery, P. 2004. Long-term decline in krill stock and increase in salps within the Southern Ocean. Nature 432:100–103.
Auel, H., and Hagen, W. 2002. Mesozooplankton community structure, abundance and biomass in the central Arctic Ocean. Mar. Biol. 140:1013–1021.
Båmstedt, U., and Tande, K. 1988. Physiological responses of *Calanus finmarchicus* and *Metridia longa* (Copepoda: Calanoida) during winter–spring transition. Mar. Biol. 99:31–38.
Bradford-Grieve, J.M., Markhaseva, E.L., Rocha, C.E.F., and Abiahy, B. 1999. Copepoda, pp. 869–1098. In D. Boltovskoy (ed.), South Atlantic Zooplankton, Vol. 2. Backhuys, Leyden.
Campbell, R.W., and Dower, J.F. 2003. Role of lipids in the maintenance of neutral buoyancy by zooplankton. Mar. Ecol. Prog. Ser. 263:93–99.

Conover, R.J., and Huntley, M. 1991. Copepods in ice-covered seas – distribution, adaptations to seasonally limited food, metabolism, growth patterns and life cycle strategies in polar seas. J. Mar. Syst. 2:1–41.
Conover, R.J., and Siferd, T.D. 1993. Dark-season survival strategies of coastal zone zooplankton in the Canadian Arctic. Arctic 46:303–311.
Cornils, A., Schnack-Schiel, S.B., Böer, M., Graeve, M., Struck, U., Al-Najjar, T., and Richter, C. 2007. Feeding of Clausocalanids (Calanoida, Copepoda) on naturally occurring particles in the northern Gulf of Aqaba (Red Sea). Mar. Biol. 151:1261–1274.
Dalsgaard, J., St. John, M., Kattner, G., Müller-Navarra, D.C., and Hagen, W. 2003. Fatty acid trophic markers in the pelagic marine environment. Adv. Mar. Biol. 46:225–340.
Dittmar, T., and Kattner, G. 2003. The biogeochemistry of the river and shelf ecosystem of the Arctic Ocean: a review. Mar. Chem. 83:103–120.
Falk-Petersen, S., Sargent, J.R., and Tande, K.S. 1987. Lipid composition of zooplankton in relation to the sub-Arctic food web. Polar Biol. 8:115–120.
Falk-Petersen, S., Hop, H., Budgell, W.P., Hegseth, E.N., Korsnes, R., Loyning, T.B., Ørbaek, J.B., Kawamura, T., and Shirasawa, K. 2000. Physical and ecological processes in the Marginal Ice Zone of the northern Barents Sea during the summer melt periods. J. Mar. Syst. 27:131–159.
Falk-Petersen, S., Timofeev, S., Pavlov, V., and Sargent, J.R. 2007. Climate variability and the possible effects on Arctic food chains. The role of *Calanus*, pp. 147–166. *In* J.B. Orbaek, T. Tombre, R. Kallenborn, E. Hegseth, S. Falk-Petersen, and A.H. Hoel (eds.), Arctic-Alpine Ecosystems and People in a Changing Environment. Springer, Berlin.
Farkas, T. 1979. Adaptation of fatty acid compositions to temperature – a study on planktonic crustaceans. Comp. Biochem. Physiol. 64B:71–76.
Flato, G.M., Boer, G.J., Lee, W.G., McFarlane, N.A., Ramsden, D., Reader, M.C., and Weaver, A.J. 2000. The Canadian centre for climate modelling and analysis global coupled model and its climate. Climate Dyn. 16:451–467.
Fraser, A.J., Sargent, J.R., and Gamble, J.C. 1989. Lipid class and fatty acid composition of *Calanus finmarchicus* (Gunnerus), *Pseudocalanus* sp. and *Temora longicornis* Müller from a nutrient-enriched seawater enclosure. J. Exp. Mar. Biol. Ecol. 130:81–92.
Gatten, R.R., Corner, E.D.S., Kilvington, C.C., and Sargent, J.R. 1979. A seasonal survey of the lipids of *Calanus helgolandicus* Claus from the English channel, pp. 275–284. *In* E. Naylor, and R.G. Hartnoll (eds.), Cyclic Phenomena in Plants and Animals. Pergamon, Oxford.
Gradinger, R. 1995. Climate change and biological oceanography of the Arctic Ocean. Phil. Trans. R. Soc. Lond. A 352:277–286.
Graeve, M., Albers, C., and Kattner, G. 2005. Assimilation and biosynthesis of lipids in Arctic *Calanus* species based on ^{13}C feeding experiments with a diatom. J. Exp. Mar. Biol. Ecol. 317:109–125.
Graeve, M., Kattner, G., and Hagen, W. 1994a. Diet-induced changes in the fatty acid composition of Arctic herbivorous copepods: experimental evidence of trophic markers. J. Exp. Mar. Biol. Ecol. 182:97–110.
Graeve, M., Hagen, W., and Kattner, G. 1994b. Herbivorous or omnivorous? On the significance of lipid compositions as trophic markers in Antarctic copepods. Deep-Sea Res. 41:915–924.
Hagen, W., and Auel, H. 2001. Seasonal adaptations and the role of lipids in oceanic zooplankton. Zoology 104:313–326.
Hagen, W., Kattner, G., and Graeve, M. 1993. *Calanoides acutus* and *Calanus propinquus*, Antarctic copepods with different lipid storage modes via wax esters or triacylglycerols. Mar. Ecol. Prog. Ser. 97:135–142.
Hagen, W., Kattner, G., and Graeve, M. 1995. On the lipid biochemistry of polar copepods: compositional differences in the Antarctic calanoids *Euchaeta antarctica* and *Euchirella rostromagna*. Mar. Biol. 123:451–457.
Hagen, W., Van Vleet, E.S., and Kattner, G. 1996. Seasonal lipid storage as overwintering strategy of Antarctic krill. Mar. Ecol. Prog. Ser. 134:85–89.
Hagen, W., Kattner, G., Terbrüggen, A., and Van Vleet, E.S. 2001. Lipid metabolism of the Antarctic krill *Euphausia superba* and its ecological implications. Mar. Biol. 139:95–104.

Hall, J.M., Parrish, C.C., and Thompson, R.J. 2002. Eicosapentaenoic acid regulates Scallop (*Placopecten magellanicus*) membrane fluidity in response to cold. Biol. Bull. 2002:201–203.

Hansen, A.S., Nielsen, T.G., Levinsen, H., Madsen, S.D., Thingstad, T.F., and Hansen, B.W. 2003. Impact of changing ice cover on pelagic productivity and foodweb structure in Disko Bay, West Greenland: a dynamic model approach. Deep-Sea Res. I 50:171–187.

Harrington, G.W., Beach, D.H., Dunham, J.E., and Holz, G.G. 1970. The polyunsaturated fatty acids of dinoflagellates. J. Protozool. 17:213–219.

Heath, M.R., Boyle, P., Gislason, A., Gurney, W., Hay, S.J., Head, E., Holmes, S., Ingvarsdóttir, A., Jónasdóttir, S.H., Lindeque, P., Pollard, R., Rasmussen, J., Richards, K., Richardson, K., Smerdon, G., and Speirs, D. 2004. Comparative ecology of over-wintering *Calanus finmarchicus* in the northern North Atlantic, and implications for life-cycle patterns. ICES J. Mar. Sci. 61:698–708.

Hirche, H.-J. 1987. Temperature and plankton II. Effect on respiration and swimming activity in copepods from the Greenland Sea. Mar. Biol. 94:347–356.

Hirche, H.-J. 1989. Spatial distribution of digestive enzyme activities of *Calanus finmarchicus* and *C. hyperboreus* in Fram Strait/Greenland Sea. J. Plankton Res. 11:431–443.

Hirche, H.-J. 1996. The reproductive biology of the marine copepod *Calanus finmarchicus* – a review. Ophelia 44:111–128.

Hopkins, T.L., and Torres, J.J. 1989. Midwater food web in the vicinity of a marginal ice zone in the western Weddell Sea. Deep-Sea Res. 36:543–560.

Hopkins, C.C.E., Tande, K.S., Grønvik, S., and Sargent, J.R. 1984. Ecological investigations of the zooplankton community of Balsfjorden, northern Norway: an analysis of growth and overwintering tactics in relation to niche and environment in *Metridia longa* (Lubbock), *Calanus finmarchicus* (Gunnerus), *Thysanoessa inermis* (Krøyer) and *Thysanoessa raschi* (M. Sars). J. Exp. Mar. Biol. Ecol. 82:77–99.

Jónasdóttir, S.H. 1999. Lipid content of *Calanus finmarchicus* during overwintering in the Faroe-Shetland channel. Fish. Oceanogr. 8 (suppl. 1):61–72.

Kattner, G. 1989. Lipid composition of *Calanus finmarchicus* from the North Sea and the Arctic. A comparative study. Comp. Biochem. Physiol. 94B:185–188.

Kattner, G., and Krause, M. 1987. Changes in lipids during the development of *Calanus finmarchius* s.l. from Copepodid I to adult. Mar. Biol. 96:511–518.

Kattner, G., and Krause, M. 1989. Seasonal variations of lipids (wax esters, fatty acids and alcohols) in calanoid copepods from the North Sea. Mar. Chem. 26:261–275.

Kattner, G., and Graeve, M. 1991. Wax ester composition of dominant calanoid copepods of the Greenland Sea/Fram Strait region. Polar Res. 10:479–487.

Kattner, G., and Hagen, W. 1995. Polar herbivorous copepods – different pathways in lipid biosynthesis. ICES J. Mar. Sci. 52:329–335.

Kattner, G., Krause, M., and Trahms, J. 1981. Lipid composition of some typical North Sea copepods. Mar. Ecol. Prog. Ser. 4:69–74.

Kattner, G., Gercken, G., and Eberlein, K. 1983. Development of lipids during a spring plankton bloom in the northern North Sea. I. Particulate fatty acids. Mar. Chem. 14:149–162.

Kattner, G., Hirche, H.-J., and Krause, M. 1989. Spatial variability in lipid composition of calanoid copepods from Fram Strait, the Arctic. Mar. Biol. 102:473–480.

Kattner, G., Graeve, M., and Hagen, W. 1994. Ontogenetic and seasonal changes in lipid and fatty acid/alcohol compositions of the dominant Antarctic copepods *Calanus propinquus*, *Calanoides acutus* and *Rhincalanus gigas*. Mar. Biol. 118:637–644.

Kattner, G., Hagen, W., Falk-Petersen, S., Sargent, J.R., and Henderson, R.J. 1996. Antarctic krill *Thysanoessa macrura* fills a major gap in marine lipogenic pathways. Mar. Ecol. Prog. Ser. 134:295–298.

Kattner, G., Hagen, W., Graeve, M., and Albers, C. 1998. Exceptional lipids and fatty acids in the pteropod *Clione limacina* (Gastropoda) from both polar oceans. Mar. Chem. 61:219–228.

Kattner, G., Albers, C., Graeve, M., and Schnack-Schiel, S.B. 2003. Fatty acid and alcohol composition of the small polar copepods, *Oithona* and *Oncaea*: indication on feeding modes. Polar Biol. 26:666–671.

Kattner, G., Hagen, W., Lee, R.F., Campbell, R., Deibel, D., Falk-Petersen, S., Graeve, M., Hansen, B.W., Hirche, H.J., Jónasdóttir, S.H., Madsen, M.L., Mayzaud, P., Müller-Navarra, D.C., Nichols, P.D., Paffenhöfer, G.-A., Pond, D., Saito, H., Stübing, D., and Virtue, P. 2007. Perspectives on marine zooplankton lipids. Can. J. Fish. Aquat. Sci. 64:1628–1639.

Laakmann, S. 2004. Abundanz und Reproduktionserfolg ausgewählter calanoider Copepoden während der Frühjahrsplanktonblüte um Helgoland. M.Sc. thesis, University of Bremen, pp. 75.

Lee, R.F. 1974. Lipid composition of the copepod *Calanus hyperboreus* from the Arctic Ocean. Changes with depth and season. Mar. Biol. 26:313–318.

Lee, R.F. 1975. Lipids of Arctic zooplankton. Comp. Biochem. Physiol. 51B:263–266.

Lee, R.F., and Hirota, J. 1973. Wax esters in tropical zooplankton and nekton and the geographical distribution of wax esters in marine copepods. Limnol. Oceanogr. 18:227–239.

Lee, R.F., Hirota, J., and Barnett, A.M. 1971. Distribution and importance of wax esters in marine copepods and other zooplankton. Deep-Sea Res. 18:1147–1165.

Lee, R.F., Nevenzel, J.C., and Paffenhöfer, G.-A. 1972. The presence of wax esters in marine planktonic copepods. Naturwissenschaften 59:406–411.

Lee, R.F., Nevenzel, J.C., and Lewis, A.G. 1974. Lipid changes during life cycle of marine copepod *Euchaeta japonica* Marukawa. Lipids 9:891–898.

Lee, R.F., Hagen, W., and Kattner, G. 2006. Lipid storage in marine zooplankton. Mar. Ecol. Prog. Ser. 307:273–306.

Lischka, S., and Hagen, W. 2005. Life histories of the copepods *Pseudocalanus minutus*, *P. acuspes* (Calanoida) and *Oithona similis* (Cyclopoida) in the Arctic Kongsfjorden (Svalbard). Polar Biol. 28:910–921.

Lischka, S., and Hagen, W. 2007. Seasonal lipid dynamics of the copepods *Pseudocalanus minutus* (Calanoida) and *Oithona similis* (Cyclopoida) in the Arctic Kongsfjorden (Svalbard). Mar. Biol. 150:445–454.

Loeb, V., Siegel, V., Holm-Hansen, O., Hewitt, R., Fraser, W., Trivelpiece, W., and Trivelpiece, S. 1997. Effects of sea-ice extent and krill or salp dominance on the Antarctic food web. Nature 387:897–900.

Mauchline, J. 1998. The biology of calanoid copepods. Adv. Mar. Biol. 33:1–660.

Mayor, D., Anderson, T., Irigoien, X., and Harris, R. 2006. Feeding and reproduction of *Calanus finmarchicus* during non-bloom conditions in the Irminger Sea. J. Plankton Res. 28:1167–1179.

Metz, C. 1995. Seasonal variation in the distribution and abundance of *Oithona* and *Oncaea* species (Copepoda, Crustacea) in the southeastern Weddell Sea, Antarctica. Polar Biol. 15:187–194.

Miller, C.B. 1993. Pelagic production processes in the Subarctic Pacific. Prog. Oceanogr. 32:1–15.

Mumm, N. 1993. Composition and distribution of mesozooplankton in the Nansen Basin, Arctic Ocean, during summer. Polar Biol. 13:451–461.

Norrbin, M.E., Olsen, R.-E., and Tande, K.S. 1990. Seasonal variation in lipid class and fatty acid composition of two small copepods in Balsfjorden, northern Norway. Mar. Biol. 105:205–211.

Paffenhöfer, G.-A. 1993. On the ecology of marine cyclopoid copepods (Crustacea, Copepoda). J. Plankton Res. 15:37–55.

Peters, J., Renz, J., van Beusekom, J., Boersma, M., and Hagen, W. 2006. Trophodynamics and seasonal cycle of the copepod *Pseudocalanus acuspes* in the central Baltic Sea (Bornholm Basin): evidence from lipid composition. Mar. Biol. 149:1417–1429.

Peters, J., Dutz, J., and Hagen, W. 2007. Role of essential fatty acids on the reproductive success of the copepod *Temora longicornis* in the North Sea. Mar. Ecol. Prog. Ser. 341:153–163.

Petersen, W. 1998. Life cycle strategies in coastal upwelling zones. J. Mar. Syst. 15:313–326.

Richardson, A.J., and Schoeman, D.S. 2004. Climate impact on plankton ecosystems in the Northeast Atlantic. Science 305:1609–1613.

Richardson, K., Jónasdóttir, S.H., Hay, S.J., and Christoffersen, A. 1999. *Calanus finmarchicus* egg production and food availability in the Faroe-Shetland Channel and northern North Sea: October-March. Fish. Oceanogr. 8:153–162.

Richter, C. 1994. Regional and seasonal variability in the vertical distribution of mesozooplankton in the Greenland Sea. Rep. Polar Res. 154:1–87.

Runge, J.A., Therriault, J., Legendre, L., Ingram, R.G., and Demers, S. 1991. Coupling between ice microalgal productivity and the pelagic, metazoan food web in the southeastern Hudson Bay: a synthesis of results. Polar Res. 10:325–338.

Saito, H., and Kotani, Y. 2000. Lipids of four boreal species of calanoid copepods: origin of monoene fats of marine animals at higher trophic levels in the grazing food chain in the subarctic ocean ecosystem. Mar. Chem. 71:69–82.

Sargent, J.R., and Henderson, R.J. 1986. Lipids, pp. 59–108. In E.D.S. Corner and S.C.M. O'Hara (eds.), The Biological Chemistry of Marine Copepods. Clarendon, Oxford.

Sargent, J.R., Eilertsen, H.C., Falk-Petersen, S., and Taasen, J.P. 1985. Carbon assimilation and lipid production in phytoplankton in northern Norwegian fjords. Mar. Biol. 85:109–116.

Sakshaug, E. 1997. Biomass and productivity distributions and their variability in the Barents Sea. ICES J. Mar. Sci. 54:341–351.

Schnack, S.B., Marschall, S., and Mizdalski, E. 1985. On the distribution of copepods and larvae of *Euphausia superba* in Antarctic waters during February 1982. Meeresforschung 30:251–263.

Schnack-Schiel, S.B., and Hagen, W. 1995. Life-cycle strategies of *Calanoides acutus*, *Calanus propinquus* and *Metridia gerlachei* (Copepoda: Calanoida) in the eastern Weddell Sea, Antarctica. ICES J. Mar. Sci. 52:541–548.

Scott, C.L., Kwasniewski, S., Falk-Petersen, S., and Sargent, J.R. 2002. Species differences, origins and functions of fatty alcohols and fatty acids in the wax esters and phospholipids of *Calanus hyperboreus*, *C. glacialis* and *C. finmarchicus* from Arctic waters. Mar. Ecol. Prog. Ser. 235:127–134.

Serreze, M.C., Holland, M., and Stroeve, J. 2007. Perspectives on the Arctic's shrinking sea-ice cover. Science 315:1533–1536.

Sewell, R.B.S. 1947. The free-swimming planktonic copepods. Systematic account. Sci. Rep. John Murray Exped. 1933–1934. Br. Mus. Nat. Hist. 8:1–303.

Shinitzky, M. 1984. Physiology of membrane fluidity. Vol. I, II. CRC Inc., Boca Raton

Siegel, V. 2005. Distribution and population dynamics of *Euphausia superba*: Summary of recent findings. Polar Biol. 29:1–22.

Smetacek, V., and Nicol, S. 2005. Polar ocean ecosystems in a changing world. Nature 437:362–368.

Smith, S.L., and Schnack-Schiel, S.B. 1990. Polar zooplankton, pp. 527–598. In W.O. Smith (ed.), Polar Oceanography, Part B: Chemistry, Biology, and Geology. Academic, San Diego.

Stillwell, W., and Wassall, S.R. 2003. Docosahexaenoic acid: membrane properties of a unique fatty acid. Chem. Phys. Lipids 126:1–27.

Verheye, H.M., Hagen, W., Auel, H., Ekau, W., Loick, N., Rheenen, I., Wencke, P., and Jones, S. 2005. Life strategies, energetics and growth characteristics of *Calanoides carinatus* (Copepoda) in the Angola-Benguela Front region. African J. Mar. Sci. 27:641–652.

Visser, A.W, and Jónasdóttir, S.H. 1999. Lipids, buoyancy and the seasonal vertical migration of *Calanus finmarchicus*. Fish. Oceanogr. 8:100–106.

Ward, P., Shreeve, R.S., and Cripps, G.C. 1996. *Rhincalanus gigas* and *Calanus simillimus*: Lipid storage patterns of two species of copepod in the seasonally ice free zone of the Southern Ocean. J. Plankton Res. 18:1439–1454.

Wu, P., Wood, R., and Stott, P. 2005. Human influence on increasing Arctic river discharges. Geophys. Res. Lett. 32:L02703.

Chapter 12
Tracing Aquatic Food Webs Using Fatty Acids: From Qualitative Indicators to Quantitative Determination

Sara J. Iverson

12.1 Introduction

Food web structure, predator–prey dynamics, foraging behavior, and consequences of these factors for individual growth, reproduction and survival are central to our understanding of ecosystem structure and functioning. Moreover, in the current context of understanding (and managing) ecosystems in the face of ongoing environmental change, important questions include: What are the critical prey of key consumers in relation to prey abundance, availability, and nutritional quality? What are the ecosystem processes responsible for food web production? And, how do these processes respond to changes in physical forcing? A fundamental requirement to understand any of these areas is an accurate assessment of trophic relationships and consumer diets. However, in aquatic, and especially marine ecosystems, such information is generally not easily or reliably obtained. In these systems, the relative inaccessibility of free-ranging organisms and the inability to directly observe species interactions make it difficult to accurately characterize diet. Traditional approaches, such as examining gut contents, have well-recognized biases in addition to representing only snapshots of recent meals and may therefore not be reliable indicators of long-term diet (Iverson et al. 2004). Thus, alternative approaches have been developed, which use various types of trophic markers. One of the most promising of these approaches is the use of lipids and fatty acids (FA) to study food web dynamics.

Lipids comprise a large group of chemically heterogeneous compounds, the majority of which include esters of FA as part of their structure. FA represent the "building blocks" of lipids and are the largest constituent of neutral lipids (NL), such as triacylglycerols (TAG) and wax esters (WE), as well as of the polar phospholipids (PL). All FA consist of carbon atom chains, which are most commonly even-numbered and straight, containing 14–24 carbons and 0–6 double bonds, with

---

S.J. Iverson
Department of Biology, Life Sciences Centre, Dalhousie University,
Halifax, Nova Scotia, Canada B3H 4J1
e-mail: Sara.Iverson@Dal.ca

a methyl (CH_3) terminal at one end and an acid (carboxyl, COOH) group at the other. The array of FA present in nature is exceptionally complex with the possibility of routinely identifying 70 FA within a given organism (e.g., Table 12.1).

Three characteristics of FA and their storage patterns make them useful tracers of diets and marine food-web structure. First, organisms are able to biosynthesize, modify chain-length, and introduce double bonds in FA, but they are subject to biochemical limitations in these processes depending on the phylogenetic group and even species. Such limitations generally increase with increasing phylogenetic order, culminating in vertebrates (Cook 1996). Second, unlike other dietary nutrients (e.g., proteins and carbohydrates), which are completely broken down during digestion, FA are released from ingested lipid molecules during digestion, but are generally not degraded, and are taken up by tissues in their basic form. The important consequences of these restrictions within plants, bacteria, and animals, and the uptake of intact FA by consumer tissues, is that individual isomers as well as "families" of FA bioaccumulate through food chains, and they can be traced back to specific food web origins. Third, unlike most other nutrients, fat is stored in animal bodies in reservoirs. These often substantial stores can later be mobilized to provide fuel for short or long-term energy demands (e.g., Pond 1998). Thus, FA accumulate over

Table 12.1 Fatty acids (FA) routinely identified in marine organisms on a polar capillary column (e.g., DB-23, Agilent Technologies; 30 m × 0.25 mm ID), listed in order of elution (Iverson et al. 1997, 2002, 2004, 2006a)

12:0	16:1n-7	18:1n-9	20:1n-7	21:5n-3
13:0	7-*methyl* 16:0	18:1n-7	20:1n-5	22:4n-9
iso-14:0	16:1n-5	18:1n-5	20:2n-11/12	22:4n-6
14:0	16:2n-6	18:2Δ5,11	20:2n-9	22:5n-6
14:1n-9	*iso*-17:0	18:2n-7	20:2n-6	22:4n-3
14:1n-7	16:2n-4	18:2n-6	20:3n-6	22:5n-3
14:1n-5	16:3n-6	18:2n-4	20:4n-6	22:6n-3
iso-15:0	17:0	18:3n-6	20:3n-3	24:1n-9
anti-15:0	16:3n-4	18:3n-4	20:4n-3	
15:0	17:1	18:3n-3	20:5n-3	NMI FA
15:1n-8	16:3n-1	18:3n-1	22:1n-11	20:2Δ5,11
15:1n-6	16:4n-3	18:4n-3	22:1n-9	20:2Δ5,13
iso-16:0	16:4n-1	18:4n-1	22:1n-7	20:3Δ5,11,14
16:0	18:0	20:0	22:2n-11/12	22:2NMID (*unknown*)
16:1n-11	18:1n-13	20:1n-11	22:2n-9	22:2Δ7,13
16:1n-9	18:1n-11	20:1n-9	22:2n-6	22:2Δ7,15

FA are named as carbon number:number of double bonds and location (*n*-x) of the double bond nearest the terminal methyl group, where all additional double bonds are separated by a –CH_2– group (i.e., "methylene-interrupted"). Non-methylene interrupted FA (NMI FA) are separated by more than one methylene group; these are generally very small peaks and require special attention in identifying, and thus are listed separately (Budge et al. 2007). Other FA that have been reported in aquatic or marine organisms include *iso*-4:0, 4:0, *iso*-5:0, 5:0, *iso*-10:0, *iso*-12:0 (Koopman et al. 1996, 2003, 2006) and minor or trace amounts of *iso*- and *anteiso*-isomers (methyl branch at the second and third carbon, respectively) of 13–18 carbon saturated FA, *trans*Δ6–16:1, 16:2n-7, 16:2n-1, 16:3n-4, 16:4n-4, 18:2n-3, 18:4n-6, 18:5n-3, 19:0, 22:3n-6 (e.g., Ackman 1980, 2002; Ackman et al. 1972; Budge et al. 2006) time and represent an integration of dietary intake over days, weeks, or months, depending on the organism and its energy intake and storage rates.

Since the mid 1930s (Lovern 1935; Klem 1935), numerous studies have demonstrated the transfer of FA from prey to predator both at the base and apex of food webs (reviewed in Dalsgaard et al. 2003; Iverson et al. 2004; Budge et al. 2006). Until recently, FA have been used largely in a qualitative or semiquantitative way to infer aspects of food webs. However, recent advances involve the development of methods that use FA to quantitatively estimate diets of individual predators (Iverson et al. 2004, 2006b, 2007). The objectives of this chapter are: (1) to provide an overview of the biochemistry, metabolism, and key assumptions that are central to understanding how and why FA can be used as trophic tracers; (2) to discuss different qualitative and quantitative ways in which lipids can be used in food web and foraging ecology studies, and the methods necessary for those applications; and (3) to consider future areas to advance this research. This review focuses primarily on marine ecosystems but also refers to some freshwater and terrestrial systems.

12.2 Characteristics and Constraints on Lipid Biosynthesis, Digestion, and Deposition as They Relate to Tracing Trophic Relationships

12.2.1 De Novo Fatty Acid Biosynthesis

Important differences in FA biosynthesis among organisms allow the original source of some FA to be identified. General and specific principles and characteristics of FA biosynthesis have been described extensively in a number of reviews (e.g., Cook 1996; Gurr and Harwood 1991; Kattner and Hagen 1995; Vance and Vance 1996; Dalsgaard et al. 2003; see Chap. 9). However, some particularly relevant points include that de novo synthesis of FA occurs from 2-carbon precursors by sequential additions of 2-carbon units to a growing chain, which is released from the enzyme complex usually at 14–18 carbons. Additional 2-carbon units may also be added, generally up to 24 carbons. During this process, double bonds can be added (i.e., desaturation) by specific enzymes. Primary producers such as unicellular phytoplankton and seaweeds (macroalgae) typically produce FA ranging from 14 to 24 carbons with various degrees of unsaturation. Alga are essentially the only organisms that possess the enzymes necessary for producing long-chain polyunsaturated FA (PUFA), such as 20:5n-3 and 22:6n-3 (e.g., Sargent and Henderson 1995; Cook 1996). These FA occur throughout the marine food web in sequentially higher trophic levels, since animals (i.e., consumers) are not capable of inserting a double bond between the terminal methyl end and the n-9 carbon. Other unusual FA, such as 16:2n-4 and 16:4n-1, are produced only by certain algae and diatoms (Viso and Marty 1993; Dunstan et al. 1994).

In contrast to primary producers, animals synthesize fewer and simpler FA. These tend to be restricted to 14:0, 16:0, and 18:0 saturated FA and their monounsaturated isomers 14:1n-5, 16:1n-7, and 18:1n-9, respectively. These monounsaturates are produced by the Δ9 desaturase enzyme present in all animals, which inserts a double bond at the ninth carbon from the carboxyl end. Animals can elongate both endogenously and exogenously produced FA to some extent, but this is generally limited, and both de novo biosynthesis and elongation/desaturation of FA are inhibited by diets containing adequate or excess fat, and long-chain PUFA (Nelson 1992). However, some invertebrates tend to have greater capacities for biosynthesis and modification of FA than higher animals. Of the zooplankton, the best studied are the calanoid copepods, which have an unusual ability to produce large amounts of long-chain monounsaturated fatty alcohols (i.e., the WE of 20:1n-11, 20:1n-9, 22:1n-11, 22:1n-9) as part of their primary storage fats (Pascal and Ackman 1976; Sargent 1978; see Chap. 6). When found in higher trophic level organisms, these FA are thought to originate from copepods (Sargent and Henderson 1986). Although less is known about many other taxa of zooplankton, they appear to biosynthesize the relatively more common FA described above (Dalsgaard et al. 2003). Certain benthic bivalve mollusks and carnivorous gastropods produce unusual FA, in which the double bonds are separated by more than one methylene group (i.e., non-methylene interrupted, NMI FA) (Joseph 1982; Budge et al. 2007). Fishes, birds, and mammals have the greatest restrictions on FA biosynthesis and follow the typical animal pattern described above. Marine predators generally have very low carbohydrate diets, but amino acids from proteins consumed in excess of immediate energy and nutrient requirements can be broken down to enter the usual FA biosynthetic pathways. An important exception to these patterns is found in some odontocetes (toothed whales), which synthesize large amounts of very short branched-chain FA (*iso*-4:0, *iso*-5:0, *iso*-10:0, or *iso*-12:0) in their cranial fats and blubber, which have no relation to diet (Koopman et al. 2003, 2006).

12.2.2 Lipid Biosynthesis

FA in nature rarely exist in free form, and in both primary producers and consumers endogenously synthesized and exogenously derived FA are generally incorporated as part of a compound. Although there are a number of lipid classes, the ones of concern in the present context are TAG, WE, and PL (Budge et al. 2006). PL are found in structural components such as cell membranes. Since fairly specific FA compositions are required for proper membrane structure and function, FA in PL tend to be fairly specific (Chap. 10) and highly conserved relative to diet. Thus, although they can be influenced by dietary FA intake, FA in PL are not particularly useful as dietary tracers (see Sect. 12.3.1). FA are most commonly stored in NL, of which TAG are by far the most common storage form. TAG can be distributed as droplets throughout an animal's body or deposited as adipose tissue (i.e., the specific fat storage tissue of vertebrates). Adipose tissue is composed of specialized cells called adipocytes, which alternately increase or decrease in volume with fattening (deposition of TAG) or fasting (mobilization of TAG), respectively. WE are another important storage form of fat in certain species of crustaceans (e.g., copepods and other zooplankton), fish (e.g., myctophids), and marine mammals (some odontocetes; Koopman 2007). In order for WE to appear in the storage lipids of an animal, that animal must synthesize them from dietary FA or from FA biosynthesized de novo, which includes the process of converting long-chain FA to their corresponding fatty alcohols (see Sect. 12.2.3).

12.2.3 *Digestion, Modification, and Deposition of Dietary Lipids and Fatty Acids*

The way in which lipids are digested by monogastric (i.e., non-ruminant) animals, and subsequently modified, has a significant impact on their utility as trophic tracers. Ingested TAG are hydrolyzed (ester bonds are broken) in the gut by lipases and esterases to its component FA, monoacylglycerol, and glycerol. If short-chain FA (<14 carbons) are ingested, they are transported to the liver and immediately oxidized (Brindley 1991). All other products are transported in the blood via various carriers (e.g., lipoproteins) to tissues where absorption takes place. At fat storage sites, FA are most commonly reesterified into TAG and sequestered. A modification of this process is found in animals that consume and that store WE, processes that are independent from one another within an organism (Budge et al. 2006). WE are prevalent in marine systems as food for higher trophic consumers where, upon ingestion, they are hydrolyzed in the gut to their component FA and fatty alcohol. Gut enzymes then oxidize the fatty alcohol to its corresponding FA (without modification of chain length and double bond positions) and then both FA enter the pool of FA available for transport and deposition (Sargent 1976; Budge and Iverson 2003). If the consumer's storage fat is in the form of TAG, then TAG are deposited regardless of whether the animal has a diet high in WE. However, according to phylogeny, some animals store their fat in part or whole as WE (see Sect. 12.2.2). In these animals, certain FA are reduced to their corresponding fatty alcohols after digestion, and these fatty alcohols are then incorporated, along with other ingested FA, into WE for storage (Sargent 1976; Sargent and Henderson 1986). Nevertheless, as with TAG digestion and deposition, FA chain lengths and double bond positions are generally conserved.

As a consequence of these digestive properties, dietary FA ≥ 14 carbons are generally deposited in animal tissue with no or minimal modification, and thus one can distinguish between FA that could be biosynthesized by the animal or those that most likely come from the diet. However, some (or even many) FA may be routed through the liver before deposition occurs, while others may be completely oxidized in tissues for immediate energy needs. Thus, although dietary FA consumed in excess of immediate energy requirements are deposited largely intact in storage reservoirs such as adipose tissue, there are several points during metabolism and transportation when there is the potential for FA to be modified by animals. For example, some marine invertebrates such as copepods (see Sect. 12.2.1) are generally thought to have a greater capacity to modify (elongate or desaturate) dietary FA than higher animals. Freshwater *Daphnia* were shown to elongate/desaturate 18:3n-3 to 20:5n-3 (Schlechtriem et al. 2006). However even in these invertebrates, direct incorporation of dietary FA has been demonstrated (Sargent and Henderson 1986; Dalsgaard et al. 2003). In contrast, fishes have a more limited ability to modify FA, but may still be better able to modify some exogenous FA than birds and mammals. However, given that their natural marine diets contain high levels of essential long-chain n-3 and n-6 PUFA (e.g., Ackman 1980), overall modification of dietary FA in fish is probably limited relative to direct dietary deposition (e.g., Kirsch et al. 1998; see Sect. 12.3.2.1).

Birds and mammals have very limited abilities to modify exogenously consumed FA by elongation and desaturation. In these species, preformed dietary FA are less likely to enter typical lipid synthetic pathways and such processes are, in any case, inhibited by diets containing adequate or excess fat, as well as those high in long-chain PUFA (Nelson 1992). Additionally, desaturation of exogenously consumed FA may be confined primarily to the $\Delta 9$ desaturase enzyme acting on some saturated 16:0 (e.g., Budge et al. 2004). Thus, FA that have been elongated and desaturated within marine birds and mammals are unlikely to make a significant contribution to their adipose FA stores. A more important process however, especially in mammals, may be peroxisomal chain-shortening of some long-chain monounsaturated FA. Thus, some ingested 20:1 and especially 22:1 isomers are likely shortened primarily to their 18-carbon isomers (Norseth and Christophersen 1978; Osmundsen et al. 1979; Cooper et al. 2005, 2006), resulting in somewhat reduced and increased deposition of these FA relative to diet, respectively.

Finally, it is clear that the biochemical pathways that animals are *capable* of performing are not necessarily the same as their *propensity* for using these pathways. Studies of FA metabolism using foreign foods and/or feeding regimes that those organisms are not accustomed to may result in forced stimulation of compensatory biochemical pathways. Examples of such studies include aquaculture fish fed artificial feeds containing terrestrial plant oil FA, rats fed fish oil FA, or species "starved" on severe nutrient depleted diets. Under natural conditions, marine fishes are adapted to and require marine FA (Ackman 1980), and carnivorous marine mammals are likewise highly adapted to efficiently digesting and depositing lipids high in marine FA without modification (e.g., Iverson et al. 1995). Additionally, some seabirds and marine mammals are well adapted to periods of prolonged fasting during which they do not enter the terminal phases of starvation, but instead maintain homeostatic regulation (Castellini and Rea 1992; Mellish and Iverson 2001). A question often asked is what happens to the FA profile of a predator when mobilizing lipid stores rather than depositing them. The issue of differential mobilization is not fully understood; however, differential release of some FA during fasting has been reported (Groscolas 1990; Raclot 2003); this is in contrast to data from natural long-term fasting studies in several phocid and otariid pinniped pups and juveniles, which have shown no temporal change in overall FA composition (D. Noren, S.J. Iverson and J.E. Mellish unpublished data). Although further studies are needed, current evidence indicates that effects of short-term fasting are unlikely to have major impacts on overall FA composition. In summary, although biosynthesis and modification of FA does occur, by far the greatest quantitative contribution to the fat stores of higher marine or aquatic predators arises from direct deposition of dietary FA (Ackman and Eaton 1966; Rouvinen and Kiiskinen 1989; Colby et al. 1993; Iverson 1993; Kirsch et al. 1998; Iverson et al. 1995, 2007).

12.3 Tracing Trophic Pathways Using Lipids and Fatty Acids

Fatty acids can be used to study trophic relationships and food webs in several ways to provide information about consumers and their diets. One approach assumes that consumers of similar phylogeny will also share similarities in their capacity to biosynthesize, digest, and modify dietary FA. Thus, finding differences or changes in FA composition allows inferences to be made about differences or changes in diets of predators, both within and between populations, without trying to specify what prey species are eaten. The second approach uses individual biomarkers to infer or possibly identify predator–prey relationships. These biomarkers tend to be relatively rare in nature, especially at higher trophic levels, but when found in consumers can indicate consumption of specific taxa at lower trophic levels. More recently, a third approach has been developed. This uses a statistical model, combined with coefficients to account for predator metabolism and a comprehensive prey FA database, to quantitatively estimate species composition of predator diets from their FA stores. All three approaches can provide valuable insight about consumer diets and foraging ecology that otherwise could not be obtained in complex aquatic ecosystems. Each of these approaches is considered next with reference to how they have been used and validated, and their limitations.

12.3.1 Tissue Sampling and Analysis

Methods for isolation, preparation, and analysis of lipids and FA have been extensively reviewed elsewhere (Christie 1982; Ackman 1986, 2002; Parrish 1999; Iverson et al. 2001a; Budge et al. 2006; and references therein). Likewise, a thorough discussion of appropriate tissue sampling and storage for FA analyses can be found in Budge et al. (2006); however, several points are specifically pertinent to trophic studies. First, not all tissues provide equal information on diet using FA. Second, different tissues are usually required depending on whether the organism is being examined as prey or predator (in some studies certain species may be both).

12.3.1.1 Predator Sampling

A metabolically active energy storage reservoir will be most readily influenced by dietary FA intake and therefore should be the tissue sampled. This reservoir will experience rapid turnover as a result of dietary intake and fat mobilization during fasting, and will be the most reflective of trophic relationships. As stated previously, PL FA reflect biosynthetic pathways and conserved membrane structural requirements, whereas TAG and WE FA largely reflect stored dietary FA. Thus, skin, muscle (but see caveat below) or other structural tissues will contain more structural PL FA and should be avoided. In vertebrates, the appropriate lipid samples will usually come from adipose tissue storage sites (or blubber in the case of pinnipeds and cetaceans). Such adipose tissue is often conveniently found subcutaneously and thus can be sampled in both live mammals and birds using relatively noninvasive biopsy techniques. However, it is important to confirm that the fat storage sites have comparable FA composition at different locations in an animal. Depth through the tissue sampled appears to be of importance only when appropriately sampling blubber (see Budge et al. 2006). Otherwise, when a true fat storage site is sampled, and not structural adipose tissues (e.g., cushions in eye sockets, tailstocks of cetaceans, skin-associated blubber in marine mammals), FA composition is uniform across body sites in various species of pinnipeds, cetaceans, polar bears (*Ursus maritimus*), and seabirds (Koopman et al. 1996; Cooper 2004; Thiemann et al. 2006; Iverson et al. 2007). However, some fish store lipid as modified adipose or lipid pockets in their muscle (e.g., salmonids, mackerels, herring, *Clupea pallasi*) or liver (e.g., gadoids such as cod, *Gadus morhua*), whereas many invertebrates such as sea urchins and jellyfishes store lipid primarily in their gonads and digestive tracts. Additionally, muscle FA in such species, including those in the PL, may reflect longer term systemic differences in diet among populations. In zooplankton, lipid pockets may be more difficult to isolate and thus whole animals must be analyzed with the recognition that there will be substantial contribution of conserved PL FA to the FA patterns analyzed; a similar issue is encountered when sampling vertebrate blood to assess diet. One solution is to fractionate NL (TAG and WE) from PL in the extracted lipid for subsequent FA analyses. However, ideally, when blood is sampled, lipoproteins that specifically carry FA from the digestive tract (e.g., chylomicrons in mammals, portomicrons in birds) should be isolated from other lipoproteins (which reflect endogenous FA conservation) and analyzed to examine the most recent meal (e.g., Cooper et al. 2005). Finally, in marine and other aquatic mammals (especially carnivores, pinnipeds and cetaceans), milk is a useful tissue to sample as it sequesters recent preformed dietary FA (Iverson and Oftedal 1995) in income breeders (females that feed during lactation, e.g., Wamberg et al. 1992; Iverson 1993; Iverson et al. 2001b). In capital breeders (females that fast throughout lactation), milk fat output relies principally on uptake of FA mobilized from adipose tissue or blubber and thus will be most influenced by diet of individuals before lactation.

12.3.1.2 Prey Sampling

Most marine and aquatic predators, from invertebrates to the highest vertebrates, consume their prey whole, thus they consume all the NL and PL contained in the prey and its digestive tract. Hence, whole prey should be homogenized and analyzed, even though including prey stomach contents may increase within-species variability in FA composition. One exception occurs if the goal is to directly link the producer of an individual FA biomarker to the consumer, in which case gut contents of the prey should be removed. Complications arise when a given organism is to be used as both a predator and prey. In this case, if the aim is to quantitatively estimate prey contribution to predator diet (see Sect. 12.3.3), sampling as both predator and prey needs to be considered. For example, to assess diets of both cod and seals, all potential prey of cod and seals must be sampled as prey (i.e., whole). However, for cod, the liver would be best isolated for direct assessment of its diet as predator (see Sect. 12.3.1.1), while the whole cod would have to be used as prey for seals. This can be solved by gravimetrically isolating the liver from the rest of the cod body, analyzing the two parts separately, and subsequently reconstructing the FA composition of the whole cod mathematically. The same would be true for determining the diet of a seal (i.e., sampling its blubber only), but analyzing the whole seal as prey for killer whales, *Orcinus orca*. However, for polar bears, which primarily consume only the blubber of seals, the same seal blubber sample can serve as both predator and prey sample (Iverson et al. 2006b).

12.3.2 Qualitative and Semiquantitative Approaches: Predators Alone and Tracers Which Infer Prey Type

12.3.2.1 Uses and Evidence

Qualitative evaluation of spatial or temporal variation in diets of predators can readily be studied by comparing profiles of FA present in consumer lipid depots. Such inferences are most informative in the context of knowing something about the FA characteristics of that ecosystem. This approach, and that of the individual FA biomarker, has been used successfully in a number of ecosystems. Much of the research studying trophic relationships, especially near the bottom of food webs, has used these two approaches. The biomarker approach can be extended by recognizing unusual levels of certain FA or of ratios among FA that can only be attributed to one or a few prey types and thus can indicate their likely importance in the diet. These methods are most successfully used in primary consumers and perhaps other lower trophic levels, as those FA originate at these lower trophic levels. In principle, this opportunity will be relatively rare in higher trophic level predators, as FA originating at the base of the food web become relatively ubiquitous throughout higher levels. Given that these biomarker approaches have been extensively reviewed (e.g., Napolitano 1999; Dalsgaard et al. 2003), they will only be summarized here, with emphasis on newer multivariate analyses of higher predator FA patterns and quantitative estimates of diet composition.

One of the earlier examples of individual FA markers, and confirmed more recently, was the finding that 16:2n-4 and 16:4n-1 are produced by only certain diatoms (Ackman and McLachlan 1977; Viso and Marty 1993; Dunstan et al. 1994). These

FA are metabolically inert in consumers and therefore their presence in the depot fats of some fish can be indicative of specific feeding habits (Ackman et al. 1975). In high latitudes, pennate diatoms, which dominate the sea ice flora, contain elevated amounts of 22:6n-3 and C_{18} PUFA compared with centric diatoms, which dominate the open water spring bloom flora (McConville 1985). This raises the possibility of investigating the relative contribution of these two groups of primary producers to their consumers (Cripps and Hill 1998), and possibly to food web production in the Arctic. The ability to trace diatom- versus flagellate-based food webs in juvenile cod was demonstrated by St. John and Lund (1996) using the ratio of 16:1n-7/16:0. Although other such "biomarkers" (e.g., 20:5n-3 and other PUFA) cannot be isolated to a specific primary producer, differing levels of these FA may characterize different producers. Calanoid copepods synthesize considerable amounts of isomers of 20:1 and 22:1, which they incorporate into their WE as fatty alcohols (see Sect. 12.2.1). Thus, these FA are not very useful for examining diet variations among copepods. Nevertheless, many studies have demonstrated the conservative incorporation of unaltered dietary FA into copepod WE, which can clearly be used as trophic tracers (Dalsgaard et al. 2003; see Chap. 6). Another example of a tracer was the discovery of $trans\Delta6$–16:1 (*trans*-6-hexadecenoic acid) indicating jellyfish in the diet of both ocean sunfish (*Mola mola*) and the leatherback turtle (*Dermochelys coriacea coriacea*) (Ackman et al. 1972; Hooper et al. 1973).

As trophic levels increase, the ability to use a unique FA to trace feeding to a specific food type is reduced. Certainly, 20:1 and 22:1 FA isomers are very useful as copepod markers in the next-higher level consumers such as larval herring (Ackman 1980). This is also true for the zooplanktivorous fin whale (*Balaenoptera physalus*), and allowed differentiation between North Atlantic and Antarctic populations (Ackman and Eaton 1966). However, in higher trophic level predators (including larger adult herring), these FA taken alone do not make it possible to determine whether the predator consumed copepods directly or consumed copepod consumers. For example, the FA 20:1n-11 and 22:1n-11, arising from copepod WE, are the dominant FA in herring from Prince William Sound, Alaska. These predictably increased in concentration from zooplanktivorous juvenile herring to large piscivorous adults (Fig.12.1), consistent with known ontogenetic changes in herring diets (Iverson et al. 2002). Here, the increase in 20:1 and 22:1 isomers could not be attributed to greater copepod consumption, but to greater consumption of fish that ate copepods and therefore concentrated copepod signatures. Likewise, at higher trophic levels such as seals, it is generally not possible to distinguish between direct consumption of zooplankton or of herring when using only these FA. Thus, in principle, inferring diets directly from one or a few FA is a risky practice for higher trophic levels.

Nevertheless, as indicated above, certain FA or combinations of FA can still serve as useful ecosystem markers at upper trophic levels. For instance, while 22:1n-11 is the dominant isomer over 22:1n-9 in most marine ecosystems, 20:1n-9 is by far the dominant isomer over 20:1n-11 in the North Atlantic (e.g., Budge et al. 2002). However, several studies have shown the reverse in the Pacific, with 20:1n-11 dominating 20:1n-9 in fish and invertebrates (Saito and Murata 1998). On this basis, North Pacific and North Atlantic harbor seals (*Phoca vitulina*) consuming

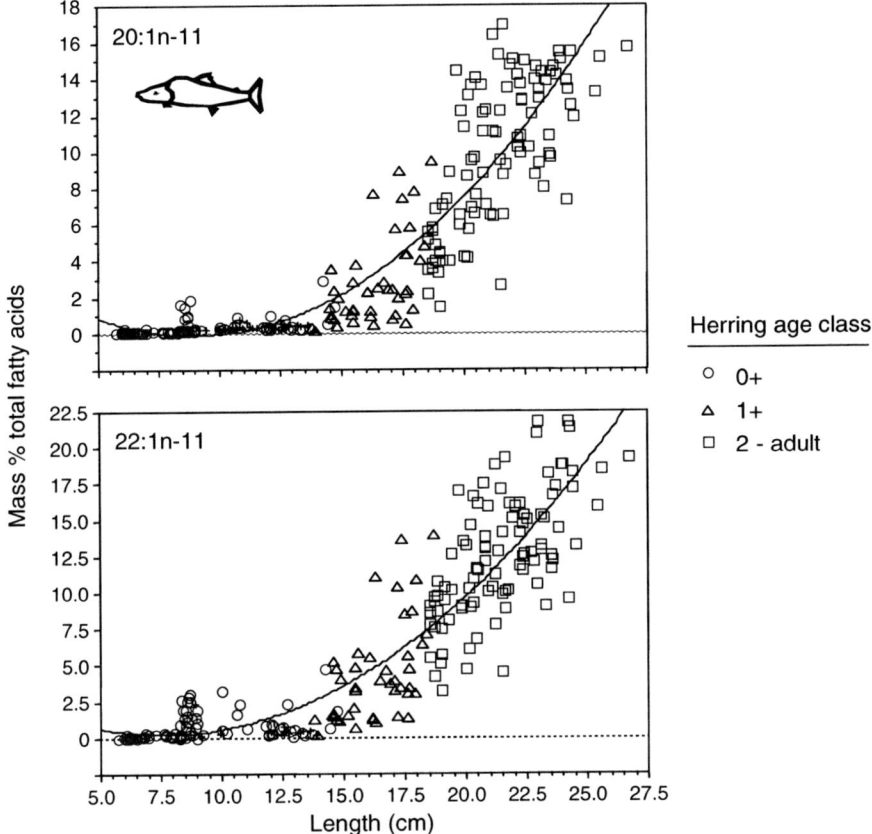

Fig. 12.1 Variation in two indicator FA in Pacific Herring ($n = 300$) from similar regions within Prince William Sound, Alaska, as a function of body length and age class. Data taken from Iverson et al. (2002)

these prey can be readily distinguished (Iverson et al. 1997). Similarly, Smith et al. (1996) used low levels of such marine-based FA, coupled with high levels of 18:2n-6, 18:3n-3, and 20:4n-6 (FA typical of primary producers in freshwater and terrestrial ecosystems, with low abundance in marine ecosystems), to distinguish freshwater from marine harbor seals. Based on the same principle, terrestrial foraging was confirmed in polar bears in late summer in Hudson Bay after measuring up to 7% 18:3n-3 in the milk of several females (which also had berry stains on their teeth; S.J. Iverson and I. Stirling unpublished data). Typically, levels of 18:3n-3 in marine ecosystems (seals or prey) are <0.5%, and have only been found at high levels in the milk of carnivores that have fed directly on plants (Iverson and Oftedal 1995; Iverson et al. 2001b).

Other types of biomarkers, or at least general trophic tracers, can include odd- and branched-chain (i.e., *iso* or *anteiso*) FA of 14–18 carbons, but these may or may not be useful depending on the phylogenetic order. These FA arise from bacteria and may indicate detrital feeding in primary consumers (Dalsgaard et al. 2003). However, in higher trophic levels, trace quantities of these FA could also largely reflect bacteria present in the gut flora of the consumer itself, rendering conclusions based on these trophic markers less certain (Iverson et al. 2004). In contrast, a potentially important group of FA are the NMI FA, which are produced only by certain benthic mollusks and gastropods (see Sect. 12.2.1), leading Paradis and Ackman (1977) to suggest that NMI FA might be useful biomarkers in food web studies. Recent work used the different types or proportions of these NMI FA produced by different mollusk species to reveal niche separation between sympatric bearded seals (*Erignathus barbatus*) and walruses (*Odobenus rosmarus rosmarus*), which both specialize on benthic mollusks (Budge et al. 2007). Furthermore, differences in the proportions of specific NMI FA have been used to estimate the relative importance of bearded seals and walruses, amongst other marine mammal prey, in the diets of large adult male polar bears; and these findings were confirmed through quantitative diet analyses (see Sect. 12.3.3) using entirely independent FA and diet estimates (Thiemann et al. 2007).

Finally, an example of another type of biomarker, but at the whole lipid level, is the case of WE, although this is in principle limited to a few predators. Only certain species synthesize and store WE (see Sect. 12.2.2), but because WE are broken down during digestion in the predator and reassembled independently as TAG or WE (see Sect. 12.2.3), they will generally not be useful tracers. However, an exception is found in Procellariiform seabirds (e.g., albatrosses and petrels), which store a significant amount of dietary lipid in stomach oils, in addition to their adipose tissue. These stomach oils do not undergo metabolic processing and can therefore contain large amounts of WE reflecting recent prey ingestion; at the same time no WE are ever present in adipose tissue of the same individuals, which instead reflects deposition of both dietary FA and WE fatty alcohols over a longer integration period (Wang et al. 2007). Thus, the presence of WE in stomach oils in Procellariiforms can be used not only as confirmation of consumption of prey species that make WE, but also provides an opportunity to examine different time scales of dietary integration.

Evaluating differences in the full array of FA among predators at similar phylogenetic levels is a promising, qualitative, way to look at trophic interactions, especially at higher trophic levels. The term "signature," or profile, refers to the relative concentration of the full array of FA identified or a subset of those FA most indicative of dietary intake, rather than just one or a few biomarkers (Iverson 1993). Such arrays will likely be most informative at higher trophic levels. Support for this approach comes from a number of experimental studies on fish, birds, and mammals, and can be illustrated clearly by a few examples. When Atlantic cod were switched to diets consisting of a single prey, i.e., low-fat squid (*Illex illecebrosus*) or high-fat mackerel (*Scomber scombrus*), the whole body FA signatures of these cod clearly reflected the FA signature of their diets (Fig.12.2). Had only liver (the major site of lipid storage in cod) been sampled and analyzed, there would likely have been an even closer correspondence with diet (see Sect. 12.3.1.1). Such findings demonstrate that FA composition of whole fish will largely be defined by their general ecology and dietary habits. This raises the possibility of using FA to characterize foraging

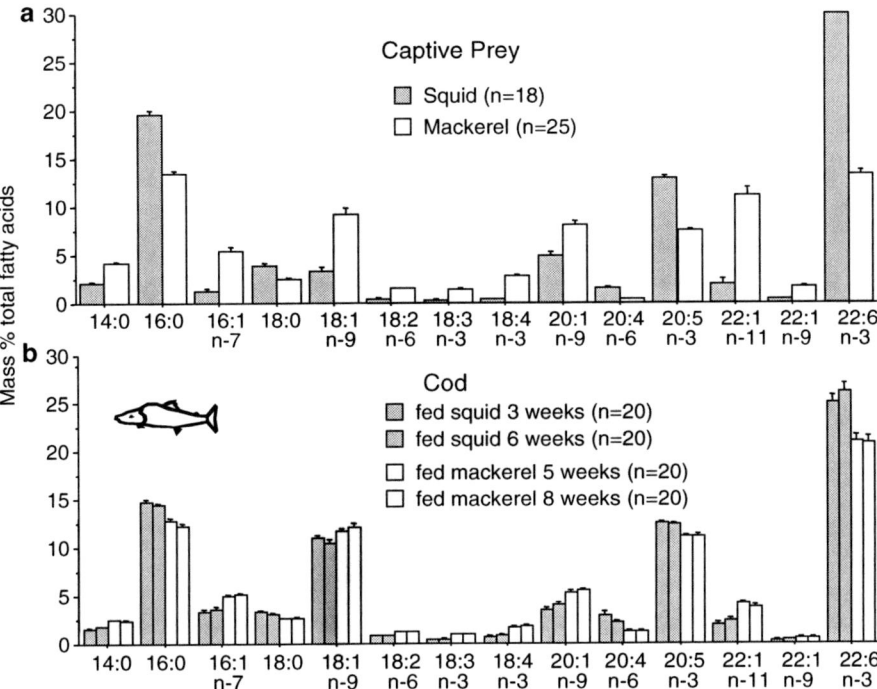

Fig. 12.2 Relative amounts (mass %) of selected abundant FA in (**a**) captive prey diets and (**b**) Atlantic cod dietary treatment groups. *Bars* are means and *vertical lines* are +1 SE. All cod were caught from the same net trawl and housed in seawater tanks at 2.8–4.0°C; all cod were fed other cod from the same lot for 4 weeks and then all were switched to the same diet treatment, but individuals were sequentially removed for whole body FA analysis. The first two *bars*: cod switched to diet of squid (2% lipid) for 3 and 6 weeks, respectively; second two *white bars*: remaining cod switched to diet of mackerel (16% lipid) for 5 and 8 weeks, respectively. Data taken from Kirsch et al. (1998)

differences within and between species and also to use their FA profiles in the evaluation of diet studies of higher predators (Budge et al. 2002; Iverson et al. 2002).

Similarly, in a study of two captive seabird species large differences in the overall FA signatures of two different prey types were reflected in synsacral adipose tissue biopsies of common murres (*Uria aalge*) and red-legged kittiwakes (*Rissa brevirostris*) (Fig. 12.3a). Although some absolute differences between diet and adipose tissue were apparent, the greatest influence on the FA composition of adipose tissue was the diet. Equally important, FA profiles of the two different species on the same diets were nearly identical and slight differences could be attributed to early feeding differences prior to the experiment. FA patterns in these same seabird species in the wild bear no resemblance to those of the captive birds (Fig.12.3b) and correspond to known foraging differences between murres and kittiwakes (Iverson et al. 2007). Thus, differences found in FA patterns of free-ranging individuals indicate differences in diet of those individuals.

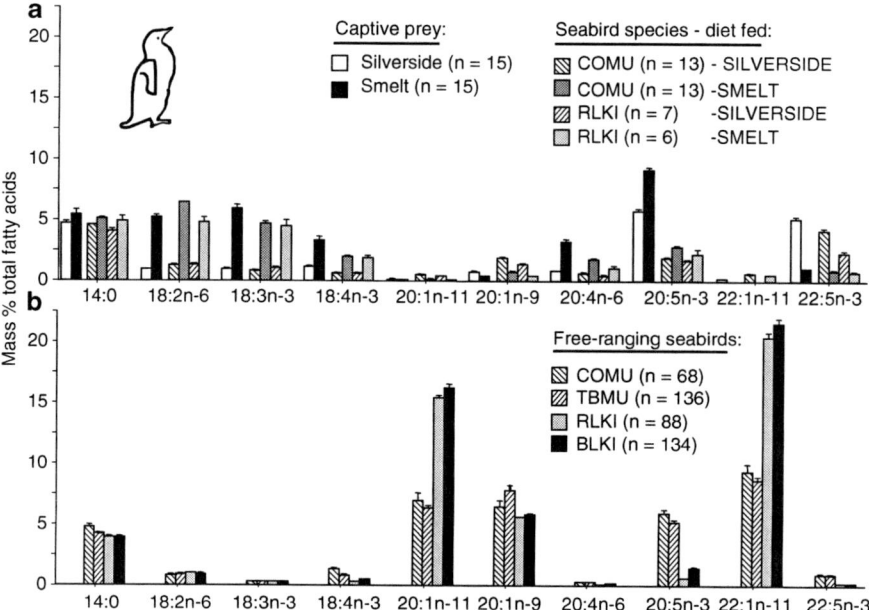

Fig. 12.3 (a) Selected abundant FA (mass %) with large overall variance illustrating characteristic differences in patterns among the two primary prey items and in the adipose tissue of captive common murre (COMU) and red-legged kittiwake (RLKI) chicks fed two prey types. *Bars* are means and *vertical lines* are +1 SE. For the diet trial, chicks were fed diets consisting of either silverside (*Menidia menidia*) or smelt (*Osmerus mordax*) from 15- to 42- or 45-day post-hatching. Prior to this, chicks had been fed either silverside (COMU) or a mixture of silverside and herring (RLKI) from 0- to 10- or 15-day post-hatching. (b) The same selected FA in free-ranging murres (COMU and thick-billed, TBMU, *Uria lomvia*) and kittiwakes (RLKI and black-legged, BLKI, *Rissa tridactyla*) in the Bering Sea. Data taken from Iverson et al. (2007)

Finally, in marine and aquatic mammals, a number of studies have documented the influence of dietary FA source on body FA composition (Iverson et al. 2004; Budge et al. 2006). An example of this is provided by another comparison of captive individuals to their free-ranging counterparts. The Hawaiian monk seal (*Monachus schauinslandi*) is one of only two pinnipeds to inhabit a tropical ecosystem. These systems contain organisms with FA profiles that are quite different from those of northern/temperate or polar ecosystems. For example, typical of tropical ecosystems (Dalsgaard et al. 2003) prey throughout the Northwestern Hawaiian Islands (NWHI) generally contain very low levels of long-chain monounsaturated isomers of 20:1 and 22:1, and relatively very high levels of long-chain n-6 PUFA. Comparison of blubber FA signatures of monk seals feeding in the NWHI versus captive monk seals maintained for an extended period on North Atlantic herring illustrates the dramatic and predictable influence that dietary FA have on lipid stores (Fig.12.4). Such comparisons of predator FA signatures alone have provided considerable insight into foraging ecology, diet shifts, and habitat segregation of many species.

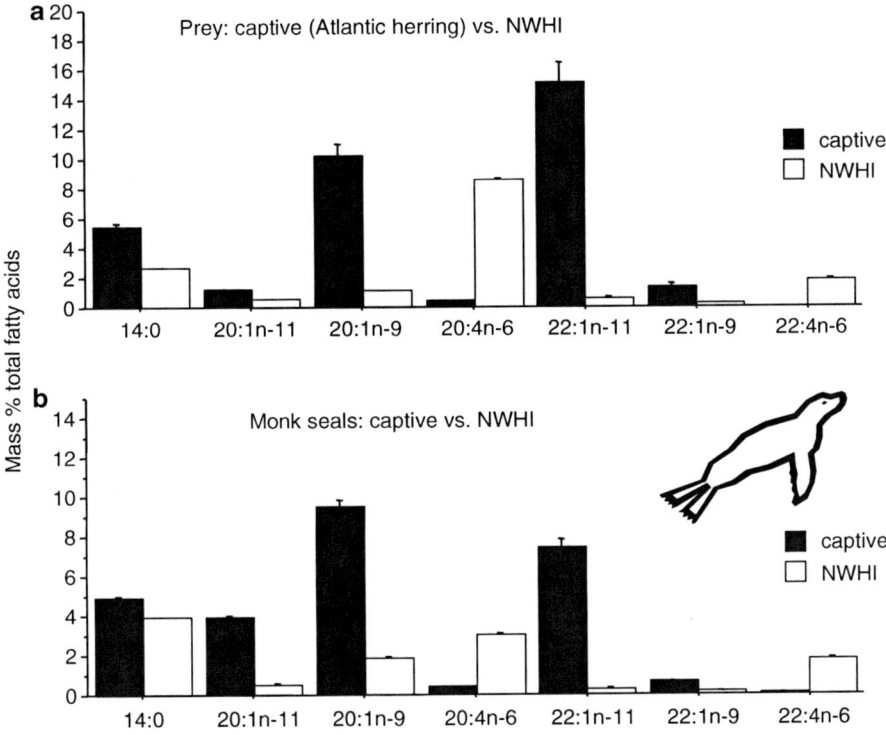

Fig. 12.4 Selected dietary FA (mean + 1 SE) in (**a**) the prey (herring) of captive monk seals in comparison to prey in the Northwestern Hawaiian Islands (NWHI) and (**b**) captive monk seals ($n = 10$) fed Atlantic herring in comparison to that of the blubber of free-ranging monk seals ($n = 157$) in the NWHI. Values for captive prey are the average of all herring analyzed ($n = 25$, from five different lots fed) and for wild prey are simply the average of all prey species previously analyzed in the NWHI data base ($n = 1,540$ individuals; S.J. Iverson and G. Antonelis, personal communication) for comparison purposes. The high levels of 14:0, 20:1n-11, 20:1n-9, 22:1n-11, and 22:1n-9 of Atlantic herring were reflected in captive seals, while much lower levels of these components and the high levels of n-6 PUFA in NWHI prey were reflected in wild seals. Data from Iverson et al. (2003)

12.3.2.2 Methods and Statistics

When evaluating different FA arrays, important issues include the choice of FA and subsequent data analyses. Finding a "significant" difference in levels of a specific FA among groups does not indicate whether this difference is biologically meaningful or whether the overall profile of FA differs between groups. Multivariate analyses, which also allow pattern recognition, are generally the most powerful as they use the maximum number of FA for differentiating predators and resolving trophic interactions (Budge et al. 2006). However, because there are generally a large number of FA identified (up to 70 in marine samples) in relation to sample size, there are restrictions on the number of FA that can be used (usually n-1 of group sample size) to meet statistical requirements. Thus, choices must be made as to which FA to use. Clearly, not all FA provide information about diet for various consumers and these can usually be removed (see also Sect. 12.3.3). For instance, any FA with <14 carbons within a consumer will have arisen from de novo biosynthesis and has no relation to diet (see Sect. 12.2.3). In mammals, 14:1n-5 arises almost entirely from biosynthesis from precursors, as demonstrated by very elevated levels of this FA found in the blubber of newborns (Iverson et al. 1995) or in the structured outer blubber of small cetaceans (Koopman et al. 1996). Whether the isomers of 20:1 and 22:1 or odd and branched-chain FA are valuable will depend on the phylogenetic order of the consumer: the former will not be useful dietary markers in calanoid copepods and perhaps other zooplankton, while the latter should be used in primary consumers but not higher orders (see Sect. 12.3.2.1). The FA 22:5n-3 is often considered a potential intermediate between 20:5n-3 and 22:6n-3; however, this may require evaluation as it can also be highly indicative of diet (Fig.12.3a; Iverson et al. 2007). Other FA may be removed as they are not indicative of diet or reliably measured. Thus, when reducing the FA set for statistical reasons, a useful practice is to choose FA that are obvious dietary markers (e.g., Appendix A in Iverson et al. 2004), as well as those that are the most abundant and/or exhibit the greatest average variance across individuals. In higher order consumers, these can also include FA that could be biosynthesized, but are likely to be most heavily influenced by dietary intake (Fig.12.2). Since FA analyses generally report values as mass % of total FA, it may be necessary to renormalize the chosen FA over 100% and/or transform the data (e.g., taking the log ratio of all chosen FA over a reference FA) to meet requirements of normality (Budge et al. 2002; Iverson et al. 2002). Proportional data can also be arcsine square root transformed, i.e., = acrsine(sqrt(proportion)), to achieve normality.

Useful multivariate techniques for these types of analyses have been reviewed extensively by Budge et al. (2006). In brief, multivariate analysis of variance (MANOVA) tests whether the mean differences among groups, based on a linear combination of response variables, could have occurred by chance, and subsequently allows identification of the FA(s) contributing to the differences. Discriminant function analysis (DFA) is used to classify samples into groups, and to describe differences among those groups, by creating a series of uncorrelated linear relationships among the original FA variables. The plot of scores derived can effectively reveal relationships among and within sample groups. Similar to DFA, principal component analysis (PCA) can describe relationships among variables, as well as reduce large numbers of variables to fewer components that represent most of the variance in the data. Complementary techniques that have fewer restrictions on numbers of variables used in relation to sample size include hierarchical cluster analysis and classification and regression trees (CART), as well as analysis of similarity (ANOSIM, Clarke 1993). All of these techniques provide the investigator with effective methods to examine trophic relationships among predators, and also allow evaluation of the ability to differentiate the prey of a predator when subsequently considering quantitative methods for estimating diet (see Sect. 12.3.3). For example, DFA analysis of 13 species of fish and invertebrates in Prince William Sound characterized similarities and differences among these consumers

12 Tracing Aquatic Food Webs Using Fatty Acids

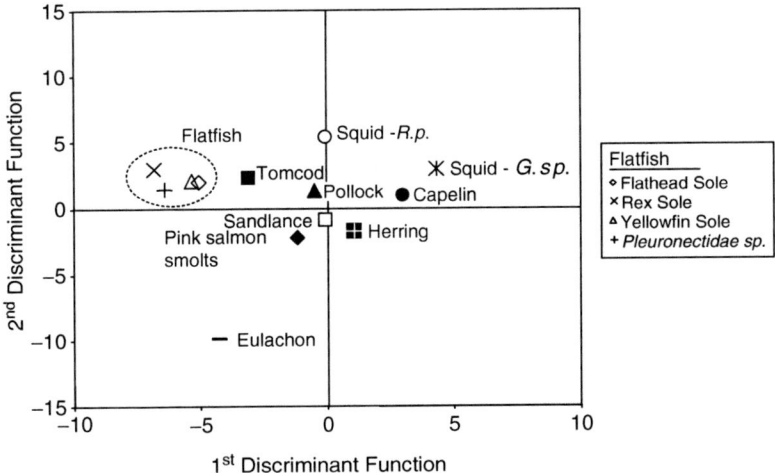

Fig. 12.5 Discriminant analysis performed on 13 common species of fish and invertebrates (each with $n \geq 17$, $n = 1{,}050$ total) in Prince William Sound, Alaska, using 17 of the most abundant FA, which together accounted for about 88% of total FA. The plot shows the group centroids (within group mean for each discriminant function) for the first and second (of 12 significant) discriminant functions, which accounted for 63% of total variance. This analysis classified individuals to species with 93.0% success rate when flatfish species were grouped as a whole. Herring (*see* Fig.12.1) were classified with 94.5% success. Reproduced from data in Iverson et al. (2002)

(Iverson et al. 2002). Overall, 93% of individuals were correctly classified to species by their FA composition. Flatfish species, which feed primarily on benthic infauna and epifauna, grouped closely together, and farther away from the more pelagic zooplanktivores and piscivores such as herring and pollock, clearly illustrating ecological differences (Fig.12.5). Despite the large FA variation in herring with ontogeny (Fig.12.1), 94.5% were still correctly classified, illustrating that between-species differences in FA signatures tend to be more pronounced than within-species differences as has been found elsewhere (Budge et al. 2002).

12.3.3 *Quantitative Estimation of Predator Diets*

The characteristics of FA biosynthesis, digestion, and deposition among organisms (see Sect. 12.2), coupled with the wide arrays of FA arising in marine food chains (Table 12.1) and the findings that characteristic FA signatures found in many prey types can be used to accurately classify individual species (e.g., Figs. 12.2a, 12.3a, 12.5) raises the possibility of using FA signatures to produce quantitative estimates of predator diets. This is perhaps the area of greatest current interest in using FA to elucidate food web relationships, especially for investigators working at higher trophic levels. Although statistical techniques described above (see Sect. 12.3.2) address important questions about patterns of FA composition among individuals and populations, they cannot be used to determine species composition of the diet of higher predators. The number and patterns of FA within a predator, as well as among and within potential prey species, are extremely complex, and there will always be absolute differences between predator and prey FA signatures due to predator metabolism. Thus, it is neither possible to visually assess species composition of diets in higher predators in a quantitative manner, nor usually even in a qualitative manner. Quantitative FA signature analysis (QFASA) is a first generation statistical tool designed to quantitatively estimate predator diet using FA signatures (Iverson et al. 2004).

The basic approach of QFASA is to determine the weighted mixture of prey species FA signatures that most closely resembles that of the predator's FA stores to thereby infer its diet. Details of the initial QFASA model are provided by Iverson et al. (2004) and are further discussed by Budge et al. (2006). Briefly, assuming appropriate sampling and analysis of predator and prey lipids (see Sect. 12.3.1), the model proceeds by applying weighting factors (calibration coefficients) to individual predator FA to account for the effects of predator metabolism on FA deposition in lipid stores. These are empirically determined from controlled long-term feeding trials. It then takes the average FA signature of each prey species (or within-species group) and estimates the mixture of prey signatures that comes closest to matching that of the predator's FA stores by minimizing the statistical distance (e.g., Kulback-Liebler) between that prey species mixture and the weighted predator FA profile. Finally, this proportional mixture is weighted by the proximate fat content (i.e., relative FA contribution) of each prey species to estimate the proportions in the predator's diet. Each of these steps requires careful consideration.

Perhaps the most important issue when using FA quantitatively is accounting for predator metabolism. Even in predators that consume high-fat diets, there will be physiological affects on FA deposition. Thus the FA composition of the predator will never exactly match that of their prey. At present, predator effects on FA deposition are accounted for using calibration coefficients, which are simple ratios for each FA in the predator fat divided by that FA in the diet. The principle is that if a predator had consumed a constant diet over a very long term, the FA signature of its lipid stores would resemble this diet as much as possible, and any differences would be attributable to metabolic processing of individual FA. These "coefficients" can be derived from studies where a predator has been fed a constant diet for a prolonged period, with the hope that this is sufficient time to have completely turned over all of its stored FA. Unfortunately, we do not yet know what the total FA turnover time is for most predators. Even so, data have been obtained from studies on captive pinnipeds, mink (*Mustela vison*), and seabirds (Fig.12.6). The key points from these data are that the calibration coefficients are, in general, similar across bird and mammal species and diets. This confirms the dominance of dietary FA deposition, as well as postulated secondary impacts from biosynthesis and modification (see Sect. 12.2). Second, the application of these coefficients in the QFASA model has been shown to be critical to accurate diet estimates (Iverson et al. 2004). To date, calibration coefficients have not been estimated or evaluated in fish or invertebrates, but could potentially be done from controlled studies such as those summarized in

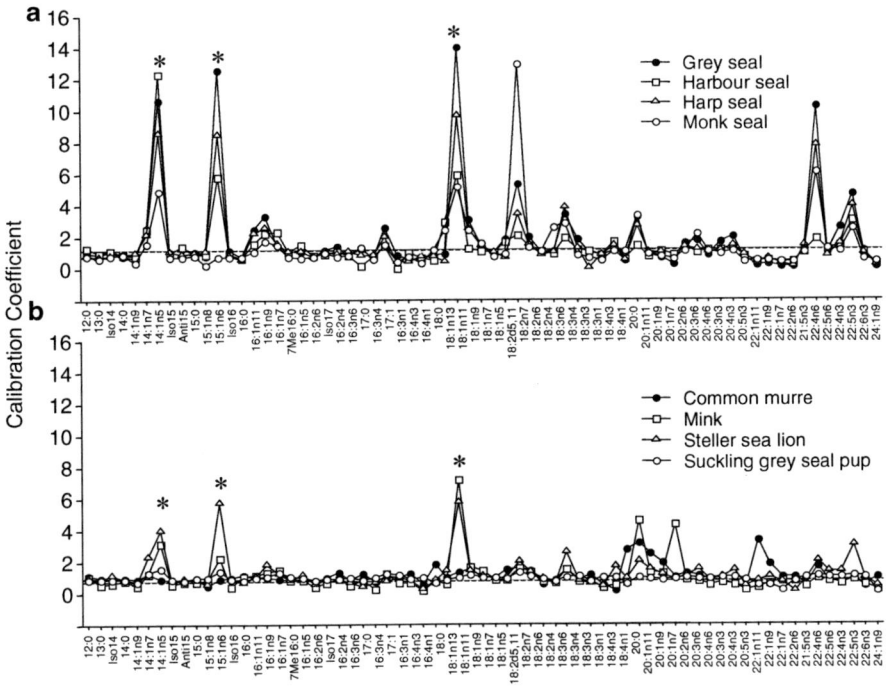

Fig. 12.6 Calibration coefficients (CCs, means of 5–16 individuals each) calculated from controlled experiments on (**a**) four species of phocid seals and (**b**) one otariid species, mink, and common murres all fed fish diets (3–10% fat), and in suckling grey seal pups consuming their mother's milk (60% fat). CCs were calculated according to Iverson et al. (2004) as: % of each FA in each predator/average %FA in diet, but without using trimmed means. The 1:1 line denotes the deviation of a given FA in the predator from that consumed in diet. *Asterisks* indicate examples of FA with large deviations from 1:1 but which occur at only minor or trace amounts in marine lipids, with contributions from biosynthesis, and routinely not used in QFASA modeling. Data taken from Iverson et al. (2004); Tollit et al. (2006), Iverson et al. (2007); and Nordstrom et al. (2008)

Chap. 6. There may be other ways to quantitatively incorporate metabolism effects on consumer FA deposition and such research should be encouraged.

There are a number of other important issues that should be recognized to successfully use QFASA (see Iverson et al. 2004 and Budge et al. 2006 for detailed discussion). These include building a representative and comprehensive prey FA database for each predator and sampling all prey sufficiently to allow quantitative evaluation of within and between-species variability to confirm the ability to reliably differentiate prey species in the model. The choice of the FA used is also critical to model outcomes and will likely also depend upon the reliability of calibration coefficients determined for specific FA. Ways to incorporate within-species variability in prey FA and fat content in estimates, as well as ways to measure goodness of fit, are still being developed. Optimization of model outputs can be assessed when there is corroborating evidence available (i.e., studies of captive animals or some free-ranging animals where other supporting diet information exists), but this is often not possible. Understanding the detection limits of prey is important, especially in cases where there is interest in prey species rather than predators. Finally, the ability to apply QFASA to lower trophic levels has not yet been explored.

Despite issues concerning its further advancement, QFASA has now been successfully validated for a variety of predators including pinnipeds, mink, and seabirds (Iverson et al. 2004, 2007; Cooper 2004; Cooper et al. 2005; Nordstrom et al. 2008). Since most validation studies are performed within the constraints of captive animal experiments, tested diets are usually composed of only one or a few prey species, potentially limiting conclusions drawn. However, two studies have allowed evaluation of QFASA in free-ranging animals in their natural habitat consuming a wide array of prey. Results from a captive validation study of seabirds were extended to the study of 235 free-ranging murres and kittiwakes by comparing QFASA diet estimates to those derived from stomach contents in the same individuals. Although one would not expect results from the two analyses to be identical, mainly because the two methods integrate diet over different time scales, both methods indicated the same dominant prey (Fig.12.7a, b) and characterized well-established differences in the known diets of murres and kittiwakes in the Bering Sea. In the Northwest Atlantic, free-ranging harbor seals, fitted with head-mounted video-cameras, allowed estimates of prey capture, which were then compared with

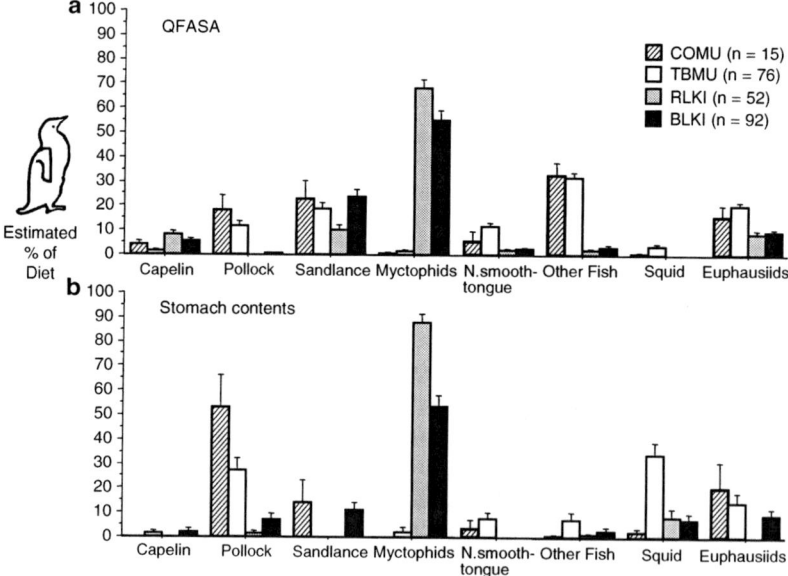

Fig. 12.7 Diet estimates of free-ranging murres (COMU, TBMU, see Fig.12.3) and kittiwakes (RLKI, BLKI) (n = 235) in the Bering Sea using (a) QFASA (modeled on 161 prey representing 15 species) in comparison with (b) stomach contents analysis in the same individuals. Bars are means of each species estimated in diets and vertical lines are +1 SE. Reproduced from data in Iverson et al. (2007)

12 Tracing Aquatic Food Webs Using Fatty Acids   301

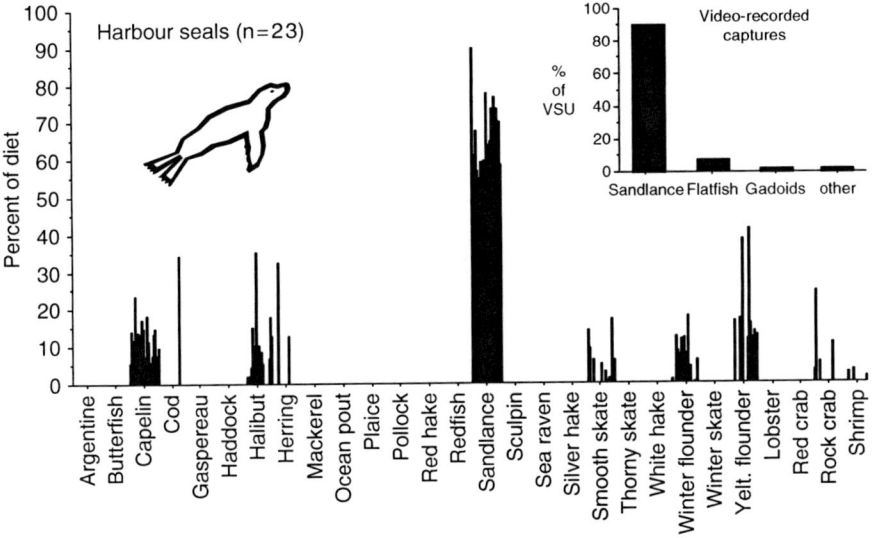

Fig. 12.8 Individual QFASA diet estimates of the contribution of prey species to diets of 23 free-ranging adult male harbour seals, at Sable Island, Nova Scotia, deployed with an animal-borne video system ("Crittercam") and filmed during natural feeding events. Seal signatures were modeled using a Scotian Shelf prey database of 954 prey representing 28 species. *Inset*: prey types consumed in video recordings of these seals, expressed as percent of all 10-min video-sampling units (VSU) that filmed prey captures and which contained identifiable items (n = 223). Reproduced from data in Iverson et al. (2004)

the diet estimates in the same individuals obtained using QFASA (Fig. 12.8). QFASA diet estimates were similar to those recorded on video and were also consistent with major prey identified in gastric lavage data and fecal analyses (Iverson et al. 2004). On the basis of these studies, QFASA is now being used to address broader ecosystem-scale processes (Bowen et al. 2006; Iverson et al. 2006b, 2007; Beck et al. 2007). The current conclusions drawn from such quantitative studies are that QFASA is a potentially powerful tool in ecological research, which has provided new insight into the foraging patterns and ecology of free-ranging predators that would otherwise not be discernable. Nevertheless, QFASA must be used with due diligence and investigators should understand that there are many issues that remain to be resolved, or need to be further investigated, to improve its reliability.

12.4 Summary, Conclusions, and the Future

In their very thorough review, Dalsgaard et al. (2003) noted that FA can only presently be used qualitatively or semiquantitatively and concluded that the state of the art for using FA as trophic tracers was essentially at the same level described 30 years ago by Sargent (1976): *"At the present state of our knowledge it would appear that fatty acid analyses represent a rather blunt tool in defining food chain inter-relationships. Until further knowledge is accumulated it would appear best to apply fatty acid analyses as a corroborative method to support prey–predator relationships already indicated on independent grounds, such as analyses of stomach contents*". In just a few short years since this review, the field of FA signature analysis has clearly advanced. With the advent of QFASA at higher trophic levels, we are now successfully using FA to understand some otherwise entirely intractable systems and ecosystem processes. Using QFASA, we are beginning to understand demographic (i.e., sex, age) and individual sources of diet variability in an abundant marine predator, the grey seal (*Halichoerus grypus*), and how it uses the North Atlantic ocean relative to prey abundance and distribution (Bowen et al. 2006; Beck et al. 2007). We have studied polar bear foraging at continental scales (i.e., across the entire Canadian Arctic), providing new insight into individual and population responses to changes in prey distribution and climate (Iverson et al. 2006b). We have also linked seabird foraging to productivity and prey abundance throughout the Bering Sea (Iverson et al. 2007). In the tropical NWHIs, we are gaining insights into the key prey of the critically endangered Hawaiian monk seal (Iverson et al. 2006a). Nevertheless, there remains much work before we understand the full potential and limitations of QFASA. Although we have advanced considerably and are somewhat beyond the "first-cut" stage, the potential to use QFASA (or other statistical models or methods) to investigate trophic ecology will continue to be an area for fruitful research.

Understanding metabolism effects on FA stored in consumers (which for simplicity is referred to as "calibration") is extremely important in using FA to study food webs using any of the methods outlined above, and in QFASA, is also likely largely tied into choosing the appropriate FA set for modeling. QFASA was obviously designed for upper trophic level endothermic vertebrates, but so far little is known about how this can be applied to ectothermic fish and reptiles, and even less about its potential application to primary consumers such as zooplankton. Certainly one way to advance this knowledge would be to conduct more rigorously controlled long-term studies to estimate calibration in a wide variety of consumers, including whether the principle of calibration coefficients could be used in lower orders. Even at the highest trophic levels, coefficients have only been evaluated in one species of seabird and none have yet been evaluated for cetaceans, polar bears, or sea otters.

A potentially powerful trophic indicator, which should be further studied and calibrated is mammalian milk. Given the metabolic dominance of the lactating mammary gland, ingested FA are directed first into milk over other tissues (Iverson 1993). For example, even though consumption of small amounts of very low-fat terrestrial vegetation could not be detected in polar bear adipose tissue using either FA or stable isotope analyses (Hobson and Stirling 1997), it was detected in milk FA (see Sect. 12.3.2.1). However, QFASA has not yet been adapted to analyses of milk.

Additionally, within higher-predator species, the influence of fat content of the diet on quantitative calibration has not yet been addressed but should be important, as higher fat diets are presumably associated with greater fat deposition and lower biosynthesis or modification of FA by predators. Similarly, the effects on predator FA deposition of consumption of prey with widely differing fat contents requires further evaluation. Future studies should also examine whether the FA composition of the control diet used to determine calibration affects the eventual modeling with prey of similar versus widely differing FA compositions. There also needs to be a better understanding of the quantitative effects of fasting and mobilization on the composition of predator FA stores, especially in species that routinely fast as part of their life history. Finally, while it has been demonstrated that QFASA estimates correctly reflect major components of diet, the ability to identify trace components may be limited. Care is also clearly required in interpreting results from prey species with similar FA signatures, which may cause false positives in estimates (e.g., Iverson et al. 2004; Budge et al. 2006; Nordstrom et al. 2008).

In theory, if all the above requirements for using QFASA (see also Sect. 12.3.3) were met for each consumer in a given ecosystem, one could model diets of each consumer/predator at each subsequently higher trophic level. Iteration backwards might allow estimates of ecosystem processes responsible for food web production and budgets. Whether this could ever be realized remains to be established, but almost certainly could not be accomplished using FA and QFASA alone. Additional tools for better understanding trophic transfer of FA in general and the performance of QFASA in particular could make use of advanced and targeted techniques, such as using radio or stable isotope-labeled FA to trace pathways from diet to deposition (Budge et al. 2004; Cooper 2004; Cooper et al. 2006). These types of methods should provide even better insight at all trophic levels but are fairly cost-intensive. Of course, combining FA and QFASA with other types of corroborative evidence will always be useful and aid in further validation of using FA as accurate trophic tracers. One particularly important area that should continue to be explored is combining FA with stable isotope analyses of tissues or of individual FA (e.g., Kharlamenko et al. 2001; Hebert et al. 2006). Recently, the convergence of diet estimates derived from FA and stable isotopes was illustrated by the correlation between estimates of the trophic position of free-ranging grey seals from QFASA and $\delta^{15}N$ values (Tucker et al. 2008). Additionally, the stable carbon isotopic composition of diatom FA has been used to investigate the contribution of ice algae to higher trophic levels and shows promise in tracing carbon flow through an Arctic marine food web (Budge et al. 2008).

In conclusion, while there are many areas of research to pursue and questions to address, it is clear that FA can be used as powerful trophic tracers of carbon flow, food webs, predator–prey interactions, and even ecosystem structure and dynamics within marine and other aquatic ecosystems. FA are particularly useful in that they are relatively easily measured, sensitive to change and responsive to change in a predictable manner, and are integrative. The fact that techniques have advanced to the point of estimating overall species composition of diets of individual predators, at time scales relevant to the ecological processes affecting survival, is exciting. Coupled with other information, FA can provide valuable insights about the ecology of individual species and ecosystem structure and functioning.

Acknowledgements The writing of this chapter was supported in part by the Natural Sciences and Engineering Research Council (NSERC), Canada. I thank all of my many collaborators and students associated with the FatLab at Dalhousie, who have contributed greatly to our current state of understanding in the field of lipids and fatty acids as food web tracers. I thank W.D. Bowen for comments on an earlier version of the manuscript.

References

Ackman, R.G. 1980. Fish lipids, Part, pp. 86–103. In J.J. Connell (ed.), Advances in Fish Science and Technology. Fishing News Books, Surrey.

Ackman, R.G. 1986. WCOT (capillary) gas-liquid chromatography, pp. 137–206. In R.J. Hamilton, and J.B. Rossell (eds.), Analysis of Oils and Fats. Elsevier, London.

Ackman, R.G. 2002. The gas chromatograph in practical analysis of common and uncommon fatty acids for the 21st century. Anal. Chim. Acta 465:175–192.

Ackman, R.G., and Eaton, C.A. 1966. Lipids of the fin whale (*Balaenoptera physalus*) from north Atlantic waters. III. Occurence of eicosenoic and docosenoic fatty acids in the zooplankter *Meganyctiphanes norvegica* (M. Sars) and their effect on whale oil composition. Can. J. Biochem. 44:1561–1566.

Ackman, R.G., and McLachlan, J. 1977. Fatty acids in some Nova Scotian marine seaweeds: a survey for octadecapentaenoic and other biochemically novel fatty acids. Proc. Nova Scotian Inst. Sci. 28:47–64.

Ackman, R.G., Hooper, S.N., and Sipos, J.C. 1972. Distribution of the *trans*-6-hexadecenoic acid and other fatty acids in tissues and organs of the Atlantic leatherback turtle, *Dermochelys coriacea coriacia*. L. Int. J. Biochem. 3:171–179.

Ackman, R.G., Eaton, C.A., and Linke, B.A. 1975. Differentiation of freshwater characteristics of fatty acids in marine specimens of the Atlantic sturgeon, *Acipenser oxyrhynchus*. Fish. Bull. US Fish. Wildl. Serv. 73:838–845.

Beck, C.A., Iverson, S.J., Bowen, W.D., and Blanchard, W. 2007. Sex differences in grey seal diet reflect seasonal variation in foraging behaviour and reproductive expenditure: evidence from quantitative fatty acid signature analysis. J. Anim. Ecol. 76:490–502.

Bowen, W.D., Beck, C.A., Iverson, S.J., Austin, D., and McMillan, J.I. 2006. Linking predator foraging behaviour and diet with variability in continental shelf ecosystems: grey seals of eastern Canada, pp. 63–81. In I.L. Boyd, S.W. Wanless, and C.J. Camphuysen (eds.), Top Predators in Marine Ecosystems. Cambridge University Press, Cambridge.

Brindley, D.N. 1991. Metabolism of triacylglycerols, pp. 171–203. In D.E. Vance and J.E. Vance (eds.), Biochemistry of Lipids, Lipoproteins and Membranes. Elsevier Science, New York.

Budge, S.M., and Iverson, S.J. 2003. Quantitative analysis of fatty acid precursors in marine samples: direct conversion of wax ester alcohols and dimethylacetals to fatty acid methyl esters. J. Lipid Res. 44:1802–1807.

Budge, S.M., Iverson, S.J., Bowen, W.D., and Ackman, R.G. 2002. Among- and within-species variability in fatty acid signatures of marine fish and invertebrates on the Scotian Shelf, Georges Bank, and southern Gulf of St. Lawrence. Can. J. Fish. Aquat. Sci. 59:886–898.

Budge, S.M., Cooper, M.H., and Iverson, S.J. 2004. Demonstration of the deposition and modification of dietary fatty acids in pinniped blubber using radiolabelled precursors. Physiol. Biochem. Zool. 77:682–687.

Budge, S.M., Iverson, S.J., and Koopman, H.N. 2006. Studying trophic ecology in marine ecosystems using fatty acids: a primer on analysis and interpretation. Mar. Mamm. Sci. 22:759–801.

Budge, S.M., Springer, A.M., Iverson, S.J., and Sheffield, G. 2007. Fatty acid biomarkers reveal niche separation in an arctic benthic food web. Mar. Ecol. Progr. Ser. 336:305–309.

Budge, S.M., Wooller, N.J., Springer, A.M., Iverson, S.J., and Divoky, G.J. 2008. Tracing carbon flow from the bottom to the top in of an Arctic marine food web using fatty acid-stable isotope analysis. Oecologia 157:117–129.

Castellini, M.A., and Rea, L.D. 1992. The biochemistry of natural fasting at its limits. Experientia 48:575–582.
Christie, W.W. 1982. Lipid Analysis. Pergamon, Oxford.
Clarke, K.R. 1993. Non-parametric multivariate analysis of changes in community structure. Aust. J. Ecol. 18:117–143.
Colby, R.H., Mattacks, C.A., and Pond, C.M. 1993. The gross anatomy, cellular structure and fatty acid composition of adipose tissue in captive polar bears (*Ursus maritimus*). Zoo Biol. 12:267–275.
Cook, H.W. 1996. Fatty acid desaturation and chain elongation in eukaryotes, pp. 129–152. *In* D.E. Vance and J.E. Vance (eds.), Biochemistry of Lipids and Membranes. Elsevier, Amsterdam.
Cooper, M.H. 2004. Fatty acid metabolism in marine carnivores: implications for quantitative estimation of predator diets. Ph.D. Thesis, Dalhousie University, Halifax, NS.
Cooper, M.H., Iverson, S.J., and Heras, H. 2005. Dynamics of blood chylomicron fatty acids in a marine carnivore: implications for lipid metabolism and quantitative estimation of predator diets. J. Comp. Physiol. B 175:133–145
Cooper, M.H., Iverson, S.J., and Rouvinen-Watt, K. 2006. Metabolism of dietary cetoleic acid (22:1n-11) in mink (*Mustela vison*) and grey seals (*Halichoerus grypus*) studied using radiolabelled fatty acids. Physiol. Biochem. Zool. 79:820–829.
Cripps, G.C., and Hill, H.J. 1998. Changes in lipid composition of copepods and *Euphasia superba* associated with diet and environmental conditions in the marginal ice zone, Bellingshausen Sea, Antarctica. Deep Sea Res. Part I. 45:1357–1381.
Dalsgaard, J., St John, M., Kattner, G., Müller-Navarra, D.C., and Hagen, W. 2003. Fatty acid trophic markers in the pelagic marine environment. Adv. Mar. Biol. 46:225–340.
Dunstan G.A., Volkman, J.K, Barrett, S.M., Leroi J-M, and Jeffrey, S.W. 1994. Essential polyunsaturated fatty acids from 14 species of diatom (*Bacillariophyceae*). Phytochemistry 35:155–161.
Groscolas, R. 1990. Metabolic adaptations to fasting in emperor and king penguins, pp. 269–295. *In* L.S. Davis and J.T. Darby (eds.), Penguin Biology. Academic, New York.
Gurr, M.I., and Harwood, J.L. 1991. Lipid Biochemistry: An Introduction. Chapman and Hall, New York.
Hebert C.E., Arts, M.T. and Weseloh, D.V.C. 2006. Ecological tracers can quantify food web structure and change. Environ. Sci. Technol. 40:5618–5623.
Hobson, K.A., and Stirling, I. 1997. Terrestrial foraging by polar bears during the ice free period in western Hudson Bay: metabolic pathways and limitations of the stable isotope method. Mar. Mamm. Sci. 13:359–367.
Hooper, S.N., Paradis, M., and Ackman, R.G. 1973. Distribution of trans-6-hexadecanoic acid, 7-methyl-7-hexadecanoic acid and common fatty acids in lipids of the ocean sunfish *Mola mola*. Lipids 8:509–516.
Iverson, S.J. 1993. Milk secretion in marine mammals in relation to foraging: can milk fatty acids predict diet? Symp. Zool. Soc. Lond. 66:263–291.
Iverson, S.J., and Oftedal, O.T. 1995. Phylogenetic and ecological variation in the fatty acid composition of milks, pp. 789–827. *In* R.G. Jensen (ed.), The Handbook of Milk Composition. Academic, Orlando.
Iverson, S.J., Oftedal, O.T., Bowen, W.D., Boness, D.J., and Sampugna, J. 1995. Prenatal and postnatal transfer of fatty acids from mother to pup in the hooded seal. J. Comp. Physiol. 165:1–12.
Iverson, S.J., Frost, K.J., and Lowry, L.L. 1997. Fatty acid signatures reveal fine scale structure of foraging distribution of harbor seals and their prey in Prince William Sound, Alaska. Mar. Ecol. Progr. Ser. 151:255–271.
Iverson, S.J., Lang, S., and Cooper, M. 2001a. Comparison of the Bligh and Dyer and Folch methods for total lipid determination in a broad range of marine tissue. Lipids 36:1283–1287.
Iverson, S.J., MacDonald, J., and Smith, L.K. 2001b. Changes in diet of free-ranging black bears in years of contrasting food availability revealed through milk fatty acids. Can. J. Zool. 79:2268–2279.
Iverson, S.J., Frost, K.J., and Lang, S. 2002. Fat content and fatty acid composition of forage fish and invertebrates in Prince William Sound, Alaska: factors contributing to among and within species variability. Mar. Ecol. Progr. Ser. 241:161–181.

Iverson, S.J., Stewart, B.S., and Yochem, P.K. 2003. Captive validation and calibration of fatty acid signatures in blubber as indicators of prey in hawaiian monk seal diet. NOAA Technical Report, San Diego, CA, 22 pp.

Iverson, S.J., Field, C., Bowen, W.D., and Blanchard, W. 2004. Quantitative fatty acid signature analysis: a new method of estimating predator diets. Ecol. Monogr. 74:211–235.

Iverson, S.J., Piche, J., and Blanchard, W. 2006a. Hawaiian monk seals and their prey in the Northwestern Hawaiian Islands: assessing characteristics of prey species fatty acid signatures and consequences for estimating monk seal diets using quantitative fatty acid signature analysis (QFAA). NOAA technical Report, Honolulu, HA, 127 pp.

Iverson, S.J., Stirling, I., and Lang, S.L.C. 2006b. Spatial and temporal variation in the diets of polar bears across the Canadian arctic: indicators of changes in prey populations and environment, pp. 98–117. In I.L. Boyd, S.W. Wanless, and C.J. Camphuysen (eds.), Top Predators in Marine Ecosystems. Cambridge University Press, Cambridge.

Iverson, S.J., Springer, A.M., and Kitaysky, A.S. 2007. Seabirds as indicators of food web structure and ecosystem variability: qualitative and quantitative diet analyses using fatty acids. Mar. Ecol. Progr. Ser. 352:235–244.

Joseph, J.D. 1982. Lipid composition of marine and estuarine invertebrates. Part II: Mollusca. Progr. Lipid Res. 21:109–153.

Kattner, G., and Hagen, W. 1995. Polar herbivorous copepods – different pathways in lipid biosynthesis. ICES J. Mar. Sci. 52:329–335.

Kharlamenko, V.I., Kiyashko, S.I., Imbs, A.B., and Vyshkvartzev, D.I. 2001. Identification of food sources of invertebrates from the seagrass *Zostera marina* community using carbon and sulfur stable isotope ratio and fatty acid analyses. Mar. Ecol. Progr. Ser. 45:17–22.

Kirsch, P.E., Iverson, S.J., Bowen, W.D., Kerr, S. and Ackman, R.G. 1998. Dietary effects on the fatty acid signatures of whole Atlantic cod (*Gadus morhua*). Can. J. Fish. Aquat. Sci. 55:1378–1386.

Klem, A. 1935. Studies in the biochemistry of whale oils. Hvalradets Skr. Nr. 11:49–108.

Koopman, H.N. 2007. Phylogenetic, ecological, and ontogenetic factors influencing the biochemical structure of the blubber of odontocetes. Mar. Biol. 151:277–291.

Koopman, H.N., Iverson, S.J., and Gaskin, D.E. 1996. Stratification and age-related differences in blubber fatty acids of the male harbour porpoise (*Phocoena phocoena*). J. Comp. Physiol. 165:628–639.

Koopman, H.N., Iverson, S.J., and Read, A.J. 2003. High concentrations of isovaleric acid in the fats of odontocetes: variation and patterns of accumulation in blubber vs. stability in the melon. J. Comp. Physiol. 173:247–261.

Koopman, H.N., Budge, S.M., Ketten, D.R., and Iverson, S.J. 2006. The topographical distribution of lipids inside the mandibular fat bodies of odontocetes: remarkable complexity and consistency. IEEE J. Oceanic Eng. 31:95–106.

Lovern, J.A. 1935. C. Fat metabolism in fishes. VI. The fats of some plankton Crustacea. Biochem. J. 29:847–849.

McConville, M.J. 1985. Chemical composition and biochemistry of sea ice microalgae, pp. 105–146. In R.A. Horner (ed.), Sea Ice Biota. CRC Press, Boca Raton.

Mellish, J.E., and Iverson, S.J. 2001. Blood metabolites as indicators of nutrient utilization in fasting, lactating phocid seals: does depletion of nutrient reserves terminate lactation? Can. J. Zool. 79:303–311.

Napolitano, G.E. 1999. Fatty acids as trophic and chemical markers in freshwater ecosystems, pp. 21–44. In M.T. Arts and B.C Wainman (eds.), Lipids in Freshwater Ecosystems. Springer, New York.

Nelson, G.J. 1992. Dietary fatty acids and lipid metabolism, pp. 437–471. In C.K. Chow (ed.), Fatty Acids in Foods and Their Health Implications. Marcel Dekker, New York.

Nordstrom, C.A., Wilson, L.J. Iverson, S.J., and Tollit, D.J. 2008. Evaluating quantitative fatty acid signature analysis (QFASA) using harbour seals (*Phoca vitulina richardsi*) in captive feeding studies. Mar. Ecol. Progr. Ser. 360:245–263

Norseth, J., and Christophersen, B.O. 1978. Chain shortening of erucic acid in isolated liver cells. FEBS Lett. 88:353–357.

Osmundsen, H., Neat, C.E., and Norum, K.R. 1979. Peroxisomal oxidation of long chain fatty acids. FEBS Lett. 99:292–296.

Paradis, M., and Ackman, R.G. 1977. Potential for employing the distribution of anomalous nonmethylene-interrupted dienoic fatty acids in several marine invertebrates as part of food web studies. Lipids 12:170–176.

Parrish, C.C. 1999. Determination of total lipid, lipid classes, and fatty acids in aquatic samples, pp. 4–20. *In* M.T. Arts and B.C. Wainman (eds.), Lipids in Freshwater Ecosystems. Springer, New York.

Pascal, J.C., and Ackman, R.G. 1976. Long chain monoethylenic alcohol and acid isomers in lipids of copepods and capelin. Chem. Phys. Lipids 16:219–223.

Pond, C.M. 1998. The Fats of Life. Cambridge University Press, Cambridge, UK.

Raclot, T. 2003. Selective mobilization of fatty acids from adipose tissue triacylglycerols. Progr. Lipid Res. 42:257–288.

Rouvinen, K., and Kiiskinen, T. 1989. Influence of dietary fat source on the body fat composition of mink (*Mustela vison*) and blue fox (*Alopex lagopus*). Acta Agric. Scand. 39:279–288.

Saito, H., and Murata, M. 1998. Origin of the monoene fats in the lipid of midwater fishes: relationship between the lipids of myctophids and those of their prey. Mar. Ecol. Progr. Ser. 168:21–33.

Sargent, J.R. 1976. The structure, metabolism and function of lipids in marine organisms, pp. 149–212. *In* D.C. Malins and J.R. Sargent (eds.), Biochemical and Biophysical Perspectives in Marine Biology, Academic, London.

Sargent, J.R. 1978. Marine wax esters. Sci. Progr. 65:437–458.

Sargent, J.R., and Henderson, R.J. 1986. Lipids, pp. 59–108. In E.D.S. Corner and S.C.M. O'Hara (eds.), The Biological Chemistry of Marine Copepods Vol I. Calrendon Press, Oxford.

Sargent, J.R., and Henderson, R.J. 1995. Marine (n-3) polyunsaturated fatty acids, pp. 32–65. *In* R.J. Hamilton (ed.), Developments in Oils and Fats. Blackie Academic and Professional, London.

Schlechtriem, C., Arts, M.T., and Zellmer, I.D. 2006. Effect of temperature on the fatty acid composition and temporal trajectories of fatty acids in fasting *Daphnia* pulex (Crustacea, Cladocera). Lipids 41:397–400.

Smith, R.J., Hobson, K.A., Koopman, H.N., and Lavigne, D.M. 1996. Distinguishing between populations of fresh- and salt-water harbour seals (*Phoca vitulina*) using stable isotope ratios and fatty acid profiles. Can. J. Fish. Aquat. Sci. 53:272–279.

St. John, M.A., and Lund, T. 1996. Lipid biomarkers: linking the utilization of frontal plankton biomass to enhanced condition of juvenile North Sea cod. Mar. Ecol. Progr. Ser. 131:75–85.

Tollit, D.J., Heaslip, S., Deagle, B., Iverson, S.J., Joy, R., Rosen, D.A.S., and Trites, A.W. 2006. Estimating diet composition in sea lions: which technique to choose?, pp. 293–308. *In* A. Trites, S. Atkinson, D. DeMaster, L. Fritz, T. Gelatt, L, Rea and K. Whynne (eds.), Sea Lions of the World. Alaska Sea Grant College Program, University of Alaska Fairbanks.

Thiemann, G.W., Iverson, S.J., and Stirling, I. 2006. Seasonal, sexual, and anatomical variability in the adipose tissue composition of polar bears (*Ursus maritimus*). J. Zool. Lond. 269:65–76.

Thiemann, G.W., Budge, S.M., Iverson, S.J. and Stirling, I. 2007. Unusual fatty acid biomarkers reveal age- and sex-specific foraging in polar bears (*Ursus maritimus*). Can. J. Zool. 85:505–517.

Tucker, S., Bowen, W.D. and Iverson, S.J. 2008. Convergence of diet estimates derived from fatty acids and stable isotopes within individual grey seals. Mar. Ecol. Progr. Ser. 354:267–276.

Vance, D.E., and Vance, J.E. 1996. Biochemistry of Lipids, Lipoproteins and Membranes. Elsevier Science, Amsterdam.

Viso, C.A., and Marty, J. 1993. Fatty acids from 28 marine microalgae. Prog. Lipid Res. 32:1521–1533.

Wamberg, S., Olesen, C.R., and Hansen, H.O. 1992. Influenece of dietary sources of fat on lipid synthesis in mink (*Mustela vison*) mammary tissue. Comp. Biochem. Physiol. A 103:199–204.

Wang, S.W., Iverson, S.J., Springer, A.M., and Hatch, S.A. 2007. Fatty acid signatures of stomach oil and adipose tissue of northern fulmars (*Fulmarus glacialis*) in Alaska: implications for diet analysis of Procellariiform birds. J. Comp. Physiol. B 177:893–903.

Chapter 13
Essential Fatty Acids in Aquatic Food Webs

Christopher C. Parrish

13.1 Introduction

Aquatic ecosystems occupy the largest part of the biosphere, and lipids in those systems provide the densest form of energy. Total lipid energy can be used to predict features of animal population dynamics such as egg production by fish stocks. Difficulties in determining the relationship between spawner biomass and the number of offspring produced (recruitment) have led researchers to look at lipids (Marshall et al. 1999). A positive association between recruitment and liver weights in cod prompted an investigation of total lipid energy as a proxy for total egg production by fish stocks. Marshall et al. (1999) found a highly significant linear relationship between total egg production and total lipid energy, and they suggested this approach should be used at other trophic levels too. Total lipid content of fish has also been connected to climate-induced community changes (Litzow et al. 2006). It is hypothesized that this relates to the dietary availability of just two fatty acids which were positively correlated with total lipid content.

The study of fatty acids in aquatic food webs has often focussed on their broad use as biomarkers in trophic transfer studies (e.g. Napolitano 1999; Dalsgaard et al. 2003; Iverson et al. 2004). By contrast the study of fatty acids in aquaculture has usually centred on only two or three fatty acids and their importance as essential dietary nutrients. The focus has been on the long-chain fatty acids, docosahexaenoic acid (DHA, 22:6n-3), eicosapentaenoic acid (EPA, 20:5n-3) and, to a lesser, extent arachidonic acid (ARA, 20:4n-6) which are required by organisms for optimal health. These polyunsaturated fatty acids (PUFA) maintain membrane structure and function and are precursors of bioactive compounds in vertebrates (Lands – Chap. 14), invertebrates, and plants. In finfish, they are required for normal somatic growth, survival, neural development, pigmentation, and reproduction (e.g. Sargent et al. 1999a). Based on a detailed examination of lipid biochemistry, this

---

C.C. Parrish
Ocean Sciences Centre, Memorial University of Newfoundland,
St. John's, Newfoundland, A1C 5S7, Canada
e-mail: cparrish@mun.ca

chapter argues for a crossover between these two approaches, i.e. a broader examination of nutritional fatty acids in aquaculture and a more thorough consideration of nutritional implications in trophic transfer studies.

13.2 Definition of Essential Fatty Acids

In 1930, linoleic acid (LIN, 18:2n-6) was termed an 'essential fatty acid' because it could eliminate acute deficiency states in rats that had been fed fat-free diets (Burr and Burr 1930). The ensuing search for fatty acids with essential fatty acid activity revealed a variety of polyunsaturated fatty acids which had the first double bond in the n-6 position. Later it was also shown that α-linolenic acid (ALA, 18:3n-3) could remove deficiency symptoms (Gurr and Harwood 1991), and now we know there to be a number of n-3 fatty acids with carbon numbers ranging from C_{14} to C_{36} that qualify as essential fatty acids (Cunnane 2000). In fact there are 23 PUFA in which the first double bond starts either 3 or 6 carbons from the methyl end that have essential fatty acid activity, some of which are shown in Figs. 13.1 and 13.2

In the synthesis of unsaturated fatty acids, the first double bond is usually inserted near the middle of the molecule in all organisms, for example in the n-9 position in stearic acid (18:0) to create oleic acid (18:1n-9). In animals, subsequent

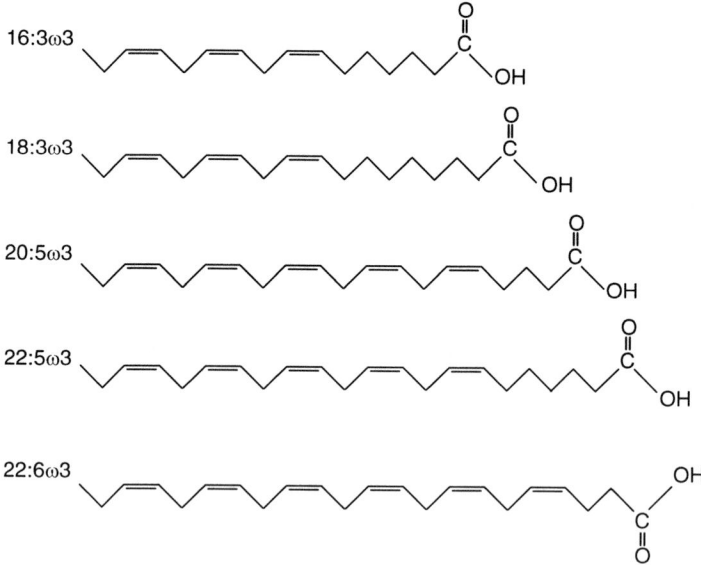

Fig. 13.1 Structures of some n-3 polyunsaturated fatty acids present in aquatic samples. Hexadecatrienoic acid (16:3n-3), α-linolenic acid (ALA, 18:3n-3), eicosapentaenoic acid (EPA, 20:5n-3), n-3 docosapentaenoic acid (n-3DPA, 22:5n-3), and docosahexaenoic acid (DHA, 22:6n-3) are all related biochemically because of the location of the first double bond 3 carbons from the methyl end of the chain

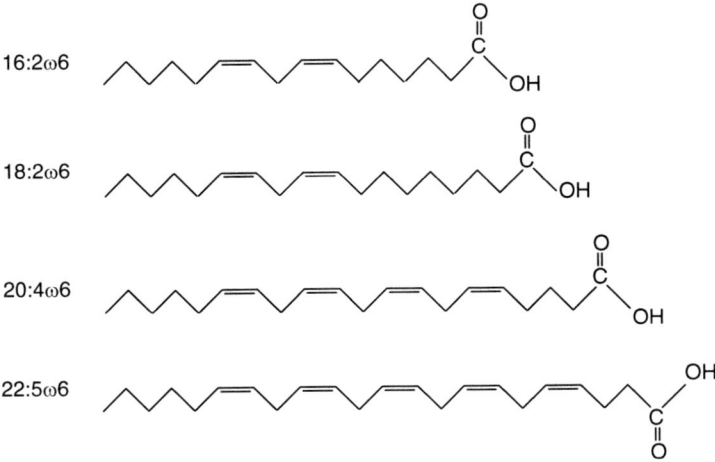

Fig. 13.2 Structures of some n-6 polyunsaturated fatty acids present in aquatic samples. Hexadecadienoic acid (16:2n-6), linoleic acid (LIN, 18:2n-6), arachidonic acid (ARA, 20:4n-6), and n-6 docosapentaenoic acid (n-6DPA, 22:5n-6) are all related biochemically because of the location of the first double bond 6 carbons from the methyl end of the chain

double bonds are introduced between an existing double bond and the carboxyl end of the molecule, while plants normally introduce a second double bond between the existing position and the terminal methyl group. The inability of animals to insert a double bond between the first one and the methyl end, combined with a requirement for fatty acids with the first double bond in the n-3 or n-6 position for disease prevention, is the basis of the essentiality of n-3 and n-6 fatty acids. When provided with sufficient n-3 and n-6 fatty acids in the diet, most animals can make other n-3 and n-6 fatty acids by desaturation and elongation or by retroconversion to shorter-chain fatty acids, but the n-3 and n-6 series are not interconvertible in vertebrates and most other animals except in the case of transgenic animals (Kang et al. 2004). The extent to which a given species can convert one n-3 fatty acid to another or one n-6 fatty acid to another leads to degrees of essentiality.

It is generally believed that ARA, EPA, and DHA are the most important long-chain ($C_{20} - C_{22}$) PUFA in mammals (Simopoulos 2002; Ruxton et al. 2004; Shahidi and Miraliakbari 2004; Wijendran and Hayes 2004) and fish (Sargent et al. 1999a, b; Montero et al. 2003, 2004). They have to be supplied to animals in their diet, although some animals can synthesize at least some of them when sufficient quantities of the LIN and ALA precursors are available. Many freshwater fish can synthesize these long-chain PUFA by a series of desaturations and elongations, although pollution decreases the ability of whitefish, *Coregonus lavaretus* to convert LIN to ARA by a factor of twofold to threefold (Toivonen et al. 2001). However, marine fish appear to have lost the ability to express a key elongase and/or desaturase gene (Sargent et al. 1999a, 2002; Tocher 2003). Gurr and Harwood (1991) make a distinction between 'essential metabolites' and 'essential nutrients', so that using this terminology, LIN and ALA would be considered to be essential nutrients in freshwater fish, while the long-chain products would be essential metabolites. In marine fish, ARA, EPA, and DHA would be essential nutrients. Similarly, rats and humans have some ability to synthesize LIN and ALA from 16:2n-6 and 16:3n-3 in green vegetables (Cunnane and Likhodii 1996), so that the essential nature of LIN and ALA depends on availability of precursors as well as the amount of these fatty acids in storage. These two short-chain PUFA (Figs. 13.1 and 13.2) also commonly occur in microalgae (e.g. Dunstan et al. 1992; Viso and Marty 1993; Nanton and Castell 1998). Borrowing terminology used in the amino acid literature, Cunnane (1996) suggested essential fatty acids should be divided into 'indispensable' or 'conditionally dispensable' fatty acids where requirements may change according to amounts in storage, age, or availability of other precursors. He subsequently modified the two categories to 'conditionally indispensable' and 'conditionally dispensable' fatty acids on the basis that there is insufficient evidence that any single PUFA is absolutely indispensable through the lifespan (Cunnane 2000).

While the mammalian literature recognizes 23 fatty acids having essential fatty acid activity, for the most part, the aquatic literature recognizes only two: EPA and DHA (e.g. Klein Breteler et al. 1999; Anderson and Pond 2000; Muller-Navarra et al. 2000, 2004; Arts et al. 2001; Tang and Taal 2005; Litzow et al. 2006; but see Kainz et al. 2004; Ahlgren et al. – Chap. 7). This despite the fact that ARA is now known to be an essential fatty acid during early development of marine finfish, albeit at lower concentrations (reviewed by Sargent et al. 1999b; Izquierdo et al. 2000).

In the aquatic literature, EPA and DHA are termed 'essential fatty acids' without qualification, but generally appear to be assumed to be essential nutrients, essential metabolites and indispensable, using the terminology of Gurr and Harwood (1991) and Cunnane (1996). It is clearly time for an examination of whether two fatty acids are enough to describe essential fatty acid status in aquatic food webs and to start to examine the degree of essentiality of these and other PUFA.

13.3 Effects of Essential Fatty Acids

Public awareness of n-3 fatty acids has increased dramatically over recent years (Lands – Chap. 14). Consumption of EPA and DHA has beneficial effects on plasma lipids and lipoproteins (Harris 1997a, b), cardiovascular disease (Kris-Etherton et al. 2002), cancer (Shahidi and Miraliakbari 2004), inflammatory and autoimmune diseases (Simopoulos 2002), brain development and function (Ruxton et al. 2004), and even adipose tissue hypertrophy (Parrish et al. 1990). For the most part, meta-analyses have confirmed these observations (He et al. 2004; Balk et al. 2006; Mozaffarian and Rimm 2006), although one recent one has not (Hooper et al. 2006). Nonetheless, Hooper et al. state that their findings 'do not rule out an important effect of omega 3 fats on total mortality' and they advise consumption of more oily fish.

There are marked differences in the effects of dietary fish oils on plasma lipid and lipoprotein concentrations in experimental animals and in humans. Harris (1997a) reviewed studies of seven species of experimental animals and compared them with human trials (Harris 1997b). The n-3 fatty acids consistently lowered serum triacylglycerol concentrations in humans but not in most species of experimental animals. Conversely there was a marked reduction in high-density lipoprotein-cholesterol concentrations in experimental animals, which is almost never seen in humans. Harris suggests these differences result not only from species differences but also from feeding experimental animals much larger concentrations of n-3 acids.

Many people know that EPA and DHA are recommended by the American Heart Association, for example, and that they are derived from fish. However, few people realize that the main source of these fatty acids is the aquatic food web, and they have to be acquired directly or indirectly by fish themselves through their diets (Gladyshev et al. - Chap. 8; Lands - Chap. 14). Even fewer know that EPA and DHA are actually very important at various trophic levels in the food web. In finfish, EPA and DHA are important in growth, immunity, and stress resistance. In sea bream, a 75% reduction in EPA + DHA resulted in a 13% reduction in growth, an increase in erythrocyte fragility, and alterations in cellular immunity and in renal morphology including renal tube degeneration (Montero et al. 2004). In another experiment a 50% reduction in dietary EPA + DHA, but with replacement this time with ALA, significantly affected stress response in sea bream (Montero et al. 2003). Plasma cortisol concentrations were significantly higher after net chasing or overcrowding.

DHA alone, which is especially rich in vertebrate neural tissue, is needed for finfish eye development and schooling behaviour. Phospholipids in the excitable membranes of the central nervous system are uniquely enriched in DHA. This n-3 fatty acid is contained in the phospholipids of the membrane bilayer, where it may produce an optimal acyl-chain packing array for the functioning of transmembrane proteins involved in the excitatory response. Animals deprived of n-3 fatty acids show reduced levels of DHA in the brain and retina and concomitant impairments in retinal function and learning ability (e.g. Moriguchi et al. 2000). DHA-deficient yellowtail, *Seriola quinqueradiata*, do not show schooling behaviour (Masuda and Tsukamoto 1999), and dietary DHA was important in terms of survival, eye development, and pigmentation in halibut larvae (Shields et al. 1999). Pigmentation may require signal transmission via the visual system to the brain which controls melanin synthesis.

Dietary DHA is also needed in invertebrates, for example for successful hatching in copepods (Arendt et al. 2005) and for somatic growth in shellfish. The growth of zebra mussel larvae was enhanced by DHA supplementation and not by EPA supplementation (Wacker et al. 2002), while in postlarval sea scallops the lowest growth rate occurred with the diet containing the lowest DHA concentrations (Milke et al. 2004).

In marine organisms, ARA, is another equally important fatty acid even though it is present in low proportions, except in echinoderms (Copeman and Parrish 2003). This n-6 acid has received extensive attention in the mammalian literature, and it is well established that ARA acts in concert with EPA to control physiological properties. Much attention has been paid to the role of these C_{20} compounds as precursors of a wide variety of short-lived hormone-like substances called eicosanoids (Lands – Chap. 14), which are a large group of cyclized compounds found in protozoa and all major animal phyla (Stanley and Howard 1998). ARA is important in sea urchin and finfish eggs (Ciapa et al. 1995; Pickova et al. 1997) and is needed for finfish growth, survival, and stress resistance (Bell and Sargent 2003).

13.4 Mechanisms of the Effects of Dietary Essential Fatty Acids

The competition among dietary fatty acids for key enzymes is central to our understanding of the importance of essential fatty acids in aquatic food webs. There is a fundamental biological difference between lipid nutrition and carbohydrate and protein nutrition (Sargent et al. 2002). The enzyme-substrate interaction in carbohydrate and protein metabolism depends mainly on strong ionic and hydrogen bond interactions resulting in high specificities in enzyme-catalysed reactions. Enzyme-substrate interaction in lipid metabolism depends much more on weak hydrophobic interactions (such as van der Waals and dispersion forces) resulting in lower specificities in enzyme-catalysed reactions. Thus, the fatty acid composition in animals depends on the low specificity of incorporation of dietary fatty acids into lipids (Sargent et al. 2002). It is for this reason that recent work in aquaculture nutrition has focussed on dietary fatty acid ratios, especially the DHA to EPA ratio (Ahlgren et al. – Chap. 7).

A multitude of studies have been undertaken on the mechanisms associated with the health effects of dietary fish oils in mammals. It appears that membrane phospholipids may be central to many of the mechanisms postulated for their effects. One way in which they exert their influence is through their effects on membrane structure. Phospholipids are by far the major lipid components of most membranes, and fatty acids are in most cases the principal determinants of a membrane's internal physical state (Thompson 1992). Highly unsaturated fatty acids have specific effects on the lipid bilayer and its dynamics, with potential influences on membrane protein functioning. Reduction in plasma triacylglycerol concentrations was one factor that early studies suggested might improve human health by reducing atherosclerosis. Now, however, the preferred mechanism for reduction in cardiovascular mortality is an antiarrhythmic effect arising from modulation of calcium ion flux (Hallaq and Leaf 1992; Goodnight 1996). High concentrations or high fluxes are believed to cause ventricular arrhythmias by producing a strong current, but in rat myocytes n-3 fatty acids reduce calcium flux into the cells. The reason for this effect is thought to be an alteration of calcium channel structure or function through inclusion of n-3 fatty acids in the plasma membrane.

Similarly, the mechanism for the dietary regulation of the activity of lipogenic enzymes (Herzberg 1989; Herzberg and Rogerson 1989) and of tumour development (Jiang et al. 1997) may relate to the fatty acid profile of the nuclear envelope phospholipids (Clandinin et al. 1992). Transport of RNA occurs only through nuclear pore complexes with lipid clustering modulating their opening and closing. Enzymes providing energy for translocation of RNA also occur near these complexes. In addition, the nuclear envelope possesses binding sites for hormones. These processes may all be affected by diet-induced changes in membrane structure. Changes in adipocyte size and function following dietary supplementation with fish oil may also relate to the modification of membrane lipid structure (Parrish et al. 1997). In rat adipocyte plasma membranes, the ratio of phosphatidylcholine to sphingomyelin was found to be increased as was the proportion of molecular species with PUFA. Changes in physicochemical properties of plasma membranes are known to affect hormone receptors and the specific activities of enzymes. As might be expected, there are studies where increased membrane fluidity can apparently be ruled out as the mechanism of the metabolic effect of fish oil feeding. For example, Dulloo et al. (1994) demonstrated suppression in calcium-dependent heat production in muscles by fish oil feeding, which is the opposite of that which would be expected if there was an increase in membrane fluidity. However, they did not refer to other ways in which membrane lipids can regulate cellular processes.

In addition to exerting their effects through their physicochemical properties, membrane phospholipids are also a source of bioactive compounds or messengers (Arts and Kohler - Chap. 10: Lands - Chap. 14). ARA at position *sn-2* of phosphatidylcholine and phosphatidylethanolamine is released by phospholipase A_2 to generate eicosanoids. Minute quantities of eicosanoids have potent physiological effects (e.g. Gurr and Harwood 1991). One group of eicosanoids, the leucotrienes, are generated by the action of lipoxygenase and are powerful mediators of the inflammatory process (e.g. Goodnight 1996). Another group of eicosanoids, the prostaglandins and thromboxanes, are generated by the action of cyclooxygenase. They are involved in reproduction, blood clotting, the circulation, kidney function, pain and stress reactions, protection of the gut from self-digestion, and regulation of immunological responses (e.g. Urich 1994; Goodnight 1996). Competition between ARA and EPA for these enzymes results in the production of different types of prostaglandins, thromboxanes and leucotrienes in varying quantities. The action of cyclooxygenase on ARA gives rise to the series 2 cascade with two double bonds in the molecules; EPA gives the series 3. These compounds are involved in regulating many types of actions as well as inflammation. Aspirin is a cyclooxygenase inhibitor that decreases the synthesis of prostaglandins.

Together, prostaglandins and leucotrienes constitute a group of extracellular mediator molecules that are part of an organism's defence system (Arts and Kohler – Chap. 10); however, reports on the effect of n-3 and n-6 fatty acids on immune response in fish are not conclusive and often contradictory, probably because of other physiological and environmental factors (Lall 2000). For example, although the feeding of vegetable oil to salmon reduced nonspecific immune parameters, there was no increase in mortality when challenged by a pathogen (Bell and Sargent 2003). Nonetheless, given that eicosanoids are clearly involved in mammalian and invertebrate immunity (Stanley and Howard 1998), further research with fish should produce a direct connection. The importance of ARA for finfish growth, survival, and stress resistance is already much clearer (Bell and Sargent 2003).

Not only are the fatty acids important in messenger formation, but different types of phospholipids are involved in different ways as well. The enzymes known as protein kinase C require the phospholipid phosphatidylserine for their activity. Protein kinase

C play a central role in a system of signal transduction through the membrane that acts on another membrane lipid: phosphatidylinositol-4,5, biphosphate (e.g. Urich 1994). These enzymes become active on binding to a lipid bilayer and in turn are influenced by the level of polyunsaturation in that bilayer (Stubbs 1992).

Although it has long been known that dietary fatty acids can alter membrane phospholipid composition (e.g. Gurr and Harwood 1991), the fact that different types of membrane phospholipids are affected differently (e.g. Aukema and Holub 1989; Zsigmond et al. 1990; Parrish et al. 1997) is less well known. This is because it is analytically more challenging to perform molecular species analyses than a simple total fatty acid profile. The importance of studying molecular species for determination of physical properties is underlined by the work of Fodor et al. (1995) who showed that changes in overall membrane fluidity in carp livers could be related to just a few molecular species (Arts and Kohler – Chap. 10).

A membrane can contain well over 100 different molecular species. The glycerophospholipids among them consist of complex mixtures of diacyl- (e.g. Fig.13.3), alkenylacyl-, and alkylacyl-glycerol derivatives. Not only does fish oil feeding replace some of the ARA in these molecular species (Aukema and Holub 1989; Careaga-Houck and Sprecher 1989), but there is also a preference for incorporation of n-3 fatty acids in ethanolamine glycerophospholipids (Yeo et al. 1989; Parrish et al. 1997), especially alkenylacyl ethanolamine glycerophospholipids (Aukema and Holub 1989). For example, after feeding rats a diet supplemented with 21% EPA + DHA, the proportion of these fatty acids in phosphatidyl ethanolamine was also 21%, more than twice the proportion in other phospholipid classes (Parrish et al. 1997). It is also interesting that dietary fish oil can create several molecular species that do not exist without it, while for the most part preserving those that do exist without fish oil feeding (Parrish et al. 1997). As well, fish oil feeding can cause a dramatic increase in the C_{24} monoenoic acid, nervonic acid, in sphingomyelin which contains no PUFA (Parrish et al. 1997). Bettger (1998) indicates that there is a neutraceutical benefit to this increase in nervonic acid in sphingomyelin which is involved in a cell signalling cascade leading to cell division or apoptosis (death). The increase in nervonic acid was not observed when an n-6 fatty acid diet was fed (Zsigmond et al. 1990).

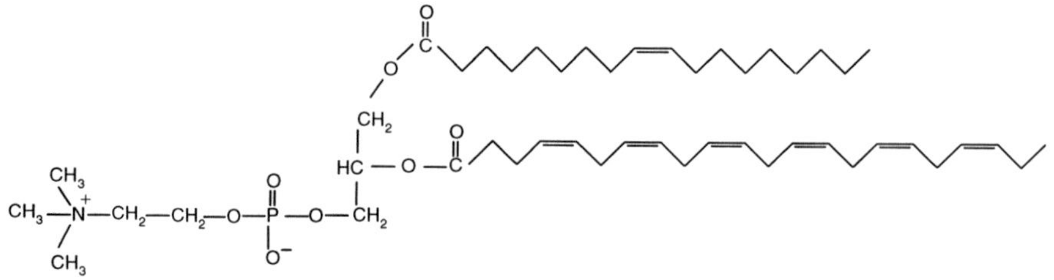

Fig. 13.3 Structure of an important phospholipid molecular species present in aquatic samples: 1-oleoyl, 2-docosahexaenoyl phosphatidyl choline

Thus, it is strongly recommended that molecular species analyses are performed in order to help define the mechanisms of the effects of essential fatty acids. However, simply cataloguing changes in molecular species distributions in membranes is also not enough: if the mechanism is believed to involve changes in physicochemical properties then fluidity determinations (Muriana and Ruiz-Gutierrez 1992; Hall et al. 2002) should be made concurrently, while phospholipase A_2 activity (Evans et al. 1998) should be determined if the effect involves eicosanoids. The rate-limiting step in the biosynthetic pathway of the eicosanoids is the release of a PUFA by phospholipase A_2 (e.g. Gurr and Harwood 1991; Urich 1994). This enzyme becomes active on binding to a membrane bilayer, and it may in turn be influenced by the level of polyunsaturation in that bilayer (Stubbs 1992). As well, it is interesting that phospholipase A_2 is also needed to produce another type of biologically active lipid: platelet-activating factor (Gurr and Harwood 1991).

The influence of dietary fatty acids on an organism's lipid composition provides a challenge for membranes since their physical characteristics are a key determinant of membrane structure and function. Although the fatty acids in phospholipids respond readily to changes in dietary content, membrane lipid composition is regulated. In fact, organisms which inhabit variable environments exploit the considerable chemical diversity among membrane lipid constituents so that lipids of appropriate physical properties are matched to the prevailing environmental conditions (Hazel and Williams 1990). If essential fatty acids are missing, animals will modify their own metabolic pathways to try and compensate, but physiological abnormalities may result (Thomson 1992).

Ectothermic animals increase the membrane content of unsaturated fatty acids in response to cold (Arts and Kohler – Chap. 10) in order to counteract the ordering effects of reduced temperature. This defence of membrane fluidity in ectotherms following thermal challenge, termed homeoviscous adaptation, was first described by Sinensky (1974) for *Escherichia coli*. Finfish membranes have high levels of *sn-1* monoenoic, *sn-2* polyenoic phospholipid molecular species (e.g. 18:1n-9/DHA, Fig.13.3) with cold adaptation, rendering the membranes less packed (Farkas et al. 2001). During winter, this adaptation may have to be performed at a time when lipid levels are low. Lipid depletion during overwintering is quite common in finfish (e.g. Arctic charr: Jorgensen et al. 1997; Jobling et al. 1998; Atlantic salmon: Morgan et al. 2002; and perch: Eckmann 2004).

Although phospholipid molecular species containing DHA are important in controlling finfish membrane fluidity, a direct correlation with DHA has not been found. In contrast, Hall et al. (2002) found a simple but very strong relationship between fluidity and a single polyunsaturated fatty acid, EPA, in gill membranes from the scallop *Placopecten magellanicus*. This suggests a possible dual function for this fatty acid in scallops as EPA's biological importance is usually associated with its role as an eicosanoid precursor. DHA, with its exceptional flexibility resulting from the electronic structure of the polyunsaturated chain (Eldho et al. 2003) has been thought to have mainly a structural function in membranes. However, it too may play a dual role as it has recently been found to be the precursor of bioactive compounds in rainbow trout brain cells (Hong et al. 2005).

13.5 Are N-6 PUFA Essential Fatty Acids in Aquatic Food Webs?

The importance of ARA for marine finfish in culture is becoming well established, but there is also evidence for the importance of ARA and LIN in crustaceans, suggesting a wider ecological role for ARA as well as other n-6 fatty acids. ARA was the most efficiently accumulated PUFA in planktonic food webs in oligotrophic lakes (Kainz et al. 2004), and suboptimal levels of ARA were implicated in the decline of copepods in a Baltic Sea mesocosm experiment (Ahlgren et al. 2005). ARA and to a lesser extent, LIN, improved survival, moulting frequency, and growth rate in the Chinese prawn (*Penaeus chinensis*: Xu et al. 1994). On the other hand, it was LIN that was highly correlated with field estimates of egg viability in the copepod, *Calanus helgolandicus* (Pond et al. 1996). However, principal components analysis also indicated that ARA content in the food and in the eggs was a potential factor involved in hatching success.

While EPA, DHA, and to a lesser extent ARA are often quantitatively dominant among the long-chain PUFA, there are a number of n-3 and n-6 long-chain PUFA that are commonly found in lipid extracts, especially from aquatic food webs, which may be nutritionally important too. For this type of investigation, fish larvae are particularly attractive because they are small and grow very quickly. In fact, fish larvae can be among the smallest of all vertebrates and they can have the highest growth rates of all vertebrates. In addition, lipid nutrition of larvae is recognized as one of the bottlenecks for mass production of many species, and the effects of essential fatty acids on growth, survival, and stress resistance of marine larvae have received increasing attention in the last decade (Sargent et al. 1999b; Izquierdo et al. 2000).

Parrish et al. (2007) took advantage of an unusual molecular and isotopic composition of a thraustochytrid protist (*Schizochytrium*) to follow n-6 docosapentaenoic acid (n-6DPA, 22:5n-6) in a short food chain leading to cod larvae. Using mass spectrometry (MS) they showed that n-6DPA had the properties of an essential fatty acid allowing them to postulate that it acts in concert with DHA, and also to test the hypothesis put forward by Metz et al. (2001) that EFA synthesized by an unusual biochemical pathway in bacteria and protists may be transferred to fish.

Known pathways of PUFA synthesis involve the processing of the saturated 16:0 or 18:0 by elongation and desaturation so that the synthesis of DHA from acetyl-CoA requires nearly 70 reactions. Polyketide synthase systems conduct the same reactions but in an abbreviated sequence. This route is found in some marine bacteria and primitive eucaryotes like the thraustochytrid protist *Schizochytrium* (Metz et al. 2001). This abbreviated pathway is thought to be responsible for providing an unusual fatty acid isotopic signature in Algamac® (Parrish et al. 2007) which consists entirely of spray-dried *Schizochytrium* sp. To determine this signature, fatty acid stable carbon isotope ratios (δ^{13}C, ‰) were measured after combustion, by continuous flow – isotope ratio MS. Compound-specific carbon isotope determinations are a refinement of bulk isotope measurements commonly used in food web studies (e.g. Canuel et al. 1995; Stapp et al. 1999; Vander Zanden et al. 1999; Post et al. 2000).

Algamac is also unusual in that it has high proportions of n-6DPA (9%) which larval cod take up rapidly when it is made available in the diet (Parrish et al. 2007). This fatty acid is also present in a few species of microalgae (Nanton and Castell 1998; Milke et al. 2004, 2006) and especially in *Pavlova* sp. where it can reach levels similar to those found in Algamac (Pernet et al. 2005). Exceptionally high amounts can be found in labyrinthulids, another type of marine microorganism which is related to the thraustochytrids. One isolate produces n-6DPA as its sole PUFA, amounting to 48% of total fatty acids (Kumon et al. 2003). Extensive bioaccumulation of this C_{22} fatty acid has been linked to improved growth in different scallop species and at different life stages (Milke et al. 2004, 2006; Pernet et al. 2005) as well as in cod larvae (Garcia et al. 2005, 2008). In sea scallop (*Placopecten magellanicus*) early veligers, the initial proportion of n-6DPA was well below dietary amounts (<2%: Pernet et al. 2005), but it increased markedly during ontongeny to become a major component (up to 20%). Marked accumulation of n-6DPA was observed in sea scallop larvae irrespective of diet and life stage: late veliger, pediveliger, or postlarvae. This fatty acid may be important in later stages as well. Alkanani et al. (2007) tested 140 variables to predict the growth of mussels starting at 2-cm length in the field. N-6DPA was among only four that were found to be significant.

The fatty acid n-6DPA may play an important structural role in membranes and/ or may be a precursor of bioactive compounds. Despite the similarity in chemical structure and some physical properties, DHA and n-6DPA chains do pack differently in bilayers (Eldho et al. 2003). The DHA chain, with one additional double bond, is more flexible which would likely have an impact on membrane proteins and their functions. Rats fed n-3-deficient diets appear to try and maintain an overall C_{22} PUFA level in brain phospholipids by replacing DHA with n-6DPA, but their behaviour is affected (Moriguchi et al. 2000).

In animals, eicosanoid bioactive compounds are derived from 20-carbon precursors (eicosa is Greek for 20), but 18-carbon compounds are also used in algae, for which the term 'oxylipin' is applied (Gerwick 1994). Recently the C_{22} DHA has been found to be a precursor of bioactive compounds termed 'docosanoids' (docosa is Greek for 22) generated via enzymatic oxygenations (Hong et al. 2003, 2005). The same enzymes could work on the C_{22} n-6DPA to form a parallel series of competitive products as found with the C_{20} eicosanoid precursors, EPA and ARA.

13.6 Ratios and Groups of Essential Fatty Acids in Food Webs

The balance required in dietary fatty acids and in membrane fatty acids has led to a focus on dietary fatty acid ratios in aquaculture nutrition (and see Ahlgren et al. – Chap. 7). A DHA:EPA ratio of 2:1 has been promoted, based on the ratio in finfish eggs (Sargent 1995; Sargent et al. 1999a; Bell et al. 2003). Most marine fish larvae feed on copepod eggs and nauplii during the first weeks after the onset of exogenous feeding and 2:1 is also the ratio found in early life stages of some copepod species (e.g. *Eurytemora velox*: Shields et al. 1999). However, Copeman et al. (2002)

demonstrated a relationship between unusually high DHA:EPA ratios and increased growth and survival in cultured yellowtail flounder, calling into question the assumption in aquaculture nutrition that 'nature knows best'. Using dietary DHA:EPA ratios ranging from 1 to 8, a strong positive correlation was obtained between the ratio in the diet and larval size and survival. High DHA:EPA ratios may be a more general requirement for cold water marine finfish larvae since Park et al. (2006) found cod larvae fed dietary ratios between 9 and 10 had the best growth and survival. It is interesting that high DHA:EPA ratios may be important for copepods themselves. Higher dietary ratios improve survival, time to maturity, maturation rate, egg production, hatching success, and female length in some copepod species (*Acartia tonsa*: Jonasdottir and Kiorboe 1996; *Gladioferens imparipes*: Payne and Rippingale 2000; *Temora longicornis*: Arendt et al. 2005) with values ranging from ~0.0 to 52.3. It is also interesting that the DHA:EPA ratio has been used as a bioindicator to distinguish dinoflagellate from diatom containing natural samples, with ratios ≥1 signalling the dominance of dinoflagellates (Budge and Parrish 1998; Dalsgaard et al. 2003). One implication from this could be that dinoflagellate-based food webs would provide more favourable essential fatty acid ratios; however, some marine protozoa (e.g. *Oxyrrhis marina*: Klein Breteler et al. 1999) and some copepods (e.g. *Eucyclops serrulatus*: Desvilletes et al. 1997; *Tisbe* sp.: Nanton and Castell 1998) can make DHA from short-chain precursors, e.g. ALA. In the first example, though, we are still talking about a dinoflagellate, albeit a heterotrophic dinoflagellate.

Other fatty acid ratios of interest include Σn-3PUFA: Σn-6PUFA (e.g. Sargent 1995; Milke et al. 2004), EPA:ARA (e.g. Sargent et al. 1999a; Milke et al. 2004), n-6DPA:DHA (e.g. Moriguchi et al. 2000), n-6DPA:ARA (e.g. Milke et al. 2004; Garcia et al. 2008), and DHA:ARA (Ahlgren et al. 2005). A statistical relationship with a ratio implies that the two fatty acids are not interchangeable and that they compete with each other biochemically. If a unique value of the ratio is important it indicates that the fatty acids within the organism should be present in a certain proportion for optimal structure and/or function.

Another approach to examining the importance of essential fatty acids is to treat them as groups and to look both at group correlations and variability. This is akin to assuming fatty acid interconversion by consumers before use, or an unlimited substitution in their functional roles (Anderson and Pond 2000). By taking groups of biochemically related fatty acids the possibilities of elongation and desaturation and/or retroconversion are implied. Egg-hatching success in the calanoid copepod *Acartia tonsa*, for example, was related to the sum of ALA, EPA, and DHA (Broglio et al. 2003). An examination of PUFA of the same chain length may reveal a mainly structural function. For example, looking at the stability in the sum of n-6DPA and DHA proportions is a way of seeing if these two C_{22} PUFA are substituting for each other (Moriguchi et al. 2000; Pernet et al. 2005). It is also possible to broaden this approach, for example by investigating the sum of LIN, ALA, ARA, EPA, and DHA (Kainz et al. 2004) which may reveal the importance and interchangeablity of n-3 and n-6 PUFA. In a field study of mussel growth, the sum of ARA, EPA, and DHA was significantly correlated with growth, while DHA and ARA individually were not (Alkanani et al. 2007), suggesting interchangeablity of C_{20} and C_{22} PUFA, presumably in a structural role.

13.7 Conclusions

Aquatic organisms need certain n-3 and n-6 fatty acids to support optimal health. An absolute requirement for these PUFA in the diet will occur only if there is not a sufficient amount already in storage, if they cannot be made from other compounds, or if they cannot be replaced by other compounds. There are probably several fatty acids in aquatic food webs with essential fatty acid activity, but these will depend on the organism, its life stage, and on environmental conditions. It is important to investigate all n-3 and n-6 PUFA, the relationships among them, and whether they may be essential nutrients or metabolites and conditionally dispensable or indispensable. Essential fatty acid activity is expressed physiologically (e.g. in terms of growth response) and biochemically (e.g. in terms of membrane fatty acid composition).

Membrane phospholipids may be central to many of the mechanisms of the health effects of essential fatty acids. Incorporation of n-3 and n-6 fatty acids in membranes affects their physicochemical characteristics altering cellular functions including the properties of certain membrane-bound enzymes and binding to receptors. Dietary PUFA also exert their influence by changing the availability of substrate fatty acids in the membrane that are used as sources of bioactive compounds or messengers. The importance of different fatty acids and phospholipids with different head groups in regulation of cellular processes, together with the fact that fluidity may be controlled by just a few compounds, suggests that molecular species analyses should be more widespread in this field. This type of analysis, together with determinations of membrane fluidity and phospholipase A_2 activity, might go a long way to defining the essentiality of individual fatty acids.

References

Ahlgren, G., Van Nieuwerburgh, L., Wanstrand, I., Pedesen, M., Boberg, M., Snoeijs, P. 2005. Imbalance of fatty acids in the base of the Baltic Sea food web – a mesocosm study. Can. J. Fish. Aquat. Sci. 62:2240–2253.

Alkanani, T., Parrish C.C., Thompson, R.J., and McKenzie C.H. 2007. Role of fatty acids in cultured mussels, *Mytilus edulis*, grown in Notre Dame Bay, Newfoundland. J. Exp. Mar. Biol. Ecol. 348:33–45.

Anderson, T.R. and Pond, D.W. 2000. Stoichiometric theory extended to micronutrients: Comparison of the roles of essential fatty acids, carbon, and nitrogen of marine copepods. Limnol. Oceanogr. 45:1162–1167.

Arendt, K.E., Jonasdottir, S.H., Hansen, P.J., and Gartner, S. 2005. Effects of dietary fatty acids on the reproductive success of the calanoid copepod *Temora longicornis*. Mar. Biol. 146:513–530.

Arts, M.T., Ackman, R.G., and Holub, B.J. 2001. "Essential fatty acids" in aquatic ecosystems: a crucial link between diet and human health and evolution. Can. J. Fish. Aquat. Sci. 58:122–137.

Aukema, H.M., and Holub, B.J. 1989. Effect of dietary supplementation with a fish oil concentrate on the alkenylacyl class of ethanolamine phospholipid in human platelets. J. Lipid Res. 30:59–64.
Balk, E.M., Lichtenstein, A.L., Chung, M., Kupelnick, B., Chew, P., and Lau, J. 2006. Effects of omega-3 fatty acids on serum markers of cardiovascular disease risk: A systematic review. Atherosclerosis 189:19–30.
Bell, J.G., McEvoy, L.A., Estevez, A., Shields, R.J., and Sargent, J.R. 2003. Optimising lipid nutrition in first-feeding flatfish larvae. Aquaculture 227: 211–220.
Bell, J.G., and Sargent, J.R. 2003. Arachidonic acid in aquaculture feeds: current status and future opportunities. Aquaculture 218:491–499.
Bettger, W. 1998. Exploring the link between very long chain fatty acids and human health. Agrifood research in Ontario. Ministry of Agriculture, Food and Rural Affairs. Vol. 21:15
Broglio, E., Jonasdottir, S.H., Calbet, S., Jakobsen, H.H., and Saiz, E. 2003. Effect of heterotrophic versus autotrophic food on feeding and reproduction of the calanoid copepod *Acartia tonsa*: relationship with prey fatty acid composition. Aquat. Microb. Ecol. 31:267–278.
Budge, S.M., and Parrish, C.C. 1998. Lipid biogeochemistry of plankton, settling matter and sediments in Trinity Bay, Newfoundland.II. Fatty acids. Org. Geochem. 29:1547–1559.
Burr, G.O., and Burr, M.M. 1930. On the nature and role of the fatty acids essential in nutrition. J. Biol. Chem. 86:587–621.
Canuel, E.A., Cloern, J.E., Ringelberg, D.B., Guckert, J.B., and Rau, G.H. 1995. Molecular and isotopic tracers used to examine sources of organic matter and its incorporation into the food webs of San Francisco Bay. Limnol. Oceanogr. 40:67–81.
Careaga-Houck, M., and Sprecher, H. 1989. Effect of a fish oil diet on the composition of rat neutrophil lipids and the molecular species of choline and ethanolamine glycerophospholipids. J. Lipid Res. 30:77–87.
Ciapa, B., Allemand, D., and De Renzis, G. 1995. Effect of arachidonic acid on Na^+/H^+ exchange and neutral amino acid transport in sea urchin eggs. Exp. Cell Res. 218:248–254.
Clandinin, M.T., Cheema, S., Field, C.J., and Baracos, V.E., 1992. Impact of dietary essential fatty acids on insulin responsiveness in adipose tissue, muscle, and liver, pp. 416–420. *In* A. Sinclair and R. Gibson [eds.], Essential fatty acids and eicosanoids. American Oil Chemists' Society, Champaign.
Copeman L.A., Parrish, C.C., Brown, J.A., and Harel, M. 2002. Effects of docosahexaenoic, eicosapentaenoic, and arachidonic acids on the early growth, survival, lipid composition and pigmentation of yellowtail flounder (*Limanda ferruginea*): a live food enrichment experiment. Aquaculture 210:285–304.
Copeman, L.A., and Parrish, C.C. 2003. Marine lipids in a cold coastal ecosystem: Gilbert Bay, Labrador. Mar. Biol. 143:1213–1227.
Cunnane, S.C. 1996. Recent studies on the synthesis, β-oxidation, and deficiency of linoleate and α-linolenate: are essential fatty acids more aptly named indispensable or conditionally dispensable fatty acids? Can. J. Physiol. Pharmacol. 74:629–639.
Cunnane, S.C., and Likhodii, S.S., 1996. ^{13}C NMR spectroscopy and gas chromatograph – combustion – isotope ratio mass spectrometry: complementary applications in monitoring the metabolism of ^{13}C-labelled polyunsaturated fatty acids. Can. J. Physiol. Pharmacol. 74:761–768.
Cunnane, S.C. 2000. The conditional nature of the dietary need for a polyunsturates: a proposal to reclassify 'essential fatty acids' as 'conditionally-indispensable' or 'conditionally-dispensable' fatty acids. British J. Nutrition 84:803–812.
Dalsgaard, J., St. John, M., Kattner, G., Müller-Navarra, D., and Hagen, W., 2003. Fatty acid trophic markers in the pelagic marine environment. Adv. Mar. Biol. 46:225–340.
Desvilettes, C., Bourdier, G., and Breton, J.C. 1997. On the occurrence of a possible bioconversion of linolenic acid into docosahexaenoic acid by the copepod *Eucylcops serrulatus* fed on microalgae. J. Plankton Res. 19:273–278.
Dulloo, A.G., Decrouy, A., and Chinet, A., 1994. Suppression of Ca^{2+}-dependent heat production in mouse skeletal muscle by high fish oil consumption. Metabolism 43: 931–934.

Dunstan, G.A., Volkman, J.K., Jeffrey, S.W., and Barrett, S.M., 1992. Biochemical composition of microalgae from green algal classes Chlorophyceae and Prasinophyceae. 2. Lipid classes and fatty acids. J. Exp. Mar. Biol. Ecol. 16:115–134.

Eckmann, R. 2004. Overwinter changes in mass and lipid content of *Perca fluviatilis* and *Gymnocephalus cernuus*. J. Fish Biol. 65:1498–1511.

Eldho, N.V., Feller, S.E., Tristram-Nagle, S., Polozov, I.V., and Gawrisch, K. 2003. Polyunsaturated docosahexaenoic vs docosapentaenoic acid – differences in lipid matrix properties from the loss of one double bond. J. Am. Chem. Soc. 125:6409–6421.

Evans, R.P., Parrish, C.C., Zhu, P., Brown, J.A., and Davis, P.J. 1998. Changes in phospholipase A^2 activity and lipid content during early development of Atlantic halibut, (*Hippoglossus hippoglossus*). Mar. Biol. 130:369–376.

Farkas T., Fodor, E., Kitajka, K., and Halver, J.E. 2001. Response of fish membranes to environmental temperature. Aquacul. Res. 32:645–655.

Fodor, E., Jones, R.H., Buba, C., Kitajka, K., Dey, I., and Farkas, T. 1995. Molecular architecture and biophysical properties of phospholipids during thermal adaptation in fish: an experimental and model study. Lipids 30:1119–1126.

Garcia, A.S., Parrish, C.C., and Brown, J.A. 2005. Effect of different live food enrichments on early growth and lipid composition of Atlantic cod larvae (*Gadus morhua*). European Aquaculture Society. Special Pub. No. 36:164–167. Belgium.

Garcia, A.S., Parrish, C.C., and Brown, J.A. 2008. Use of enriched rotifers and *Artemia* during larviculture of Atlantic cod (*Gadus morhua*, Linnaeus, 1758): Effects on early growth, survival and lipid composition. Aquacul. Res. 39:406–419.

Gerwick, W.H. 1994. Structure and biosynthesis of marine algal oxylipins. Biochim. Biophys. Acta 1211:243–255.

Goodnight, S.H. 1996. The fish oil puzzle. Sci. Med. September/October: 42–51.

Gurr, M.I., and Harwood, J.L. 1991. Lipid Biochemistry. An introduction. 4th edition. London, England: Chapman-Hall. 406 pp.

Hall, J.M., Parrish, C.C. and Thompson, R.J., 2002. Eicosapentaenoic acid regulates scallop (*Placopecten magellanicus*) membrane fluidity in response to cold. Biol. Bull. 202:201–203.

Hallaq, H., and Leaf, A., 1992. Stabilization of cardiac arrhythmias by ω-3 polyunsaturated fatty acids, pp. 245–247. *In* A. Sinclair and R. Gibson [eds.], Essential fatty acids and eicosanoids. American Oil Chemists' Society, Champaign.

Harris, W.S. 1997a. N-3 Fatty acids and serum lipoproteins: animal studies. Am. J. Clin. Nutr. 65:1611S–1616S.

Harris, W.S. 1997b. N-3 fatty acids and serum lipoproteins: human studies. Am. J. Clin. Nutr. 65:1645S–1654S.

Hazel, J.R., and Williams, E.E. 1990. The role of alterations in membrane lipid composition in enabling physiological adaptation of organisms to their physical environment. Prog. Lipid Res. 29:167–227.

He, K., Song, Y., Daviglus, M.L., Liu, K., Van Horn, L., Dyer, A.R., and Greenland, P. 2004. Accumulated evidence on fish consumption and coronary heart disease mortality: A meta-analysis of cohort studies. Circulation 109:2705–2711.

Herzberg, G.R. 1989. The mechanism of serum triacylglycerol lowering by dietary fish oil, pp. 143–158. *In* R.K. Chandra [ed.], Health effects of fish and fish oils. ARTS Biomedical Publishers and Distributors, St. John's.

Herzberg, G.R. and Rogerson, M. 1989. The effect of dietary fish oil on muscle and adipose tissue lipoprotein lipase. Lipids 24:351–353.

Hong, S., Gronert, K., Devchand, P.R., Moussignac, R.-L., and Serhan, C.N. 2003. Novel docosatrienes and 17S-resolvins generated from docosahexaenoic acid in murine brain, human blood, and glial cells. J. Biol. Chem. 278:14677–14687.

Hong, S., Tjonahen, E., Morgan, E.L., Lu, Y., Serhan, C.N., and Rowley, A.F. 2005. Rainbow trout (*Oncorhynchus mykiss*) brain cells biosynthesize novel docosahexaenoic acid-derived resolvins and protectins – Mediator lipidomic analysis. Prostaglandins other Lipid Mediat. 78:107–116.

Hooper, L., Thompson, R.L., Harrison, R.A., Summerbell, C.D., Ness, A.R., Moore, H.J., Worthington, H.V., Durrington, P.N., Higgins, J.P.T., Capps, N.E., Riemersma, R.A., Ebrahim, S.B.J., and Smith, G.D. 2006. Risks and benefits of omega 3 fats for mortality, cardiovascular disease, and cancer: systematic review. British Medical J. 332:752–760.

Iverson, S.J., Field, C., Bowen, W.D., and Blanchard, W. 2004. Quantitative fatty acid signature analysis: a new method of estimating predator diets. Ecol. Monogr. 74:211–235.

Izquierdo, M.S., Socorro, J., Arantzamendi, L., and Hernandez-Cruz, C.M. 2000. Recent advances in lipid nutrition in fish larvae. Fish Physiol. Biochem. 22:97–107.

Jiang, Y.-H., Lupton, J.R., and Chapkin, R.S. 1997. Dietary fish oil blocks carcinogen-induced down-regulation of colonic protein kinase C isozymes. Carcinogen. 18:351–357.

Jobling, M., Johansen, S.J.S., Foshaug, H., Burkow, I.C., and Jorgensen, E.H. 1998. Lipid dynamics in anadromous Arctic charr, *Salvelinus alpinus* (L.): seasonal variations in lipid storage depots and lipid class composition. Fish Physiol. Biochem. 18:225–240.

Jonasdottir, S.H, and Kiorboe, T. 1996. Copepod recruitment and food composition: do diatoms affect hatching success? Mar. Biol. 125:743–750.

Jorgensen E.H., Johansen, S.J.S., and Jobling, M 1997. Seasonal patterns of growth, lipid deposition and lipid depletion in anadromous Arctic charr. J. Fish Biol. 51:312–326.

Kainz, M., Arts, M.T., and Mazumder, A. 2004. Essential fatty acids in the planktonic food web and their ecological role for higher trophic levels. Limnol. Oceanogr. 49:1784–1793.

Kang, J.X., Wang, J., Wu, L., and Kang, Z.B. 2004. *Fat-1* mice convert n-6 to n-3 fatty acids. Nature 427:504.

Klein Breteler, W.C.M., Schogt, N., Baas, M., Schouten, S., and Kraay, G. W. 1999. Trophic upgrading of food quality by protozoans enhancing copepod growth: role of essential lipids. Mar. Biol. 135:191–198.

Kris-Etherton, P.M., Harris, W.S., and Appel, L.J. 2002. Fish consumption, fish oil, omega-3 fatty acids, and cardiovascular disease. Circulation 106:2747–2757.

Kumon, Y., Yokoyama, R., Yokochi, T., Honda, D., and Nakahara, T. 2003. A new labyrinthulid isolate, which solely produces n-6 docosapentaenoic acid. Appl. Microbiol. Biotechnol. 63:22–28.

Lall, S.P. 2000. Nutrition and health of fish. *In* L.E. Cruz-Suarez, D. Ricque-Marie, M.Tapia-Salazar, M.A. Olvera-Novoa and R. Civera-Cerecedo [eds.], Avances en nutricion acuicola V. Proceedings of V Simposium Internacional de Nutricion Acuicola. November 2000. Mérida, Yucatán, Mexico.

Litzow, M., Bailey, K.M., Prahl, F.G., and Heintz, R. 2006. Climate regime shifts and reorganization of fish communities: the essential fatty acid limitation hypothesis. Mar. Ecol. Prog. Ser. 315:1–11.

Marshall, C.T., Yaragina, N.A., Lambert, Y., and Kjesbu, O.S. 1999. Total lipid energy as a proxy for total egg production by fish stocks. Nature 402:288–290.

Masuda R., and Tsukamoto, K. 1999. School formation and concurrent developmental changes in carangid fish with reference to dietary conditions. Environ. Biol. Fishes. 56:243–252.

Metz, J.G., Roessler, P., Facciotti, D., Levering, C., Dittrich, F., Lassner, M., Valentine, R., Lardizabal, K., Domergue, F. Yamada, A., Yazawa, K., Knauf, V., and Browse, J. 2001. Production of polyunsaturated fatty acids by polyketide synthase in both prokaryotes and eukaryotes. Science 293:290–293.

Milke, L.M., Bricelj, V.M, and Parrish C.C. 2004. Growth of postlarval sea scallops, *Placopecten magellanicus*, on microalgal diets, with emphasis on the nutritional role of lipids and fatty acids. Aquaculture 234:293–317.

Milke, L.M., Bricelj, V.M. and Parrish, C.C. 2006. Comparison of early life history stages of the bay scallop, *Argopecten irradians*: Effects of microalgal diets on growth and biochemical composition. Aquaculture 260:272–289.

Montero, D., Kalinowski, T., Obach, A., Robaina, L., Tort, L., Caballero, M.J., and Izquierdo, M.S. 2003. Vegetable lipid sources for gilthead seabream (*Sparus aurata*): effects on fish health. Aquaculture 225:353–370.

Montero, D., Socorro, J., Tort, L., Caballero, M.J., Robaina, L., Vergara, J.M., and Izquierdo, M.S. 2004. Glomerulonephritis and immunosuppression associated with dietary essential fatty acid deficiency in gilthead sea bream, *Sparus aurata* L., juveniles. J. Fish Diseases. 27:297–306.

Morgan, I. J., McCarthy, I.D., and Metcalfe, N.B. 2002. The influence of life-history strategy on lipid metabolism in overwintering juvenile Atlantic salmon. J. Fish Biol. 60:674–686.

Moriguchi, T., Greiner, R.S., and Salem, Jr., N. 2000. Behavioral deficits associated with dietary induction of decreased brain docosahexaenoic acid concentration. J. Neurochem. 75:2563–2573.

Mozaffarian, D., and Rimm, E.B. 2006. Fish intake, contaminants, and human health. J. Amer. Medical Assoc. 296:1885–1899.

Müller-Navarra, D.C., Brett, M.T., Liston, A.M., and Goldman, C.R. 2000. A highly unsaturated fatty acid predicts carbon transfer between primary producers and consumers. Nature 403:74–77.

Müller-Navarra, D.C., Brett, M.T., Park, S., Chandra, S., Ballantyne, A.P., Zorita, E., and Goldman, C.R. 2004. Unsaturated fatty acid content in seston and tropho-dynamic coupling in lakes. Nature 427:69–72.

Muriana, F.J.G., and Ruiz-Gutierrez, V. 1992. Effect of n-6 and n-3 polyunsaturated fatty acids ingestion on rat liver membrane-associated enzymes and fluidity. J. Nutr. Biochem. 3:659–663.

Nanton, D.A., and Castell, J.D., 1998. The effects of dietary fatty acids on the fatty acid composition of the harpacticoid copepod, *Tisbe* sp., for use as a live food for marine fish larvae. Aquaculture 163:251–261.

Napolitano, G.E. 1999. Fatty acids as trophic and chemical markers in freshwater ecosystems, pp. 21–44. In M.T. Arts, and B.C. Wainman [eds.], Lipids in freshwater ecosystems. Springer, New York.

Park, H.G., Puvanendran, V., Kellett, A., Parrish, C.C., Brown, J.A. 2006. Effect of enriched rotifers on growth and survival of Atlantic cod (*Gadus morhua* L.) larvae. ICES J. Mar. Sci. 63:285–295.

Parrish, C.C., Pathy, D.A., and Angel, A. 1990. Dietary fish oils limit adipose tissue hypertrophy in rats. Metabolism 39:217–219.

Parrish, C.C., Myher, J.J., Kuksis, A., and Angel, A. 1997. Lipid structure of rat adipocyte plasma membranes following dietary lard and fish oil. Biochim. Biophys. Acta. 1323:253–262.

Parrish, C.C., Whiticar, M., and Puvanendran V. 2007. Is ω6 docosapentaenoic acid an essential fatty acid during early ontogeny in marine fauna? Limnol. Oceanogr. 52:476–479.

Payne, M.F., and Rippingale, R.J. 2000. Evaluation of diets for culture of the calanoid copepod *Gladioferens imparipes*. Aquaculture 187:85–96.

Pernet, F., Bricelj, V.M., and Parrish, C.C. 2005. Effect of varying dietary levels of ω6 polyunsaturated fatty acids during the early ontogeny of the sea scallop, *Placopecten magellanicus*. J. Exp. Mar. Biol. Ecol. 327:115–133.

Pickova, J., Dutta, P.C., Larsson, P.-O., and Kiessling, A. 1997. Early embryonic cleavage pattern, hatching success, and egg-lipid fatty acid composition: comparison between two cod (*Gadus morhua*) stocks. Can. J. Fish. Aquat. Sci. 54:2410–2416.

Pond, D., Harris, R., Head, R., and Harbour, D. 1996. Environmental and nutritional factors determining seasonal variability in the fecundity and egg viability of *Calanus helgolandicus* in coastal waters off Plymouth, UK. Mar. Ecol. Prog. Ser. 143:45–63.

Post, D.M., Pace, M. L., and Hairston, N.G. Jr. 2000. Ecosystem size determines food-chain length in lakes. Nature 405:1047–1049.

Ruxton, C.H.S., Reed, S.C., Simpson, M.J.A. and Millington, K.J. 2004. The health benefits of omega-3 polyunsaturated fatty acids: a review of the evidence. J. Hum. Nutr. Dietet. 17:449–459.

Sargent, J.R. 1995. Origins and functions of egg lipids: nutritional implications, pp. 353–372. In N.R. Bromage and R.J. Roberts [eds.], Broodstock management and egg and larval quality. Blackwell Science, Oxford.

Sargent, J.R., Bell, G., McEvoy, L., Tocher, D., and Estevez, A. 1999a. Recent developments in the essential fatty acid nutrition of fish. Aquaculture 177:191–199.

Sargent, J.R., McEvoy, L., Estevez, A., Bell, G., Bell, M., Henderson, J., and Tocher, D. 1999b. Lipid nutrition of marine fish during early development: current status and future directions. Aquaculture 179:217–229.

Sargent, J.R., Tocher, D.R, and Bell, J.G. 2002. The lipids, pp 181–257. *In* J.E. Halver and R.W. Hardy [eds.], Fish nutrition, 3rd edition. Elsevier, New York.

Shahidi, F., and Miraliakbari, H. 2004. Omega-3 (n–3) fatty acids in health and disease: Part 1 -Cardiovascular disease and cancer. J. Med. Food 7:387–401.

Shields, R.J., Bell, J.G., Luizi, F.S., Gara, B., Bromage, N.R., and Sargent, J.R. 1999. Natural copepods are superior to enriched *Artemia* nauplii as feed for halibut larvae (*Hippoglossus hippoglossus*) in terms of survival, pigmentation and retinal morphology: relation to dietary essential fatty acids. J. Nutr. 129:1186–1194.

Simopoulos, A.P. 2002. Omega-3 fatty acids in inflammation and autoimmune diseases. J. Amer. Coll. Nutr. 21:495–505.

Sinensky, M. 1974. Homeoviscous adaptation – a homeostatic process that regulates the viscosity of membrane lipids in *Escherichia coli*. Proc. Nat. Acad. Sci. USA 71:522–525.

Stanley, D.W., and Howard, R.W. 1998. The biology of prostaglandins and related eicosanoids in invertebrates: cellular, organismal and ecological actions. Amer. Zool. 38:369–381.

Stapp, P., Polis, G.A., and Piñero F.S. 1999. Stable isotopes reveal strong marine and El Niño effects on island food webs. Nature 401:467–469.

Stubbs, C.D. 1992. The structure and function of docosahexaenoic acid in membranes, pp. 116–121. *In* A. Sinclair and R. Gibson [eds.], Essential fatty acids and eicosanoids. American Oil Chemists' Society, Champaign.

Tang, K.W. and Taal, M. 2005. Trophic modification of food quality by heterotrophic protists: species-specific effects on copepod egg production and egg hatching. J. Exp. Mar. Biol. Ecol. 318:85–98.

Tocher, D.R. 2003. Metabolism and functions of lipids and fatty acids in teleost fish. Rev. Fish. Sci. 11:107–184.

Toivonen L.V., Nefedova, Z.A., Sidorov, V.S., Sharova, Y.N. 2001 Adaptive changes in fatty acid compositions of whitefish *Coregonus lavaretus* L. tissue lipids caused by anthropogenic factors. Applied Biochem. Microbiol. 37:314–317.

Thompson, G. A., Jr. 1992. The regulation of membrane lipid metabolism. 230 pgs. Boca Raton: CRC Press.

Urich, K., 1994. Comparative animal biochemistry. 782 pgs. Springer, New York.

Viso, C.A., and Marty, J.-C. 1993. Fatty acids from 28 marine microalgae. Phytochemistry 34:1521–1533.

Wacker, A., Becher, P., and von Elert, E. 2002. Food quality effects of unsaturated fatty acids on larvae of the zebra mussel *Dreissena polymorpha*. Limnol. Oceanogr. 47:1242–1248.

Wijendran, V., and Hayes, K.C. 2004. Dietary n-6 and n-3 fatty acid balance and cardiovascular health. Annu. Rev. Nutr. 24:597–615.

Xu, X. L., Ji, W. J., Castell, J.D., O'Dor, R.K. 1994. Essential fatty acid requirement of the Chinese prawn, *Pinaeus chinensis*. Aquaculture 127:29–40.

Yeo, Y.K., Philbrick, D.-J., and Holub, B.J. 1989. Altered acyl chain composition of alkylacyl, alkenylacyl, and diacyl subclasses of choline and ethanolamine glycerophospholipids in rat heart by dietary fish oil. Biochim. Biophys. Acta 1001:25–30.

Vander Zanden, M.J., Casselman, J.M., Rasmussen, J.B. 1999. Stable isotope evidence for the food web consequences of species invasions in lakes. Nature 401:464–466.

Zsigmond, E., Parrish, C., Fong, B., and Angel, A. 1990. Changes in dietary lipid saturation modify fatty acid composition and high-density-lipoprotein binding of adipocyte plasma membrane. Am. J. Clin. Nutr. 52:110–119.

Chapter 14
Human Life: Caught in the Food Web

William E. M. Lands

14.1 Life in the web

14.1.1 Connections Make Complexity

Ecology develops awareness of dynamic interacting parts (whether enzymes and substrates, consumers and autotrophs, or predators and prey) which transfer energy, biomass, and information along chains of cause-and-effect connected events. This review will focus on polyunsaturated lipids that act in sequential steps that link cause to consequence, including some of the crosslinks that weave the chains into a complex web of interactions within habitats. In coastal areas, humans eat abundant supplies of finfish and shellfish that contain polyunsaturated fatty acids (PUFA) that originated from other life-forms. The PUFA are predominantly synthesized by phytoplankton. Phytoplankton production is mostly consumed by zooplankton or benthic invertebrates such as shellfish which are, in turn, consumed by fish and ultimately by humans when they eat fish.

In marine and freshwater systems, thousands of gene-defined organisms interact with a highly dynamic aquatic environment. Each species, cell, or enzyme transfers energy, biomass, and information into new forms consumed by other species, cells, or enzymes. Most animals survive by eating other life-forms that have already succeeded in gathering needed energy and biomass (minerals, vitamins, carbon, nitrogen, etc.). The chain of events connecting food supply to each species' survival involves enzyme-catalyzed molecular transformations within the cells of each participating organism. This in turn, depends, on DNA-defined proteins catalyzing and signaling different processes at different times in environment-triggered responses. The space and time occupied by each transient event differs greatly in terms of energy, biomass, or information. This review summarizes some information on the health consequences of these molecular events to build understanding that may change future human eating behaviors.

W.E.M. Lands
6100 Westchester Park Drive, Apt. #1219, College Park, MD, USA
e-mail: wemlands@att.net

Complex organisms develop their differentiated functions over time, accumulating long-term consequences from earlier short-term adaptations. Timing is vital because of limited storage ability at every stage of life. Each link in the chain of events depends on the arrival of supplies that influence the processes and consequences that lead to the next-step supplies. Whenever a species lacks metabolic machinery to make an essential component, this affects its nutritional physiology. Cumulative effects of successful or failed food transfers over time, both in quantity and/or in quality, can cause very different cognitive and physical health states in individuals with common genetic blueprints. For example, we see evidence linking essential PUFA and its critical subset of 20- and 22-carbon chain (C_{20} and C_{22}) highly unsaturated fatty acids (HUFA) with healthy cognitive functioning in the humans caught in a complex food web. There are harmful consequences in turning away from the aquatic foodstuffs that facilitated the transformation of hominids to *Homo sapiens* (Walter et al. 2000; Marean et al. 2007). We need researchers to carefully evaluate the past 100,000 years of adaptations in light of the food web that now supports human needs.

Choice of food consumed often seems based on proximity, convenience, taste/odor, and ingestability rather than on specific nutritional requirements. Similarly, enzyme catalysts seem unlikely to sense a need for the substrates they consume or the products they produce. Rather, most life events seem sustained by whatever food is available. Life thrives, survives, or dies within a changing web of energy, biomass, and information transfers where changed supplies can cause new patterns. Survivors fit in complex webs of interdependencies from which few on earth can be "independent." To avoid unintended outcomes that follow unrecognized causes, this review informs readers about the PUFA and HUFA they eat (examples in Table 14.1) and the resulting outcomes on human health. This transferred information may alter readers' perception and understanding of the importance of PUFA and HUFA in the aquatic and human food web and the future choices people may make as a species with respect to how to manage aquatic resources and how to select food.

14.1.2 Transfer of Essential Fatty Acids

The phytoplankton drifting in marine and freshwaters have genes that define molecular machinery to make and use chlorophyll, which uses solar energy to break water's H-O bonds and form molecular oxygen. The relocated energy-rich electrons move through various molecular intermediates in multistep pathways catalyzed by enzymes that couple this energy to permit the fixation of CO_2 into carbohydrate biomass. From this beginning, other steps couple the transfer of energy into new amino acids, proteins, nucleic acids, lipids, and specific cellular structures. Bacteria and both plant and animal eukaryotes convert metabolic fragments into energy-rich saturated and monoenoic fatty acids (see Table 14.1). In addition, some cyanobacteria (and eukaryotes containing chloroplasts with related DNA; Clegg et al. 1994) have gene-defined enzymes that create n-3 and n-6 PUFA structures from monoenoic acids. However, animals lack the DNA-encoded enzymes that permit the manufacture

Table 14.1 PUFA and HUFA in different life-forms

Author	Life-form	Genus	Sat	Mono	18:2n-6	18:3n-3	18:4n-3	20:4n-6	20:5n-3	22:5n-3	22:6n-3	HUFA	%n-6HUFA
Ferr	Algae	Scenedesmus	44.4	15.3	2.1	1.3	0.0	0.0	0.0	0.0	0.0	0.0	0.0
Keny	Algae	Synechococcus	49.0	44.0	0.0	0.0	0.0	0.0	0.0	0.0	0.0	0.0	0.0
Keny	Algae	Synechococcus	33.1	56.1	1.0	0.0	0.0	0.0	0.0	0.0	0.0	0.0	0.0
Keny	Algae	Synechococcus	32.0	29.0	18.0	10.0	0.0	0.0	0.0	0.0	0.0	0.0	0.0
Keny	Algae	Chlorogloea	42.0	34.0	2.0	0.0	0.0	0.0	0.0	0.0	0.0	0.0	0.0
Otle	Microalgae	Spirulina	55.7	11.2	17.4	0.0	0.0	0.0	0.0	0.0	0.0	0.0	0.0
Otle	Microalgae	Chlorella	25.3	28.0	12.0	15.8	0.0	0.0	0.0	0.0	0.3	0.3	0.0
Colo	Macroalgae	Nereocystis	27.0	15.8	6.0	7.6	12.2	16.8	12.6	0.0	0.2	29.5	56.8
Colo	Macroalgae	Ulva	30.9	14.5	3.7	24.1	17.1	1.1	3.4	0.0	0.5	5.0	22.2
Colo	Macroalgae	Gloiopeltis	70.8	22.0	1.0	1.7		1.0	1.9	0.0	1.1	3.9	25.5
Colo	Macroalgae	Soliera	64.6	17.0	7.8	7.6		0.8	1.0	0.0	0.7	2.4	32.9
Wen	Diatoms	Nitzschia	33.4	40.1	2.6	1.7	0.0	4.1	17.3	0.0	0.0	21.4	19.1
Wen	Diatoms	Nitzschia	31.7	45.6	3.4	0.7	0.0	3.5	14.5	0.0	0.0	18.1	19.5
Aren	Diatom	Thalassiosira	35.7	25.0	1.9	1.5	1.5	0.8	8.4	0.0	1.9	11.1	7.2
Aren	Chlorophyte	Dunaliella	39.5	21.7	2.5	0.8	4.2	0.8	1.6	1.0	4.7	8.2	10.0
Aren	Haptophyte	Phaeocystis	31.4	15.8	3.8	3.2	14.2	0.2	1.0	0.2	7.7	9.0	1.9
Aren	Haptophyte	Isochrysis	20.4	10.6	5.8	22.6	2.3	0.0	0.0	0.1	0.0	0.1	0.0
Fern	Microalgae	Isochrysis	43.5	23.2	2.8	3.2	11.7	0.8	0.5	0.3	8.0	13.4	34.3
Velo	Algae	Dunaliella	23.2	16.1	9.1	51.4	0.2	0.0	0.0	0.0	0.0	0.0	0.0
Velo	Algae	Rhodomonas	32.8	5.9	8.1	21.0	19.6	0.0	6.0	0.0	6.4	12.5	0.4
Velo	Protist	Oxyrrhis	34.9	11.0	4.5	12.0	0.0	0.0	0.2	0.0	37.4	37.6	0.0
Velo	Dinoflaggelate	Oxyrrhis	34.5	7.2	5.1	6.5	0.2	0.2	4.4	0.2	41.8	46.5	0.4
Velo	Copepod	Acartia	56.3	14.5	0.0	7.5	6.6	0.0	8.1	0.0	7.0	15.2	0.3
Velo	Copepod	Acartia	57.5	8.5	0.1	23.6	0.1	0.1	4.5	0.1	5.7	10.3	0.7
Velo	Copepod	Acartia	42.2	7.9	0.0	13.8	13.9	0.0	9.8	0.0	12.3	22.2	0.2
Velo	Copepod	Acartia	57.8	7.0	0.1	3.8	0.1	0.1	5.4	0.1	25.8	31.3	0.2
Shie	Fungus	Schizochytrium	57.1	11.0	0.4	1.2	0.2	0.5	0.6	0.1	18.1	27.9	32.6
Shie	Brachiopod	Artemia	19.4	35.2	4.0	19.1	2.1	1.5	5.3	0.1	5.1	14.5	27.6
Shie	Copepod	Eurytemora	28.9	29.1	2.0	1.3	0.6	1.8	10.8	0.2	21.8	35.5	7.6
Pers	Copepod	Heterocope	34.3	10.3	3.6	4.1	6.7	3.7	10.9	0.2	20.7	35.3	10.5

(continued)

Table 14.1 (continued)

Author	Life-form	Genus	Sat	Mono	18:2n-6	18:3n-3	18:4n-3	20:4n-6	20:5n-3	22:5n-3	22:6n-3	HUFA	%n-6HUFA
Pers	Copepod	*Arctodiaptomus*	32.8	12.6	4.8	7.4	14.4	1.7	6.9		13.6	22.2	7.7
Pers	Cladoceran	*Bythothrephes*	30.8	19.2	3.6	5.3	4.1	9.3	23.0		2.1	34.4	27.0
Pers	Cladoceran	*Holopedium*	33.6	16.0	4.1	6.7	11.0	6.8	17.5		2.1	26.4	25.8
Pers	Cladoceran	*Bosmina*	32.6	20.8	5.3	7.6	10.4	4.7	14.4		2.6	21.7	21.7
Pers	Cladoceran	*Daphnia*	51.8	20.2	4.5	5.7	6.0	1.6	3.0		2.1	6.7	23.9
Bara	Cladoceran	*Daphnia*	21.7	26.4	7.1	8.3	0.5	1.3	0.5		0.0	1.8	71.3
Bara	Cladoceran	*Daphnia*	24.4	25.7	11.6	22.2	0.1	0.6	0.2		0.1	0.9	65.0
Hess	Cladoceran	*Daphnia*	33.9	25.4	6.7	5.9	11.9	0.8	8.6		1.5	11.0	7.6
Hess	Cladoceran	*Daphnia*	30.6	32.3	5.8	8.0	7.4	0.9	10.1		0.3	11.3	8.2
Hess	Cladoceran	*Daphnia*	29.7	23.4	4.3	6.7	11.1	1.2	12.5		1.0	14.7	7.9
Hess	Cladoceran	*Daphnia*	28.9	32.9	5.2	19.6	3.8	1.2	5.7		0.0	6.9	17.1
Hess	Cladoceran	*Daphnia*	31.6	29.8	5.3	5.3	7.3	5.3	12.9		0.5	18.7	28.4
Phle	Crustacea-krill	*Euphasia*	27.8	20.4	3.2	1.3	5.2	0.0	20.5		22.6	43.1	0.0
Dela	Oysters	*Crassostrea*	18.2	19.9	1.5	0.9	1.3	3.7	13.5	1.4	21.2	40.8	11.5
Soud	Mix of algae + diatoms for oyster diet		33.4	23.1	3.4	4.1	5.4	0.7	10.4	0.1	3.2	14.9	8.1
Soud	Oysters	*Crassostrea*	28.4	11.1	0.9	0.6	1.2	2.3	12.3	1.6	17.3	34.3	9.0
Soud	Oysters	*Crassostrea*	34.6	20.4	2.8	2.4	0.5	1.4	12.3	0.6	11.9	26.6	6.8
Fern	Clams	*Rudiapes*	28.1	31.6	1.9	2.6	7.0	1.5	1.1	0.4	16.0	20.5	14.6
Toch	Arctic charr	*Salvalinus*	22.1	36.7	4.4	3.3	2.7	0.7	6.7	1.5	18.5	27.4	2.6
Mour	Sea bass	*Dicentrarchus*	24.5	26.0	3.8	1.1	1.2	0.9	9.3	1.5	17.3	29.0	3.1
Cast	Smelt	*Osmerus*	22.3	49.7	0.2	0.0	0.0	4.8	3.5	0.3	15.0	23.6	20.3
Cast	Alewife	*Alosa*	24.1	34.9	3.7	3.6	2.9	2.4	8.2	1.5	6.0	19.4	19.1
Cast	Cod	*Gadus*	18.6	39.2	1.9	0.6	0.5	1.4	12.9	1.1	12.7	28.4	6.0
Cast	Trout	*Oncorhynchus/Salmo*	18.1	32.3	4.6	5.2	1.5	2.2	5.0	2.6	19.0	28.8	7.6
Cast	Herring	*Clupea*	17.2	60.5	0.7	0.3	1.5	0.4	7.4	1.1	3.9	12.8	3.1
Cast	Herring	*Clupea*	28.1	46.2	1.6	0.6	2.8	0.4	6.6	1.3	7.6	15.9	2.5
Cast	Menhaden	*Brevoortia*	40.9	27.2	1.1	0.9	1.9	1.2	10.2	1.6	12.8	25.8	4.7
Cast	Salmon	*Salmo*	22.4	40.3	2.0	1.0	2.0	0.9	6.7	2.3	16.1	26.0	3.5
Huyn	Herring	*Clupea*	28.7	46.5	1.0	0.3	0.6	0.9	7.5	0.6	10.0	19.1	4.9
Mora	Sole	*Solea*	29.1	36.8	10.1	10.3	1.2	0.8	2.0	1.3	3.0	7.1	11.6
Roll	Salmon	*Salmo*	22.8	41.5	7.1	8.9	3.8	0.8	2.2	1.0	7.7	11.6	7.0

Code	Sample	Genus											
Sta2	Greenland	*Homo*	43.6	19.2	14.0	0.1	5.2	4.9	1.6	7.9	21.3	32.5
Koba	Japan, 57-year	*Homo*	47.6	13.6	17.3	0.2	6.4	3.7	1.2	7.3	18.6	34.4
DE1b	Quebec Inuits	*Homo*			22.2	0.0	6.2	3.0		5.0	14.2	43.9
Toku	Nagoya, Japan	*Homo*	30.4	21.3	31.7	0.2	6.6	2.6	0.7	5.4	16.3	46.8
Kuri	Japan dietitians	*Homo*	32.1	21.0	27.8	0.9	4.9	1.7	0.4	3.2	10.9	51.4
Sand	England	*Homo*	30.7	20.1	33.3	0.8	7.2	1.5	1.2	3.0	15.1	62.5
Raat	Minnesota	*Homo*	40.9	12.3	21.6	0.2	12.2	0.7	1.0	4.5	18.4	66.2
Chaj	Spain	*Homo*	47.1	19.7	15.9	0.1	6.8	0.6	0.4	5.5	22.7	71.7
DE2	James Bay Cree	*Homo*			21.1	0.0	9.3	0.7		3.0	13.0	71.5
DE1a	Quebec	*Homo*			22.1	0.0	6.4	0.5		1.3	8.2	78.0
BEYD	ARIC study	*Homo*	49.3	9.2	22.0	0.1	11.5	0.6		2.9	14.9	76.9
STA5	Detroit, 25 year-old	*Homo*	31.1	24.1	26.5	0.4	7.6	0.2	0.3	1.8	12.0	81.8
Kiec	Ohio elderly	*Homo*	45.0	15.0	25.7	0.6	10.2	0.7	0.2	0.8	13.2	86.8

Information on the fatty acids in diverse samples is arrayed with literature citations noted by the first four letters of the first author's last name,[a] a general description of the life-form, and the taxonomic genus name. *Sat* saturated fatty acids; *Mono* monoenoic fatty acids. Values are the percent in total fatty acids, except for the %n-6 in HUFA listed in the last column on the right. Not all reports listed every fatty acid, and some less common n-6 fatty acids are in the calculated HUFA and %n-6 in HUFA, but omitted to keep the table simple

[a]Except for citations: (*DE1a* Dewailly et al. 2001a; *DE1b* Dewailly et al. 2001b; *DE2* Dewailly et al. 2002; *STA2* Stark et al. 2002; *STA5* Stark et al. 2005)

of these structures, and they therefore must obtain PUFA by eating plants (or herbivorous animals). Because the n-3 and n-6 PUFA and the HUFA have important consequences for the healthy development of animals, they are termed dietary essential fatty acids (EFA; see also Parrish – Chap. 13). The supply of EFA, especially the C_{20} and C_{22} HUFA, in the food web and its link to human health is the primary focus of this review.

Examples of suppliers and consumers of PUFA and HUFA are listed in Table 14.1. For example, different species of *Synechococcus* form and accumulate linoleate (18:2n-6) ranging from 0 to 18% of cellular fatty acids with no 20 or 22 carbon acids. On the other hand, diatoms (e.g., in the genus *Nitzschia*) form and accumulate appreciable amounts of HUFA, mostly with the n-3 structure. Veloza et al. (2006) showed transfer and accumulation of PUFA and HUFA when they fed the crustacean copepod, *Acartia tonsa*, with protist dinoflagellates (*Oxyrrhis marina*) that had eaten algae (*Dunaliella* or *Rhodomonas*). *Oxyrrhis* elongated and desaturated the PUFA 18:3n-3, abundant in *Dunaliella*, accumulating the HUFA 22:6n-3. It also accumulated PUFA and HUFA obtained from *Rhodomonas*. In turn, the *A. tonsa* accumulated PUFA and HUFA obtained from eating *Oxyrrhis*. In these cases, n-3 forms predominated in HUFA, and n-6 HUFA were less than 1% of the accumulated HUFA (although n-6 HUFA are abundant in some algae (Bigogno et al. 2002)). Shields et al. (1999) fed brine shrimp (*Artemia salina*) grown with the marine algae (*Schizochytrium sp.*) or calanoid copepods (*Eurytemora velox*) to halibut larvae (*Hippoglossus hippoglossus*), which then accumulated the PUFA and HUFA that they could not make from simple metabolic fragments.

When Persson and Vrede (2006) compared PUFA and HUFA in herbivorous and carnivorous zooplankton genera of *Daphnia*, *Bosmina*, *Holopedium*, *Bythotrephes*, *Arctodiaptomus*, and *Heterocope*, the accumulated HUFA ranged from 7 to 35% of total fatty acids with 8–27% n-6 in HUFA. Other collections of the small crustacean genus, *Daphnia*, contained 8–71% n-6 in HUFA depending on their local food supplies (Hessen and Leu 2006; Barata et al. 2005). In studying eutrophication with a cyanobacterial bloom, Müller-Navarra et al. (2000) suggested that an undesirable reduction in growth of *Daphnia* could be due to a relative deficit of diatoms and their 20:5n-3. In a similar way, reproductive success of the calanoid copepod, *Temora longicornis*, was related to availability of n-3 HUFA in the food web (Arendt et al. 2005). Different critical life events may require different specific EFA, and negative effects may come from either too little supply or too much supply. Researchers can productively examine mechanisms by which the just-noted impact of HUFA supply on crustacean success has parallels for vertebrate, mammal, and human success.

Table 14.1 shows about half of the fatty acids in diverse living systems are saturated and monoenoic acids, whereas HUFA can be from 0 to 40% of tissue fatty acids. Transfers of PUFA and HUFA along food chains and around the food web likely affect how different animal species thrive or survive. Human life in coastal regions has been well supported by easy access to crustacean and molluskan shellfish and abundant, oil-rich fish such as herrings, salmonids, and tunas. We need to know how important zooplankton HUFA are in sustaining harvests of these species. The HUFA in marine species are predominately n-3 HUFA (see Table 14.1), with

14 Human Life: Caught in the Food Web

somewhat higher proportions of n-6 HUFA in some freshwater species (e.g., smelt and alewife). For optimally sustaining pike (*Esox lucius*), access to adequate levels of both n-3 and n-6 HUFA is required (Engstrom-Ost et al. 2005). Optimal conditions for sustaining humans (Lands 2003c; Hibbeln et al. 2006a) and crustaceans seem to require about 20–40% n-6 HUFA in the total HUFA. However, biomass transfers to *Homo sapiens* from the worldwide food web currently lead to different ethnic populations having anywhere from 20% to more than 80% n-6 in HUFA (see Table 14.1). These differences in HUFA proportions are not as much determined by different genes as by culture-oriented food choices with little attention to n-3 or n-6 contents. The differences are also clear from the composition of fatty acids in human mother's milk, which varies worldwide and is influenced by diet composition (Kuipers et al. 2005, 2007) Thus, humans are caught in a food web that has serious outcomes. Evidence from observational and interventional studies indicates that low dietary intakes of marine foods and their 20- and 22-carbon n-3 HUFA are associated with serious human disabilities worldwide.

14.1.3 Prevalent Problems for Humans

The consequences of moving vitamin-like essential fatty acids along the food web that supports human health will be discussed in later sections. They can viewed in the context of the two most prevalent human disabilities worldwide, cardiovascular disease and major depression (Murray and Lopez 1997a, b). These may account for 23% of disability-adjusted life years (DALY) among developed nations (Table 14.2). Additionally, cancers, traffic accidents, alcohol abuse, osteoarthritis,

Table 14.2 The most abundant forms of human disability

	Worldwide			Developed nations		
Rank	Disease or Injury	DALYs	Cum. (%)	Disease or Injury	DALYs	Cum. (%)
	All causes	1,389	6	All causes	161	
1	Ischemic Heart disease	82	12	Ischemic Heart disease	18	11
2	Unipolar major depression	79	17	Cerebrovascular disease	10	17
3	Road traffic accidents	71	21	Unipolar major depression	10	23
4	Cerebrovascular disease	61	25	Lung and throat cancers	7	28
5	Chronic pulmonary disease	58	28	Road traffic accidents	7	32
6	Lower respiratory infections	43	31	Alcohol abuse	6	36
7	Tuberculosis	43	34	Osteoarthritis	5	40
8	War injuries	41	37	Dementia & CNS disorders	5	43
9	Diarrheal diseases	37	40	Chronic pulmonary disease	5	46
10	HIV-AIDS	36		Self-inflicted injuries	4	48

The disability-adjusted life years (DALY) were calculated by Murray and Lopez (1997a, b)

dementia and central nervous system (CNS) disorders, chronic pulmonary disease, and self-inflicted injuries account for another 25% of DALY, and these events may also have important associations to food selections. We need more research to identify the chains of molecular events causing these consequences so that we can design effective preventive interventions.

14.2 Evidence of Impaired Neural Development

Low intakes of n-3 HUFA in foods are associated with low levels in tissues and with impaired development of healthy neurobehavioral acts in humans. For example, results from thousands of subjects in the Avon Longitudinal Study of Parents and Children (ALSPAC) show maternal seafood intakes of more than 340 g per week have beneficial effects on child development, suggesting that advice to limit seafood consumption may be detrimental (Hibbeln et al. 2007). Seafood intake less than 340 g per week during pregnancy was associated with higher risk of children being in the lowest quartile for verbal intelligence quotient (IQ) (overall trend, $p = 0.004$), and with higher risk of suboptimum score for prosocial behavior, fine motor, communication, and social development. Unfortunately, the US Environmental Protection Agency currently advises pregnant women to limit their seafood intake to 12 ounces (i.e., ~340 g per week; US 2004). Careful revision of that advice intended to decrease exposure to mercury is now needed.

14.2.1 Stress, Corticotrophins, and Aggression

Fear and anxiety, components of defensive and violent behaviors, accompany elevated levels of corticotrophin-releasing hormone in the cortical–hippocampal–amygdala pathway. A small observational study correlated higher levels of cerebrospinal fluid corticotrophin-releasing hormone with lower percentage of docosahexaenoic acid (DHA; 22:6n-3) in total plasma fatty acids (Hibbeln et al. 2004a). Placebo-controlled trials may determine if dietary omega-3 fatty acid interventions can reduce excessive corticotrophin-releasing hormone levels and psychiatric illnesses. One pilot study of psychiatric patients with alcoholism, depression, or both, showed lower n-3 HUFA status was associated with higher concentrations of neuroactive steroids (Nieminen et al. 2006).

Deficiencies in DHA and eicosapentaenoic acid (EPA, 20:5n-3) at critical periods of neurodevelopment may lower serotonin levels and result in a cascade of suboptimal development of neurotransmitter systems, limiting regulation of the limbic system by the frontal cortex (Hibbeln et al. 2006b). Maladaptations to transient environmental stress and elevated corticoids in young animals can lead to "glucocorticoid programming" and altered neurophysiological and neuropathological status of adult laboratory animals and humans (Seckl and Meaney 2006). Accumulated effects of earlier transient trauma may give permanent disorders of depression, aggression, and hostility in adults. More data are needed to distinguish influences of diet on long-term accumulated neurodevelopmental defects from short-term reversible influences on adult pathology (e.g., Gesch et al. 2002).

In a sample of 3,581 urban white and black young adults (Iribarren et al. 2004), the multivariate odds ratios of scoring in the upper quartile of hostility were significantly associated (0.90; $p = 0.02$) with one standard deviation decrease in DHA intake. Prior consumption of any fish rich in n-3 fatty acids was independently associated with lower odds of high hostility (0.82; $p = 0.02$). Also, aggressive cocaine addicts (Buydens-Branchey et al. 2003) had significantly lower levels of DHA and the n-6 HUFA docosapentaenoic acid (DPA; 22:5n-6). Finally, the shifts that occurred in HUFA contents of those people during therapy support considering possible links between aggression in humans and a deficit in n-3 relative to n-6 nutrients.

14.2.2 How Food Helps

Food provides the n-3 PUFA precursors as well as the fully formed HUFA, DHA, which uniquely promotes neurite growth in hippocampal neurons, and its deficiency may contribute to cognitive dysfunction (Calderon and Kim 2004). Deficiency of DHA during development accompanies impaired learning and memory. Supplementing with DHA increased the population of neurons with longer neurite length per neuron and with higher number of branches, whereas added oleic (18:1n-9), arachidonic (20:4n-6), or n-6 DPA did not. DHA may promote neuronal integrity by facilitating membrane translocation/activation of Akt[1] while increasing phosphatidylserine (PS) in cell membranes (Akbar et al. 2005). The n-6 DPA which replaces DHA during n-3 fatty acid deficiency is less effective in accumulating PS, in translocating Akt and in preventing loss of neurons.

Depletion of DHA from neuronal tissues may also have a compounding effect on Raf-1[2] translocation in growth factor signaling (Kim et al. 2003). G protein-coupled signaling is impaired when a deficiency of n-3 EFA causes replacement of tissue n-3 DHA with n-6 DPA and gives suboptimal function in learning, memory, olfactory-based discrimination, spatial learning, and visual acuity (Niu et al. 2004). Replacement of n-3 by n-6 HUFA is correlated with desensitization of visual signaling in rod outer segments as evidenced by reduced rhodopsin activation, rhodopsin-transducin (G(t)) coupling, cGMP phosphodiesterase activity, and slower formation of metarhodopsin II (MII) and the MII-G(t) complex relative to rod outer

---

[1] Akt or protein kinase B (PKB) is an important molecule in mammalian cellular signaling. The name Akt does not refer to its function. Presumably, the "Ak" in Akt was a temporary classification name for a mouse strain developing spontaneous thymic lymphomas. The "t" stands for "transforming" the letter was added when a transforming retrovirus was isolated from the Ak strain.

[2] Raf-1 is a serine/threonine-specific kinase.

segments of n-3 FA-adequate animals. Reduced amplitude and delayed response of the electro-retinogram a-wave observed during n-3 EFA deficiency is attributed to impaired signal transduction by G protein-coupled receptors. Thus, adequate transfer of visual information from the environment to an individual depends on prior transfer of n-3 biomass to support optimal retinal function.

Neural cells exposed to ethanol accumulated considerably less phosphatidyl serine (PS) in response to the DHA enrichment and were less effective at phosphorylating Akt and suppressing caspase-3 activity (Akbar et al. 2006). Reduction of PS and a resulting neuronal cell death are undesirably enhanced by ethanol exposure during fetal development. These impairments may contribute to the long-recognized cumulative impairments that characterize fetal alcohol syndrome. The impairments illustrate the importance of maintaining adequate supplies of the essential n-3 HUFA in neural and retinal tissues. Levels of PS were consistently reduced in brain cortices of pups from ethanol-exposed dams, mainly due to the depletion of the n-3 PS18:0/22:6 species (Wen and Kim 2007).

14.2.3 Fish Also have Neural Development

Castel et al. (1972) reported that feeding rainbow trout (*Oncorhynchus mykiss*) the n-3 alpha-linolenic acid (ALA; 18:3n-3) at 0.5% or more of dietary calories prevented poor growth, fin erosion, heart myopathy, and a shock syndrome seen with PUFA-deficient diets. That report contrasted with many others that assigned an essential role only to n-6 linoleic acid (LIN; 18:2n-6) for man and other animals (e.g., Holman 1958) while denying assignment of n-3 nutrients as "essential." Rejecting an essential role for n-3 acids was then based on the limited view that n-3 acids provide no more benefit than n-6 acids (Tinoco et al. 1971). However, in trout, neuropathological shock symptoms (Sinnhuber 1969) began to appear 4 weeks to 3 months after feeding a PUFA-deficient diet, depending upon the age of the fish when the diet was started. Eating 1% LIN made this symptom appear sooner and more severely than any other diet tested. The shock syndrome resembles transportation shock reported as a "common experience of the fish culturist" (Black and Barrett 1957). In those times, commercial trout feed was usually high in LIN and low in n-3 fatty acids (Sinnhuber 1969). Such proportions in the food supply are often controlled by financial priorities of suppliers rather than physiologic outcomes of consumers.

Trout had highly variable individual somatic growth responses when fed LIN (compare photos in Fig. 3, Castel et al. 1972) suggesting diverse individual responses to transient environmental stimuli had accumulated during development (which did not occur when diets contained n-3 EFA). Maladaptations to transient environmental stress and elevated corticoids in young individuals can lead to altered neurophysiological and neuropathological status of adult laboratory animals and humans (Seckl and Meaney 2006). Such nongenetic developmental differences are well known for individual response levels of corticotrophin-releasing hormone and plasma corticosteroids, biomarkers related to aggression, homicide, suicide, and other serious human disorders. Readers might wonder if cumulative maladaptations of corticoid responses will occur in mammals with less frequency or intensity when daily food during development has higher proportions of n-3 EFA.

Because so many life-forms share related DNA-determined events, the ability of a dietary balance in n-3 and n-6 EFA to diminish or prevent maladaptations during development could be important for many species in our food web. We may wonder how n-3 and n-6 EFA affect developmental stages of zooplankton or other crustacea, which also carry echoes of our shared DNA-defined responses. Constructive research can determine how HUFA imbalances in the food web exacerbate developmental maladaptations (see Arts and Kohler – Chap. 10) and whether accumulated outcomes in adults can be reversed.

14.2.4 Self-Harm is a Depressing Adaptation

Depression, impulsivity, and suicidal intent were measured in patients with self-harm and matched controls, together with plasma lipids and EFA. Patients presenting with self-harm had more pathology on psychometric measures of depression, impulsivity, and suicidal intent than did controls (Garland et al. 2007). They also had lower mean total EFA levels (88 vs. 106 µg ml^{-1}, $p = <$ 0.001), total n-3 and n-6 EFA levels, and a higher %n-6 in HUFA (73% vs. 66%). Impulsivity and depression scores were inversely correlated with both n-6 and n-3 EFA. Supplementation with n-3 HUFA for 12 weeks led to substantial reductions in surrogate markers of suicidal behavior and to improvements in well-being for patients recruited after acts of repeated self-harm (Hallahan et al. 2007). Low levels of DHA (22:6n-3) and elevated ratios of n-6/n-3 acids are associated with major depression and, possibly, suicidal behavior (Sublette et al. 2006). A lower DHA percentage in plasma PUFA and a higher n-6/n-3 ratio predicted suicide attempt over a 2-year period.

Higher concentrations of DHA in mothers' milk ($r = -0.84$, $p < 0.0001$, $n = 16$ countries) and greater seafood consumption ($r = -0.81$, $p < 0.0001$, $n = 22$ countries) both predicted lower prevalence rates of postpartum depression (Hibbeln 2002). However, ARA and EPA contents of mothers' milk were unrelated to postpartum depression prevalence. To test the efficacy of supplemental n-3 fat, subjects received 0.5 g day^{-1} ($n = 6$), 1.4 g day^{-1} ($n = 3$), or 2.8 g day^{-1} ($n = 7$) in an 8-week trial. Across groups, pretreatment Edinburgh Postnatal Depression Scale (EPDS) and Hamilton Rating Scale for Depression (HRSD) mean scores were 18.1 and 19.1, respectively; post-treatment mean scores were 9.3 and 10.0 (Freeman et al. 2006a). Percent decreases on the EPDS and HRSD were 51.5% and 48.8%, respectively; changes from baseline were significant within each group and when combining groups. However, groups did not significantly differ in pre- or post-test scores, or change in scores. Nevertheless, greater seafood consumption predicted lower lifetime prevalence rates of bipolar I disorder, bipolar II disorder, and bipolar spectrum disorder (Noaghiul and Hibbeln 2003). Bipolar II disorder and bipolar spectrum disorder had an apparent vulnerability threshold below 50 lb year^{-1} (i.e., ~430 g week^{-1}) of seafood/person.

14.2.5 Recognizing Cognitive Benefits from Seafood

Recently, the American Psychiatric Association appointed a subcommittee to prepare a report that was reviewed and approved by its Committee on Research on Psychiatric Treatments, Council on Research, and Joint Reference Committee (Freeman et al. 2006b). The report concluded that the preponderance of epidemiologic and tissue compositional studies supports a protective effect on mood disorders of n-3 EFA intake, particularly EPA and DHA. Randomized controlled trials showed EPA and DHA appear to have negligible risks and statistically significant benefit in unipolar and bipolar depression ($p = 0.02$). A recent literature survey regarded fish as a food with unique psychotropic properties (Reis and Hibbeln. 2006). It described how fish have been culturally labeled as symbols of emotional well-being and social healing in religious and medical practices among independent cultures for at least six millennia. Recent reports about dietary HUFA preventing cognitive decline in older adults (Connor and Connor 2007; Baydoun et al. 2007; vanGelder et al. 2007) continue the transfer of hopeful healing information.

14.3 Evidence of Impaired Cardiovascular Development

Evidence of vascular injury developing progressively in young Americans has been documented repeatedly over the past 50 years (Enos et al. 1953; Newman et al. 1986; Rainwater et al. 1999; Zieske et al. 2002), but prevention of its primary dietary causes remains neglected. Risk scores predict advanced coronary artery atherosclerosis in middle-aged persons as well as youth (McMahan et al. 2007). Thus, each generation follows earlier ones in a tragic chain of preventable, cumulative errors that cause disability and death of adults. As with psychiatric disorders that have cumulative effects on individuals during development, clinical cardiovascular disease has cumulative maladaptations that eventually become recognized long after the time when they were reversible. A recent epidemiological study (Robinson and Stone 2006) noted that lifetimes of eating n-3 HUFA showed more consistent cardiovascular benefit than did clinical intervention trials with limited times of eating n-3 HUFA. Primary prevention of coronary heart disease (CHD) needs to begin in childhood.

14 Human Life: Caught in the Food Web

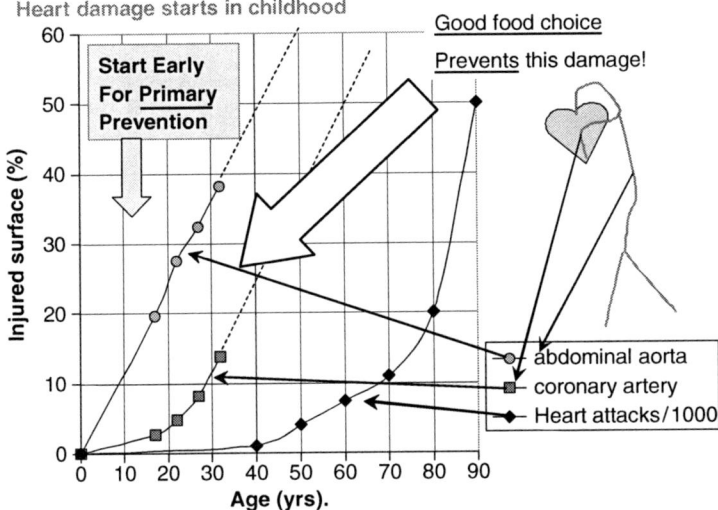

Fig. 14.1 Evidence of CHD pathology begins in childhood. Results of progressive vascular injury from the Pathobiological Determinants of Atherosclerosis in the Youth (PDAY Research Group 1990) program are in many reports, including (McGill et al. 2000; McMahan et al. 2007; Rainwater et al. 1999.)

From a US perspective, greater awareness of worldwide human living conditions developed after 1946 when the USA transferred energy and attention from war to new biomedical research programs at the National Institutes of Health. The Seven Countries Study introduced researchers to different mortality rates for CHD among different populations with different food habits. Epidemiologists at the time associated death with ingesting saturated fats and total fat energy, but said little about seafood or n-3 HUFA. Subsequently, much attention went to the hypothesis that cholesterol in the diet and the bloodstream of individuals may cause CHD death, but a molecular mechanism was never proved.

Wilber and Levine (1950) noted a near absence of cardiovascular disease among Alaskan Inuits who had consistently high serum cholesterol. Other reports showed clearly that dietary cholesterol was NOT causing death, although elevated cholesterol in the blood was associated with death in the USA. In 1984, an appointed committee substituted political consensus for rigorous logic and asserted that elevated plasma cholesterol was a *cause* of death (Consensus 1985). This assertion was also accompanied by stern advice to lower caloric imbalances in daily life. Unfortunately, the advice about food energy was not effectively transferred or implemented in primary prevention programs, and an epidemic of obesity developed over the next twenty years.

The HUFA proportions for diverse human populations in Table 14.1 illustrate diverse impacts of the food web on various human groups. A lower incidence of CHD deaths among people eating more marine food (with higher n-3 HUFA contents) is illustrated in Fig. 14.2.

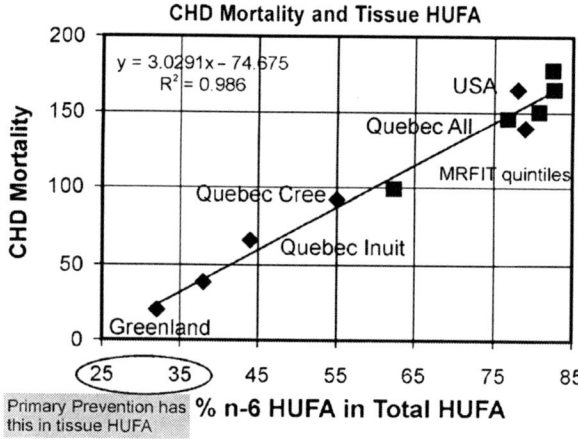

Fig. 14.2 Higher proportions of n-6 in HUFA predict higher CHD mortality rates. The figure is derived from an earlier version (Lands 2003a). *Squares* indicate quintiles in the MRFIT Study in the USA, where 80% individuals clustered closely

14.4 Which DNA-Coded Proteins Discriminate N-3 and N-6 Structures?

14.4.1 Putting EFA into Triacylglycerols and Phospholipids

Many enzymes are like animal species that consume readily available food without regard to downstream consequences. They are promiscuous and relatively indiscriminate in handling fatty acids with different structural features. Enzymes converting nonesterified fatty acids (NEFA) of 14–24 carbons to acyl-CoA esters consume whatever fatty acids are supplied without appreciably selecting for chain length or for numbers and locations of double bonds. Nevertheless, a selective transfer of palmitic acid (16:0) from its CoA ester to the 1-position of glycerol-3-phosphate begins the "*de novo*" lipid pathway that includes transfer to the 2-position of readily available unsaturated 16- or 18-carbon acids (e.g., palmitoleate (16:1n-7), oleate (18:1n-9), or linoleate (18:2n-6)). This leads to diacylglycerols with a saturated acid at the 1-position and an unsaturated acid at the 2-position (Hill et al. 1968).

Enzymes converting diacylglycerols (DAG) to triacylglycerols (TAG) consume acyl-CoA mostly in accord with supply abundance and not with chain length or number and location of double bonds. The nonessential saturated and monoenoic acids are steadily maintained in their CoA esters by synthesis and transport. However, a greater daily dietary supply of an EFA gives linearly greater amounts of EFA accumulated in tissue acyl-CoA and TAG. Rearrangements occur as lipases cleave TAG to DAG and provide mixtures of diverse DAG that can be acylated to TAG with little specificity for acyl-CoA structure or for the acyl content of the DAG

(Slakey and Lands 1968). Thus, tissue TAG provides a flexible, nonselective expandable storage pool of unlimited size. Its energy-rich biomass contains mainly whatever is accessible in the local microenvironment. Little selectivity for n-3 or n-6 acyl structures occurs when forming or cleaving triacylglycerols. The 18-carbon PUFA, 18:2n-6 and 18:3n-3, accumulate in tissue TAG in linear response to their percent abundance in daily dietary calories (Lands et al. 1990; Lands et al. 1992).

Cellular 1,2-diacylglycerols are reversibly transformed into phospholipids by enzymes that attach phosphate derivatives to the 3-position. These enzymes also seem fairly promiscuous and able to form diverse molecular species that reflect the DAG units available (although transfer of ethanolamine phosphate slightly favors DAG units with 22-carbon HUFA at the 2-position). Subsequent transacylase and acyltransferase actions "retailor" phospholipids with acyl chains available from acyl-CoA esters. Highly specific selections occur when placing acyl chains into phospholipids (Lands 2000, 2005b). The DNA-coded selectivities favor accumulation of HUFA at the 2-position of phospholipids. However, the number of enzymes involved and the precise acyl group interactions of each remain unknown. The resulting accumulated phospholipids retain a dominant pattern of saturated acids at the 1-position and unsaturated acids at the 2-position. These diacylglycerol units of phospholipids can form DAG (by reversible actions of enzymes like CDPcholine:DAG cholinephosphotransferase) that are acylated and accumulate nonselectively in tissue TAG. The proportions of n-3 and n-6 in HUFA of all accumulated tissue glycerolipids show little evidence of an ability of acyl-CoA transferases to discriminate between n-3 and n-6 structures.

The mixture of acyl groups in cellular acyl-CoA esters includes newly synthesized and newly imported acids plus those released from TAG and phospholipids by hydrolases. Dietary PUFA that enter the acyl-CoA mixture compete with each other when interacting with enzymes catalyzing the elongation:desaturation reactions that form longer and more highly unsaturated acyl-CoA esters. The paths involve 18:2n-6 > 18:3n-6 > 20:3n-6 > 20:4n-6 > 22:4n-6 > 24:4n-6 > 24:5n-6 > 22:5n-6 and 18:3n-3 > 18:4n-3 > 20:4n-3 > 20:5n-3 > 22:5n-3 > 24:5n-3 > 24:6n-3 > 22:6n-3. The n-3 and n-6 acids compete for the enzymes more effectively than n-7 and n-9 acids, and some transient intermediates (e.g., 20:4n-3 and 24:5n-6) do not accumulate appreciably in glycerolipids. The limited conversion rates through this multistep system make dynamic handling of 18-carbon PUFA differ appreciably from that for the derived 20- and 22-carbon HUFA.

Limited space and time become evident whenever discriminating "consumer" enzymes rapidly transform *only* preferred supply items into next-step supplies, while nondiscriminating "consumers" transform all supply items *in proportion to their abundance* to next-step supplies. Also, limited space for next-step supplies allows only rapidly formed product items to accumulate, whereas unlimited space for next-step supplies allows all product items to accumulate indiscriminately. Quantitative studies related accumulated tissue HUFA to amounts of competing PUFA supplied in daily food (Mohrhauer and Holman 1963a, b; Lands et al. 1990), giving an empirical competitive, hyperbolic "saturable" equation (Lands et al. 1992). Revised constants (see http://efaeducation.nih.gov/sig/hufacalc.html) fit later data sets (Lands 2003b) and better predict diet:tissue outcomes. The equation predicts the nonlinear hyperbolic impact of dietary EFA abundance on the proportions of n-6 in HUFA accumulated competitively in mammalian tissues (*Mus, Rattus, Canis, Homo*) and allows rational interpretations of the impact of food choices on health.

14.4.2 Hormone-Like Signaling By N-3 and N-6 Eicosanoids

HUFA-mediated adaptive responses to environmental stimuli have many steps. Activation of cellular phospholipase A_2 (PLA_2) increases release of HUFA from the 2-position of phospholipids where they accumulate in relatively high amounts. Although PLA_2 appears to interact with 20-carbon HUFA better than with 18-carbon PUFA, it discriminates little between n-3 and n-6 structures when producing nonesterified HUFA. Fatty acid oxygenases (cyclooxygenase and lipoxygenase) act on nonesterified n-3 and n-6 HUFA to form hydroperoxide derivatives that stimulate faster oxygenation. Producing the hydroperoxide activator gives explosive formation of eicosanoids in response to stimuli. Predatory cellular peroxidases remove the hydroperoxide and slow this amplifying step during synthesis of prostaglandin H and leukotriene A.

Although peroxidases and lipoxygenase discriminate little between n-3 and n-6 structures, the cyclooxygenase activity of PGH synthase forms hydroperoxide product several times faster with n-6 than n-3 HUFA (Kulmacz et al. 1994; Chen et al. 1999; Laneuville et al. 1995). This is one of the few DNA-coded protein-catalyzed events known to discriminate between n-3 and n-6 acids. One consequence of the faster explosive positive feedback in formation of n-6 eicosanoids, PGG_2 and PGH_2, is that levels of peroxidase activity that remove enough n-3 product/activator (PGG_3) to prevent its explosive positive feedback do not appreciably hinder the faster n-6 amplified actions. This dynamic difference gives more vigorous overall formation and action of n-6 than n-3 prostaglandins (Lands 2005a; Smith 2005; Liu et al. 2006). As a result, the proportion of n-6 in HUFA of tissues, a biomarker that predicts likelihood of CHD mortality (see Fig. 14.2), also predicts the likely intensity of n-6 eicosanoid-mediated events. A strong preference of trout PGH synthase for n-6 over n-3 structures allows even small proportions of n-6 arachidonate in fish HUFA to form sufficient active n-6 hormone (Liu et al. 2006). However, that preference may also cause a too-intense eicosanoid action during the earlier-noted "transportation shock" in rainbow trout (when imbalanced diets had much more n-6 than n-3 fats).

Isomerases form active hormone-like prostaglandins (PGD, PGE, PGF, PGI, TXA) from the intermediate PGH_2 and PGH_3 formed by PGH synthase. We still know little of the selectivity of these isomerizing enzymes for n-3 and n-6 structures. Slow formation of transient active hormone can allow continual degradation by predatory dehydrogenases to inactivate slowly formed eicosanoids. In this situation, significant amounts of active hormone may be formed during a 24-h period but never accumulate sufficiently at tissue receptors at any point in time to give significant signals. For thromboxane, TXA (which mediates thrombosis), the active hormone spontaneously decomposes to inert TXB within seconds. Thus, TP receptors will give appreciable signals only when a very rapid pulse of TXA is being formed to create the needed local abundance.

At this time, we know little of the n-3:n-6 selectivity of the diverse cellular receptors that bind active eicosanoid and stimulate inside the cell signaling actions by kinases and phosphatases that regulate cell physiology. A slower receptor response to a bound hormone could allow the continual removal of transient signal mediators to dissipate the signal and prevent sufficient response. Although little selectivity is known for many leukotriene-mediated events, an important selective response by the leukotriene B (LTB) receptor provides much greater chemotactic signaling with the n-6 LTB_4 than for the n-3 analog, LTB_5. This produces more vigorous inflammatory events with n-6 than n-3 mediators. Wherever cellular systems respond differently to n-3 and n-6 eicosanoid signals, we can begin understanding the chain of events by which the food web impacts human health. Important insights were recently provided by Wada et al. (2007), confirming that many DNA-coded proteins act more vigorously in n-6 eicosanoid signaling than in n-3 eicosanoid-mediated events. Over-reactions mediated by n-6 eicosanoids are targets for inhibition by highly profitable therapeutic patented drugs. Perhaps primary prevention with greater proportions of n-3 in HUFA might decrease the need for such medication.

14.4.3 How Food Hurts

When humans eat food, the carbohydrates, proteins, and nucleic acids are hydrolyzed (digested) in the gut to simple sugars, amino acids, and nucleosides that diffuse during the next hour into intestinal cells. They then enter portal blood flowing to the liver and onward to the general bloodstream flowing throughout the body. Hydrolyzed fats are reassembled in intestinal cells and secreted into lymph as lipoprotein complexes which enter the general bloodstream during the next few hours. In this way, each meal (like a frequent periodic flood) provides the bloodstream and associated tissues with more energy and biomass than can be fully metabolized to CO_2 as it arrives. The transient postprandial surplus of energy-rich biomass is handled by different tissues in different ways: forming CO_2, forming fat, forming isoprenoids, and forming eicosanoids. In the process, transient oxidant stress creates temporary tissue insults that mostly reverse over time, but occasionally become the long-lasting vascular injuries noted in Fig. 14.3.

Within cells, the carbon and electrons of fats, carbohydrates, and proteins readily transform in multistep paths to acetyl-CoA (Fig. 14.3), CO_2 plus cofactor-carried electrons. The limited capacity to store electrons is met in part by transferring them to oxygen, forming water. This transfer is strongly coupled with formation of energy-transmitting ATP from ADP. However, without coupled work returning ATP to ADP, the stalled system cannot form much CO_2 and "leaks" some of the flood of electrons to oxygen, forming reactive molecules that cause transient oxidative stress and amplifies it into harmful inflammatory tissue insults (Lands 2003c).

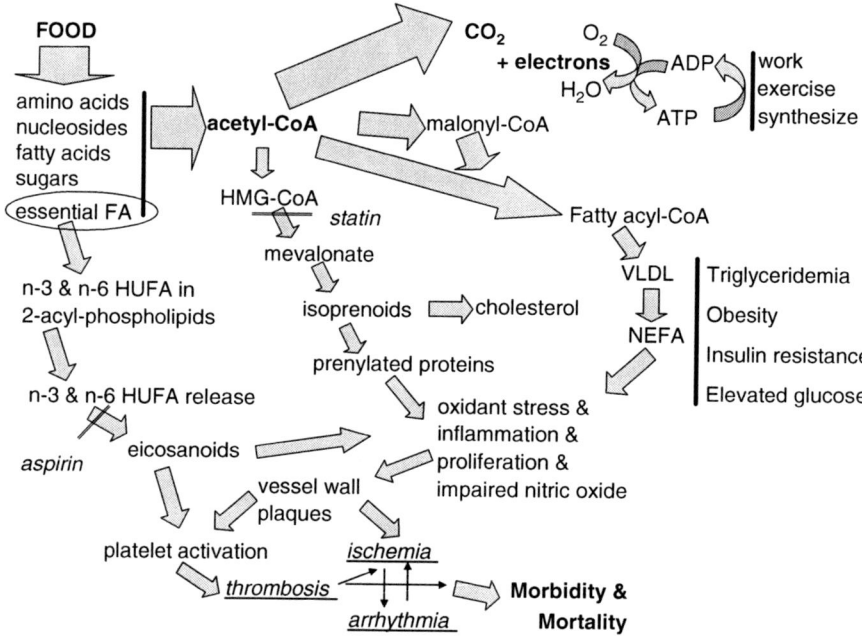

Fig. 14.3 Pathways to morbidity and death. Associated biomarkers are not always causal mediators. Transient inputs of food energy and biomass move to plasma lipids and are stored for later useful work. However, transient postprandial inflammatory states are amplified by n-6 eicosanoids.

In cells unable to convert the transient flood of metabolites to CO_2, excess acetyl-CoA forms malonyl-CoA (which polymerizes in combination with excess electrons in multistep paths that form energy-rich fats) and some hydroxymethyl glutaryl-CoA (which forms mevalonate and polymerizes in multistep paths that that form diverse isoprenoid products). Many transient intermediates (e.g., eicosanoids, prenylated proteins, and NEFA) do not accumulate in large amounts, but have important roles in mediating overall pathophysiology (Lands 2003c). Figure 14.3 shows how the transient postprandial flood of biomass "pushes" acetyl-CoA through steps that increase plasma NEFA and cellular prenylated proteins, which lead to oxidant stress and inflammation. These transient conditions can be amplified by n-6 eicosanoid actions and accumulate long-term inflammatory vessel wall plaques. Thus, two food imbalances act together to create chronic conditions that ultimately cause CHD death: imbalances in food energy which are then amplified by elevated n-6 eicosanoid-mediated inflammation, thrombosis, and arrhythmia (Fig. 14.3). Elevated plasma TAG and cholesterol are indicators of transient food energy imbalance, and the % n-6 in HUFA of plasma indicates imbalances in intakes of n-3 and n-6 fats.

The much-discussed clinical indicator of CHD risk, plasma total cholesterol (TC), poorly predicted the absolute death rates observed in a 25-year follow-up of the Seven Countries Study (Verschuren et al. 1995). In fact, it had no clear ability

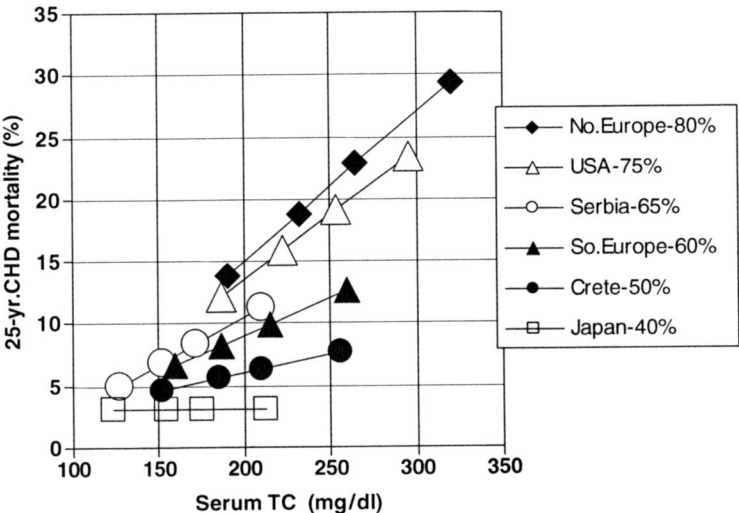

Fig. 14.4 Food energy impact on mevalonate products is fatal only to the degree that n-6 exceeds n-3 in HUFA. Deaths (as % of subjects studied) reported by Verschuren et al. (1995) are predicted quantitatively using mg total cholesterol (TC) per deciliter plasma and %n-6 in HUFA: death from CHD = 3 + 3 × (%n-6 in HUFA − 40) × (TC − 100)/1,000.

to predict death in Japan. However, the observed CHD mortality was well predicted when the likely proportion of n-6 in HUFA for the different populations was combined in the following way: CHD death = 3 + 3 × (%n-6 in HUFA − 40) × (TC − 100)/1,000. Figure 14.4 shows that higher values for total cholesterol (TC) predicted higher death rates only to the degree that the %n-6 exceeded the %n-3 in HUFA. This result suggests that higher tissue proportions of n-3 HUFA may diminish vigorous inflammatory and thrombotic signals of n-6 eicosanoids and prevent the transient food energy-induced increases in NEFA and isoprenoids (for which TC is a biomarker) from being amplified into fatal outcomes. In this way, the toxicity of food energy imbalances depends on the context of ambient tissue n-3 HUFA.

14.4.4 Choosing Healthy Foods

Most transfers of fatty acids among individual organisms, cells, or enzymes are somewhat indifferent to n-3 or n-6 structures, allowing both to compete with each other for many processes. However, when physiologically important events discriminate n-3 from n-6 structures, the relative abundance of these two forms in food supplies may cause unintended consequences. The wide range of disorders linked to imbalances in n-3 and n-6 are discussed in detail in Lands (2005a). Humans may eat indiscriminately whatever is available, much like most species do in the food web within which they survive. However, the high proportion of n-3 in the food web of estuaries is not matched by food in the current commercial agriculture-based food web for Americans. Financial and marketing priorities that are controlling the supply of the food web may depend more on aspects of storage, transport, and profits rather than biological sustainability, nutrient content, or human physiology (Nestle 2002). Humans and their companion animals now are in a web of widely publicized and marketed foods that gives little information or interpretation of the PUFA and HUFA content.

One possible unintended consequence might be the higher rates of homicide (Hibbeln et al. 2004b) that are correlated with a greater apparent consumption of n-6 linoleic acid over a 20-fold range (0.51–10.2/100,000) for the years 1961–2000 in Argentina, Australia, Canada, the United Kingdom, and the United States ($r = 0.94$, $p < 0.00001$). The apparent linoleic acid intake from seed oil sources ranged from 0.29 en% (percentage of daily food energy) (Australia in 1962) to 8.3 en% (US in 1990s). This supply may be seen, in hindsight, to be an oversupply. Scientists have long known that an adequate intake of linoleate for humans may be less than 0.5% of daily food energy (Cuthbertson 1976; see http://efaeducation.nih.gov/sig/dri1.html). However, multibillion dollar profits from selling vegetable oils rich in n-6 fats help fund marketing messages that inform the public of benefits from replacing saturated with unsaturated fats, but fail to describe clinical conditions linked to excessive n-6 eicosanoid actions. The public could benefit from wider distribution of balanced information from which to make food choices, but it provides few financial incentives to fund such distribution.

To estimate a healthy dietary allowance for n-3 HUFA that would meet nutrient requirements for 97–98% of the world's population, a deficiency in n-3 HUFA was defined as attributable risk from 13 morbidity and mortality outcomes (Hibbeln et al. 2006a). The outcomes included death from all causes, coronary heart disease, stroke, cardiovascular disease, homicide, bipolar disorder, and major and postpartum depressions. The potential attributable burden of disease ranged from 20.8% (all-cause mortality in men) to 99.9% (bipolar disorder). The n-3 HUFA intake for Japan (0.37% of energy, or 750 mg/d) met criteria for uniformly protecting > 98% of the populations worldwide.

Table 14.3 provides information on average PUFA and HUFA intakes for some countries in the context of how much n-3HUFA would be needed daily to balance the other EFA and maintain a tissue biomarker level of 40% n-6 in HUFA. The low intake of linoleate in the Philippines would need only small intakes of n-3 HUFA to sustain that tissue biomarker level. With the current US food web, a healthy dietary allowance for n-3 HUFA was estimated to be 3.5 g/d for a 2,000-kcal diet. Because of the well-established competition between n-3 and n-6 structures in lipid metabolism (Mohrhauer and Holman 1963b, Lands et al. 1992), this allowance can likely be lowered to one-tenth of that amount by consuming fewer n-6 fats.

Interactive computer software (http://efaeducation.nih.gov/sig/kim.html) informs users of n-3, n-6 and caloric contents of nearly 12,000 different servings of food.

Table 14.3 PUFA and HUFA in the food web

Country	Current avg dietary intake (en%)			n-3 HUFA needed for 40% n-6 in HUFA	
	LIN	ALA	ARA	(en%)	(mg/d)
Philippines	0.80	0.08	0.06	0.125	278
Denmark	2.23	0.33	0.09	0.45	1,000
Iceland	2.48	0.33	0.10	0.54	1,200
Colombia	3.21	0.24	0.04	0.51	1,133
Ireland	3.57	0.42	0.06	0.62	1,378
UK	3.91	0.77	0.07	0.72	1,600
Netherlands	4.23	0.28	0.08	0.88	1,956
Australia	4.71	0.49	0.07	0.90	2,000
Italy	5.40	0.51	0.06	0.95	2,111
Germany	5.57	0.62	0.06	1.00	2,222
Bulgaria	7.02	0.06	0.05	1.25	2,778
Israel	7.79	0.67	0.07	1.45	3,222
USA	8.91	1.06	0.08	1.65	3,667

The results were provided by Hibbeln et al. 2006a,b. The major dietary PUFA are: LIN = 18:2n-6, ALA = 18:3n-3, and the major dietary n-6 HUFA is ARA = 20:4n-6. The daily intake of n-3 HUFA predicted to maintain tissue biomarker value of 40% n-6 in HUFA (an average value for Japan) is indicated in the two right-hand columns as percent of daily food energy (en%) and mg d^{-1}

It predicts the calorie balance and likely tissue proportions of n-6 in HUFA that result from eating the servings. The freely available information allows people to plan individual daily menu plans to meet their personalized sense of taste and risk aversion. Examples of PUFA and HUFA contents of various types of food and of diverse daily menu plans can be found in Lands (2005a).

14.5 Making a Better Future

Human interactions with the food web have shifted from hunter-gatherer to domestic farmer to functional food fabricator. Hunters ate species that carried genes permitting survival in the competitive food web of nature, whereas farmers selectively bred species that carried genes permitting optimal financial profits. Now, functional foods may come from mixing gene products of diverse species or transferring genes to provide new species with previously unknown mixtures of nutrients. Human concerns for sustainability involve questions of quality of life for individuals and their environment. Biomedical research thrived with the transfer of information about the homology of DNA coding patterns among genera of bacteria (*Escherichia* and *Cyanobacteria*), simple eukaryotes (*Saccharomyces*), and more complex plants (*Arabidopsis*) and animals (*Caenorabditis, Drosophila, Danio, Mus* and *Homo*). The research provided deep insight into molecular events that cause health and disease. Converting that information into daily practice has been vigorously pursued by biomedical corporations to develop profitable treatments for wealthy people in distress with evident disease. Unfortunately, a sense of need or will to use that information to prevent disease and avoid too expensive treatments is not rewarded by current market forces and is in little demand.

Sustaining the diverse food web that once provided humans with foods needed for good health and quality of life will be a challenge. Hopefully, efforts to provide those supplies will carefully avoid the unintended consequences noted in this review.

The simplicity of management effort favors "monoculture" agricultural programs that return well-recognized profits and contrast sharply with the priority ecological programs that sustain diverse species and are supported by no traditionally defined profits. The interdependence of diverse life-forms that evolved over time led to some species thriving and some vanishing. The genus *Homo* continues to increase in abundance and adds greater pressure on existing supplies in the estuarine (and aquatic in general) food web. The current commercial food web in which *Homo sapiens* is currently "trapped" depends much upon easily stored grains, cereals, and nuts that contain much n-6 PUFA and little or no n-3 HUFA. Ironically, *Vigna mungo* (black gram), rich in n-3 PUFA, provides major food energy and biomass to poor people in India and southeast Asia. However, *Vigna radiata* (green gram) commercially developed in richer nations has much more n-6 than n-3 PUFA. Awareness of the contents of PUFA and HUFA may some day be a priority in efforts to sustain a healthy food web for humans. Rational selections of food can be aided with the interactive planning software noted earlier (http://efaeducation.nih.gov/sig/kim.html). However, the future food web will need more items containing more n-3 PUFA and HUFA if humans are to sustain healthy proportions of tissue HUFA.

Current experiments in gene transfer are considering ways to "short-circuit" the historic stepwise transfer of PUFA and HUFA through diverse species of phytoplankton, crustaceans, mollusks, and fish to humans. Inserting DNA-coded machinery into domesticated plant and animal species (Lai et al. 2006) may increase their tissue contents of n-3 PUFA and HUFA and help humans thrive with less food from estuarine food webs. This domestic agricultural alternative of mixing genes among previously wild species may compensate for the greatly increased human consumption of the species that have survived till now in an estuarine food web that supplied so much n-3 HUFA for human health. The diverse chapters in this book may further help readers better understand what can be done to improve human life caught in the food web.

References

Akbar, M., Calderon, F., Wen, Z., and Kim, H.Y. 2005. Docosahexaenoic acid: a positive modulator of Akt signaling in neuronal survival. Proc. Natl. Acad. Sci. USA. 102:10858–10863.

Akbar, M., Baick, J., Calderon, F., Wen, Z., and Kim, H.Y. 2006. Ethanol promotes neuronal apoptosis by inhibiting phosphatidylserine accumulation. J. Neurosci. Res. 83:432–440.

Arendt, K.E., Jonasdotter, S.H., Hansen, P.J., and Gartner, S. 2005. Effects of dietary fatty acids on the reproductive success of the calanoid copepod *Temora longicornis*. Marine Biol. 146:513–530.
Barata, C., Navarro, J.C., Varo, I., Riva, M.C., Arun, S., and Porte, C. 2005. Changes in antioxidant enzyme activities, fatty acid composition and lipid peroxidation in *Daphnia magna* during the aging process. Comp. Biochem. Physiol. B. 140:81–90.
Baydoun, M.A., Kaufman, J.S., Satia, J.A., Rosamond, W., and Folsom, A.R. 2007. Plasma n-3 fatty acids and the risk of cognitive decline in older adults: the Atherosclerosis Risk in Communities Study. Am. J. Clin. Nutr. 85:1103–1111.
Bigogno, C., Khozin-Goldberg, I., Boussiba, S., Vonshak, A., and Cohen, Z. 2002. Lipid and fatty acid composition of the green oleaginous alga *Parietochloris incisa*, the richest plant source of arachidonic acid. Phytochem. 60:497–503.
Black, E.C., and Barrett, X. 1957. Increase in levels of lactic acid in the blood of cutthroat and steelhead trout following handling and live transport. Can. Fish Cult. 20:13.
Buydens-Branchey, L., Branchey, M., McMakin, D.L., and Hibbeln, J.R. 2003. Polyunsaturated fatty acid status and aggression in cocaine addicts. Drug Alcohol Depend. 71:319–323.
Calderon, F., and Kim, H.Y. 2004. Docosahexaenoic acid promotes neurite growth in hippocampal neurons. J. Neurochem. 90:979–988.
Castel, J.D., Sinnhuber, R.O., Wales, J.H., and Lee, D.J. 1972. Essential Fatty Acids in the Diet of Rainbow Trout *(Salmo gairdneri):* Growth, Feed Conversion and some Gross Deficiency Symptoms. J. Nutr. 102:77–86.
Castell, J.D. 1979. Review of lipid requirements of finfish. In Finfish nutrition and fishfeed technology, edited by J.E. Halver and K. Trews. Proceedings of a World Symposium sponsored and supported by EIFAC/FAO/ICES/IUNS, Hamburg, 20–23 June, 1978. Schr. Bundesforschungsanst. Fisch., Hamb., (14/15) vol.1: 59–84. (http://www.fao.org/docrep/X5738E/x5738e1a.gif; accessed April, 2007).
Chajes, V., Elmstahl, S., Martinez-Garcia, C., Van Kappel, A.L., Bianchini, F., Kaaks, R., and Riboli E. 2001. Comparison of fatty acid profile in plasma phospholipids in women from Granada (southern Spain) and Malmo (southern Sweden). Int. J. Vitam. Nutr. Res. 71:237–242.
Chen, W., Pawelek, T.R., and Kulmacz, R.J. 1999. Hydroperoxide dependence and cooperative cyclooxygenase kinetics in prostaglandin H synthase-1 and -2. J. Biol. Chem. 274:20301–20306.
Clegg, M.T., Gaut, B.S., Learn, G.H. Jr., and Morton, B.R. 1994. Rates and patterns of chloroplast DNA evolution (chloroplast DNA/ribulose-1,5-bisphosphate carboxylase/plant evolution/evolutionary rates). Proc. Nat. Acad. Sci. USA 1994; 91:6795–6801.
Colombo, M.L., Rise, P., Giavarini, F., DeAngelis, L., Galli, C., and Bolis, C.L. 2006. Plant Foods Hum. Nutr. 61:67–72.
Connor, W.E., and Connor, S.L. 2007. The importance of fish and docosahexaenoic acid in Alzheimer disease. Am. J. Clin. Nutr. 85:929–930.
Consensus conference. 1985. Lowering blood cholesterol to prevent heart disease. J. Am. Med. Assoc. 253:2080–2086.
Cuthbertson, W.F.J. 1976. Essential fatty acid requirements in infancy. Am. J. Clin. Nutr. 29:559–568.
Delaporte, M., Soudant, P., Moal, J., Giudicelli, E., Lambert, C., Seguineau, C., and Samain, J.F. 2006. Impact of 20:4n-6 supplementation on the fatty acid composition and hemocyte parameters of the Pacific oyster *Crassostrea gigas*. Lipids 41:567–576.
Dewailly, E., Blanchet, C., Gingras, S., Lemieux, S., Sauve, L., Bergeron, J., and Holub, B.J. 2001a. Relations between n-3 fatty acid status and cardiovascular disease risk factors among Quebecers. Am. J. Clin. Nutr.74:603–611.
Dewailly, E., Blanchet, C., Lemieux, S., Sauve, L., Gingras, S., Ayotte, P., and Holub, B.J. 2001b. N-3 fatty acids and cardiovascular disease risk factors among the Inuit of Nunavik. Am. J. Clin. Nutr. 74:464–473.
Dewailly, E., Blanchet, C., Gingras, S., Lemieux, S., and Holub, B.J. 2002.Cardiovascular disease risk factors and n-3 fatty acid status in the adult population of James Bay Cree. Am. J. Clin. Nutr. 76:85–92.

Engstrom-Ost, J., Lehtiniemi, M., Jonasdottir, S.H., and Viitasalo, M. 2005. Growth of pike larvae (*Esox lucius*) under different conditions of food quality and salinity. Ecol. Freshw. Fish.14:385–393.

Enos, W.F., Holmes, R.H., and Beyer, J. 1953. Coronary disease among United States soldiers killed in action in Korea. J. Am. Med. Assoc. 152:1090–1093.

Fernandez-Reiriz, M.J., Labarta, U., Albentosa, M., and Perez-Camacho, A. 1999. Lipid profile and growth of the clam spat, *Ruditapes decussatus* (L), fed with microalgal diets and cornstarch. Comp. Biochem. Physiol. B. 124:309–318.

Ferrao-Filho, A.S., Fileto, C., Lopes, N.P., and Arcifa, M.S. 2003. Effects of essential fatty acids and N- and P-limited algae on the growth rate of tropical cladocerans. Freshw. Biol. 48:759–767.

Freeman, M.P., Hibbeln, J.R., Wisner, K.L., Brumbach, B.H., Watchman, M., and Gelenberg, A.J. 2006a. Randomized dose-ranging pilot trial of omega-3 fatty acids for postpartum depression. Acta Psychiatr. Scand. 113:31–35.

Freeman, M.P., Hibbeln, J.R., Wisner, K.L., Davis, J.M., Mischoulon, D., Peet, M., Keck, P.E. Jr, Marangell, L.B., Richardson, A.J., Lake, J., and Stoll, A.L. 2006b. Omega-3 fatty acids: evidence basis for treatment and future research in psychiatry. J. Clin. Psychiatry. 67:1954–1967.

Garland, M.R., Hallahan, B., McNamara, M., Carney, P.A., Grimes, H., Hibbeln, J.R., Harkin, A., and Conroy, R.M. 2007. Lipids and essential fatty acids in patients presenting with self-harm. Br. J. Psychiatry. 190:112–117.

Gesch, C.B., Hammond, S.M., Hampson, S.E., Eves, A., Crowder, M.J. 2002. Influence of supplementary vitamins, minerals and essential fatty acids on the antisocial behaviour of young adult prisoners. Randomised, placebo-controlled trial. Br. J. Psychiatry. 181:22–28.

Hallahan, B., Hibbeln, J.R., Davis, J.M., and Garland, M.R. 2007. Omega-3 fatty acid supplementation in patients with recurrent self-harm: Single-centre double-blind randomized controlled trial. Br. J. Psychiatry. 190:118–122.

Hessen, D.O., and Leu E. 2006. Trophic transfer and trophic modification of fatty acids in high Arctic lakes. Freshw. Biol. 51:1987–1998.

Hibbeln, J.R. 2002. Seafood consumption, the DHA content of mothers' milk and prevalence rates of postpartum depression: a cross-national, ecological analysis. J. Affect Disord. 69:15–29.

Hibbeln, J.R., Bissette, G., Umhau, J.C., and George, D.T. 2004a. Omega-3 status and cerebrospinal fluid corticotrophin releasing hormone in perpetrators of domestic violence. Biol. Psychiatry. 56:895–897.

Hibbeln, J.R., Nieminen, L.R., and Lands W.E.M. 2004b. Increasing homicide rates and linoleic acid consumption among five Western countries, 1961–2000. Lipids 39:1207–1213.

Hibbeln, J.R., Nieminen, L.R., Blasbalg, T.L., Riggs, J.A., and Lands, W.E.M. 2006a. Healthy intakes of n-3 and n-6 fatty acids: estimations considering worldwide diversity. Am. J. Clin. Nutr. 83:1483S–1493S.

Hibbeln, J.R., Ferguson, T.A., Blasbalg, T.L. 2006b. Omega-3 fatty acid deficiencies in neurodevelopment, aggression and autonomic dysregulation: opportunities for intervention. Int. Rev. Psychiatry 18:107–118.

Hibbeln, J.R., Davis, J.M., Steer, C., Emmett, P., Rogers, I., Williams, C., and Golding, J. 2007. Maternal seafood consumption in pregnancy and neurodevelopmental outcomes in childhood (ALSPAC study): an observational cohort study. Lancet 369:578–585.

Hill, E.E., Lands, W.E.M., and Slakey, P.M., Sr. 1968. The incorporation of 14C-Glycerol into different species of diglycerides and triglycerides in rat liver slices. Lipids 3:411–416.

Holman, R.T. Essential fatty acids.1958. Nutr. Rev.16:33–35.

Huynh, M.D., Kitts, D.D., Hu, C., and Trites, A.W. 2007. Comparison of fatty acid profiles of spawning and non-spawning Pacific herring, *Clupes harengus pallasi*. Comp. Biochem. Physiol. B. 146:504–511.

Iribarren, C., Markovitz, J.H., Jacobs, D.R. Jr, Schreiner, P.J., Daviglus, M., and Hibbeln, J.R. 2004. Dietary intake of n-3, n-6 fatty acids and fish: relationship with hostility in young adults – the CARDIA study. Eur. J. Clin. Nutr. 58:24–31.

Kenyon, C.N. 1972. Fatty acid composition of unicellular strains of blue-green algae. J. Bacteriol. 109:827–834.

Kiecolt-Glaser, J.K., Belury, M.A., Porter, K., Beversdorf, D.Q., Lemeshow, S., and Glaser, R. 2007. Depressive symptoms, omega-6:omega-3 fatty acids, and inflammation in older adults. Psychosom. Med. 69:217–224.

Kim, H.Y., Akbar, M., and Lau, A. 2003. Effects of docosapentaenoic acid on neuronal apoptosis. Lipids 38:453–457.

Kobayashi, M., Sasaki, S., Kawabata, T., Hasegawa, K., Akabane, M., and Tsugane, S. 2001. Single measurement of serum phospholipid fatty acid as a biomarker of specific fatty acid intake in middle aged Japanese men. Eur. J. Clin. Nutr. 55:643–650.

Kulmacz, R.J., Pendleton, R.B., and Lands, W.E.M. 1994. Interaction between peroxidase and cyclooxygenase activities in prostaglandin-endoperoxide synthase. J. Biol. Chem. 269:5527–5536.

Kuipers, R.S., Fokkema, M.R., Smit, E.N., van der Meulen, J., Boersma, E.R., and Muskiet, F.A. 2005. High contents of both docosahexaenoic and arachidonic acids in milk of women consuming fish from Lake Kitangiri (Tanzania): targets for infant formulae close to our ancient diet? Prostaglandins Leukot. Essent. Fatty Acids. 72:279–288.

Kuipers, R.S., Smit, E.N., van der Meulen, J., Janneke Dijck-Brouwer, D.A., Rudy Boersma, E., and Muskiet, F.A. 2007. Milk in the island of Chole [Tanzania] is high in lauric, myristic, arachidonic and docosahexaenoic acids, and low in linoleic acid reconstructed diet of infants born to our ancestors living in tropical coastal regions. Prostaglandins Leukot. Essent. Fatty Acids 76:221–233.

Kuriki K., Tajima, K., and Tokudome, S. 2006. Accelerated solvent extraction for quantitative measurement of fatty acids in plasma and erythrocytes. Lipids 41:605–614.

Lai, L., Kang, J.X., Li, R., Wang, J., Witt, W.T., Yong, H.Y., Hao, Y., Wax, D.M., Murphy, C.N., Rieke, A., Samuel, M., Linville, M.L., Korte, S.W., Evans, R.W., Starzl, T.E., Prather, R.S., and Dai, Y. 2006. Generation of cloned transgenic pigs rich in omega-3 fatty acids. Nat. Biotechnol. 24:435–436.

Lands, W.E.M. 2000. Stories about acyl chains. Biochim. Biophys. Acta. 1483:1–15.

Lands, W.E.M. 2003a. Diets could prevent many diseases. Lipids 18:317–321.

Lands, W.E.M. 2003b. Functional foods in primary prevention or nutraceuticals in secondary prevention? Curr. Top. Nutraceutical Res. 1:113–120.

Lands, W.E.M. 2003c. Primary prevention in cardiovascular disease: Moving out of the shadows of the truth about death. Nutri. Metabol. Cardiovascular Disease 13:154–164.

Lands, W.E.M. 2005a. Fish, omega-3 and human health, 2nd edition, AOCS Press, Champaign IL. 235 pgs.

Lands, W.E.M. 2005b. Dietary fat and health: the evidence and the politics of prevention: careful use of dietary fats can improve life and prevent disease. Ann. NY Acad. Sci. 1055:179–192.

Lands, W.E.M., Morris, A.J., and Libelt, B. 1990. Quantitative effects of dietary polyunsaturated fats on the composition of fatty acids in rat tissues. Lipids 25:505–516.

Lands, W.E.M., Libelt, B., Morris, A., Kramer, N.C., Prewitt, T.E., Bowen, P., Schmeisser, D., Davidson, M.H., and Burns, J.H. 1992. Maintenance of lower proportions of n-6 eicosanoid precursors in phospholipids of human plasma in response to added dietary n-3 fatty acids. Biochem. Biophys. Acta. 1180:147–162.

Laneuville, O., Breuer, D.K., Xu, N., Huang, Z.H., Gage, D.A., Watson, J.T., Lagarde, M., DeWitt, D.L., and Smith, W.L. 1995. Fatty acid substrate specificities of human prostaglandin-endoperoxide H synthase-1 and -2. Formation of 12-hydroxy-(9Z, 13E/Z, 15Z)-octadecatrienoic acids from alpha-linolenic acid. J. Biol. Chem. 270:19330–19336.

Liu, W., Cao, D., Oh, S.F., Serhan, C.N., and Kulmacz, R.J. 2006. Divergent cyclooxygenase responses to fatty acid structure and peroxide level in fish and mammalian prostaglandin H synthases. Fed. Amer. Soc. Exp. Biol. J. 20:1097–108.

Marean, C.W., Bar-Matthews, M., Bernatchez, J., Fisher, E., Goldberg, P., Herries, A.I., Jacobs, Z., Jerardino, A., Karkanas, P., Minichillo, T., Nilssen, P.J., Thompson, E., Watts, I., and Williams, H.M. 2007. Early human use of marine resources and pigment in South Africa during the Middle Pleistocene. Nature 449:905–908.

McGill, H.C. Jr., McMahan, C.A., Zieske, A.W., Sloop, G.D., Walcott, J.V., Troxclair, D.A., Malcom, G.T., Tracy, R.E., Oalmann, M.C., and Strong, J.P. 2000. Associations of coronary heart disease risk factors with the intermediate lesion of atherosclerosis in youth. The Pathobiological Determinants of Atherosclerosis in Youth (PDAY) Research Group. Arterioscler. Thromb. Vasc. Biol. 20:1998–2004.

McMahan, C.A., McGill, H.C., Gidding, S.S., Malcom, G.T., Newman, W.P., Tracy, R.E., and Strong, J.P. 2007. Pathobiological Determinants of Atherosclerosis in Youth (PDAY) Research Group. PDAY risk score predicts advanced coronary artery atherosclerosis in middle-aged persons as well as youth. Atherosclerosis 190:370–377.

Mohrhauer, H., and Holman, R.T. 1963a. The effect of dose level of essential fatty acids upon fatty acid composition of the rat liver. J. Lipid Res. 4:151–159.

Mohrhauer, H., and Holman, R.T. 1963b. Effect of linolenic acid upon the metabolism of linoleic acid. J. Nutr. 81:67–74.

Morais, S., Caballero, M.J., Conceicao, L.E.C., Izquierdo, M.S., and Dinis, M.T. 2006. Dietary neutral lipid level and source in Senegalese sole (*Solea senegalensis*) larvae: Effect on growth, lipid metabolism and digestive capacity. Comp. Biochem. Physiol. B. 144:57–69.

Mourente, G., and Bell, J.G. 2006. Partial replacement of dietary fish oil with blends of vegetable oils (rapeseed, linseed and palm oils) in diets for European sea bass (*Dicentrarchus labrax* L.) over a long term growth study: Effects on muscle and liver fatty acid composition and effectiveness of a fish oil finishing diet. Comp. Biochem. Physiol. B. 143:389–399.

Müller-Navarra, D.C., Brett, M.T., Liston, A.M., and Goldman, C.R. 2000. A highly unsaturated fatty acid predicts carbon transfer between primary producers and consumers. Nature 403:74–77.

Murray, C.J., and Lopez, A.D. 1997a. Global mortality, disability, and the contribution of risk factors: Global Burden of Disease Study. Lancet 349:1436–1442.

Murray, C.J., and Lopez, A.D. 1997b. Alternative projections of mortality and disability by cause 1990–2020: Global Burden of Disease Study. Lancet 349:1498–1504.

Nestle, M. 2002. Food Politics: How the food industry influences nutrition and health. University of California Press, Berkeley, CA. 469 pgs.

Newman, W.P. 3rd, Freedman, D.S., Voors, A.W., Gard, P.D., Srinivasan, S.R., Cresanta, J.L., Williamson, G.D., Webber, L.S., and Berenson, G.S. 1986. Relation of serum lipoprotein levels and systolic blood pressure to early atherosclerosis. The Bogalusa Heart Study. N. Engl. J. Med. 314:138–144.

Nieminen, L.R., Makino, K.K., Mehta, N., Virkkunen, M., Kim, H.Y., and Hibbeln, J.R. 2006. Relationship between omega-3 fatty acids and plasma neuroactive steroids in alcoholism, depression and controls. Prostaglandins Leukot Essent Fatty Acids. 75:309–314.

Niu, S.L., Mitchell, D.C., Lim, S.Y., Wen, Z.M., Kim, H.Y., Salem, N. Jr, and Litman, B.J. 2004. Reduced G protein-coupled signaling efficiency in retinal rod outer segments in response to n-3 fatty acid deficiency. J. Biol. Chem. 279:31098–31104.

Noaghiul, S., and Hibbeln, J.R. 2003. Cross-national comparisons of seafood consumption and rates of bipolar disorders. Am. J. Psychiatry. 160:2222–2227.

Otles, S., and Pire, R. 2001. Fatty acid composition of *Chlorella* and *Spirulina* microalgae species. J. Am. Oil. Chem. Soc. 84:1708–1714.

PDAY Research Group. 1990. Relationship of atherosclerosis in young men to serum lipoprotein cholesterol concentrations and smoking. A preliminary report from the Pathobiological Determinants of Atherosclerosis in Youth (PDAY) Research Group. J. Am. Med. Assoc. 264:3018–3024.

Persson, J., and Vrede, T. 2006. Polyunsaturated fatty acids in zooplankton: variation due to taxonomy and trophic position. Freshw Biol. 51:887–900.

Phleger, C.F., Nelson, M.W., Mooney, B.D., and Nichols, P.D. 2002. Comp. Biochem. Physiol. B. 131:733–747.

Raatz, S.K., Bibus, D., Thomas, W., and Kris-Etherton, P. 2001. Total fat intake modifies plasma fatty acid composition in humans. J. Nutr. 131:231–234.

Rainwater, D.L., McMahan, C.A., Malcom, G.T., Scheer, W.D., Roheim, P.S., McGill, H.C. Jr, and Strong, J.P. 1999. Lipid and apolipoprotein predictors of atherosclerosis in youth: apolipoprotein concentrations do not materially improve prediction of arterial lesions in PDAY subjects. The PDAY Research Group. Arterioscler. Thromb. Vasc. Biol. 19:753–761.

Reis, L.C., and Hibbeln, J.R. 2006. Cultural symbolism of fish and the psychotropic properties of omega-3 fatty acids. Prostaglandins Leukot. Essent. Fatty Acids. 75:227–236.

Robinson, J.G., and Stone, N.J. 2006. Antiatherosclerotic and antithrombotic effects of omega-3 fatty acids. Am. J. Cardiol. 98:39i–49i.

Rollin, X., Peng, J., Pham, D., Ackman, R.G., and Larondelle, Y. 2003. The effects of dietary lipid and strain difference on polyunsaturated fatty acid composition and conversion in andromous and landlocked salmon (*Salmo salas* L.) parr. Comp. Biochem. Physiol. B. 134:349–366.

Sanders, T.A., Gleason, K., Griffin, B., and Miller, G.J. 2006. Influence of an algal triacylglycerol containing docosahexaenoic acid (22:6n-3) and docosapentaenoic acid (22:5n-6) on cardiovascular risk factors in healthy men and women. Br. J. Nutr. 95:525–531.

Seckl, J.R., and Meaney, M.J. 2006. Glucocorticoid "programming" and PTSD risk. Ann. NY Acad. Sci. 1071:351–378.

Shields, R.J., Bell, J.G., Luizi, F.S., Gara, B., Bromage, N.R., and Sargent, J.R. 1999. Natural copepods are superior to enriched Artemia nauplii as feed for halibut larvae (*Hippoglossus hippoglossus*) in terms of survival, pigmentation and retinal morphology: Relation to dietary essential fatty acids. J. Nutr. 129:1186–1194.

Sinnhuber, RO. 1969 The role of fats, p. 245. In O.W. Neuhaus and J.E. Halver [eds], Fish in research, Academic Press, NY.

Slakey, P.M., Sr., and Lands, W.E.M. 1968. The structure of rat liver triglycerides. Lipids 3:30–36.

Smith, W.L. 2005. Cyclooxygenases, peroxide tone and the allure of fish oil. Curr. Opin. Cell Biol. 17:174–82.

Soudant, P., Van Ryckeghem, K., Marty, Y., Moal, J., Samain, J.F., and Sorgeloos, P. 1999. Comparison of the lipid class and fatty acid composition between a reproductive cycle in nature and a standard hatchery conditioning of the Pacific Oyster *Crassostrea gigas*. Comp. Biochem. Physiol. B. 123:209–222.

Stark, K.D., Mulvad, G., Pedersen, H.S., Park, E.J., Dewailly, E., and Holub, B.J. 2002. Fatty acid compositions of serum phospholipids of postmenopausal women: a comparison between Greenland Inuit and Canadians before and after supplementation with fish oil. Nutrition 18:627–630.

Stark, K.D., Beblo, S., Murthy, M., Buda-Abela, M., Janisse, J., Rockett, H., Whitty, J.E., Martier, S.S., Sokol, R.J., Hannigan, J.H., and Salem, N. Jr. 2005. Comparison of bloodstream fatty acid composition from African-American women at gestation, delivery, and postpartum. J. Lipid Res. 46:516–525.

Sublette, M.E., Hibbeln, J.R., Galfalvy, H., Oquendo, M.A., and Mann, J.J. 2006. Omega-3 polyunsaturated essential fatty acid status as a predictor of future suicide risk. Am. J. Psychiatry. 163:1100–1102.

Tinoco, J., Williams, M.A., Hincenbergs, I., and Lyman, R.L. 1971. Evidence for nonessentiality of linolenic acid in the diet of the rat. J. Nutr. 101:937–45.

Tocher, D.R., Dick, J.R., MacGlaughlin, P., Bell, J.G. 2006. Effect of diets enriched in delta-6 desaturated fatty acids (18:3n-6 and 18:4n-3), on growth fatty acid composition and highly unsaturated fatty acid synthesis in two populations of Arctic charr (*Salvelinus alpinus* L.). Comp. Biochem. Physiol. B. 144:245–253.

Tokudome, Y., Kuriki, K., Imaeda, N., Ikeda, M., Nagaya, T., Fujiwara, N., Sato, J., Goto, C., Kikuchi, S., Maki, S., and Tokudome, S. 2003. Seasonal variation in consumption and plasma concentrations of fatty acids in Japanese female dietitians. Eur. J. Epidemiol. 18:945–53.

US Department of Health and Human Services, US Environmental Protection Agency. What you need to know about mercury in fish and shellfish. EPA and FDA advice for women who might become pregnant, women who are pregnant, nursing mothers, and young children. March, 2004: http://www.cfsan.fda.gov/~dms/admehg3.html (accessed April 5, 2007).

van Gelder, B.M., Tijhuis, M., Kalmijn, S., and Kromhout, D. 2007. Fish consumption, n-3 fatty acids, and subsequent 5-y cognitive decline in elderly men: the Zutphen Elderly Study. Am. J. Clin. Nutr. 85:1142–1147.

Velosa, A.J., Chu, F-L.E., and Tang, K.W. 2006. Trophic modification of essential fatty acids by heterotrophic protists an dits effects on the fatty acid composition of the copepod, *Acartia tonsis*. Marine Biol. 148:779–788.

Verschuren, W.M., Jacobs, D.R., Bloemberg, B.P., Kromhout, D., Menotti, A., Aravanis, C., Blackburn, H., Buzina, R., Dontas, A.S., Fidanza, F., Karvonen, M.J., Nedeljkovic, S., Nissinen, A. and Toshima, H. 1995. Serum total cholesterol and long-term coronary heart disease mortality in different cultures. Twenty-five-year follow-up of the seven countries study. J. Am. Med. Assoc. 274:131–6.

Wada, M., DeLong, C.J., Hong, Y.H., Rieke, C.J., Song, I., Sidhu, R.S., Yuan, C., Warnock, M., Schmaier, A.H., Yokoyama, C., Smyth, E.M., Wilson, S.J., FitzGerald, G.A., Garavito, R.M., Sui, X., Regan, J.W., and Smith, W.L. 2007. Enzymes and receptors of prostaglandin pathways with arachidonic acid-derived versus eicosapentaenoic acid-derived substrates and products. J. Biol. Chem. 282:22254–22266.

Walter, R.C., Buffler, R.T., Bruggemann, J.H., Guillaume, M.M., Berhe, S.M., Negassi, B., Libsekal, Y., Cheng, H., Edwards, R.L., von Cosel, R., Neraudeau, D., and Gagnon, M. 2000. Early human occupation of the Red Sea coast of Eritrea during the last interglacial. Nature 405:65–69.

Wen, Z-Y., and Chen, F. 2002.Continuous Cultivation of the diatom *Nitzschia laevis* for eicosapentaenoic acid production: Physiological study and process optimization. Biotechnol. Prog. 18:21–28.

Wen, Z., and Kim, H.Y. 2007. Inhibition of phosphatidylserine biosynthesis in developing rat brain by maternal exposure to ethanol. J. Neurosci. Res. 85:1568–1578.

Wilber, C.G., and Levine, V.E. 1950. Fat metabolism in Alaskan eskimos. Exp. Med. Surg. 8:422–425.

Zieske, A.W., Malcolm, G.T., and Strong, J.P. 2002. Natural history and risk factors of atherosclerosis in children and youth: the PDAY study. Pediatr. Pathol. Mol. Med. 21: 213–237.

Name Index

A
Abba, C., 199
Abbadi, A., 231
Abrusán, G., 138, 159, 160
Acharya, K., 157, 158, 173
Ackefors, A., 160
Ackman, R.G., 116, 258, 284, 286, 287, 289, 290, 292
Adams, S.M., 115, 240
Adlerstein, D., 13
Adolf, J.E., 34, 36, 112
Adolph, S., 72, 77
Adrian, R., 33–35, 51
Agaba, M., 223
Ahlgren, G., 28, 33, 103, 106, 115, 117, 122, 123, 150, 154, 156, 157, 159, 160, 170, 172, 178, 318, 320
Akbar, M., 335, 336
Aki, T., 222
Aktas, H., 180
Al-Fadhli, A., 5
Albers, C.S., 261, 271
Alekseev, V., 78
Alimov, A.F., 186, 189, 194, 195, 197, 198
Alimuddin, Y.G., 231
Alkanani, T., 319, 321
Alonso, D.L., 10, 17
Amblard, C., 33, 126, 128
Andersen, R.J., 5
Anderson, R.O., 237
Anderson, T.R., 312, 320
Andersson, M.X., 18
Andrikovics, S., 196
Antia, N.J., 27
Appleby, R.S., 32
Arendt, K.E., 313, 332
Arhonditsis, G.B., 129
Arisz, S.A., 3, 17
Arndt, H., 26, 27

Arts, M.T., 36, 115, 139, 148, 180, 231, 243, 270, 312
Atkinson, A., 133, 275
Auel, H., 95, 258, 264
Aukema, H.M., 316
Avery, S.V., 35
Azachi, M., 15, 16
Azam, F., 27

B
Balfry, S., 245, 249
Balk, E.M., 312
Ballantine, J.A., 57
Ballantyne, A.P., 118, 128
Ballinger, A., 194, 201
Båmstedt, U., 261
Ban, S.H., 79, 82
Barata, C., 332
Barclay, W.R., 215
Barofsky, A., 75
Barreiro, A., 78
Barrett, S.M., 51
Barrett, X., 336
Baxter, C.V., 194
Baydoun, M.A., 338
Beamish, R.J., 192
Beauchamp, D.A., 99
Beaudoin, F., 222
Bec, A., 28, 29, 32, 34, 35, 37, 46, 48, 52, 56, 94
Beck, C.A., 301, 302
Becker, C., 149
Behmer, S.T., 49, 51, 55, 56, 58, 59
Behrouzian, B., 30
Bell, J.G., 154, 155, 160, 245, 248, 249, 314, 315, 319
Bell, M.V., 70, 147–149, 154, 213, 219, 221, 229, 230

Bence, (2003), 244
Benesh, D.P., 195
Benjamin, (2003), 244
Benning, C., 18
Bergen, B.J., 106
Bernays, E.A., 56
Bettarel, Y., 26
Bettger, W., 316
Bigogno, C., 9, 10, 13, 332
Bisseret, P., 7
Bjarnov, N., 162
Black, E.C., 336
Blair, T.A., 182
Blazer, V.S., 245, 250
Boëchat, I.G., 28, 32–34, 35, 48, 51
Boersma, M., 149, 157, 182
Boland, W., 66, 71, 74
Borgå, K., 95, 96, 99, 100, 104, 106
Bottrell, H.H., 79
Bourdier, G., 33, 126, 128
Bowden, L.A., 242
Bowen, W.D., 301, 302
Bowman, J.P., 212
Bradford-Grieve, J.M., 269
Bradow, J., 83
Bransden, M.P., 248
Brett, M.T., 2, 28, 33, 117, 121, 123, 126, 127, 136, 149, 155, 156, 159, 160, 244
Bricelj, V.M., 81
Brindley, D.N., 285
Broadhurst, C.L., 148, 180
Brockerhoff, H., 243
Broglio, E., 34, 35, 37, 123, 320
Broman, D., 104
Brooks, S., 241
Brown, M.R., 14
Browse, J., 29, 219, 222
Brugerolle, G., 26
Buckman, A.H., 97, 100
Buda, C.I., 241, 242
Budge, S.M., 283–287, 288, 290, 292–298, 299, 303, 320
Burkholder, J.M., 26, 27
Burr, G.O., 310
Burr, M.M., 310
Burreau, S., 97
Burton, T.M., 201
Bury, N.R., 70
Buydens-Branchey, L., 335
Buzzi, M., 221

C

Caballero, M.J., 248
Cabana, G., 101, 104

Calbet, A., 36
Calderon, F., 335
Caldwell, G.S., 69, 77, 79, 80
Callieri, C., 28
Camazine, S.M., 77
Campbell, L.M., 101, 102, 104
Campbell, R.W., 258
Campfens, J., 95
Canfield, D.E., 195
Canuel, E.A., 318
Capuzzo, J.M., 55
Caramujo, M.-J., 123, 128
Careaga-Houck, M., 316
Caron, D.A., 28, 36
Carotenuto, Y., 78, 80, 81
Carrias, J-F., 27
Casotti, R., 69, 76, 80
Castell, J.D., 48, 123, 131, 136, 139, 148, 149, 219, 312, 319, 320, 336
Castellini, M.A., 286
Cavill, G.W.K., 77
Cembella, A.D., 67
Champagne, D.E., 56
Chapman, P.J., 46
Charnov, E.L., 82
Chatton, E., 25
Chaudron, Y., 80, 82
Chauvet, E., 46
Chen, W.H., 79, 342
Chew, E.Y., 180
Chiang, I.Z., 70
Cho, H.P., 222, 228
Christie, W.W., 287
Christophersen, B.O., 286
Chu, F.L.E., 106, 214
Chuecas, L., 71
Ciapa, B., 314
Clandinin, M.T., 314
Clare, A.S., 71
Clarke, K.R., 296
Clegg, M.T., 328
Cohen, Z., 19, 181
Colby, R.H., 287
Conklin, D.E., 60
Connel, D.W., 98
Conner, R.L., 48, 54
Connick, J.W., 83
Connor, S.L., 338
Connor, W.E., 338
Conover, R.J., 259, 260
Cook, A., 83
Cook, H.W., 65, 282, 283
Cooksey, K.E., 16
Coolbear, K.P., 241
Cooper, M.H., 286, 288, 300, 303

Name Index

Cooper, R.A., 242
Copeman, L.A., 106, 153, 180, 313, 319
Cornils, A., 269
Countway, R.E., 105
Coutteau, P., 131
Crawford, M.A., 147–149, 160
Cripps, C., 219
Cripps, G.C., 120, 133, 290
Crockett, E.L., 53, 58
Csengeri, I., 116
Cunnane, S.C., 94, 179, 310, 312
Cuthbertson, W.F.J., 346
Cutignano, A., 72

D

D'Adamo, R., 97, 98
D'Agostino, A., 117
D'Andrea, S., 221
D'Ippolito, G., 3, 72, 73
D'Abramo, L.R., 126
Dahl, C.E., 47
Dahl, J., 160
Dalsgaard, J., 95, 115, 117, 118, 133, 156, 249, 257, 258, 262, 266, 283, 284, 286, 289, 290, 292, 294, 301, 309, 320
Damude, H.G., 181
Danielsdottir, M.G., 182
Darimont, C.T., 191
Davis, B.C., 183
Dawczynski, C., 213
De Antueno, R.J., 221
Dembitsky, V.M., 7, 48
DeMott, W.R., 105, 155, 158, 182
Derenbach, J.B., 83
Desvilettes, C., 34, 35, 123, 219, 302
Dey, I., 242
Di Marzo, V., 71
Dick, J.R., 221, 230
Dick, T.A., 154
Dicke, M., 66
Dietz, R., 101
Diez, B., 27
Dittberner, U., 83
Dittmar, T., 273
Dobey, S., 190
Dolan, J.R., 36
Domergue, F., 29–31
Dong, F.M., 245
Dower, J.F., 258
Downing, J.A., 198
Dulloo, A.G., 315
Dunstan, G.A., 29, 123, 213, 283, 289, 312

Dürst, U., 73
Dutz, J., 78, 80

E

Eaton, C.A., 116, 287
Ederington, M.C., 59, 103, 138
Eichenberger, W., 5, 7, 10
Einarsson, A., 195
Einicker-Lamas, M., 20
El-Sheek, M.M., 18
Eldho, N.V., 317, 319
Elendt, B.-P, 126, 149, 157
Elias, D.O., 49, 55
Eltgroth, M.L., 11
Elvidge, J., 83
Engstrom-Ost, J., 106, 333
Enos, W.F., 338
Erwin, J.A., 29, 32, 34, 35
Estévez, A., 244
Evans, R.P., 317
Evjemo, J.O., 106

F

Fabregas, J., 13
Falk-Petersen, S., 132, 259, 261, 274
Farkas, T., 116–119, 139, 241, 271, 317
Farrell, D.J., 182
Fauré-Fremiet, E., 27
Fedorova, I., 182
Felicetti, L.A., 190
Feller, S.E., 240
Ferguson, H.W., 250
Feussner, I., 71, 73
Findlay, J.A., 70
Finizio, A., 97
Fink, P., 65, 81
Fisher, N.S., 95, 101, 105, 106
Fisk, A.T., 97–101
Flato, G.M., 273
Fleurence, J., 200
Floreto, E.A.T., 15, 17
Flynn, K.J., 67
Fodor, E., 242, 316
Fogg, G.E., 28
Folt, C.L., 79
Fontana, A., 78
Fox, K., 97, 98
Fracalossi, D.M., 249
Fraser, A.J., 126, 268, 271
Freeman, M.P., 337, 338
Frolov, A.V., 51
Fu, M., 67

G

Ganf, G.G., 54
Gao, K., 12, 20
Garcia, A.S., 319, 320
Gardarsson, A., 195
Gardner, W.S., 161
Garg, M.L., 199
Garland, M.R., 337
Gatenby, C.M., 123
Gatten, R.R., 260, 267
Geiser, F., 182
Gende, S.M., 191
Gere, G., 196
Gerster, H., 179
Gerwick, W.H., 71, 319
Gesch, C.B., 335
Gessner, M.O., 46
Ghioni, C., 221
Gifford, D.J., 28
Gilbert, L.I., 59
Giner, J.-L., 46, 48, 51, 55
Gladyshev, M.I., 182, 191, 192
Glasgow, H.B.J., 26
Glencross, B.D., 248
Goad, L.J., 44, 49
Gobas, F.A.P.C., 96, 97, 99
Gockel, G., 30
Goedkoop, W., 160, 161
Gombos, Z., 20
Goodnight, S.H., 314, 315
Goulden, C.E., 93, 141
Gouy, M., 225
Gradinger, R., 273
Graeve, M., 95, 117, 118, 130, 133, 137, 258, 261, 266, 276
Green, K.H., 199
Greenfield, B.K., 99
Gribi, C., 5
Grieneisen, M.L., 47, 49, 59
Groscolas, R., 286
Guckert, J.B., 16
Guisande, C., 60, 79
Gulati, R.D., 157, 182
Gundersen, P., 101
Gunstone, F.D., 4
Gurr, M.I., 1, 283, 310–312, 315–317
Guschina, I.A., 1, 3, 11, 18, 95

H

Hachtel, W., 30
Haeckel, E., 25
Hagen, W., 95, 99, 118, 120, 132, 141, 258, 259, 261, 262, 264, 266, 267, 271, 273, 274, 283
Hagve, T.A., 240
Haigh, W.G., 9
Haines, T.H., 47, 55, 242, 243
Hairston, N.G., 81
Hall, J.M., 240, 271, 317
Hallahan, B., 337
Hallaq, H., 314
Halperin, J.A., 180
Halver, J.E., 116
Hansen, A.S., 274
Hansen, E., 77, 82
Hansen, L.R., 70
Hanson, B.J., 181
Harrington, G.W., 258
Harris, R.P., 213, 219
Harris, W.S., 312, 313
Harrison, K.E., 70, 147, 149
Härtel, H., 18
Harvey, H.R., 48, 51, 54, 55, 58
Harwood, J.L., 1, 3–5, 9, 11, 13, 95, 181, 283, 310–312, 315–317
Hashimoto, K., 29
Hashimoto, M., 240
Hassett, R.P., 50, 51, 53
Hastings, N., 222, 223
Hauvermale, A., 214, 215
Hawker, D.W., 98
Hayes, K.C., 311
Hazel, J.R., 139, 240, 317
He, K., 312
Heath, M.R., 266
Hebert, C.E., 99, 104, 181, 183, 184, 243, 303
Heinz, E., 181
Helfield, J.M., 190–192
Henderson, R.J., 118, 119, 147, 154, 218, 257, 283, 284
Hennessey, T.M., 244
Herodek, S., 116, 139
Herzberg, G.R., 314
Hessen, D.O., 28, 120, 155, 156, 332
Hibbeln, J.R., 180, 333, 334, 337, 338, 346
Higgs, D.A., 155, 245, 249
Hilderbrand, G.V., 184, 191
Hill, E.E., 340
Hill, H.J., 290
Hinterberger, H., 77
Hirche, H.-J., 259, 260, 274
Hirota, J., 257, 258, 264, 268

Name Index

Hobson, K.A., 302
Hoffman, L.C., 182
Holey, M.E., 244
Holman, R.T., 336, 341, 346
Holub, B.J., 316
Hombeck, M., 73, 74
Hong, S., 317, 319
Hooper, L., 312
Hooper, S.N., 290
Hop, H., 99
Hopkins, C.C.E., 261
Hopkins, T.L., 264
Hossain, M.S., 240
Howard, R.W., 181, 313, 315
Hoyer, M.V., 195, 196
Hoyle, R.J., 243
Hu, H., 20
Huber, T., 240
Hulbert, A.J., 199
Hull, M.C., 241
Huntley, M.E., 81, 259
Hurd, S.D., 199, 200

I

Ianora, A., 67, 72, 77, 79
Ikawa, M., 70
Inagaki, K., 222, 226
Incardona, J.P., 98
Iribarren, C., 335
Irigoien, X., 67, 77
Iverson, S.J., 95, 281–283, 285–303, 309
Izquierdo, M.S., 248, 318

J

Jeckel, W.H., 70
Jeffries, H.P., 116
Jensen, J., 79
Jiang, H., 12
Jiang, Y.-H., 316
Jobling, M., 136, 248
Joh, T., 12, 13
Johns, R.B., 212
Johnson, M.D., 26
Johnston, T.A., 100
Jónasdóttir, S.H., 79, 117, 258, 260
Jones, A.L., 3
Jones, R.H., 67, 78, 80
Jordal, A.-E.O., 228, 229
Joseph, J.D., 213, 217, 284
Jøstensen, J.P., 212
Jüttner, F., 65, 66, 69, 70, 72–74, 81–83, 106

K

Kainz, M., 65, 95, 101, 103–105, 106, 128, 150, 156, 184, 191, 312, 318, 320
Kajikawa, K.T., 29–32
Kamiya, H., 70
Kanazawa, A., 48
Kaneda, T., 103
Kaneshiro, E.S., 33, 34
Kang, J.X., 311
Katan, M.B., 154
Kato, M., 7
Katoh, S., 67
Kattner, G., 130, 258–262, 264, 265, 267, 268, 273, 275, 276, 283
Kelly, A.M., 240, 248
Kelly, B.C., 97
Kelly, E.N., 95
Kennleyside, K.A., 99
Keusgen, M., 3
Kharlamenko, V.I., 32, 34, 35, 303
Khotimchenko, S.V., 14
Khozin-Goldberg, I., 9, 11, 13, 19, 30
Kidd, K.A., 99, 100
Kieber, R., 67
Kiiskinen, T., 287
Kim, H.Y., 335, 336
Kinney, A.J., 181
Kiørboe, T., 79
Kiron, V., 93
Kirsch, P.E., 286, 287, 293
Kitajka, K., 240
Klein Breteler, W.C.M., 28, 37, 48, 51, 56, 57, 94, 312, 320
Klem, A., 283
Kleppel, G.S., 79
Kohler, C.C., 240, 247, 248
Konwick, B.J., 96, 97
Koopman, H.N., 284, 285, 288, 296
Koski, M., 67, 77
Kotani, Y., 267, 275
Koussoroplis, A.M., 148, 181, 184
Kraemer, L.D., 101
Krause, M., 260, 262, 264, 267, 268
Kreutzer, C., 182
Kris-Etherton, P.M., 183, 312
Kroon, B.M.A., 15
Kubo, I., 77
Kuipers, R.S., 333
Kulmacz, R.J., 342
Kumon, Y., 319
Kwon, T.D., 97

L

Laabir, M., 79, 80
Laakmann, S, 264, 265
Lai, L., 348
Lake, P.S., 194
Lall, S.P., 315
LaMontagne, J.M., 79
Lampert, W., 53, 78, 81
Landfald, B., 212
Landry, M.R., 36, 37
Lands, W.E.M., 246, 247, 249, 333, 340–344, 346
Lane, R.L., 248, 249
Laneuville, O., 342
Langdon, C.J., 126
Lauritzen, L., 180
Laybourn-Parry, J., 27, 36
Leaf, A., 314
Leblond, J.D., 46, 48
Lee, R.F., 117, 118, 128, 130, 132, 141, 257, 258, 260, 261, 264, 267–269
Leeper, D.A., 70
Lefèvre, E., 27
Legendre, L., 36
Legrande, C., 70
Lenz, P.H., 119
Leonard, A.E., 181, 182, 222, 226
Leppimäki, P., 58
Leu, E., 120, 155, 332
Levine, V.E., 339
Lewis, C., 77
Lewis, H.A., 248
Lewis, R.W., 116, 126
Li, M.H., 250
Li, W.K.W., 27
Likens, G.E., 201
Likhodii, S.S., 312
Lim, S.-Y., 182, 183
Lin, Y.-H., 249
Lischka, S., 258, 259, 264, 266
Litzow, M., 309, 312
Liu, W., 342
Lodemel, J.B., 250
Loeb, V., 275
Lopez, A.D., 333
Los, D.A., 238
Loureiro, A.P.M., 80
Lourenco, S.O., 123
Lovell, R.T., 249
Lovern, J.A., 116, 283
Lozano, R., 48
Lu, C., 182
Lubzens, E., 219
Lund, E.K., 240
Lund, T., 290
Lürling, M., 159
Lynch, M., 58
Lynn, S.G., 17

M

Maazouzi, C., 160
Maberly, S.C., 197
Mackay, D., 95, 96, 97
Mackinlay, E.E., 218
MacLatchy, D.L., 52
Madenjian, C.P., 244
Makewicz, A., 11
Makhutova, O., 160
Mann, D., 83
Manning, B.B., 248
Mansour, M.P., 32
Marean, C.W., 328
Marsh, D., 148, 149
Marshall, C.T., 309
Marsot, P., 81
Martin-Creuzburg, D., 28, 48–55, 58, 59, 95, 137
Marty, J.C., 105, 283, 312
Mason, R.P., 105
Masuda, R., 244, 313
Matsui, K., 76
Matsutani, T., 71
Mauchline, J., 267, 269
Mayor, D., 260
Mazumder, A., 103
McAdam, A.G., 199
McCauley, E., 79
McConville, M.J., 290
McDowell Capuzzo, J., 28
McGill, H.C., 339
McGrattan, D., 67
McIntyre, J.K., 99, 100
McKean, M.L., 44
McLachlan, J., 289
McLarnon-Riches, C.J., 12, 15
McMahan, C.A., 338, 339
McManus, G.B., 28, 48
McMeans, B.C., 101
Meaney, M.J., 335
Meijer, L., 71
Mellish, J.E., 286
Merris, M., 48
Merritt, J.F., 199
Metz, C., 264, 266
Metz, J.G., 30, 214, 215, 318
Meyer, A., 31, 32, 213, 225
Michaelson, L.V., 222

Mignot, J-P., 30
Milke, L.M., 160, 313, 319, 320
Millar, J.S., 199
Miller, C.B., 267
Mills, G.L., 181
Miralto, A., 65, 67, 69, 72, 77, 78, 81, 82, 93
Mitchell, T.W., 182
Mock, T., 15
Moffat, C.F., 243
Mohrhauer, H., 341, 346
Montero, D., 250, 311, 313
Moreau, R.A., 49, 56, 57
Morel, F.M.M., 96
Morgan-Kiss, R.M., 11
Moriguchi, T., 313, 319, 320
Morimoto, M., 77
Morrison, H.A., 96
Mourente, G., 131, 223
Mozaffarian, D., 312
Müller, H., 73
Müller, M., 26
Müller-Navarra, D.C., 2, 28, 54, 93, 95, 105, 108, 115, 117, 123, 124, 126, 127, 129, 135–138, 149, 155, 158, 160, 182, 183, 244, 312, 332
Mumm, N., 260
Muradyan, E.A., 19, 20
Murakami, M., 70
Murata, M., 290
Murata, N., 12, 20, 238
Muriana, F.J.G., 317
Murray, C.J., 333

N
Naiman, R.J., 190, 191
Nakashima, S., 29, 33
Nalepa, T.F., 243
Nanton, D.A., 123, 131, 136, 139, 219, 312, 319, 320
Napier, J.A., 222
Napolitano, G.E., 14, 104, 289, 309
Navas, J.M, 152
Nelson, G.J., 284, 286
Nelson, J.S., 223
Nes, W.D., 49
Nes, W.R., 44
Nestle, M., 346
Neumann, R.M., 237
Newman, M.C., 95, 101
Newman, W.P., 338
Nichols, B.W., 32
Nichols, D.S., 213

Nichols, P.D., 213
Nicholson, G.L., 238
Nicol, S., 259
Nieminen, L.R., 334
Niu, S.L., 335
Noaghiul, S., 338
Nomura, T., 71
Nor Aliza, A.R., 181
Nordstrom, C.A., 299, 300, 303
Normén, L., 57
Norrbin, M.E., 258, 264, 266, 271
Norseth, J., 286
Norsker, N.H., 131
Nosawa, Y., 35

O
O'Hagan, D., 218
O'Neal, C.C., 248
Oftedal, O.T., 288
Ohvo-Rekilä, H., 47, 242
Okamoto, T., 67
Okuyama, H., 213
Olive, P.J.W., 82
Oliver, R.L., 54
Olsen, Y., 115, 148, 155
Opperhuizen, A., 106
Orihel, D.M., 104
Osmundsen, H., 286
Otwell, W.S., 154
Ourisson, G., 44

P
Paffenhöfer, G.A., 67, 72, 266
Palmtag, M.R., 131
Paradis, M., 292
Park, H.G., 320
Park, J.S., 26
Parrish, C.C., 30, 106, 126, 160, 287, 312, 313, 315, 316, 318–320
Parry, J., 27
Pascal, J.C., 284
Paterson, G., 100
Paterson, S., 96
Patil, A.D., 70
Patil, V., 123
Patterson, G.W., 1, 44, 52
Pauly, D., 192
Pawlosky, R.J., 183
Pereira, S.L., 105
Pernet, F., 319, 320
Perrière, G., 225
Perry, G.J., 212

Persson, J., 103, 106, 117–120, 122, 129, 134, 150, 155, 156, 158, 171, 332
Pesando, D., 83
Peters, J., 120, 130, 258, 264, 265, 268
Petersen, W., 269, 270
Pickhardt, P.C., 105, 106, 108
Pickova, J., 154, 314
Piironen, V., 49, 56
Place, A.R., 93, 141
Plourde, M., 179
Poerschmann, J., 17, 34, 35
Pohnert, G., 67, 71–75, 83
Polis, G.A., 199, 200
Politia, L., 183
Pomeroy, L.R., 27
Pond, C.M., 282
Pond, D.W., 133, 213, 219, 312, 320
Popov, S., 48
Porter, J.A., 47
Porter, K.G., 70
Post, D.M., 318
Poulet, S.A., 77, 79–81, 82
Prahl, F.G., 51, 58
Pronina, N.A., 19
Provasoli, L., 60, 117
Pruitt, N.L., 182

Q
Qiu, X., 214, 215
Quinn, T.P., 192

R
Raclot, T., 286
Rady, A.A., 18
Raederstorff, D., 48
Rainwater, D.L., 338, 339
Rangan, V.S., 105
Rasmussen, J.B., 100, 101
Ratledge, C., 29
Raven, J.A., 197, 198
Ravet, J.L., 115, 149, 159, 160
Rea, L.D., 286
Regnault, A., 17
Regost, C., 248
Reimchen, T.E., 191, 192
Reinikainen, M., 70
Reis, L.C., 338
Reitan, K.I., 18, 123
Renaud, S.M., 12, 123
Rengefors, K., 70
Ribalet, F., 76, 77

Richards, W.L., 154
Richardson, A.J., 274
Richardson, K., 260
Richoux, N.B., 120
Richter, C., 264
Riekhof, W.R., 30
Riley, J.P., 71
Rimm, E.B., 312
Rivkin, R.B., 36
Robert, S.S., 181, 231
Robin, J.H., 248
Robinson, J.G., 338
Rogerson, M., 314
Rohmer, M., 46, 48
Romano, G., 77, 80
Rouvinen, K., 287
Rowley, A.F., 245
Ruiz-Gutierrez, V., 317
Rukmini, C., 70
Runge, J.A., 274
Russell, N.J., 213
Russell, R.W., 95
Ruxton, C.H.S., 311, 312

S
Sabelis, M.W., 66
Saito, H., 267, 275, 290
Sakshaug, E., 259
Sanchez-Machado, D.I., 213
Sanchez-Puerta, M.V., 30
Sanders, R.W., 28
Sandgren, C., 67, 78, 83
SanGiovanni, J.P., 180
Sargent, J.R., 30, 33, 106, 115, 118, 147–150, 152, 221, 222, 231, 244, 245, 249, 257, 258, 283–285, 286, 301, 309, 311, 314, 315, 318–320
Satchwill, T., 77
Sato, N., 13, 18, 19
Sayanova, O., 29–31, 33
Schaller, H., 47
Schlechtriem, C., 139, 140, 161, 202, 219, 245, 286
Schmidt, K., 120
Schnack, S.B., 264
Schnack-Schiel, S.B., 259, 261, 264
Schnitzler, I., 74, 77
Schoeman, D.S., 274
Schuel, H., 71
Schwarzenbach, R.P., 97
Scott, F.J., 26
Scott, C.L., 99, 118, 119, 130, 260
Scott, G.R., 95

Name Index

Scott, K.M., 180
Scutt, J., 67
Seckl, J.R., 335, 336
Seiliez, I., 222, 223
Sekiguchi, H., 30
Sellem, F., 70
Serreze, M.C., 273
Sewell, R.B.S., 264
Shahidi, F., 311, 312
Shearer, K.D., 248
Sheen, S.-S, 126
Sheldon, W.M. Jr., 250
Sherr, B.F., 26–28, 37
Sherr, E.B., 26–28, 37
Shiau, S.-Y., 249
Shields, R.J., 313, 319, 332
Shilo, M., 70
Shim, Y.-H., 48
Shimizu, Y., 218
Shinitzky, M., 271
Shorland, F.B., 181
Shortreed, K.S., 28
Siegel, V., 275
Sifferd, T.D., 260
Silva, C.J., 48
Silvers, K.M., 180
Simopoulos, A.P., 115, 180, 181, 231, 311, 312
Sinensky, M., 238, 317
Singer, S.J., 238
Sinnhuber, R.O., 336
Skoglund, R.S., 99
Slakey, P.M. Sr., 341
Sloman, K.A., 95
Smetacek, V., 259
Smith, A.D., 241
Smith, R.J., 291
Smith, S.L., 105, 259
Smith, W.L., 342
Smyntek, P.M., 119, 120, 129
Snyder, R.J., 244
Son, B.W., 5
Soudant, P., 48
Southgate, D.A.T, 245
Sprecher, H., 217, 220, 316
St. John, M.A., 290
Stagliano, D.M., 194
Stampfl, P., 159
Stanley, D.W., 182, 313, 315
Stanley-Samuelson, D.W., 71, 149, 181, 182
Stapp, P., 199, 200, 318
Stevens, C.J., 130
Stillwell, W., 239, 240, 271
Stirling, I., 302
Stockner, J.G., 27, 28
Stoecker, D.K., 25, 28, 55
Stone, N.J., 338
Støttrup, J.G., 79, 131
Straile, D., 36
Stubbs, C.D., 241, 316, 317
Stübing, D., 120, 132, 134, 141, 273
Sublette, M.E., 337
Sukenik, A., 10
Sul, D., 34, 35
Sullivan, T.P., 199
Summons, R.E., 44, 46
Sun, M.-Y., 105
Sushchik, N.N., 12, 106, 160, 183, 194
Suter, W., 195
Swackhamer, D.L., 99
Szepanski, M.M., 191

T

Taal, M., 312
Taglialatela-Scafati, O., 5
Takagi, M., 16
Takizawa, E., 13
Tanabe, S., 95
Tande, K., 261
Tang, K.W., 312
Tatsuzawa, H., 13, 16
Taylor, F.J.R., 70
Teshima, S.-I., 15, 49
Thiemann, G.W., 288, 292
Thomann, R.V., 97, 99
Thompson, G.A.J., 9, 11, 12, 16, 17, 19, 35, 46, 314, 317
Thompson, S.N., 161
Thrush, M.A, 152
Tinoco, J., 336
Tocher, D.R., 93–95, 181, 221–223, 229–231, 245, 311
Toivonen, L.V., 311
Tollit, D.J., 299
Tonon, T., 29, 213
Torres, J.J., 264
Torres-Ruiz, M., 149, 160
Torstensen, B.E., 248
Tort, L., 246
Tosti, E., 77
Trautwein, E.A., 56, 58
Tremblay, R., 123
Tremolieres, A., 7
Trider, D.J., 48
Tripodi, K., 29–32
Trombetta, D., 80
Trubetskova, I., 53

Trudel, M., 100
Trueman, R.J., 243
Trushenski, J.T., 245, 247
Tsitsa-Tzardis, S.E., 52
Tsubo, Y., 83
Tsukamoto, K., 313
Tucker, S., 303

U
Ulbricht, T.L.V., 245
Urich, K., 315–317
Uscian, J.M., 181

V
Van der Kraak, G., 52
Van Donk, E., 159
van Gelder, B.M., 338
Van Horne, B., 199
Van Pelt, C.K., 32
Van Vliet, T., 154
van Wezel, A.P., 106
Vance, D.E., 283
Vance, J.E., 283
Vanderploeg, H.A., 162, 178
Vander Zanden, M.J., 318
Vanni, M.J., 191
VanWagtendonk, W.J., 48
Vardi, A., 76
Veloza, A.J., 35, 120, 123, 332
Venegas-Calerón, M., 29, 214
Vera, A., 32, 34–36
Verheye, H.M., 270
Véron, B., 46, 49, 59
Verschuren, W.M., 344, 345
Vismara, R., 131
Viso, A.C., 105, 283, 289, 312
Visser, A.W., 258
Volkman, J.K., 1, 29, 44, 46, 57, 105, 106, 115, 122, 123, 213, 217
Von Elert, E., 50, 54, 59, 95, 108, 149, 157–159, 161, 162
Voogt, P.A., 48
Vørs, N., 27
Voss, A., 147, 160
Vrede, T., 103, 117–120, 122, 129, 150, 155, 171, 332

W
Wacker, A., 58, 123, 137, 161, 162, 176–178, 313
Wada, M., 343
Wakeham, S.G., 105

Waldock, M.J., 126
Walker, G., 71
Wallis, J.G., 29, 32
Walter, R.C., 328
Walton, W., 67
Wamberg, S., 288
Wang, S.W., 292
Wang, W.X., 95, 105
Ward, O.P., 71
Ward, P., 264, 267
Wassall, S.R., 239, 240, 242, 271
Wasternack, C., 71
Watson, S.B., 66, 69, 70, 72, 77, 81–83
Watts, J.L., 219, 222
Wee, J.L., 77
Weers, P.M.M., 126, 157
Weissburg, M.J., 155
Weisse, T., 27
Wen, Z., 336
Wendel, T., 65, 72, 73
Wetzel, R.G., 197
Wichard, T., 69, 73, 76–80, 82, 84
Wickham, S.A., 28
Wieltschnig, C., 27
Wijendran, V., 311
Wilber, C.G., 339
Williams, B.L., 46, 48
Williams, E.E., 240, 317
Willson, M.F., 190, 191
Winder, M., 191, 192
Wolff, R.L., 181
Wolffrom, T., 149, 159
Wong, C.S., 97
Wonnacott, E.J., 248
Wood, J.D., 182
Wright, D.C., 57
Wu, J.T., 67
Wu, P., 273

X
Xu, X., 149

Y
Yakovleva, I.M., 14
Yamada, N., 67
Yang, X., 154
Yano, Y., 212
Yasumoto, T., 70
Yazawa, K., 214
Yen, J., 155
Yeo, Y.K., 316
Yongmanitchai, W., 71

Z

Zenebe, T., 152, 154
Zheng, X., 222, 223, 228, 229
Zhu, C.J., 12
Zhukova, N.V., 32, 34, 35
Zieske, A.W., 338
Zsigmond, E., 316

Subject Index

A
Adaptation mechanisms, 13
Adaptive immune response, 246–247
ALA. *See* α-Linolenic acid
Algal lipids
    environmental effects
        growth conditions, 11
        light, 13–15
        pH, 16
        salinity, 15–16
        temperature, 11–13
    nonpolar glycerolipids, 9–11
    nutrient regimes
        carbon availability, 19–20
        limitations, 16–18
        phosphorus and sulphur effects, 18–19
    polar glycerolipids
        betaine lipids, 7–9
        glycolipids, 2–5
        phospholipids, 5–7
α-Linolenic acid (ALA), 179
    Acartia tonsa, 320
    cryptophytes, 124
    cyano-mix diet, 159
    definition, 310
    HUFA export, 199
    n-3 and n-6 PUFA synthesis, 147
Allelopathy, 67, 78
ALSPAC. *See* Avon longitudinal study of parents and children
Amphipods, 161
Anthropogenic factors
    CO_2 concentration, 19–20
    phosphorus/sulphur effects, 18–19
    trophic transfer, elements/metals, 101
Aquaculture, 117, 122
Arachidonic acid (ARA)
    bioaccumulation, 103
    diatoms, 124
    eicosanoids, 249
    food web
        long-chain PUFA synthesis, 311
        marine finfish, 318
    membrane fluidity, 240
    physiological role, 149
    prostaglandins, 249
    PUFA synthesis, 179
    taxonomic differences, 118, 120
    teleost immunity, 249
    vertebrates, 147
    zooplankton, 118
Arctic charr *(Salvelinus alpinus)*, 154
Atlantic salmon *(Salmo salar)*, 155
Avon Longitudinal Study of Parents and Children (ALSPAC), 334

B
Bacillariophyceae/Bacillariophyta. *See* Diatoms
Behavioral avoidance, 81, 82
Benthic animals, 149
Betaine lipids, 7–9
Bioaccumulation
    lipids, 103
    organic contaminants
        body size and age, 99–100
        temperature and reproduction, 100
        xenobiotics, 95
Biomagnification, 97, 99
Biomarkers, 57, 281, 309
Brassicasterol, 45

C
Campesterol, 45
Carbon
    availability, 19–20

Carbon (cont.)
  loss, 37
  dietary, 49–50
  transfer
    crustacean grazers, 54–55
    cyanobacteria, 54
    dinoflagellates, 55
    grasshoppers, 55–56
Carrion drift, 200
Chemical tracers, 104
Chemoreceptors, 119
Chilling susceptibility, 12
Chlorophyceae, heterotrophic protists
  biosynthesis pathyways, 32
  lipid composition variability, 33
Chlorophyta
  D^7 sterols content, 57
  light effects, 15
  PUFA synthesis, 31
Chloroplast
  carbon availability, 19
  FA and FA esters, 67
  heterotrophic protist lipid composition variability, 34
  light effects, 13–15
  nutrient effects, 17
  phospholipids, 6
  phosphorus and sulphur effects, 18
  PUA biosynthesis, 73
  PUFA biosynthesis pathways, 29–31
  temperature effects, 11–12
Cladocerans, 155–156. See also Zooplankton
CO_2-concentrating mechanism, 19
Cold-water adaptation, 117, 139
Compensatory mechanisms, 18
Consumptional effects, food webs
  ARA, 313–314
  EPA and DHA, 312, 313
  plasma lipids and lipoproteins, 312
Contaminants, 93–97, 99, 104, 106–108
Copepods. See also Zooplankton
  behavioral responses, 81–82
  dietary impacts, 128
  diffusion mechanism, 80
  FA and phospholipids
    FA composition, 272, 273
    position-specific distribution, FA, 271, 272
    PUFA distribution, 270, 271
  global warming
    Antarctic ocean, 274
    poleward shift, 273
    Southern and Arctic ocean, 274–275

high-latitude
  herbivorous, 259–261
  omnivorous and carnivorous, 261–266
low-latitude, 268–269
metal concentrations, 101
mid-latitude
  large copepods, 266–267
  small copepods, 268
PUA toxicity, 78, 79
reproductive anomalies, 79
sterols
  assimilation efficiency, 55
  compensatory mechanism, 60
  egg production, 50–51, 53
  homeostasis, 58
  tetrahymanol assimilation, 59
  trophic upgrading, 56
  upwelling-system, 269–270
Coronary heart disease (CHD), 338–340
Corticotrophins, 334–335
Chrysophyceae
  glycolipids, 5
  heterotrophic protists, 27
  PUFA biosynthesis reaction, 30, 32
Chrysophyte(s)
  algal producers, 82–83
  FA-mediated semiochemical interactions, 83
  FA-related toxic events, 70
  semiochemistry, oxylipins, 77–78
Cryptomonads
  Crustaceans fed Cyanophytes, 159–160
  heterotrophic protists, 27
  temperature effects, 12
Cryptophyte(s)
  environmental effects, HUFA supply, 244
  feeding experiments
    Crustaceans fed Cyanophytes, 159
    zoobenthos, 162
  HUFA content, 180
  lipid composition variability, 33
Cyanobacteria
  carbon transfer, 54
  chlorophytes, 156
  cyanophytes, 158–159
  dietary impacts, 128, 130
  DNA coding patterns, 347
  environmental effects, HUFA supply, 244
  essential fatty acids transfer, 328
  FA-related toxic events, 70
  homeostatic fatty acid composition response, 135–136
  HUFA production, 183
  occurrence of sterols, 44

Subject Index

phytoplankton FA composition, 122–124
small copepods, 268
temperature effects, biochemistry, 13
zoobenthos, 161–162
zooplankton, reproductive function, 137

D

DALY. *See* Disability-adjusted life years
Dealkylation, 51
Depression, 333, 337
Desaturases, 147
    PUFA synthesis, 29–33
Detoxification, 80, 81
Deuterated linolenic acid, 230
DHA. *See* Docosahexaenoic acid
Diacylglycerols (DAG), 340
Diacylglycerylcarboxyhydroxymethylcholine (DGCC), 8
Diacylglycerylhydroxymethyltrimethylalanine (DGTA), 8
Diacylglyceryltrimethylhomoserine (DGTS), 8
Diatoms
    algal producers, 82–83
    behavioral responses, 81
    copepods
        high-latitude copepods, 259
        mid-latitude copepods, 266, 268
    eutrophication, 244
    FA markers, 289–290
    FA-related semiochemicals, 83
    grazer defense mechanism, 70
    heterotrophic protist
        lipid composition variability, 33
        PUFA biosynthesis, 29–30
    HUFA export, 180, 183
    lipid and fatty acid composition, 14–15
    oxylipins, 76–79
    phosphatidylcholine, 7
    PUA-biosynthesis, 72–76
    sterol
        ergosterol, 46
        invertebrates, 49
    zoobenthos, 161
    zooplankton FA
        dietary impacts, 127, 130
        phytoplankton, 122–125
        reproductive function, 138
Dietary mechanism, essential fatty acids
    bioactive compounds, 315
    enzyme-substrate interaction, 314
    finfish membranes, 9
    fish oil feeding, 315, 316
    phospholipase A_2 activity, 317

phospholipid structure, 316
physicochemical properties, 315
protein kinase C, 315–316
regulation, 314–315
Digalactosyldiacylglycerol (DGDG), 2, 4
Dinoflagellates, carbon transfer
    Karenia brevis, 55
    Oxyrrhis marina, 56
    Scrippsiella trochoidea, 51
Diphosphatidylglycerol (DPG), 5
Disability-adjusted life years (DALY), 333–334
Discriminant function analysis (DFA), 296
Docosahexaenoic acid (DHA)
    bacteria, 212
    bioactive compound synthesis, 319
    consumptional effects, 312
    cryptophytes, 124
    heterotrophic protist
        lipid composition variability, 33
        PUFA biosynthesis, 32
    HUFA export, 179
        aquatic insects, 194
        fish, 188
        HUFA requirements, 182
        marine organisms, 189
        Pacific salmon, 191
        water birds, 196
        zoobenthos, 186–187
        zooplankton, 186
    membrane fluidity, 239–240
    neurodevelopment, 334–335
    Rana ridibunda, 197
    synthesis, 217, 230
    vertebrates, 147
Docosapentaenoic acid (DPA)
    bioaccumulation, 319
    long-chain PUFA synthesis, 311
    thraustochytrid protist, 318
DPA. *See* Docosapentaenoic acid

E

Ecdysteroids
    biosynthesis, 59
    Δ^7 sterol, 51
    structure, 47
Ecosystem services, 203
Ecotoxicology, 104
Edible leafy plant oils, 181
Edinburgh Postnatal Depression Scale (EPDS), 337
Egg production, 58, 138

Eicosapentaenoic acid (EPA)
  aquatic insect emergence, 194
  consumptional effects, 312
  cryptophytes and diatoms, 124
  dietary ratios, 320
  eicosanoids, 248, 249
  food web
    self-harm, 337
    unipolar and bipolar depression, 338
  heterotrophic protist
    lipid composition variability, 33
    PUFA biosynthesis, 30, 32
  HUFA concentration, 182
  light effects, 15
  membrane fluidity, 240
  Pacific salmon, 191
  PUFA biosynthesis, 212
  teleost immunity, 249
  tropic transfer, lipids, 103
  water birds, 196
Elongase, PUFA synthesis, 29–32
Elongation of very long chain fatty acids (ELOVL), 226
Endogenous fatty acid production, 141
Energy storage, 9
Environmental effects, algal lipid
  growth conditions, 11
  light
    dark treatment, 15
    light/dark cycles, 14
    metabolism, 13
  pH, 16
  salinity, 15–16
  temperature effects, 11–13
Environmental perturbations, 243
EPA. See Eicosapentaenoic acid
EPDS. See Edinburgh Postnatal Depression Scale
Essential fatty acids (EFA). See also Polyunsaturated fatty acids (PUFA)
  consumptional effects
    ARA, 313–314
    EPA and DHA, 312, 313
    finfish, 313–314
    plasma lipids and lipoproteins, 312
  definition
    conditionally indispensable and dispensable, 312
    α-linolenic acid, 310
    metabolites and nutrients, 311–312
  dietary mechanism
    bioactive compounds, 315
    enzyme-substrate interaction, 314
    finfish membranes, 9
    fish oil feeding, 315, 316
    phospholipase A_2 activity, 317
    phospholipid structure, 316
    physicochemical properties, 315
    protein kinase C, 315–316
    regulation, 314–315
  dietary ratios, 319–320
  n-6 PUFA
    Algamac®, isotopic signature, 318, 319
    n-6 DPA, 319
    thraustochytrid protist
      Schizochytrium, 318
  neural development, 337
  phytoplankton drifting, 328
  tissue biomarker level maintainance, 346
  trophic transfer, 103
Essential metabolite, 312
Essential nutrient, 312
Estuarine food web, 348
Euphausiids, 120, 121
Eutrophication, 183, 202, 244

F
Fatty acid derivatives, 83
Fatty acids (FA). See also Essential fatty acids (EFA)
  allelopathic agents, 67
  biomarkers, 289, 290
  biosynthesis, 283
  composition, 262, 265
  deposition, 285
  derivatives, 83
  distribution, 10
  elongation, 16, 284, 311
  modification, 285, 286
  polyketide pathway, 216
  ratios, 151
  semiochemical interactions, 84
  signatures, 292
  synthesis, 180
  temperature effect, 12
  toxigenic capacity, 70
  trafficking, 31
  turnover, 136
  unsaturation, 19
Fatty acid trophic markers (FATM), 133, 258
Feeding experiments
  fish, 150, 153–155
  zoobenthos, 161–162
  zooplankton, 156–159
Fish
  homeoviscous adaptation
    definition, 237–238

Subject Index

DHA incorporation, 239–240
temperature changes, 238
HUFA biosynthesis
  biosynthetic pathway, 220–222
  elongases, 225–228
  fatty acyl desaturase, 222–224
  regulation, 228–229
  in vivo studies, stable isotopes, 229–230
HUFA supply, environmental effects, 243–245
membrane fluidity
  lipid compositional changes, 241–243
  measurement, 240–241
n-3/n-6 and DHA/ARA ratios
  ARA treatment, 152–153
  DHA-enriched rotifers, 153
  LIN and ALA addition, 154
  wild fish, 150–152
  wild *vs.* cultured, 154–155
teleost dietary FA, modulatory effects
  dietary effects, 248–249
  disease susceptibility, 250
  eicosanoids, 247–248
  immune system, 246
  innate and adaptive immune systems, 246–247
  physical and chemical barriers, 246
  teleost immunity, 249
Food webs
  chain events
    essential fatty acids transfer
      phytoplankton drifting, 328, 332
      PUFA and the HUFA, 332–333
    food chain events
      earlier short-term adaptation, 328
      food choice, 328
      molecular events, 327
  functional foods, 347–348
Food web tracing, FA and lipids
  characteristics, 282–283
  de novo FA biosynthesis, 283–284
  dietary lipids and fatty acids
    digestion, 285–286
    modification and deposition, 286–287
  FA signatures
    Atlantic cod, 292–293
    captive seabird, 293–294
    Hawaiian monk seal, 294–295
    MANOVA test and DFA, 296–297
    multivariate analysis, 295–296
  lipid biosynthesis, 284–285
  predator diet
    16:2n-4 and 16:4n-1 FA, 289–290
    20:1 and 22:1 FA isomers, 290

marine-based and 18:3n-3 FA, 291
odd and branched C_{14}–C_{18} FA, 291–292
wax esters, 292
quantitative FA signature analysis (QFASA)
  predator metabolism, 298
  predator species, 302–303
  prey FA database, 299–300
  trophic FA transfer, 303
  validation studies, 300–301
trophic pathway tracing
  predator sampling, 287–288
  prey sampling, 288–289
Fucoserratene, 83
Functional food, 57
Fungal fatty acids, 181

G

Galactolipids, 17, 72
Global warming, copepod lipids
  Antarctic ocean, 274
  poleward shift, 273
  Southern and Arctic ocean, 274–275
Glycerolipids
  nonpolar glycerolipids, 9–11
  polar glycerolipids
    betaine lipids, 7–9
    glycolipids, 2–5
    phospholipids, 5–7
Glycerophospholipids, 316
Grazer counter-defense
  behavioral responses, 81
  detoxification mechanism, 82
  protective mechanism, 81–82
Gross growth efficiency (GGE), 36
Growth cycle, 79

H

Halogenated hydrocarbon, 77
Hamilton Rating Scale for Depression (HRSD), 337
Homeostasis, 58
Heterotrophic protists (HP)
  abundance, 26
  feeding behaviour, 26–27
  gross growth efficiency (GGE), 36
  lipid composition variability
    growth and survival, 36
    habitats and diet, 34–35
    lipid sources, 33–34
    membrane components, 35
    predator–prey interactions, 36

Heterotrophic protists (HP) (cont.)
    mixotrophy, 25
    multiple trophic transfers, 37
    organic matter transfer, 36–37
    picoflagellates, 27
    PUFA biosynthetic pathway
        bifunctional $\Delta 12/\Delta 15$ desaturase, 33
        DHA, 30, 32
        eukaryotic pathway, 30
        genetics, 29
        polyketide synthase (PKS) pathway, 29–30
        prokaryotic pathway, 30
    taxonomy, 27
    trophic interfaces
        metazooplankton, HP nutritional value, 28
        picoplankton, 27–28
        trophic upgrading concept, 28–29
Hexadecatrienoic acid, 310
High-latitude copepods
    Calanus species
        composition, 261
        lipid accumulation, 259–260
        reproduction, 260
    Metridia longa and *M. gerlachei*, 261
    small and micro-copepods, 264–266
Highly unsaturated fatty acids (HUFA), 33, 36
    aquatic and terrestrial ecosystems
        algal-derived, 183
        anthropogenic impact, 180
        aquatic-derived, 180
        aquatic insect emergence, 193–195
        fluxes, 184–185
        HUFA production, 183
        humans and fisheries, 192
        landmass *vs.* aquatic surface area, 197–198
        measurements, 185
        ocean contribution, 199–200
        physiological benefits, 182–183
        salmon consumption by bears, 190–192
        sea bird colonies, 182–183
        terrestrial biomass, 198–199
        water birds, 195–196
    aquatic organisms
        fish, 188–190
        zoobenthos, 186–187, 190
        zooplankton, 186
    biosynthesis
        regulation, 228–229
        in vivo studies, stable isotopes, 229–230
    biosynthetic pathway, 220–222
    food web
        coronary heart disease (CHD), 338–340
        EFA transfer, 329–333
        impaired neural development, 334–338
        n-3 and n-6 structure discrimination, 341–347
    marine organisms, 189, 190
    molecular biology
        elongases, 225–228
        fatty acyl desaturase, 222–224
    n-3 polyunsaturated fatty acids synthesis, 179–180
    terrestrial ecosystems
        C18 compounds, 181
        critical assumptions, 180–181
        edible oils, 181
        invertebrate species, 181
        PUFA elongation and desaturation, 182
        terrestrial predators, 181–182
Homeoviscous adaptation, fish
    definition, 237–238
    DHA incorporation, 239–240
    temperature changes, 238
Hopanoids, 48, 54, 59
Hormone-like signaling, eicosanoids
    hormone-like prostaglandins, 342–343
    leukotriene B (LTB) receptor, 343
    PGH synthase, 342
HRSD. *See* Hamilton Rating Scale for Depression
HUFA. *See* Highly unsaturated fatty acids
Humoral secretions, 246
Hydroperoxy fatty acid transformation, 75–77

I
Immunomodulatory effects, 245
Immunostimulation, 249
Impaired neural development
    aggression, 335
    ALSPAC, 334
    corticotrophins, 334–335
    human disability, 333–334
    phosphatidyl serine (PS) reduction, 335–336
    rainbow trout, 336–337
    seafood, cognitive benefits, 338
    self-harm, 337–338
    stress, 334
Inflammation, 248, 343, 344
Innate immune responses, 246

Subject Index

K
Krill, 131–133, 274

L
Leukotriene B (LTB), 343
Light/dark cycle, 14
Linoleic acid (LIN), 147, 149, 159, 310, 332, 346
  chlorophytes, 124
Lipids and contaminants
  aquatic food web, 105–106
  chemical tracers, 104–105
  composition and contaminants, 106–107
  ecological and ecotoxicological relevance, 107
  omega-3 (n-3) PUFA
  organic contaminants
    bioaccumulation, 97–98
    biomagnification, 99
    biotransformation, 97
    body size and age, 99–100
    fugacity, 96
    lipids, 99
    passive diffusion, 97
    temperature and reproduction, 100
  tropic transfer, 103
  xenobiotics
    chemicals, 95
    compounds, 93
    MeHg, 101
Lipoxygenases, 74

M
Membrane fluidity
  homeoviscous adaptation
    definition, 237–238
    DHA incorporation, 239–240
    temperature changes, 238
  HUFA supply, environmental effects
    Chinook salmon and alewife, 244
    cultural eutrophication and climate change, 244–245
    Diporeia sp., 243–244
    triacylglycerols and phospholipids, 243
  lipid compositional changes
    condensing effect, 242–243
    mono and polyunsaturated FA modification, 241–242
    structural and chemical properties, 241
  measurement, 240–241
Membrane phospholipids, 18
Mevalonate (MVA pathway), 46

MGDG. *See* Monogalactosyldiacylglycerol
Microalgae, 329
  PUFA synthesis, 213
Microbial food web, heterotrophic protists fatty acid
  ecological roles, 26
  feeding behaviour, 26–27
  gross growth efficiency (GGE), 36
  heterotrophic flagellates, 27
  lipid composition variability
    growth and survival, 36
    habitats and diet, 34–35
    lipid sources, 33–34
    membrane components, 35
    predator–prey interactions, 36
  mixotrophy, 25
  multiple trophic transfers, 37
  organic matter transfer, 36–37
  picoflagellates, 27
  PUFA biosynthetic pathway
    bifunctional $\Delta12/\Delta15$ desaturase, 33
    DHA, 30, 32
    eukaryotic pathway, 30
    genetics, 29
    polyketide synthase (PKS) pathway, 29–30
    prokaryotic pathway, 30
  trophic interfaces
    metazooplankton, HP nutritional value, 28
    picoplankton, 27–28
    trophic upgrading concept, 28–29
Mid-latitude copepod
  large copepods, 266–267
  small copepods, 268
Molting hormones, 47
Monoenoic fatty acids, 328, 331
Monogalactosyldiacylglycerol (MGDG)
  carban availability, 19
  diatom *Skeletonema costatum*, 3
  light effects, 14–15
  nutrient-limitation, 17
  phosphorus effect, 17
  PUA synthesis, 72–73
  structure, 2–3
  temperature shift, 13
Monounsaturated fatty acids (MUFA)
  algal toxins, 67
  copepods, 261, 264
  cryptophytes and diatoms, 124
  fluidity control, 242
  zooplankton fatty acids
    Acartia spp., 116
    dietary impacts, 126–130

Monounsaturated fatty acids (MUFA) (cont.)
  krill, 132–133
  phytoplankton, 124–125
  quasi-homeostatic response, 136
  reproductive function, 137–138
  starvation impacts, 140
  taxonomic differences, 118–122
Multivariate analysis of variance (MANOVA), 296

N

n-3 and n-6 structure discrimination
  food energy imbalance, 344–345
  hormone-like signaling, 342–343
  linoleic acid, 346
  n-3 HUFA deficiency, 346
  PUFA and HUFA intakes, 346–347
  TAG and PL, 340–342
  transient oxidant stress, 343
n-3/n-6 and DHA/ARA ratios
  fish
    ARA treatment, 152–153
    DHA-enriched rotifers, 153
    LIN and ALA addition, 154
    wild fish, 150–152
    wild vs. cultured, 154–155
  zoobenthos
    ephemerids and chironomids, 160–161
    green algal diet, 162
    oats and *Spirulina* diet, 161–162
  zooplankton
    cladocerans and copepods, 155–156
    cyanobacteria, 158–160
    N-and P limited *Chlamydomonas*, 157
    Scenedesmus acutus diet, 157–158
    Scenedesmus quadricauda diet, 157
NEFA. See Nonesterified fatty acids
Neutral storage lipids, 117
Nitrogen
  limitation, 9
  starvation, 17
Nonesterified fatty acids (NEFA), 340, 344, 345
Non-methylene interrupted fatty acids, 292
Nutrient limitation, algal lipids, 16–18

O

Octanol–water partition coefficient, 97
Oleic acid, 310
Omnivorous and carnivorous copepods, 261–266

Organic contaminants
  bioaccumulation, factors affecting
    body size and age, 99–100
    lipids, 99
    temperature and reproduction, 100
  trophic transfer
    bioaccumulation, 97–98
    biomagnification, 99
    biotransformation, 97
    fugacity, 96
    passive diffusion, 97
Osmotic stress, 16
Over-wintering strategy, 116, 139, 141
Oxylipins
  algal producers, 82–83
  eicosanoids, 71
  grazer counter-defense
    behavioral responses, 81
    detoxification mechanism, 82
    protective mechanism, 81–82
  polyunsaturated aldehydes
    alcohol production, 75
    aquatic signaling, 71–72
    diatom *Thalassiosira rotula*, 72
    hydrocarbon production, 74–75
    hydroperoxy FA transformation, 73–74
    LOX activity, 72–73
    LOXmediated transformation, 71
  semiochemistry
    diatom diet and nonviable egg production, 80
    planktonic diatom species, 77–78
    PUA production, 76–77
    reproduction and development, 77
    survivorship and growth, 78–79

P

Phagotrophy, 26
pH effects, 16
Pheromones, 66, 73, 83
Phosphatidylcholine (PC), 6, 241, 270, 272, 315
Phosphatidylethanolamine (PE), 6, 241, 270, 272
Phosphatidylglycerol (PG), 6, 13, 14
Phosphatidylinositol (PI), 6, 241
Phosphatidyl serine (PS), 336
Phosphatidylsulfocholine, 7
Phospholipase, 248, 315, 317
Phospholipids, 239, 241, 263, 267–269, 313
Photosynthesis, 2, 20
Phytoplankton
  copepods, energy transfer, 257

Subject Index

fatty acid composition, 123, 125
production, 327
reproduction, 260
Polybrominated biphenyls, 95
Polychlorinated biphenyls (PCB), 95, 106
Polycyclic aromatic hydrocarbons, 95
Polyketide pathway
    dinoflagellates, 218
    enzymes, 215
    n-3 PUFA, 215–217
    Schizochytrium cDNA library, 215
    Shewanella ORFs, 214–215
Polyunsaturated aldehydes (PUA)
    alcohol production, 75
    aquatic signaling, 71–72
    diatom *Thalassiosira rotula*, 72
    hydrocarbon production, 74–75
    hydroperoxy FA transformation, 73–74
    LOX activity, 72–73
    LOXmediated transformation, 71
Polyunsaturated fatty acid (PUFA) ratios
    chlorophytes, 172–173
    cyanophytes, 173–174
    experimental designs, 163–164
    fish, 169
        ARA treatment, 152–153
        DHA-enriched rotifers, 153
        LIN and ALA addition, 154
        wild fish, 150–152
        wild *vs.* cultured, 154–155
    salmonids, 170
    zoobenthos, 175–178
        ephemerids and chironomids, 160–161
        green algal diet, 162
        oats and *Spirulina* diet, 161–162
    zooplankton, 171
        cladocerans and copepods, 155–156
        N-and P limited *Chlamydomonas*, 157
        Scenedesmus acutus diet, 157–158
        Scenedesmus quadricauda diet, 157
Polyunsaturated fatty acids (PUFA). *See also*
    α-Linolenic acid (ALA);
    Arachidonic acid (ARA);
    Docosahexaenoic acid (DHA);
    Eicosapentaenoic acid (EPA);
    Linoleic acid (LIN)
    biosynthesis
        bacteria, 212–213
        microalgae, 213
        PKS and PUFA synthase, 214–218
        protozoans and heterotrophic organisms, 214
    copepods, 258

food web
    EFA transfer, 329–333
    n-3 and n-6 structure discrimination, 341, 342, 346–347
    self-harm, 337
    heterotrophic protists, biosynthetic pathway
        bifunctional Δ12/Δ15 desaturase, 33
        DHA, 30, 32
        eukaryotic pathway, 30
        genetics, 29
        polyketide synthase (PKS) pathway, 29–30
        prokaryotic pathway, 30
    invertebrates, metabolism, 219
    trace elements, 93
Predator diet, FA signatures. *See also* Food web tracing, FA and lipids
    16:2n-4 and 16:4n-1 FA, 289–290
    20:1 and 22:1 FA isomers, 290
    Atlantic cod, 292–293
    captive seabird, 293–294
    Hawaiian monk seal, 294–295
    MANOVA test and DFA, 296–297
    marine-based and 18:3n-3 FA, 291
    multivariate analysis, 295–296
    odd and branched 14-18C FA, 291–292
    quantitative FA signature analysis (QFASA)
        higher-predator species, 302–303
        predator metabolism, 298
        prey FA database, 299–300
        trophic FA transfer, 303
        validation studies, 300–301
    wax esters, 292
Prostacyclins, 247
Prostaglandin, 247, 342
Protein kinase C, 315–316
Protozoans
    PUFA synthesis, 214
PUFA. *See* Polyunsaturated fatty acids (PUFA)

Q

Quantitative FA signature analysis (QFASA)
    higher-predator species, 302–303
    predator metabolism, 298
    prey FA database, 299–300
    trophic FA transfer, 303
    validation studies, 300–301

R

Rainbow trout *(Oncorhynchus mykiss)*, 154, 221

S

Salinity effects, 15–16
Saturated fatty acids (SAFA)
    Acartia, 116–117
    algal toxins, 67
    cell membranes, 238
    PUA-biosynthesis, 72
    zooplankton
        cyanobacteria, 124
        dietary impacts, 126–130
        n-3 HUFA content, 129–130
        phytoplankton, 124–125
        quasi-homeostatic response, 135–136
        reproductive function, 137–138
        starvation impacts, 140
        taxonomic differences, 118
        temperature impacts, 139
        trophic markers, 130
Seabird colonies, 184, 201
Seafood, cognitive benefits, 338
Semiochemicals. *See also* Oxylipins
    fatty acids and FA esters
        algal toxins, 67
        allelopathic agents, 67, 70
        food web interaction, 70
    PUFA derivatives
        pheromones and grazer defense, 66
        properties, 68–69
Stearic acid, 310
Stearidonic acid (SDA), 124
Sterols
    biomarker, 57
    biosynthesis, 46
    carbon transfer
        crustacean grazers, 54–55
        cyanobacteria, 54
        dinoflagellates, 55
        grasshoppers, 55–56
    compensatory mechanism, 59–60
    dietary sterol
        eukaryotic cell membranes, 59
        requirements, 57–58
        sterol limitation and structural features, 58–59
    human nutrition, 56–57
    nutritional requirements
        heterotrophic protists, 47–48
        invertebrates, 48–51
        vertebrates, 52
    occurrence
        eukaryotes, 44–46
        heterotrophic protists, 46
        prokaryotes, 43–44
    physiological properties, 46–47
    trophic upgrading, 56
    zooplankton, ecological implications
        dietary sterols, 53
        somatic growth rates, 52–53
Steryl ester, 58
Stigmasterol, 45
Sulfolipid, 3
Sulphur, 18

T

TAG. *See* Triacylglycerols (TAG)
Taste-odor, 66
Teleost FA, modulatory effects
    dietary effects, 248–249
    disease susceptibility, 250
    eicosanoids, 247–248
    immune system, 246
    innate and adaptive immune systems, 246–247
    physical and chemical barriers, 246
    teleost immunity, 249
Temperature effects, 11–12
Tetrahymanol, 48, 54, 59, 60
Thermoadaptation mechanism, 13
Thylakoid membrane, 15
Toxigenic effects, 67, 70
Triacylglycerols (TAG)
    copepods and euphausids, 142
    FA distribution, 9–10
    fish, FA assembly, 243
    n-3 fatty acids, 313
    PUFA-rich TAG, 9–11
    small copepods, 268
    storage, 264
    structure, 9
    vs. wax esters, 257
    zooplankton, 117
Triterpenoid alcohols, 44
Trophic efficiency, 36
Trophic interactions, 43
Trophic levels, 94, 108
Trophic position, 100
Trophic repackaging, 28
Trophic tracers, 283
Trophic transfer, 36, 94, 96, 100, 103, 117
    elements or metals, 100–102
    lipids, 103
    organic contaminants
        bioaccumulation, 97–98
        biomagnification, 99
        biotransformation, 97
        fugacity, 96
        passive diffusion, 97

Subject Index

Trophic upgrading, 28, 37, 56, 60, 94
Tropical ecosystems, 294, 295
Tropical oceans, 268
Tropics, 262, 265, 271

U
Upwelling-system copepods, 269–270

V
Volatile hydrocarbon, 72

W
Wax esters
　accumulation, 266
　fatty acid and alcohol compositions, 263
　geographical distribution, marine zooplankton, 257
　large copepod species, 260
　metabolism, 285
　small copepod species, 264, 268
　storage lipids, 117
　temperate species, 267
Wound-activated synthesis, 73

X
Xenobiotics
　bioaccumulation, 104
　chemicals, 95
　compounds, 93
　MeHg, 101

Z
Zoobenthos
　n-3/n-6 and DHA/ARA ratios
　　ephemerids and chironomids, 160–161
　　green algal diet, 162
　　oats and *Spirulina* diet, 161–162
　PUFA ratios, 175–178

Zooplankton. *See also* Zooplankton fatty acid (FA) composition
　n-3/n-6 and DHA/ARA ratios
　　cladocerans and copepods, 155–156
　　N-and P limited *Chlamydomonas*, 157
　　Scenedesmus acutus diet, 157–158
　　Scenedesmus quadricauda diet, 157
　PUFA metabolism, 219
　PUFA ratios, 171
　sterols, ecological implications
　　dietary sterols, 53
　　somatic growth rates, 52–53
Zooplankton fatty acid (FA) composition
　dietary impacts
　　Daphnia spp., 126
　　euphausiids (krill), 131–133
　　FATM approach, 133
　　freshwater copepods and *C. dubia*, 128
　　harpacticoid copecods and artemia spp, 131
　　laboratory and field studies, 134
　　Lake Washington, 129–130
　　marine calanoid copecods, 130
　　oligotrophic alpine lakes, 129
　FA turnover, 136–137
　historical analysis, 116–117
　HUFA, 115
　phytoplankton
　　bivariate plot, 125
　　mean composition, 122–125
　quasi-homeostatic response, 135–136
　reproductive investment, 137–138
　starvation impacts, 140
　taxonomic differences
　　copepods and cladoceran, 118–119
　　discriminant function analysis, 121–122
　　mean FA composition, 119–121
　　wax ester synthesis, 118
　temperature impacts, 139–140
　vs. marine and freshwater dietary, 117

Printed in the United States
147841LV00003B/2/P